Lecture Notes in Physics

Volume 830

The Lecture Notes in Physics

The series Lecture Notes in Physics (LNP), founded in 1969, reports new developments in physics research and teaching—quickly and informally, but with a high quality and the explicit aim to summarize and communicate current knowledge in an accessible way. Books published in this series are conceived as bridging material between advanced graduate textbooks and the forefront of research and to serve three purposes:

- to be a compact and modern up-to-date source of reference on a well-defined topic
- to serve as an accessible introduction to the field to postgraduate students and nonspecialist researchers from related areas
- to be a source of advanced teaching material for specialized seminars, courses and schools

Both monographs and multi-author volumes will be considered for publication. Edited volumes should, however, consist of a very limited number of contributions only. Proceedings will not be considered for LNP.

Volumes published in LNP are disseminated both in print and in electronic formats, the electronic archive being available at springerlink.com. The series content is indexed, abstracted and referenced by many abstracting and information services, bibliographic networks, subscription agencies, library networks, and consortia.

Proposals should be sent to a member of the Editorial Board, or directly to the managing editor at Springer:

Christian Caron
Springer Heidelberg
Physics Editorial Department I
Tiergartenstrasse 17
69121 Heidelberg/Germany
christian.caron@springer.com

Stefan Scherer · Matthias R. Schindler

A Primer for Chiral Perturbation Theory

 Springer

Dr. Stefan Scherer
Johannes Gutenberg-Universität Mainz
Institut für Kernphysik
Johann-Joachim-Becher-Weg 45
55099 Mainz
Germany
e-mail: scherer@kph.uni-mainz.de

Dr. Matthias R. Schindler
Department of Physics and Astronomy
University of South Carolina
712 Main Street
Columbia, SC 29208
USA

and

Department of Physics
The George Washington University
Washington, DC
USA
e-mail: mschindl@mailbox.sc.edu

ISSN 0075-8450
ISBN 978-3-642-19253-1
DOI 10.1007/978-3-642-19254-8
Springer Heidelberg Dordrecht London New York

e-ISSN 1616-6361
e-ISBN 978-3-642-19254-8

Cover design: eStudio Calamar, Berlin/Figueres

Printed on acid-free paper

Springer is part of Springer Science+Business Media (www.springer.com)

Preface

Chiral perturbation theory (ChPT) is the effective field theory of quantum chromodynamics (QCD) at energies well below typical hadron masses. This means that it is a systematic and model-independent approximation of QCD, based on the symmetries of the underlying theory and general principles of quantum field theory. Starting from early work on the interaction of pions, ChPT has grown to become a valuable tool to analyze and interpret a host of low-energy experiments involving the lowest-mass meson and baryon octets and decuplets. The application to $\pi\pi$ scattering and pion photoproduction are just two of the large number of remarkable successes of ChPT.

This monograph is based on lectures on chiral perturbation theory given by one of us (S.S.) on various occasions, supplemented with additional material. It is aimed at readers familiar with elementary concepts of field theory and relativistic quantum mechanics. The goal of these lecture notes is to provide a *pedagogical introduction* to the basic concepts of chiral perturbation theory (ChPT) in the mesonic and baryonic sectors. We therefore also derive and explain those aspects that are considered well known by "experts." In particular, we often include intermediate steps in derivations to illuminate the origin of our final results. We have also tried to keep a reasonable balance between mathematical rigor and illustrations by means of simple examples. Numerous exercises throughout the text cover a wide range of difficulty, from very easy to quite difficult and involved. Ideally, at the end of the course, the reader should be able to perform simple calculations in the framework of ChPT and to read the current literature. Solutions to all exercises are provided for readers to check their own work.

These lecture notes include the following topics: Chapter 1 deals with QCD and its global symmetries in the chiral limit, explicit symmetry breaking in terms of the quark masses, and the concept of Green functions and Ward identities reflecting the underlying chiral symmetry. In Chap. 2, the idea of a spontaneous breakdown of a global symmetry is discussed and its consequences in terms of the Goldstone theorem are demonstrated. Chapter 3 deals with mesonic chiral perturbation theory. The principles entering the construction of the chiral Lagrangian are outlined and a number of elementary applications are discussed. In Chap. 4, these methods are extended to include the interaction between Goldstone bosons and baryons in the single-baryon sector. Chapter 5 discusses more advanced applications and topics that are closely related to chiral perturbation theory.

This work is not intended as a comprehensive review of the status of chiral perturbation theory. This also means that we cannot cite all of the vast literature, especially on advanced applications. Readers interested in the present status of applications are encouraged to consult the numerous available lecture notes, review articles, and conference proceedings. A list of suggested references is provided at the end of Chap. 5.

While the number of people who have contributed to our understanding of the topics discussed in this monograph is too large to acknowledge each of them individually, we would like to thank H.W. Fearing, J. Gegelia, H.W. Grießhammer, and D. R. Phillips for numerous interesting and stimulating discussions that have most directly influenced us. We are grateful to A. Neiser for the careful reading of and helpful comments on the manuscript. We would also like to thank all students who participated in previous classes on ChPT and gave important feedback. The support and patience of our editor C. Caron is gratefully acknowledged. S.S. would like to thank M. Hilt for extensive technical support. M.R.S. would like to thank the Lattice and Effective Field Theory group at Duke University for their hospitality. This work was carried out in part with financial support from the Center for Nuclear Studies at the George Washington University, National Science Foundation CAREER award PHY-0645498, and US-Department of Energy grant DE-FG02-95ER-40907.

Mainz and Columbia, SC, April 2011   Stefan Scherer
                                                                 Matthias R. Schindler

Contents

Chapter 1
Quantum Chromodynamics and Chiral Symmetry

1.1 Some Remarks on SU(3)

The special unitary group SU(3) plays an important role in the context of the strong interactions, because

1. it is the gauge group of quantum chromodynamics (QCD);
2. flavor SU(3) is approximately realized as a global symmetry of the hadron spectrum, so that the observed (low-mass) hadrons can be organized in approximately degenerate multiplets fitting the dimensionalities of irreducible representations of SU(3);
3. the direct product $SU(3)_L \times SU(3)_R$ is the chiral-symmetry group of QCD for vanishing u-, d-, and s-quark masses.

Thus, it is appropriate to first recall a few basic properties of SU(3) and its Lie algebra su(3) [8, 34, 43].

The group SU(3) is defined as the set of all unitary, unimodular, 3×3 matrices U, i.e. $U^\dagger U = \mathbb{1}$,[1] and $\det(U) = 1$. In mathematical terms, SU(3) is an eight-parameter, simply-connected, compact Lie group. This implies that any group element can be parameterized by a set of eight independent real parameters $\Theta = (\Theta_1, \ldots, \Theta_8)$ varying over a continuous range. The Lie-group property refers to the fact that the group multiplication of two elements $U(\Theta)$ and $U(\Psi)$ is expressed in terms of eight *analytic* functions $\Phi_i(\Theta; \Psi)$, i.e. $U(\Theta)U(\Psi) = U(\Phi)$, where $\Phi = \Phi(\Theta; \Psi)$. It is simply connected because every element can be connected to the identity by a continuous path in the parameter space and compactness requires the parameters to be confined in a finite volume. Finally, for compact Lie groups, every finite-dimensional representation is equivalent to a unitary one and

---

[1] Throughout this monograph we adopt the convention that $\mathbb{1}$ stands for the unit matrix in n dimensions. It should be clear from the respective context which dimensionality actually applies.

S. Scherer and M. R. Schindler, *A Primer for Chiral Perturbation Theory*,
Lecture Notes in Physics 830, DOI: 10.1007/978-3-642-19254-8_1,
© Springer-Verlag Berlin Heidelberg 2012

can be decomposed into a direct sum of irreducible representations (Clebsch-Gordan series).

Elements of SU(3) are conveniently written in terms of the exponential representation[2]

$$U(\Theta) = \exp\left(-i\sum_{a=1}^{8} \Theta_a \frac{\lambda_a}{2}\right) = \exp\left(-i\Theta_a \frac{\lambda_a}{2}\right), \tag{1.1}$$

with Θ_a real numbers, and where the eight linearly independent matrices λ_a are the so-called Gell-Mann matrices, satisfying

$$\frac{\lambda_a}{2} = i\frac{\partial U}{\partial \Theta_a}(0,\ldots,0), \tag{1.2}$$

$$\lambda_a = \lambda_a^\dagger, \tag{1.3}$$

$$\text{Tr}(\lambda_a \lambda_b) = 2\delta_{ab}, \tag{1.4}$$

$$\text{Tr}(\lambda_a) = 0. \tag{1.5}$$

The Hermiticity of Eq. 1.3 is responsible for $U^\dagger = U^{-1}$. On the other hand, since $\det[\exp(C)] = \exp[\text{Tr}(C)]$, Eq. 1.5 results in $\det(U) = 1$. An explicit representation of the Gell-Mann matrices is given by [25]

$$\lambda_1 = \begin{pmatrix} 0 & 1 & 0 \\ 1 & 0 & 0 \\ 0 & 0 & 0 \end{pmatrix}, \quad \lambda_2 = \begin{pmatrix} 0 & -i & 0 \\ i & 0 & 0 \\ 0 & 0 & 0 \end{pmatrix}, \quad \lambda_3 = \begin{pmatrix} 1 & 0 & 0 \\ 0 & -1 & 0 \\ 0 & 0 & 0 \end{pmatrix},$$

$$\lambda_4 = \begin{pmatrix} 0 & 0 & 1 \\ 0 & 0 & 0 \\ 1 & 0 & 0 \end{pmatrix}, \quad \lambda_5 = \begin{pmatrix} 0 & 0 & -i \\ 0 & 0 & 0 \\ i & 0 & 0 \end{pmatrix}, \quad \lambda_6 = \begin{pmatrix} 0 & 0 & 0 \\ 0 & 0 & 1 \\ 0 & 1 & 0 \end{pmatrix},$$

$$\lambda_7 = \begin{pmatrix} 0 & 0 & 0 \\ 0 & 0 & -i \\ 0 & i & 0 \end{pmatrix}, \quad \lambda_8 = \sqrt{\frac{1}{3}}\begin{pmatrix} 1 & 0 & 0 \\ 0 & 1 & 0 \\ 0 & 0 & -2 \end{pmatrix}. \tag{1.6}$$

The set $\{i\lambda_a | a = 1,\ldots,8\}$ constitutes a basis of the Lie algebra su(3) of SU(3), i.e., the set of all complex, traceless, skew-Hermitian, 3×3 matrices. The Lie product is then defined in terms of ordinary matrix multiplication as the commutator of two elements of su(3). Such a definition naturally satisfies the Lie properties of anticommutativity

---

[2] Most of the time, we will make use of the repeated-index summation convention, i.e., wherever in any term of an expression a literal index occurs twice, this term is to be summed over all possible values of the index. However, sometimes we will find it instructive to explicitly keep the summation symbol including its range of summation.

Table 1.1 Totally antisymmetric nonvanishing structure constants of SU(3)

abc	123	147	156	246	257	345	367	458	678
f_{abc}	1	$\frac{1}{2}$	$-\frac{1}{2}$	$\frac{1}{2}$	$\frac{1}{2}$	$\frac{1}{2}$	$-\frac{1}{2}$	$\frac{1}{2}\sqrt{3}$	$\frac{1}{2}\sqrt{3}$

$$[A, B] = -[B, A] \tag{1.7}$$

as well as the Jacobi identity

$$[A, [B, C]] + [B, [C, A]] + [C, [A, B]] = 0. \tag{1.8}$$

In accordance with Eqs. 1.1 and 1.2, elements of su(3) can be interpreted as tangent vectors in the identity of SU(3).

The structure of the Lie group is encoded in the commutation relations of the Gell-Mann matrices,

$$\left[\frac{\lambda_a}{2}, \frac{\lambda_b}{2}\right] = if_{abc}\frac{\lambda_c}{2}, \tag{1.9}$$

where the totally antisymmetric real structure constants f_{abc} are obtained from Eq. 1.4 as

$$f_{abc} = \frac{1}{4i}\text{Tr}([\lambda_a, \lambda_b]\lambda_c). \tag{1.10}$$

Exercise 1.1 Verify Eq. 1.10.

Exercise 1.2 Show that f_{abc} is totally antisymmetric.
Hint: Consider the symmetry properties of $\text{Tr}([A, B]C)$.

The independent nonvanishing values are explicitly summarized in Table 1.1 [25]. Roughly speaking, these structure constants are a measure of the non-commutativity of the group SU(3).

The anticommutation relations of the Gell-Mann matrices read

$$\{\lambda_a, \lambda_b\} = \frac{4}{3}\delta_{ab}\mathbb{1} + 2d_{abc}\lambda_c, \tag{1.11}$$

where the totally symmetric d_{abc} are given by

$$d_{abc} = \frac{1}{4}\text{Tr}(\{\lambda_a, \lambda_b\}\lambda_c), \tag{1.12}$$

and are summarized in Table 1.2 [25].

Exercise 1.3 Verify Eq. 1.12 and show that d_{abc} is totally symmetric.

Clearly, the anticommutator of two Gell-Mann matrices is not necessarily a Gell-Mann matrix. For example, the square of a (nontrivial) skew-Hermitian matrix is not skew Hermitian.

Table 1.2 Totally symmetric nonvanishing d symbols of SU(3)

abc	118	146	157	228	247	256	338	344	355	366	377	448	558	668	778	888
d_{abc}	$\frac{1}{\sqrt{3}}$	$\frac{1}{2}$	$\frac{1}{2}$	$\frac{1}{\sqrt{3}}$	$-\frac{1}{2}$	$\frac{1}{2}$	$\frac{1}{\sqrt{3}}$	$\frac{1}{2}$	$\frac{1}{2}$	$-\frac{1}{2}$	$-\frac{1}{2}$	$-\frac{1}{2\sqrt{3}}$	$-\frac{1}{2\sqrt{3}}$	$-\frac{1}{2\sqrt{3}}$	$-\frac{1}{2\sqrt{3}}$	$-\frac{1}{\sqrt{3}}$

Exercise 1.4 Using

$$\lambda_a\lambda_b = \frac{1}{2}\{\lambda_a, \lambda_b\} + \frac{1}{2}[\lambda_a, \lambda_b] = \frac{2}{3}\delta_{ab}\mathbb{1} + h_{abc}\lambda_c, \quad h_{abc} \equiv d_{abc} + if_{abc},$$

in combination with Eqs. 1.4 and 1.5, traces of products of Gell-Mann matrices may be evaluated recursively. Verify

$$\mathrm{Tr}(\lambda_a\lambda_b\lambda_c) = 2h_{abc},$$

$$\mathrm{Tr}(\lambda_a\lambda_b\lambda_c\lambda_d) = \frac{4}{3}\delta_{ab}\delta_{cd} + 2h_{abe}h_{ecd},$$

$$\mathrm{Tr}(\lambda_a\lambda_b\lambda_c\lambda_d\lambda_e) = \frac{4}{3}h_{abc}\delta_{de} + \frac{4}{3}\delta_{ab}h_{cde} + 2h_{abf}h_{fcg}h_{gde}.$$

Hint: h_{abc} is invariant under cyclic permutations, i.e. $h_{abc} = h_{bca} = h_{cab}$.

Moreover, it is convenient to introduce as a ninth matrix

$$\lambda_0 = \sqrt{\frac{2}{3}}\mathbb{1},$$

such that Eqs. 1.3 and 1.4 are still satisfied by the nine matrices λ_a. In particular, the set $\{i\lambda_a | a = 0, \ldots, 8\}$ constitutes a basis of the Lie algebra u(3) of U(3), i.e., the set of all complex, skew-Hermitian, 3×3 matrices. Many useful properties of the Gell-Mann matrices can be found in Sect. 8 of Ref. [12].

Finally, an *arbitrary* 3×3 matrix M can be written as

$$M = \sum_{a=0}^{8} M_a\lambda_a, \qquad (1.13)$$

where M_a are complex numbers given by

$$M_a = \frac{1}{2}\mathrm{Tr}(\lambda_a M). \qquad (1.14)$$

1.2 Local Symmetries and the QCD Lagrangian

The gauge principle has proven to be a tremendously successful method in elementary particle physics to generate interactions between matter fields through the exchange of massless gauge bosons (for a detailed account see, e.g., Refs. [1, 13, 15, 43, 54]).

1.2.1 The QED Lagrangian

The best-known example is quantum electrodynamics (QED) which is obtained from promoting the global U(1) symmetry of the Lagrangian describing a free electron,[3]

$$\Psi \mapsto \exp(i\Theta)\Psi : \mathscr{L}_{\text{free}} = \bar{\Psi}\left(i\gamma^{\mu}\partial_{\mu} - m\right)\Psi \mapsto \mathscr{L}_{\text{free}}, \qquad (1.15)$$

to a local symmetry. In this process, the parameter $0 \leq \Theta < 2\pi$ describing an element of U(1) is allowed to vary smoothly in space-time, $\Theta \to \Theta(x)$, which is referred to as gauging the U(1) group. To keep the invariance of the Lagrangian under local transformations one introduces a four-vector potential $\mathscr{A}_{\mu}$ into the theory which transforms under the gauge transformation $\mathscr{A}_{\mu} \mapsto \mathscr{A}_{\mu} + \partial_{\mu}\Theta/e$. The method is referred to as gauging the Lagrangian with respect to U(1):

$$\mathscr{L}_{\text{QED}} = \bar{\Psi}\left[i\gamma^{\mu}\left(\partial_{\mu} - ie\mathscr{A}_{\mu}\right) - m\right]\Psi - \frac{1}{4}\mathscr{F}_{\mu\nu}\mathscr{F}^{\mu\nu}, \qquad (1.16)$$

where $\mathscr{F}_{\mu\nu} = \partial_{\mu}\mathscr{A}_{\nu} - \partial_{\nu}\mathscr{A}_{\mu}$ denotes the electromagnetic field-strength tensor.[4] The covariant derivative of Ψ,

$$D_{\mu}\Psi \equiv (\partial_{\mu} - ie\mathscr{A}_{\mu})\Psi,$$

is defined such that under a so-called gauge transformation of the second kind

$$\Psi(x) \mapsto \exp[i\Theta(x)]\Psi(x), \quad \mathscr{A}_{\mu}(x) \mapsto \mathscr{A}_{\mu}(x) + \partial_{\mu}\Theta(x)/e, \qquad (1.17)$$

it transforms in the same way as Ψ itself:

$$D_{\mu}\Psi(x) \mapsto \exp[i\Theta(x)]D_{\mu}\Psi(x). \qquad (1.18)$$

In Eq. 1.16, the term containing the squared field strength makes the gauge potential a dynamical degree of freedom as opposed to a pure external field. A mass term $M^2\mathscr{A}^2/2$ is not included since it would violate gauge invariance and thus the gauge principle requires massless gauge bosons.[5] In the present case we identify $\mathscr{A}_{\mu}$ with the electromagnetic four-vector potential and $\mathscr{F}_{\mu\nu}$ with the field-strength tensor containing the electric and magnetic fields. The gauge principle has (naturally) generated the interaction of the electromagnetic field with matter:

$$\mathscr{L}_{\text{int}} = -(-e)\bar{\Psi}\gamma^{\mu}\Psi\mathscr{A}_{\mu} = -J^{\mu}\mathscr{A}_{\mu}, \qquad (1.19)$$

---

[3] We use the standard representation for the Dirac matrices (see, e.g., Ref. [11]).

[4] We use natural units, i.e., $\hbar = c = 1$, $e > 0$, and $\alpha = e^2/4\pi \approx 1/137$.

[5] Masses of gauge fields can be induced through a spontaneous breakdown of the gauge symmetry.

Table 1.3 Quark flavors and their charges and masses

Flavor	u	d	s
Charge [e]	2/3	$-1/3$	$-1/3$
Mass [MeV]	1.7–3.3	4.1–5.8	101^{+29}_{-21}
Flavor	c	b	t
Charge [e]	2/3	$-1/3$	2/3
Mass [GeV]	$1.27^{+0.07}_{-0.09}$	$4.19^{+0.18}_{-0.06}$	$172.0 \pm 0.9 \pm 1.3$

See Ref. [39] for details

where J^μ denotes the electromagnetic current (density). If the underlying gauge group is non-Abelian, the gauge principle associates an independent gauge field with each independent continuous parameter of the gauge group.

1.2.2 The QCD Lagrangian

QCD is the gauge theory of the strong interactions [18, 29, 51] with color SU(3) as the underlying gauge group. For a comprehensive guide to the literature on QCD, see Ref. [35]. Historically, the color degree of freedom was introduced into the quark model to account for the Pauli principle in the description of baryons as three-quark states [28, 30]. The matter fields of QCD are the so-called quarks which are spin-1/2 fermions, with six different flavors (u, d, s, c, b, t) in addition to their three possible colors (see Table 1.3). Since quarks have not been observed as asymptotically free states, the meaning of quark masses and their numerical values are tightly connected with the method by which they are extracted from hadronic properties (see Ref. [39] for a thorough discussion). Regarding the so-called current-quark-mass values of the light quarks, one should view the quark-mass terms merely as symmetry breaking parameters with their magnitude providing a measure for the extent to which chiral symmetry is broken [46].[6]

The QCD Lagrangian can be obtained from the Lagrangian for free quarks by applying the gauge principle with respect to the group SU(3). Denoting the quark field components by $q_{\alpha,f,A}$, where $\alpha = 1, \ldots, 4$ refers to the Dirac-spinor index, $f = 1, \ldots, 6$ to the flavor index, and $A = 1, 2, 3$ to the color index, respectively, the "free" quark Lagrangian without interaction may be regarded as the sum of $6 \times 3 = 18$ free fermion Lagrangians:

$$\mathscr{L}_{\text{free quarks}} = \sum_{A=1}^{3} \sum_{f=1}^{6} \sum_{\alpha,\alpha'=1}^{4} \bar{q}_{\alpha,f,A} \left(\gamma^\mu_{\alpha\alpha'} i\partial_\mu - m_f \delta_{\alpha\alpha'} \right) q_{\alpha',f,A}. \qquad (1.20)$$

---

[6] The expression *current-quark masses* for the light quarks is related to the fact that they appear in the divergences of the vector and axial-vector currents (see Sect. 1.3.6).

Suppressing the Dirac-spinor index, we introduce for each quark flavor f a color triplet

$$q_f = \begin{pmatrix} q_{f,1} \\ q_{f,2} \\ q_{f,3} \end{pmatrix}. \tag{1.21}$$

The gauge principle is applied with respect to the group SU(3), i.e., the transformations are flavor independent as all q_f are subject to the same local SU(3) transformation:

$$q_f(x) \mapsto q_f'(x) = \exp\left(-i\Theta_a(x)\frac{\lambda_a^c}{2}\right)q_f(x) \equiv U(x)q_f(x). \tag{1.22}$$

The eight λ_a^c denote Gell-Mann matrices acting in color space and the Θ_a are smooth, real functions in Minkowski space. Technically speaking, each quark field q_f transforms according to the fundamental representation of color SU(3). For the adjoint quark fields, Eq. 1.22 implies the transformation behavior

$$q_f^\dagger(x) \mapsto q_f^\dagger(x)U^\dagger(x). \tag{1.23}$$

Because of the partial derivatives acting on the quark fields, the Lagrangian of Eq. 1.20 is not invariant under the transformations of Eqs. 1.22 and 1.23. In order to keep the invariance of the Lagrangian under local transformations, one introduces eight four-vector gauge potentials $\mathscr{A}_{a\mu}$ into the theory, transforming as[7]

$$\mathscr{A}_\mu \equiv \mathscr{A}_{a\mu}\frac{\lambda_a^c}{2} \mapsto \mathscr{A}_\mu' = U\mathscr{A}_\mu U^\dagger + \frac{i}{g_3}\partial_\mu U U^\dagger. \tag{1.24}$$

The ordinary partial derivative $\partial_\mu q_f$ is replaced by the covariant derivative

$$D_\mu q_f \equiv \left(\partial_\mu + ig_3\mathscr{A}_\mu\right)q_f, \tag{1.25}$$

which, by construction, transforms as the quark field. In Eqs. 1.24 and 1.25, g_3 denotes the strong coupling constant. We note that the interaction between quarks and gluons is independent of the quark flavors which can be seen from the fact that only one coupling constant g_3 appears in Eq. 1.25.

Exercise 1.5 Show that the covariant derivative $D_\mu q_f$ transforms as q_f, i.e., $D_\mu q_f \mapsto (D_\mu q_f)' = D_\mu' q_f' = UD_\mu q_f$.

So far we have only considered the matter-field part of $\mathscr{L}_{\text{QCD}}$ including its interaction with the gauge fields. In order to treat the gauge potentials $\mathscr{A}_{a\mu}$ as

---

[7] Under a gauge transformation of the first kind, i.e., a global SU(3) transformation, the second term on the right-hand side of Eq. 1.24 would vanish and the gauge fields would transform according to the adjoint representation.

dynamical degrees of freedom, one defines a generalization of the field-strength tensor to the non-Abelian case as

$$\mathscr{G}_{a\mu\nu} \equiv \partial_\mu \mathscr{A}_{a\nu} - \partial_\nu \mathscr{A}_{a\mu} - g_3 f_{abc} \mathscr{A}_{b\mu} \mathscr{A}_{c\nu}, \qquad (1.26)$$

with the SU(3) structure constants given in Table 1.1. Given Eq. 1.24, the field-strength tensor transforms under SU(3) as

$$\mathscr{G}_{\mu\nu} \equiv \mathscr{G}_{a\mu\nu} \frac{\lambda_a^c}{2} \mapsto U\mathscr{G}_{\mu\nu} U^\dagger. \qquad (1.27)$$

Exercise 1.6 Verify Eq. 1.27.
Hint: Equation 1.26 is equivalent to $\mathscr{G}_{\mu\nu} = \partial_\mu \mathscr{A}_\nu - \partial_\nu \mathscr{A}_\mu + ig_3[\mathscr{A}_\mu, \mathscr{A}_\nu]$.

The QCD Lagrangian obtained by applying the gauge principle to the free Lagrangian of Eq. 1.20, finally, reads [7, 40]

$$\mathscr{L}_{QCD} = \sum_{\substack{f=u,d,s, \\ c,b,t}} \bar{q}_f \left(i\slashed{D} - m_f \right) q_f - \frac{1}{4} \mathscr{G}_{a\mu\nu} \mathscr{G}_a^{\mu\nu}. \qquad (1.28)$$

Using Eq. 1.4, the purely gluonic part of $\mathscr{L}_{QCD}$ can be written as

$$-\frac{1}{2} \mathrm{Tr}_c \left(\mathscr{G}_{\mu\nu} \mathscr{G}^{\mu\nu} \right),$$

which, using the cyclic property of traces, $\mathrm{Tr}(AB) = \mathrm{Tr}(BA)$, together with $UU^\dagger = \mathbb{1}$, is easily seen to be invariant under the transformation of Eq. 1.27.

In contradistinction to the Abelian case of quantum electrodynamics, the squared field-strength tensor gives rise to gauge-field self interactions involving vertices with three and four gauge fields of strength g_3 and g_3^2, respectively. Such interaction terms are characteristic of non-Abelian gauge theories and make them much more complex than Abelian theories.

From the point of view of gauge invariance, the strong-interaction Lagrangian could also involve a term of the type

$$\mathscr{L}_\theta = \frac{g_3^2 \bar\theta}{64\pi^2} \varepsilon_{\mu\nu\rho\sigma} \mathscr{G}_a^{\mu\nu} \mathscr{G}_a^{\rho\sigma}, \quad \varepsilon_{0123} = 1, \qquad (1.29)$$

where $\varepsilon_{\mu\nu\rho\sigma}$ denotes the totally antisymmetric Levi-Civita tensor.[8] The so-called θ term of Eq. 1.29 implies an explicit P and CP violation of the strong interactions which, for example, would give rise to an electric dipole moment of the neutron.

---

[8]
$$\varepsilon_{\mu\nu\rho\sigma} = \begin{cases} +1 & \text{if } \{\mu, \nu, \rho, \sigma\} \text{ is an even permutation of } \{0,1,2,3\} \\ -1 & \text{if } \{\mu, \nu, \rho, \sigma\} \text{ is an odd permutation of } \{0,1,2,3\} \\ 0 & \text{otherwise} \end{cases}.$$

The present empirical information indicates that the θ term is small [44] and, in the following, we will omit Eq. 1.29 from our discussion.

1.3 Accidental, Global Symmetries of the QCD Lagrangian

1.3.1 Light and Heavy Quarks

The six quark flavors are commonly divided into the three light quarks u, d, and s and the three heavy flavors c, b, and t,

$$
\begin{pmatrix} m_u = (1.7 - 3.3)\,\text{MeV} \\ m_d = (4.1 - 5.8)\,\text{MeV} \\ m_s = (80 - 130)\,\text{MeV} \end{pmatrix} \ll 1\,\text{GeV} \le \begin{pmatrix} m_c = 1.27^{+0.07}_{-0.09}\,\text{GeV} \\ m_b = 4.19^{+0.18}_{-0.06}\,\text{GeV} \\ m_t = (172.0 \pm 0.9 \pm 1.3)\,\text{GeV} \end{pmatrix},
$$

$$(1.30)$$

where the scale of 1 GeV is associated with the masses of the lightest hadrons containing light quarks, e.g., $m_\rho = 770\,\text{MeV}$, which are not Goldstone bosons resulting from spontaneous symmetry breaking. The scale associated with spontaneous chiral symmetry breaking, $4\pi F_\pi \approx 1,170\,\text{MeV}$, is of the same order of magnitude. A nonvanishing pion-decay constant F_π is a necessary and sufficient criterion for spontaneous chiral symmetry breaking (see Sect. 3.2.2).

The masses of the lightest meson and baryon containing a charmed quark, $D^+ = c\bar{d}$ and $\Lambda_c^+ = udc$, are $(1,869.5 \pm 0.4)\,\text{MeV}$ and $(2,286.46 \pm 0.14)\,\text{MeV}$, respectively [41]. The threshold center-of-mass energy to produce, say, a D^+D^- pair in e^+e^- collisions is approximately 3.74 GeV, and thus way beyond the low-energy regime which we are interested in. In the following, we will approximate the full QCD Lagrangian by its light-flavor version, i.e., we will ignore effects due to (virtual) heavy quark-antiquark pairs $h\bar{h}$.

Comparing the proton mass, $m_p = 938\,\text{MeV}$, with the sum of two up and one down current-quark masses (see Table 1.3),

$$
m_p \gg 2m_u + m_d, \tag{1.31}
$$

shows that an interpretation of the proton mass in terms of current-quark-mass parameters must be very different from, say, the situation in the hydrogen atom, where the mass is essentially given by the sum of the electron and proton masses, corrected by a small amount of binding energy. In this context we recall that the current-quark masses must not be confused with the constituent-quark masses of a (nonrelativistic) quark model, which are typically of the order of 350 MeV. In particular, Eq. 1.31 suggests that the Lagrangian $\mathscr{L}^0_{\text{QCD}}$, containing only the light-flavor quarks in the so-called chiral limit $m_u, m_d, m_s \to 0$, might be a good starting point in the discussion of low-energy QCD:

$$\mathscr{L}^0_{\text{QCD}} = \sum_{l=u,d,s} \bar{q}_l i \not{D} q_l - \frac{1}{4} \mathscr{G}_{a\mu\nu} \mathscr{G}^{\mu\nu}_a. \tag{1.32}$$

We repeat that the covariant derivative D_μ in Eq. 1.25 acts on color indices only, but is independent of flavor.

1.3.2 Left-Handed and Right-Handed Quark Fields

In order to fully exhibit the global symmetries of Eq. 1.32, we consider the chirality matrix $\gamma_5 = \gamma^5 = i\gamma^0\gamma^1\gamma^2\gamma^3 = \gamma_5^\dagger, \{\gamma^\mu, \gamma_5\} = 0, \gamma_5^2 = \mathbb{1}$, and introduce the projection operators

$$P_R = \frac{1}{2}(\mathbb{1} + \gamma_5) = P_R^\dagger, \quad P_L = \frac{1}{2}(\mathbb{1} - \gamma_5) = P_L^\dagger, \tag{1.33}$$

where the subscripts R and L refer to right-handed and left-handed, respectively, as will become clearer below. The 4×4 matrices P_R and P_L satisfy a completeness relation,

$$P_R + P_L = \mathbb{1}, \tag{1.34}$$

are idempotent,

$$P_R^2 = P_R, \quad P_L^2 = P_L, \tag{1.35}$$

and respect the orthogonality relations

$$P_R P_L = P_L P_R = 0. \tag{1.36}$$

Exercise 1.7 Verify the properties of Eqs. 1.33–1.36.

The combined properties of Eqs. 1.33–1.36 guarantee that P_R and P_L are indeed projection operators which project from the Dirac field variable q to its chiral components q_R and q_L,

$$q_R = P_R q, \quad q_L = P_L q. \tag{1.37}$$

We recall in this context that a chiral (field) variable is one which under parity is transformed into neither the original variable nor its negative.[9] Under parity, the quark field is transformed into its parity conjugate,

$$P : q(t, \vec{x}) \mapsto \gamma_0 q(t, -\vec{x}),$$

and hence

---

[9] In case of fields, a transformation of the argument $\vec{x} \mapsto -\vec{x}$ is implied.

$$q_R(t, \vec{x}) = P_R q(t, \vec{x}) \mapsto P_R \gamma_0 q(t, -\vec{x}) = \gamma_0 P_L q(t, -\vec{x}) = \gamma_0 q_L(t, -\vec{x}) \neq \pm q_R(t, -\vec{x}),$$

and similarly for q_L.[10]

The terminology right-handed and left-handed fields can easily be visualized in terms of the solution to the free Dirac equation. For that purpose, let us consider an extreme relativistic positive-energy solution to the free Dirac equation with three-momentum $\vec{p}$,[11]

$$u(\vec{p}, \pm) = \sqrt{E + m} \begin{pmatrix} \chi_\pm \\ \frac{\vec{\sigma} \cdot \vec{p}}{E+m} \chi_\pm \end{pmatrix} \xrightarrow{E \gg m} \sqrt{E} \begin{pmatrix} \chi_\pm \\ \pm \chi_\pm \end{pmatrix} \equiv u_\pm(\vec{p}),$$

where we assume that the spin in the rest frame is either parallel or antiparallel to the direction of momentum

$$\vec{\sigma} \cdot \hat{p} \chi_\pm = \pm \chi_\pm.$$

In the standard representation of Dirac matrices[12] we find

$$P_R = \frac{1}{2} \begin{pmatrix} \mathbb{1}_{2 \times 2} & \mathbb{1}_{2 \times 2} \\ \mathbb{1}_{2 \times 2} & \mathbb{1}_{2 \times 2} \end{pmatrix}, \quad P_L = \frac{1}{2} \begin{pmatrix} \mathbb{1}_{2 \times 2} & -\mathbb{1}_{2 \times 2} \\ -\mathbb{1}_{2 \times 2} & \mathbb{1}_{2 \times 2} \end{pmatrix}.$$

Exercise 1.8   Show that

$$P_R u_+ = u_+, \quad P_L u_+ = 0, \quad P_R u_- = 0, \quad P_L u_- = u_-.$$

In the extreme relativistic limit (or better, in the zero-mass limit), the operators P_R and P_L project onto the positive and negative helicity eigenstates, i.e., in this limit chirality equals helicity.

Our goal is to analyze the symmetry of the QCD Lagrangian with respect to independent global transformations of the left- and right-handed fields. There are 16 independent 4×4 matrices Γ, which can be expressed in terms of the unit matrix $\mathbb{1}$, the Dirac matrices γ^μ, the chirality matrix γ_5, the products $\gamma^\mu \gamma_5$, and the six matrices $\sigma^{\mu\nu} = i[\gamma^\mu, \gamma^\nu]/2$. In order to decompose the corresponding 16 quadratic forms into their respective projections to right- and left-handed fields, we make use of

$$\bar{q} \Gamma q = \begin{cases} \bar{q}_R \Gamma q_R + \bar{q}_L \Gamma q_L & \text{for } \Gamma \in \Gamma_1 \equiv \{\gamma^\mu, \gamma^\mu \gamma_5\}, \\ \bar{q}_R \Gamma q_L + \bar{q}_L \Gamma q_R & \text{for } \Gamma \in \Gamma_2 \equiv \{\mathbb{1}, \gamma_5, \sigma^{\mu\nu}\}, \end{cases} \tag{1.38}$$

where[13]

---

[10]   Note that in the above sense, also q is a chiral variable. However, the assignment of handedness does not have such an intuitive meaning as in the case of q_L and q_R.

[11]   Here we adopt a covariant normalization of the spinors, $u^{(\alpha)\dagger}(\vec{p}) u^{(\beta)}(\vec{p}) = 2E \delta_{\alpha\beta}$, etc.

[12]   Unless stated otherwise, we use the convention of Bjorken and Drell [11].

[13]   For notational convenience we write $\bar{q}_L$ and $\bar{q}_R$ instead of $\overline{q_L}$ and $\overline{q_R}$.

$$\bar{q}_R = q_R^\dagger \gamma_0 = q^\dagger P_R^\dagger \gamma_0 = q^\dagger P_R \gamma_0 = q^\dagger \gamma_0 P_L = \bar{q} P_L,$$
$$\bar{q}_L = \bar{q} P_R.$$

Exercise 1.9 Verify Eq. 1.38.
Hint: Insert unit matrices as

$$\bar{q}\Gamma q = \bar{q}(P_R + P_L)\Gamma(P_R + P_L)q,$$

and make use of $\{\Gamma, \gamma_5\} = 0$ for $\Gamma \in \Gamma_1$ and $[\Gamma, \gamma_5] = 0$ for $\Gamma \in \Gamma_2$ as well as the properties of the projection operators derived in Exercise 1.7.

We stress that the validity of Eq. 1.38 is general and does not refer to "massless" quark fields.

We now apply Eq. 1.38 to the term in the Lagrangian of Eq. 1.32 containing the contraction of the covariant derivative with γ^μ. This quadratic quark form decouples into the sum of two terms which connect only left-handed with left-handed and right-handed with right-handed quark fields. The QCD Lagrangian in the chiral limit can then be written as

$$\mathscr{L}^0_{\text{QCD}} = \sum_{l=u,d,s} \left(\bar{q}_{R,l}\, i\slashed{D}q_{R,l} + \bar{q}_{L,l}\, i\slashed{D}q_{L,l}\right) - \frac{1}{4}\mathscr{G}_{a\mu\nu}\mathscr{G}_a^{\mu\nu}. \qquad (1.39)$$

Due to the flavor independence of the covariant derivative, $\mathscr{L}^0_{\text{QCD}}$ is invariant under

$$\begin{pmatrix} u_L \\ d_L \\ s_L \end{pmatrix} \mapsto U_L \begin{pmatrix} u_L \\ d_L \\ s_L \end{pmatrix} = \exp\left(-i\sum_{a=1}^{8}\Theta_{La}\frac{\lambda_a}{2}\right)e^{-i\Theta_L}\begin{pmatrix} u_L \\ d_L \\ s_L \end{pmatrix},$$

$$\begin{pmatrix} u_R \\ d_R \\ s_R \end{pmatrix} \mapsto U_R \begin{pmatrix} u_R \\ d_R \\ s_R \end{pmatrix} = \exp\left(-i\sum_{a=1}^{8}\Theta_{Ra}\frac{\lambda_a}{2}\right)e^{-i\Theta_R}\begin{pmatrix} u_R \\ d_R \\ s_R \end{pmatrix}, \qquad (1.40)$$

where U_L and U_R are independent unitary 3×3 matrices and where we have extracted the factors $e^{-i\Theta_L}$ and $e^{-i\Theta_R}$ for future convenience. We have thus decomposed the U(3) × U(3) transformations into SU(3) × SU(3) × U(1) × U(1) transformations. Note that the Gell-Mann matrices act in flavor space. We will refer to the invariance of $\mathscr{L}^0_{\text{QCD}}$ under SU(N)$_L$ × SU(N)$_R$ ($N = 2$ or 3) as chiral symmetry.

$\mathscr{L}^0_{\text{QCD}}$ is said to have a classical *global* U(3)$_L$ × U(3)$_R$ symmetry. Applying Noether's theorem one would expect a total of $2 \times (8 + 1) = 18$ conserved currents from such an invariance.

1.3.3 Noether Theorem

Noether's theorem [6, 24, 31, 42] establishes the connection between continuous symmetries of a dynamical system and conserved quantities (constants of the motion). For simplicity we consider only internal symmetries. (The method can also be used to discuss the consequences of Poincaré invariance.)

In order to identify the conserved currents associated with the transformations of Eqs. 1.40, we briefly recall the method of Gell-Mann and Lévy [24], which we will then apply to Eq. 1.39.

We start with a Lagrangian $\mathscr{L}$ depending on n independent fields Φ_i and their first partial derivatives $\partial_\mu \Phi_i$ $(i = 1, \ldots, n)$, collectively denoted by the symbols Φ and $\partial_\mu \Phi$,[14]

$$\mathscr{L} = \mathscr{L}(\Phi, \partial_\mu \Phi), \tag{1.41}$$

from which one obtains n equations of motion:

$$\frac{\partial \mathscr{L}}{\partial \Phi_i} - \partial_\mu \frac{\partial \mathscr{L}}{\partial \partial_\mu \Phi_i} = 0, \quad i = 1, \ldots, n. \tag{1.42}$$

Suppose the Lagrangian of Eq. 1.41 to be invariant under a continuous, global transformation of the fields depending smoothly on r real parameters. The method of Gell-Mann and Lévy [24] now consists of promoting this global symmetry to a *local* one, from which we will then be able to identify the Noether currents. To that end we consider transformations which depend on r real local parameters $\varepsilon_a(x)$,[15]

$$\Phi_i(x) \mapsto \Phi_i'(x) = \Phi_i(x) + \delta\Phi_i(x) = \Phi_i(x) - i\varepsilon_a(x)F_{ai}[\Phi(x)], \tag{1.43}$$

and obtain, neglecting terms of order ε^2, as the variation of the Lagrangian,

$$\begin{aligned}
\delta\mathscr{L} &= \mathscr{L}(\Phi', \partial_\mu \Phi') - \mathscr{L}(\Phi, \partial_\mu \Phi) \\
&= \frac{\partial \mathscr{L}}{\partial \Phi_i}\delta\Phi_i + \frac{\partial \mathscr{L}}{\partial \partial_\mu \Phi_i} \underbrace{\partial_\mu \delta\Phi_i}_{= -i\partial_\mu \varepsilon_a F_{ai} - i\varepsilon_a \partial_\mu F_{ai}} \\
&= \varepsilon_a \left(-i\frac{\partial \mathscr{L}}{\partial \Phi_i}F_{ai} - i\frac{\partial \mathscr{L}}{\partial \partial_\mu \Phi_i}\partial_\mu F_{ai} \right) + \partial_\mu \varepsilon_a \left(-i\frac{\partial \mathscr{L}}{\partial \partial_\mu \Phi_i}F_{ai} \right) \\
&\equiv \varepsilon_a \partial_\mu J_a^\mu + \partial_\mu \varepsilon_a J_a^\mu.
\end{aligned} \tag{1.44}$$

According to this equation we define for each infinitesimal transformation a four-current density as

---

[14] The extension to higher-order derivatives is also possible.

[15] Note that the transformation need not be realized linearly on the fields.

$$J_a^\mu = -i\frac{\partial\mathscr{L}}{\partial\partial_\mu\Phi_i}F_{ai}. \tag{1.45}$$

By calculating the divergence $\partial_\mu J_a^\mu$ of Eq. 1.45

$$\partial_\mu J_a^\mu = -i\left(\partial_\mu\frac{\partial\mathcal{L}}{\partial\partial_\mu\Phi_i}\right)F_{ai} - i\frac{\partial\mathscr{L}}{\partial\partial_\mu\Phi_i}\partial_\mu F_{ai}$$

$$= -i\frac{\partial\mathscr{L}}{\partial\Phi_i}F_{ai} - i\frac{\partial\mathscr{L}}{\partial\partial_\mu\Phi_i}\partial_\mu F_{ai},$$

where we made use of the equations of motion, Eq. 1.42, we explicitly verify the consistency with the definition of $\partial_\mu J_a^\mu$ according to Eq. 1.44. From Eq. 1.44 it is straightforward to obtain the four-currents[16] as well as their divergences as

$$J_a^\mu = \frac{\partial\delta\mathscr{L}}{\partial\partial_\mu\varepsilon_a}, \tag{1.46}$$

$$\partial_\mu J_a^\mu = \frac{\partial\delta\mathscr{L}}{\partial\varepsilon_a}. \tag{1.47}$$

We chose the parameters of the transformation to be local. However, the Lagrangian of Eq. 1.41 was only assumed to be invariant under a *global* transformation. In that case, the term $\partial_\mu\varepsilon_a$ disappears, and since the Lagrangian is invariant under such transformations, we see from Eq. 1.44 that the current J_a^μ is conserved, $\partial_\mu J_a^\mu = 0$. For a conserved current the charge

$$Q_a(t) = \int d^3x J_a^0(t,\vec{x}) \tag{1.48}$$

is time independent, i.e., a constant of the motion.

Exercise 1.10 By applying the divergence theorem for an infinite volume, show that $Q_a(t)$ is a constant of the motion for $\delta\mathscr{L} = 0$. Assume that the fields and thus the current density vanish sufficiently rapidly for $|\vec{x}| \to \infty$.

Exercise 1.11 Consider the Lagrangian of two real scalar fields Φ_1 and Φ_2 of equal masses m with a so-called $\lambda\Phi^4$ interaction:

$$\mathscr{L} = \frac{1}{2}[\partial_\mu\Phi_1\partial^\mu\Phi_1 + \partial_\mu\Phi_2\partial^\mu\Phi_2 - m^2(\Phi_1^2 + \Phi_2^2)] - \frac{\lambda}{4}(\Phi_1^2 + \Phi_2^2)^2, \tag{1.49}$$

where $m^2 > 0$ and $\lambda > 0$.

(a) Determine the variation $\delta\mathscr{L}$ under the infinitesimal, local transformation of the fields

---

[16] Most of the time, we follow common practice and speak of four-currents instead of four-current densities.

Table 1.4 Different versions of conservation laws

Invariant quantity	Current density or charge
$\delta\mathscr{L} = 0$	$J^\mu = \frac{\partial\mathscr{L}}{\partial\partial_\mu\Phi}\delta\tilde\Phi$
$\delta\mathscr{L} = \varepsilon\partial_\mu\mathscr{J}^\mu$	$J^\mu = \frac{\partial\mathscr{L}}{\partial\partial_\mu\Phi}\delta\tilde\Phi - \mathscr{J}^\mu$
$\delta L = 0$	$Q = \int d^3x \frac{\partial\mathscr{L}}{\partial\partial_0\Phi}\delta\tilde\Phi$
$\delta L = \varepsilon\frac{d\mathscr{Q}(t)}{dt}$	$Q = \int d^3x \frac{\partial\mathscr{L}}{\partial\partial_0\Phi}\delta\tilde\Phi - \mathscr{Q}$
$\delta S = 0$	Explicit form of J^μ not known

The transformation of the fields is symbolically written as $\Phi \mapsto \Phi + \delta\Phi = \Phi + \varepsilon\delta\tilde\Phi$. L and S refer to the Lagrange function and the action, respectively. The second column denotes which quantity can be *explicitly* obtained from the Lagrangian

$$\Phi'_1 = \Phi_1 + \delta\Phi_1 = \Phi_1 - \varepsilon(x)\Phi_2, \quad \Phi'_2 = \Phi_2 + \delta\Phi_2 = \Phi_2 + \varepsilon(x)\Phi_1. \quad (1.50)$$

(b) Apply the method of Gell-Mann and Lévy to determine the corresponding current J^μ and show that J^μ is conserved.

In the above discussion, we have assumed that the Lagrangian is invariant under a global transformation of the type Eq. 1.43 which is sufficient for the present purposes. However, we would like to mention that, demanding less restrictive assumptions, it is still possible to derive conservation laws of the type $\partial_\mu J^\mu = 0$ (see Ref. [52]). The various possibilities are summarized in Table 1.4.

So far we have discussed Noether's theorem on the classical level, implying that the charges $Q_a(t)$ can have any continuous real value. However, we also need to discuss the implications of a transition to a quantum theory.

To that end, let us first recall the transition from classical mechanics to quantum mechanics. Consider a point mass m in a central potential $V(r)$, i.e., the corresponding Lagrange and Hamilton functions are rotationally invariant. As a result of this invariance, the angular momentum $\vec{l} = \vec{r} \times \vec{p}$ is a constant of the motion which, in classical mechanics, can have any continuous real value. In the transition to quantum mechanics, the components of $\vec{r}$ and $\vec{p}$ turn into Hermitian, linear operators, satisfying the commutation relations

$$[\hat{x}_i, \hat{p}_j] = i\delta_{ij}, \quad [\hat{x}_i, \hat{x}_j] = 0, \quad [\hat{p}_i, \hat{p}_j] = 0.$$

The components of the angular momentum operator are given by

$$\hat{l}_i = \varepsilon_{ijk}\hat{x}_j\hat{p}_k$$

which, for later comparison with the results in quantum field theory, we express in terms of the 3×3 matrices L_i^{ad} of the adjoint representation,

$$\hat{l}_i = -i\hat{p}_j \underbrace{(-i\varepsilon_{ijk})}_{= (L_i^{\text{ad}})_{jk}} \hat{x}_k. \quad (1.51)$$

Both the matrices of the adjoint representation and the components of the angular momentum operator satisfy the angular momentum commutation relations,

$$[L_i^{\text{ad}}, L_j^{\text{ad}}] = i\varepsilon_{ijk}L_k^{\text{ad}}, \quad [\hat{l}_i, \hat{l}_j] = i\varepsilon_{ijk}\hat{l}_k.$$

Since the components of the angular momentum operator cannot simultaneously be diagonalized, the states are organized as eigenstates of $\hat{l}_i\hat{l}_i$ and $\hat{l}_3$ with eigenvalues $l(l+1)$ and $m = -l, \ldots, l$ ($l = 0, 1, 2, \ldots$). Also note that the angular momentum operators are the generators of rotations. The rotational invariance of the quantum system implies that the components of the angular momentum operator commute with the Hamilton operator,

$$[\hat{H}, \hat{l}_i] = 0,$$

i.e., they are still constants of the motion. One then simultaneously diagonalizes $\hat{H}, \hat{l}_i\hat{l}_i$, and $\hat{l}_3$. For example, the energy eigenvalues of the hydrogen atom are given by

$$E_n = -\frac{\alpha^2 m}{2n^2} \approx -\frac{13.6}{n^2}\,\text{eV},$$

where $n = n' + l + 1, n' \geq 0$ denotes the principal quantum number, and the degeneracy of an energy level is given by n^2 (spin neglected). The value E_1 and the spacing of the levels are determined by the *dynamics* of the system, i.e., the specific form of the potential, whereas the multiplicities of the energy levels are a consequence of the underlying rotational *symmetry*.[17]

Having the example from quantum mechanics in mind, let us turn to the analogous case in quantum field theory. After canonical quantization, the fields Φ_i and their conjugate momenta $\Pi_i = \partial\mathscr{L}/\partial(\partial_0\Phi_i)$ are considered as linear operators acting on a Hilbert space which, in the Heisenberg picture, are subject to the equal-time commutation relations

$$\begin{aligned}
[\Phi_i(t, \vec{x}), \Pi_j(t, \vec{y})] &= i\delta^3(\vec{x} - \vec{y})\delta_{ij}, \\
[\Phi_i(t, \vec{x}), \Phi_j(t, \vec{y})] &= 0, \\
[\Pi_i(t, \vec{x}), \Pi_j(t, \vec{y})] &= 0.
\end{aligned} \qquad (1.52)$$

As a special case of Eq. 1.43 let us consider infinitesimal transformations that are *linear* in the fields,

$$\Phi_i(x) \mapsto \Phi_i'(x) = \Phi_i(x) - i\varepsilon_a(x)t_{a,ij}\Phi_j(x), \qquad (1.53)$$

where the $t_{a,ij}$ are constants generating a mixing of the fields. The angular-momentum analogue reads

---

[17] In fact, the accidental degeneracy for $n \geq 2$ is a result of an even higher symmetry of the $1/r$ potential, namely an SO(4) symmetry (see, e.g., Ref. [34]).

$$\hat{x}_i \mapsto \hat{x}_i - i\varepsilon_k(-i\varepsilon_{kij})\hat{x}_j.$$

From Eq. 1.45 we then obtain

$$J_a^\mu(x) = -i\frac{\partial \mathscr{L}}{\partial \partial_\mu \Phi_i}t_{a,ij}\Phi_j, \tag{1.54}$$

$$Q_a(t) = -i\int d^3x\Pi_i(x)t_{a,ij}\Phi_j(x), \tag{1.55}$$

where $J_a^\mu(x)$ and $Q_a(t)$ are now operators. Note the perfect analogy to the angular momentum case of Eq. 1.51.

In order to interpret the charge operators $Q_a(t)$, let us make use of the equal-time commutation relations, Eqs. 1.52, and calculate their commutators with the field operators,

$$\begin{aligned}[Q_a(t), \Phi_k(t, \vec{y})] &= -it_{a,ij}\int d^3x[\Pi_i(t, \vec{x})\Phi_j(t, \vec{x}), \Phi_k(t, \vec{y})]\\ &= -t_{a,kj}\Phi_j(t, \vec{y}),\end{aligned} \tag{1.56}$$

which corresponds to

$$[\hat{l}_k, \hat{x}_i] = i\varepsilon_{kij}\hat{x}_j.$$

Note that we did not require the charge operators to be time independent.

Exercise 1.12 Using the equal-time commutation relations of Eqs. 1.52, verify Eq. 1.56.

On the other hand, for the transformation behavior of the Hilbert space associated with a global infinitesimal transformation, we make an ansatz in terms of an infinitesimal unitary transformation[18]

$$|\alpha'\rangle = [1 + i\varepsilon_a G_a(t)]|\alpha\rangle, \tag{1.57}$$

with Hermitian operators G_a. Demanding

$$\langle\beta|A|\alpha\rangle = \langle\beta'|A'|\alpha'\rangle \quad \forall \quad |\alpha\rangle, |\beta\rangle, \varepsilon_a, \tag{1.58}$$

in combination with Eq. 1.53 yields the condition

$$\begin{aligned}\langle\beta|\Phi_i(x)|\alpha\rangle &= \langle\beta'|\Phi_i'(x)|\alpha'\rangle\\ &= \langle\beta|[1 - i\varepsilon_a G_a(t)][\Phi_i(x) - i\varepsilon_b t_{b,ij}\Phi_j(x)][1 + i\varepsilon_c G_c(t)]|\alpha\rangle.\end{aligned}$$

By comparing the terms linear in ε_a on both sides,

---

[18] We have chosen to have the fields (field operators) rotate actively and thus must transform the states of the Hilbert space in the opposite direction.

$$0 = -i\varepsilon_a[G_a(t), \Phi_i(x)] \quad \underbrace{-i\varepsilon_a t_{a,ij}\Phi_j(x)}_{} \quad , \qquad (1.59)$$
$$= i\varepsilon_a[Q_a(t), \Phi_i(x)]$$

we see that the infinitesimal generators $G_a(t)$, acting on the Hilbert space states, that are associated with the transformation of the fields are identical with the charge operators $Q_a(t)$ of Eq. 1.55.

Finally, evaluating the commutation relations for the case of several generators,

$$[Q_a(t), Q_b(t)] = -i(t_{a,ij}t_{b,jk} - t_{b,ij}t_{a,jk}) \int d^3x \Pi_i(t,\vec{x})\Phi_k(t,\vec{x}), \qquad (1.60)$$

we find the right-hand side of Eq. 1.60 to be again proportional to a charge operator, if

$$t_{a,ij}t_{b,jk} - t_{b,ij}t_{a,jk} = iC_{abc}t_{c,ik}, \qquad (1.61)$$

i.e., in that case the charge operators $Q_a(t)$ form a Lie algebra

$$[Q_a(t), Q_b(t)] = iC_{abc}Q_c(t) \qquad (1.62)$$

with structure constants C_{abc}.

Exercise 1.13 Using the canonical commutation relations of Eqs. 1.52, verify Eq. 1.60.

From now on we assume the validity of Eq. 1.61 and interpret the constants $t_{a,ij}$ as the entries in the ith row and jth column of an $n \times n$ matrix T_a,

$$T_a = \begin{pmatrix} t_{a,11} & \cdots & t_{a,1n} \\ \vdots & & \vdots \\ t_{a,n1} & \cdots & t_{a,nn} \end{pmatrix}.$$

Because of Eq. 1.61, these matrices form an n-dimensional representation of a Lie algebra,

$$[T_a, T_b] = iC_{abc}T_c.$$

The infinitesimal, linear transformations of the fields Φ_i may then be written in a compact form,

$$\begin{pmatrix} \Phi_1(x) \\ \vdots \\ \Phi_n(x) \end{pmatrix} = \Phi(x) \mapsto \Phi'(x) = (\mathbb{1} - i\varepsilon_a T_a)\Phi(x). \qquad (1.63)$$

In general, through an appropriate unitary transformation, the matrices T_a may be decomposed into their irreducible components, i.e., brought into block-diagonal form, such that only fields belonging to the same multiplet transform into each other under the symmetry group.

Exercise 1.14 In order to also deal with the case of fermions, we discuss the isospin invariance of the strong interactions and consider, in total, five fields. The commutation relations of the isospin algebra su(2) read

$$[Q_i, Q_j] = i\varepsilon_{ijk}Q_k. \tag{1.64}$$

A basis of the so-called fundamental representation ($n = 2$) is given by

$$T_i^f = \frac{1}{2}\tau_i \quad (\text{f: fundamental}) \tag{1.65}$$

with the Pauli matrices

$$\tau_1 = \begin{pmatrix} 0 & 1 \\ 1 & 0 \end{pmatrix}, \quad \tau_2 = \begin{pmatrix} 0 & -i \\ i & 0 \end{pmatrix}, \quad \tau_3 = \begin{pmatrix} 1 & 0 \\ 0 & -1 \end{pmatrix}. \tag{1.66}$$

We replace the fields Φ_4 and Φ_5 by the nucleon doublet containing the proton and neutron fields,

$$\Psi = \begin{pmatrix} p \\ n \end{pmatrix}. \tag{1.67}$$

A basis of the so-called adjoint representation ($n = 3$) is given by

$$T_i^{\text{ad}} = \begin{pmatrix} t_{i,11}^{\text{ad}} & t_{i,12}^{\text{ad}} & t_{i,13}^{\text{ad}} \\ t_{i,21}^{\text{ad}} & t_{i,22}^{\text{ad}} & t_{i,23}^{\text{ad}} \\ t_{i,31}^{\text{ad}} & t_{i,32}^{\text{ad}} & t_{i,33}^{\text{ad}} \end{pmatrix}, \quad t_{i,jk}^{\text{ad}} = -i\varepsilon_{ijk} \quad (\text{ad: adjoint}), \tag{1.68}$$

i.e.

$$T_1^{\text{ad}} = \begin{pmatrix} 0 & 0 & 0 \\ 0 & 0 & -i \\ 0 & i & 0 \end{pmatrix}, \quad T_2^{\text{ad}} = \begin{pmatrix} 0 & 0 & i \\ 0 & 0 & 0 \\ -i & 0 & 0 \end{pmatrix}, \quad T_3^{\text{ad}} = \begin{pmatrix} 0 & -i & 0 \\ i & 0 & 0 \\ 0 & 0 & 0 \end{pmatrix}. \tag{1.69}$$

With $\Phi_{1,2,3} \to \vec{\Phi}$ we consider the pseudoscalar pion-nucleon Lagrangian

$$\mathscr{L} = \bar{\Psi}(i\,\slashed{\partial} - m_N)\Psi + \frac{1}{2}\left(\partial_\mu\vec{\Phi}\cdot\partial^\mu\vec{\Phi} - M_\pi^2\vec{\Phi}^2\right) - ig\bar{\Psi}\gamma_5\vec{\Phi}\cdot\vec{\tau}\Psi, \tag{1.70}$$

where $g = g_{\pi N} = 13.2$ denotes the pion-nucleon coupling constant. As a specific application of the infinitesimal transformation of Eq. 1.53 we take

$$\begin{pmatrix} \vec{\Phi} \\ \Psi \end{pmatrix} \mapsto [\mathbb{1} - i\varepsilon_a(x)T_a]\begin{pmatrix} \vec{\Phi} \\ \Psi \end{pmatrix}, \quad T_a = \begin{pmatrix} T_a^{\text{ad}} & 0_{3\times2} \\ 0_{2\times3} & T_a^f \end{pmatrix}, \tag{1.71}$$

(T_a block-diagonal), i.e.

$$\Psi \mapsto \Psi' = \left(\mathbb{1} - i\vec{\varepsilon}(x)\cdot\frac{\vec{\tau}}{2}\right)\Psi, \tag{1.72}$$

$$\vec{\Phi} \mapsto \left(\mathbb{1} - i\vec{\varepsilon}(x) \cdot \vec{T}^{\mathrm{ad}}\right)\vec{\Phi} = \vec{\Phi} + \vec{\varepsilon} \times \vec{\Phi}. \tag{1.73}$$

(a) Show that the variation of the Lagrangian is given by

$$\delta\mathscr{L} = \partial_\mu \vec{\varepsilon} \cdot \left(\bar{\Psi}\gamma^\mu \frac{\vec{\tau}}{2}\Psi + \vec{\Phi} \times \partial^\mu \vec{\Phi}\right). \tag{1.74}$$

From Eqs. 1.46 and 1.47 we find

$$\vec{J}^\mu = \frac{\partial \delta\mathscr{L}}{\partial \partial_\mu \vec{\varepsilon}} = \bar{\Psi}\gamma^\mu \frac{\vec{\tau}}{2}\Psi + \vec{\Phi} \times \partial^\mu \vec{\Phi}, \tag{1.75}$$

$$\partial_\mu \vec{J}^\mu = \frac{\partial \delta\mathscr{L}}{\partial \vec{\varepsilon}} = 0. \tag{1.76}$$

We obtain three time-independent charge operators

$$\vec{Q} = \int d^3x \left[\Psi^\dagger(x)\frac{\vec{\tau}}{2}\Psi(x) + \vec{\Phi}(x) \times \vec{\Pi}(x)\right]. \tag{1.77}$$

These operators are the infinitesimal generators of transformations of the Hilbert space states. The generators decompose into a fermionic and a bosonic piece, which commute with each other. Using the anticommutation relations for the fermion fields

$$\{\Psi_{\alpha,r}(t,\vec{x}), \Psi^\dagger_{\beta,s}(t,\vec{y})\} = \delta^3(\vec{x} - \vec{y})\delta_{\alpha\beta}\delta_{rs}, \tag{1.78}$$

$$\{\Psi_{\alpha,r}(t,\vec{x}), \Psi_{\beta,s}(t,\vec{y})\} = 0, \tag{1.79}$$

$$\{\Psi^\dagger_{\alpha,r}(t,\vec{x}), \Psi^\dagger_{\beta,s}(t,\vec{y})\} = 0, \tag{1.80}$$

where α and β denote Dirac indices, and r and s denote isospin indices, and the commutation relations for the boson fields

$$[\Phi_r(t,\vec{x}), \Pi_s(t,\vec{y})] = i\delta^3(\vec{x} - \vec{y})\delta_{rs}, \tag{1.81}$$

$$[\Phi_r(t,\vec{x}), \Phi_s(t,\vec{y})] = 0, \tag{1.82}$$

$$[\Pi_r(t,\vec{x}), \Pi_s(t,\vec{y})] = 0, \tag{1.83}$$

together with the fact that fermion fields and boson fields commute, we will verify:

$$[Q_i, Q_j] = i\varepsilon_{ijk}Q_k. \tag{1.84}$$

We start from

$$[Q_i, Q_j] = \int d^3x d^3y \Big[\Psi^\dagger(t, \vec{x}) \frac{\tau_i}{2} \Psi(t, \vec{x}) + \varepsilon_{ikl} \Phi_k(t, \vec{x}) \Pi_l(t, \vec{x}),$$

$$\Psi^\dagger(t, \vec{y}) \frac{\tau_j}{2} \Psi(t, \vec{y}) + \varepsilon_{jmn} \Phi_m(t, \vec{y}) \Pi_n(t, \vec{y}) \Big]$$

$$= \int d^3x d^3y \Big(\Big[\Psi^\dagger(t, \vec{x}) \frac{\tau_i}{2} \Psi(t, \vec{x}), \Psi^\dagger(t, \vec{y}) \frac{\tau_j}{2} \Psi(t, \vec{y}) \Big]$$

$$+ \Big[\varepsilon_{ikl} \Phi_k(t, \vec{x}) \Pi_l(t, \vec{x}), \varepsilon_{jmn} \Phi_m(t, \vec{y}) \Pi_n(t, \vec{y}) \Big] \Big)$$

$$= A_{ij} + B_{ij}.$$

For the evaluation of A_{ij} we make use of

$$\Big[\Psi^\dagger_{\alpha,r}(t, \vec{x}) \widehat{\mathcal{O}}_{1\alpha\beta,rs} \Psi_{\beta,s}(t, \vec{x}), \Psi^\dagger_{\gamma,t}(t, \vec{y}) \widehat{\mathcal{O}}_{2\gamma\delta,tu} \Psi_{\delta,u}(t, \vec{y}) \Big]$$

$$= \widehat{\mathcal{O}}_{1\alpha\beta,rs} \widehat{\mathcal{O}}_{2\gamma\delta,tu} \Big[\Psi^\dagger_{\alpha,r}(t, \vec{x}) \Psi_{\beta,s}(t, \vec{x}), \Psi^\dagger_{\gamma,t}(t, \vec{y}) \Psi_{\delta,u}(t, \vec{y}) \Big]. \qquad (1.85)$$

(b) Verify

$$[ab, cd] = a\{b, c\}d - ac\{b, d\} + \{a, c\}db - c\{a, d\}b \qquad (1.86)$$

and express the commutator of fermion fields in terms of anticommutators as

$$\Big[\Psi^\dagger_{\alpha,r}(t, \vec{x}) \Psi_{\beta,s}(t, \vec{x}), \Psi^\dagger_{\gamma,t}(t, \vec{y}) \Psi_{\delta,u}(t, \vec{y}) \Big]$$

$$= \Psi^\dagger_{\alpha,r}(t, \vec{x}) \Psi_{\delta,u}(t, \vec{y}) \delta^3(\vec{x} - \vec{y}) \delta_{\beta\gamma} \delta_{st} - \Psi^\dagger_{\gamma,t}(t, \vec{y}) \Psi_{\beta,s}(t, \vec{x}) \delta^3(\vec{x} - \vec{y}) \delta_{\alpha\delta} \delta_{ru}.$$

In a compact notation:

$$\Big[\Psi^\dagger(t, \vec{x}) \Gamma_1 F_1 \Psi(t, \vec{x}), \Psi^\dagger(t, \vec{y}) \Gamma_2 F_2 \Psi(t, \vec{y}) \Big]$$

$$= \delta^3(\vec{x} - \vec{y}) \Big[\Psi^\dagger(t, \vec{x}) \Gamma_1 \Gamma_2 F_1 F_2 \Psi(t, \vec{y}) - \Psi^\dagger(t, \vec{y}) \Gamma_2 \Gamma_1 F_2 F_1 \Psi(t, \vec{x}) \Big], \quad (1.87)$$

where Γ_i is one of the sixteen 4×4 matrices

$$\mathbb{1}, \gamma^\mu, \gamma_5, \gamma^\mu \gamma_5, \sigma^{\mu\nu} = \frac{i}{2} [\gamma^\mu, \gamma^\nu],$$

and F_i one of the four 2×2 matrices

$$\mathbb{1}, \tau_i.$$

(c) Apply Eq. 1.87 and integrate $\int d^3y \ldots$ to obtain

$$A_{ij} = i\varepsilon_{ijk} \int d^3x \Psi^\dagger(x) \frac{\tau_k}{2} \Psi(x).$$

(d) Verify

$$[ab, cd] = a[b, c]d + ac[b, d] + [a, c]db + c[a, d]b. \tag{1.88}$$

(e) Apply Eq. 1.88 in combination with the equal-time commutation relations to obtain

$$[\Phi_k(t, \vec{x})\Pi_l(t, \vec{x}), \Phi_m(t, \vec{y})\Pi_n(t, \vec{y})]$$
$$= -i\Phi_k(t, \vec{x})\Pi_n(t, \vec{y})\delta^3(\vec{x} - \vec{y})\delta_{lm} + i\Phi_m(t, \vec{y})\Pi_l(t, \vec{x})\delta^3(\vec{x} - \vec{y})\delta_{kn}. \tag{1.89}$$

(f) Apply Eq. 1.89 and integrate $\int d^3y\ldots$ to obtain

$$B_{ij} = i\varepsilon_{ijk} \int d^3x \varepsilon_{klm}\Phi_l(x)\Pi_m(x).$$

Adding the results for A_{ij} and B_{ij} we obtain

$$[Q_i, Q_j] = i\varepsilon_{ijk}\left[\int d^3x \Psi^\dagger(x)\frac{\tau_k}{2}\Psi(x) + \int d^3x \varepsilon_{klm}\Phi_l(x)\Pi_m(x)\right]$$
$$= i\varepsilon_{ijk}Q_k.$$

1.3.4 Global Symmetry Currents of the Light-Quark Sector

The method of Gell-Mann and Lévy can easily be applied to the QCD Lagrangian by calculating the variation under the infinitesimal, local form of Eqs. 1.40,

$$\delta\mathcal{L}^0_{\text{QCD}} = \bar{q}_R\left(\sum_{a=1}^8 \partial_\mu\varepsilon_{Ra}\frac{\lambda_a}{2} + \partial_\mu\varepsilon_R\right)\gamma^\mu q_R + \bar{q}_L\left(\sum_{a=1}^8 \partial_\mu\varepsilon_{La}\frac{\lambda_a}{2} + \partial_\mu\varepsilon_L\right)\gamma^\mu q_L, \tag{1.90}$$

from which, by virtue of Eqs. 1.46 and 1.47, one obtains the currents associated with the transformations of the left-handed or right-handed quarks,

$$L^\mu_a = \frac{\partial\delta\mathcal{L}^0_{\text{QCD}}}{\partial\partial_\mu\varepsilon_{La}} = \bar{q}_L\gamma^\mu\frac{\lambda_a}{2}q_L, \quad \partial_\mu L^\mu_a = \frac{\partial\delta\mathcal{L}^0_{\text{QCD}}}{\partial\varepsilon_{La}} = 0,$$

$$R^\mu_a = \frac{\partial\delta\mathcal{L}^0_{\text{QCD}}}{\partial\partial_\mu\varepsilon_{Ra}} = \bar{q}_R\gamma^\mu\frac{\lambda_a}{2}q_R, \quad \partial_\mu R^\mu_a = \frac{\partial\delta\mathcal{L}^0_{\text{QCD}}}{\partial\varepsilon_{Ra}} = 0,$$

$$L^\mu = \frac{\partial\delta\mathcal{L}^0_{\text{QCD}}}{\partial\partial_\mu\varepsilon_L} = \bar{q}_L\gamma^\mu q_L, \quad \partial_\mu L^\mu = \frac{\partial\delta\mathcal{L}^0_{\text{QCD}}}{\partial\varepsilon_L} = 0,$$

$$R^\mu = \frac{\partial\delta\mathcal{L}^0_{\text{QCD}}}{\partial\partial_\mu\varepsilon_R} = \bar{q}_R\gamma^\mu q_R, \quad \partial_\mu R^\mu = \frac{\partial\delta\mathcal{L}^0_{\text{QCD}}}{\partial\varepsilon_R} = 0. \tag{1.91}$$

Note that a summation over color indices is implied in Eqs. 1.90 and 1.91. For example, the detailed expression for L_a^μ reads

$$L_a^\mu = \bar{q}_{L\alpha,f,A}\gamma_{\alpha\alpha'}^\mu \frac{\lambda_{aff'}}{2}\delta_{AA'}q_{L\alpha',f',A'}.$$

The eight currents L_a^μ transform under $SU(3)_L \times SU(3)_R$ as an $(8,1)$ multiplet, i.e., as octet and singlet under transformations of the left- and right-handed fields, respectively. Similarly, the right-handed currents transform as a $(1,8)$ multiplet under $SU(3)_L \times SU(3)_R$. Instead of these chiral currents one often uses linear combinations,

$$V_a^\mu = R_a^\mu + L_a^\mu = \bar{q}\gamma^\mu \frac{\lambda_a}{2}q, \tag{1.92}$$

$$A_a^\mu = R_a^\mu - L_a^\mu = \bar{q}\gamma^\mu\gamma_5\frac{\lambda_a}{2}q, \tag{1.93}$$

transforming under parity as vector and axial-vector currents, respectively,

$$P : V_a^\mu(t,\vec{x}) \mapsto V_{a\mu}(t,-\vec{x}), \tag{1.94}$$

$$P : A_a^\mu(t,\vec{x}) \mapsto -A_{a\mu}(t,-\vec{x}). \tag{1.95}$$

Exercise 1.15   Verify Eqs. 1.92 and 1.93.

From Eq. 1.91 one also obtains a conserved singlet vector current resulting from a transformation of all left-handed and right-handed quark fields by the *same* phase,

$$V^\mu = R^\mu + L^\mu = \bar{q}\gamma^\mu q,$$
$$\partial_\mu V^\mu = 0. \tag{1.96}$$

The singlet axial-vector current,

$$A^\mu = R^\mu - L^\mu = \bar{q}\gamma^\mu\gamma_5 q, \tag{1.97}$$

originates from a transformation of all left-handed quark fields with one phase and all right-handed quark fields with the *opposite* phase. However, such a singlet axial-vector current is only conserved on the *classical* level. Quantum corrections destroy the singlet axial-vector current conservation and there are extra terms, referred to as anomalies [3, 4, 10], resulting in

$$\partial_\mu A^\mu = \frac{3g_3^2}{32\pi^2}\varepsilon_{\mu\nu\rho\sigma}\mathcal{G}_a^{\mu\nu}\mathcal{G}_a^{\rho\sigma}, \quad \varepsilon_{0123} = 1. \tag{1.98}$$

The factor of three originates from the number of flavors. In the large N_c (number of colors) limit of Ref. [49] the singlet axial-vector current is conserved, because the strong coupling constant behaves as $g_3^2 \sim N_c^{-1}$.

1.3.5 The Chiral Algebra

The invariance of $\mathscr{L}^0_{\text{QCD}}$ under global $SU(3)_L \times SU(3)_R \times U(1)_V$ transformations implies that also the QCD Hamilton operator in the chiral limit, H^0_{QCD}, exhibits a global $SU(3)_L \times SU(3)_R \times U(1)_V$ symmetry. As usual, the "charge operators" are defined as the space integrals of the charge densities,

$$Q_{La}(t) = \int d^3x q_L^\dagger(t,\vec{x})\frac{\lambda_a}{2}q_L(t,\vec{x}) = \int d^3x q^\dagger(t,\vec{x})P_L\frac{\lambda_a}{2}q(t,\vec{x}), \qquad (1.99)$$

$$Q_{Ra}(t) = \int d^3x q_R^\dagger(t,\vec{x})\frac{\lambda_a}{2}q_R(t,\vec{x}) = \int d^3x q^\dagger(t,\vec{x})P_R\frac{\lambda_a}{2}q(t,\vec{x}), \qquad (1.100)$$

$$Q_V(t) = \int d^3x \left[q_L^\dagger(t,\vec{x})q_L(t,\vec{x}) + q_R^\dagger(t,\vec{x})q_R(t,\vec{x})\right] = \int d^3x q^\dagger(t,\vec{x})q(t,\vec{x}). \quad (1.101)$$

For conserved symmetry currents, these operators are time independent, i.e., they commute with the Hamiltonian,

$$[Q_{La},H^0_{\text{QCD}}] = [Q_{Ra},H^0_{\text{QCD}}] = [Q_V,H^0_{\text{QCD}}] = 0. \qquad (1.102)$$

The commutation relations of the charge operators with each other are obtained by using Eq. 1.87 applied to the quark fields,

$$[q^\dagger(t,\vec{x})\Gamma_1 F_1 q(t,\vec{x}), q^\dagger(t,\vec{y})\Gamma_2 F_2 q(t,\vec{y})]$$
$$= \delta^3(\vec{x}-\vec{y})[q^\dagger(t,\vec{x})\Gamma_1\Gamma_2 F_1 F_2 q(t,\vec{y}) - q^\dagger(t,\vec{y})\Gamma_2\Gamma_1 F_2 F_1 q(t,\vec{x})], \qquad (1.103)$$

where Γ_i and F_i are 4×4 Γ matrices and 3×3 flavor matrices, respectively.[19] After inserting appropriate projectors $P_{L/R}$, Eq. 1.103 is easily applied to the charge operators of Eqs. 1.99–1.101, showing that these operators indeed satisfy the commutation relations corresponding to the Lie algebra of $SU(3)_L \times SU(3)_R \times U(1)_V$,

$$[Q_{La}, Q_{Lb}] = if_{abc}Q_{Lc}, \qquad (1.104)$$

$$[Q_{Ra}, Q_{Rb}] = if_{abc}Q_{Rc}, \qquad (1.105)$$

$$[Q_{La}, Q_{Rb}] = 0, \qquad (1.106)$$

$$[Q_{La}, Q_V] = [Q_{Ra}, Q_V] = 0. \qquad (1.107)$$

For example (recall $P_L^2 = P_L$)

---

[19] Strictly speaking, we should also include the color indices. However, since we are only discussing color-neutral quadratic forms a summation over such indices is always implied, with the net effect that one can completely omit them from the discussion.

$$
\begin{aligned}
[Q_{La}, Q_{Lb}] &= \int d^3x d^3y \left[q^\dagger(t,\vec{x}) P_L \frac{\lambda_a}{2} q(t,\vec{x}), q^\dagger(t,\vec{y}) P_L \frac{\lambda_b}{2} q(t,\vec{y}) \right] \\
&= \int d^3x d^3y \delta^3(\vec{x} - \vec{y}) q^\dagger(t,\vec{x}) \underbrace{P_L P_L}_{= P_L} \frac{\lambda_a}{2} \frac{\lambda_b}{2} q(t,\vec{y}) \\
&\quad - \int d^3x d^3y \delta^3(\vec{x} - \vec{y}) q^\dagger(t,\vec{y}) P_L \frac{\lambda_b}{2} \frac{\lambda_a}{2} q(t,\vec{x}) \\
&= i f_{abc} \int d^3x q^\dagger(t,\vec{x}) P_L \frac{\lambda_c}{2} q(t,\vec{x}) = i f_{abc} Q_{Lc}.
\end{aligned}
$$

Exercise 1.16 Verify the remaining commutation relations, Eqs. 1.105–1.107.

It should be stressed that, even without being able to explicitly solve the equation of motion of the quark fields entering the charge operators of Eqs. 1.104–1.107, we know from the equal-time commutation relations and the symmetry of the Lagrangian that these charge operators are the generators of infinitesimal transformations of the Hilbert space associated with H_{QCD}^0. Furthermore, their commutation relations with a given operator specify the transformation behavior of the operator in question under the group $\text{SU}(3)_L \times \text{SU}(3)_R \times \text{U}(1)_V$.

1.3.6 Chiral Symmetry Breaking by the Quark Masses

So far we have discussed an idealized world with massless light quarks. The finite u-, d-, and s-quark masses in the QCD Lagrangian explicitly break the chiral symmetry, resulting in divergences of the symmetry currents. As a consequence, the charge operators are, in general, no longer time independent. However, as first pointed out by Gell-Mann, the equal-time commutation relations still play an important role even if the symmetry is explicitly broken [22]. As will be discussed later on in more detail, the symmetry currents give rise to chiral Ward identities relating various QCD Green functions to each other. Equation 1.47 allows one to discuss the divergences of the symmetry currents in the presence of quark masses. To that end, let us consider the quark-mass matrix of the three light quarks and project it onto the nine λ matrices of Eq. 1.13,

$$
\mathcal{M} = \begin{pmatrix} m_u & 0 & 0 \\ 0 & m_d & 0 \\ 0 & 0 & m_s \end{pmatrix}. \tag{1.108}
$$

Exercise 1.17 Express the quark-mass matrix in terms of the λ matrices λ_0, λ_3, and λ_8.

In particular, applying Eq. 1.38 we see that the quark-mass term mixes left- and right-handed fields,

$$\mathscr{L}_{\mathscr{M}} = -\bar{q}\mathscr{M}q = -(\bar{q}_R\mathscr{M}q_L + \bar{q}_L\mathscr{M}q_R). \tag{1.109}$$

The symmetry-breaking term transforms under $SU(3)_L \times SU(3)_R$ as a member of a $(3,3^*) \oplus (3^*,3)$ representation, i.e.,

$$\bar{q}_{R,i}\mathscr{M}_{ij}q_{L,j} + \bar{q}_{L,i}\mathscr{M}_{ij}q_{R,j} \mapsto U_{L,jk}U_{R,il}^*\bar{q}_{R,l}\mathscr{M}_{ij}q_{L,k} + (L \leftrightarrow R),$$

where $(U_L, U_R) \in SU(3)_L \times SU(3)_R$. Such symmetry-breaking *patterns* were already discussed in the pre-QCD era in Refs. [26, 27].

From $\mathscr{L}_{\mathscr{M}}$ one obtains the variation $\delta\mathscr{L}_{\mathscr{M}}$ under the infinitesimal transformations corresponding to Eqs. 1.40,

$$\delta\mathscr{L}_{\mathscr{M}} = -i\left[\bar{q}_R\left(\sum_{a=1}^{8}\varepsilon_{Ra}\frac{\lambda_a}{2} + \varepsilon_R\right)\mathscr{M}q_L - \bar{q}_R\mathscr{M}\left(\sum_{a=1}^{8}\varepsilon_{La}\frac{\lambda_a}{2} + \varepsilon_L\right)q_L\right.$$
$$+\bar{q}_L\left(\sum_{a=1}^{8}\varepsilon_{La}\frac{\lambda_a}{2} + \varepsilon_L\right)\mathscr{M}q_R - \bar{q}_L\mathscr{M}\left(\sum_{a=1}^{8}\varepsilon_{Ra}\frac{\lambda_a}{2} + \varepsilon_R\right)q_R\right]$$
$$= -i\left[\sum_{a=1}^{8}\varepsilon_{Ra}\left(\bar{q}_R\frac{\lambda_a}{2}\mathscr{M}q_L - \bar{q}_L\mathscr{M}\frac{\lambda_a}{2}q_R\right) + \varepsilon_R(\bar{q}_R\mathscr{M}q_L - \bar{q}_L\mathscr{M}q_R)\right.$$
$$\left. + \sum_{a=1}^{8}\varepsilon_{La}\left(\bar{q}_L\frac{\lambda_a}{2}\mathscr{M}q_R - \bar{q}_R\mathscr{M}\frac{\lambda_a}{2}q_L\right) + \varepsilon_L(\bar{q}_L\mathscr{M}q_R - \bar{q}_R\mathscr{M}q_L)\right], \tag{1.110}$$

which results in the following divergences,[20]

$$\partial_\mu L_a^\mu = \frac{\partial\delta\mathscr{L}_{\mathscr{M}}}{\partial\varepsilon_{La}} = -i\left(\bar{q}_L\frac{\lambda_a}{2}\mathscr{M}q_R - \bar{q}_R\mathscr{M}\frac{\lambda_a}{2}q_L\right),$$

$$\partial_\mu R_a^\mu = \frac{\partial\delta\mathscr{L}_{\mathscr{M}}}{\partial\varepsilon_{Ra}} = -i\left(\bar{q}_R\frac{\lambda_a}{2}\mathscr{M}q_L - \bar{q}_L\mathscr{M}\frac{\lambda_a}{2}q_R\right),$$

$$\partial_\mu L^\mu = \frac{\partial\delta\mathscr{L}_M}{\partial\varepsilon_L} = -i(\bar{q}_L\mathscr{M}q_R - \bar{q}_R\mathscr{M}q_L),$$

$$\partial_\mu R^\mu = \frac{\partial\delta\mathscr{L}_M}{\partial\varepsilon_R} = -i(\bar{q}_R\mathscr{M}q_L - \bar{q}_L\mathscr{M}q_R). \tag{1.111}$$

The anomaly has not yet been considered. Applying Eq. 1.38 to the case of the vector currents and inserting projection operators for the axial-vector current, the corresponding divergences read

---

[20] The divergences are proportional to the mass parameters which is the origin of the expression current-quark mass.

$$\partial_\mu V_a^\mu = -i\bar{q}_R\left[\frac{\lambda_a}{2}, \mathcal{M}\right]q_L - i\bar{q}_L\left[\frac{\lambda_a}{2}, \mathcal{M}\right]q_R \overset{(1.38)}{=} i\bar{q}\left[\mathcal{M}, \frac{\lambda_a}{2}\right]q,$$

$$\partial_\mu A_a^\mu = -i\left(\bar{q}_R\frac{\lambda_a}{2}\mathcal{M}q_L - \bar{q}_L\mathcal{M}\frac{\lambda_a}{2}q_R\right) + i\left(\bar{q}_L\frac{\lambda_a}{2}\mathcal{M}q_R - \bar{q}_R\mathcal{M}\frac{\lambda_a}{2}q_L\right)$$

$$= i\left(\bar{q}_L\left\{\frac{\lambda_a}{2}, \mathcal{M}\right\}q_R - \bar{q}_R\left\{\frac{\lambda_a}{2}, \mathcal{M}\right\}q_L\right)$$

$$= i\left(\bar{q}\frac{1}{2}(\mathbb{1} + \gamma_5)\left\{\frac{\lambda_a}{2}, \mathcal{M}\right\}q - \bar{q}\frac{1}{2}(\mathbb{1} - \gamma_5)\left\{\frac{\lambda_a}{2}, \mathcal{M}\right\}q\right)$$

$$= i\bar{q}\gamma_5\left\{\frac{\lambda_a}{2}, \mathcal{M}\right\}q,$$

$$\partial_\mu V^\mu = 0,$$

$$\partial_\mu A^\mu = 2i\bar{q}\gamma_5\mathcal{M}q + \frac{3g_3^2}{32\pi^2}\varepsilon_{\mu\nu\rho\sigma}\mathcal{G}_a^{\mu\nu}\mathcal{G}_a^{\rho\sigma}, \quad \varepsilon_{0123} = 1, \tag{1.112}$$

where the axial anomaly has also been taken into account.

We are now in the position to summarize the various (approximate) symmetries of the strong interactions in combination with the corresponding currents and their divergences.

1. In the limit of massless quarks, the sixteen currents L_a^μ and R_a^μ or, alternatively, V_a^μ and A_a^μ are conserved. The same is true for the singlet vector current V^μ, whereas the singlet axial-vector current A^μ has an anomaly.
2. For any values of quark masses, the individual flavor currents $\bar{u}\gamma^\mu u$, $\bar{d}\gamma^\mu d$, and $\bar{s}\gamma^\mu s$ are always conserved in the strong interactions reflecting the flavor independence of the strong coupling and the diagonal form of the quark-mass matrix. Of course, the singlet vector current V^μ, being the sum of the three flavor currents, is always conserved.
3. In addition to the anomaly, the singlet axial-vector current has an explicit divergence due to the quark masses.
4. For equal quark masses, $m_u = m_d = m_s$, the eight vector currents V_a^μ are conserved, because $[\lambda_a, \mathbb{1}] = 0$. Such a scenario is the origin of the SU(3) symmetry originally proposed by Gell-Mann and Ne'eman [25]. The eight axial-vector currents A_a^μ are not conserved. The divergences of the octet axial-vector currents of Eq. 1.112 are proportional to pseudoscalar quadratic forms. This can be interpreted as the microscopic origin of the PCAC relation (partially conserved axial-vector current) [5, 23] which states that the divergences of the axial-vector currents are proportional to renormalized field operators representing the lowest-lying pseudoscalar octet (for a comprehensive discussion of the meaning of PCAC see Refs. [5, 6, 23, 50]).
5. Taking $m_u = m_d \neq m_s$ reduces SU(3) flavor symmetry to SU(2) isospin symmetry.
6. Taking $m_u \neq m_d$ leads to isospin-symmetry breaking.
7. Various symmetry-breaking patterns are discussed in great detail in Ref. [45].

1.4 Green Functions and Ward Identities

For conserved currents, the spatial integrals of the charge densities are time independent, i.e., in a quantized theory the corresponding charge operators commute with the Hamilton operator. These operators are generators of infinitesimal transformations on the Hilbert space of the theory. The mass eigenstates should organize themselves in degenerate multiplets with dimensionalities corresponding to irreducible representations of the Lie group in question.[21] Which irreducible representations ultimately appear, and what the actual energy eigenvalues are, is determined by the dynamics of the Hamiltonian. For example, SU(2) isospin symmetry of the strong interactions reflects itself in degenerate SU(2) multiplets such as the nucleon doublet, the pion triplet, and so on. Ultimately, the actual masses of the nucleon and the pion should follow from QCD.

It is also well-known that symmetries imply relations between S-matrix elements. For example, applying the Wigner-Eckart theorem to pion-nucleon scattering, assuming the strong-interaction Hamiltonian to be an isoscalar, it is sufficient to consider two isospin amplitudes describing transitions between states of total isospin $I = 1/2$ or $I = 3/2$ (see, for example, Ref. [16]). All the dynamical information is contained in these isospin amplitudes and the results for physical processes can be expressed in terms of these amplitudes together with geometrical coefficients, namely, the Clebsch-Gordan coefficients.

In quantum field theory, the objects of interest are the Green functions which are vacuum expectation values of time-ordered products.[22] Pictorially, these Green functions can be understood as vertices and are related to physical scattering amplitudes through the Lehmann-Symanzik-Zimmermann (LSZ) reduction formalism [36]. Symmetries provide strong constraints not only for scattering amplitudes, i.e. their transformation behavior, but, more generally speaking, also for Green functions and, in particular, *among* Green functions. The famous example in this context is, of course, the Ward identity of QED associated with U(1) gauge invariance [53],

$$\Gamma^\mu(p, p) = -\frac{\partial}{\partial p_\mu} \Sigma(p), \qquad (1.113)$$

which relates the electromagnetic vertex of an electron at zero momentum transfer, $\gamma^\mu + \Gamma^\mu(p, p)$, to the electron self-energy, $\Sigma(p)$.

Such symmetry relations can be extended to nonvanishing momentum transfer and also to more complicated groups and are referred to as Ward-Fradkin-Takahashi identities [17, 48, 53] (or Ward identities for short). Furthermore, even

---

[21] Here we assume that the dynamical system described by the Hamiltonian does not lead to a spontaneous symmetry breakdown. We will come back to this point later.

[22] Later on, we will also refer to matrix elements of time-ordered products between states other than the vacuum as Green functions.

if a symmetry is broken, i.e., the infinitesimal generators are time dependent, conditions related to the symmetry-breaking terms can still be obtained using equal-time commutation relations [22].

1.4.1 Ward Identities Resulting from U(1) Invariance: An Example

In this section we will show how to derive Ward identities for Green functions in the framework of canonical quantization on the one hand, and quantization via the Feynman path integral on the other hand, by means of an explicit example. In order to keep the discussion transparent, we will concentrate on a simple scalar field theory with a global SO(2) or U(1) invariance. To that end, let us consider the Lagrangian of Exercise 1.11,

$$
\begin{aligned}
\mathscr{L} &= \frac{1}{2}\left(\partial_\mu \Phi_1 \partial^\mu \Phi_1 + \partial_\mu \Phi_2 \partial^\mu \Phi_2\right) - \frac{m^2}{2}\left(\Phi_1^2 + \Phi_2^2\right) - \frac{\lambda}{4}\left(\Phi_1^2 + \Phi_2^2\right)^2 \\
&= \partial_\mu \Phi^\dagger \partial^\mu \Phi - m^2 \Phi^\dagger \Phi - \lambda (\Phi^\dagger \Phi)^2,
\end{aligned}
\tag{1.114}
$$

where

$$
\Phi(x) = \frac{1}{\sqrt{2}}[\Phi_1(x) + i\Phi_2(x)], \quad \Phi^\dagger(x) = \frac{1}{\sqrt{2}}[\Phi_1(x) - i\Phi_2(x)],
$$

with real scalar fields Φ_1 and Φ_2. Furthermore, we assume $m^2 > 0$ and $\lambda > 0$, so there is no spontaneous symmetry breaking (see Chap. 2) and the energy is bounded from below. Equation 1.114 is invariant under the global (or rigid) transformations

$$
\Phi_1' = \Phi_1 - \varepsilon \Phi_2, \quad \Phi_2' = \Phi_2 + \varepsilon \Phi_1,
\tag{1.115}
$$

or, equivalently,

$$
\Phi' = (1 + i\varepsilon)\Phi, \quad \Phi'^\dagger = (1 - i\varepsilon)\Phi^\dagger,
\tag{1.116}
$$

where ε is an infinitesimal real parameter. Applying the method of Gell-Mann and Lévy, we obtain for a *local* parameter $\varepsilon(x)$,

$$
\delta \mathscr{L} = \partial_\mu \varepsilon(x)\left(i\partial^\mu \Phi^\dagger \Phi - i\Phi^\dagger \partial^\mu \Phi\right),
\tag{1.117}
$$

from which, via Eqs. 1.46 and 1.47, we derive for the current corresponding to the global symmetry,

$$
J^\mu = \frac{\partial \delta \mathscr{L}}{\partial \partial_\mu \varepsilon} = i\partial^\mu \Phi^\dagger \Phi - i\Phi^\dagger \partial^\mu \Phi,
\tag{1.118}
$$

$$\partial_\mu J^\mu = \frac{\partial \delta \mathscr{L}}{\partial \varepsilon} = 0. \qquad (1.119)$$

Recall that the identification of Eq. 1.47 as the divergence of the current is only true for fields satisfying the Euler-Lagrange equations of motion.

We now extend the analysis to a *quantum* field theory. In the framework of canonical quantization, we first define conjugate momenta,

$$\Pi_i = \frac{\partial \mathscr{L}}{\partial \partial_0 \Phi_i}, \quad \Pi = \frac{\partial \mathscr{L}}{\partial \partial_0 \Phi} = \dot{\Phi}^\dagger, \quad \Pi^\dagger = \frac{\partial \mathscr{L}}{\partial \partial_0 \Phi^\dagger} = \dot{\Phi}, \qquad (1.120)$$

and interpret the fields and their conjugate momenta as operators which, in the Heisenberg picture, are subject to the equal-time commutation relations

$$[\Phi_i(t, \vec{x}), \Pi_j(t, \vec{y})] = i\delta_{ij}\delta^3(\vec{x} - \vec{y}), \qquad (1.121)$$

and

$$[\Phi(t, \vec{x}), \Pi(t, \vec{y})] = [\Phi^\dagger(t, \vec{x}), \Pi^\dagger(t, \vec{y})] = i\delta^3(\vec{x} - \vec{y}). \qquad (1.122)$$

The remaining equal-time commutation relations vanish. For the quantized theory, the current operator then reads

$$J^\mu(x) =: \left(i\partial^\mu \Phi^\dagger \Phi - i\Phi^\dagger \partial^\mu \Phi\right) :, \qquad (1.123)$$

where : : denotes normal or Wick ordering, i.e., annihilation operators appear to the right of creation operators. For a conserved current, the charge operator, i.e., the space integral of the charge density, is time independent and serves as the generator of infinitesimal transformations of the Hilbert space states,

$$Q = \int d^3x J^0(t, \vec{x}). \qquad (1.124)$$

Applying Eq. 1.122, it is straightforward to calculate the equal-time commutation relations[23]

$$\begin{aligned}
[J^0(t, \vec{x}), \Phi(t, \vec{y})] &= \delta^3(\vec{x} - \vec{y})\Phi(t, \vec{x}), \\
[J^0(t, \vec{x}), \Pi(t, \vec{y})] &= -\delta^3(\vec{x} - \vec{y})\Pi(t, \vec{x}), \\
[J^0(t, \vec{x}), \Phi^\dagger(t, \vec{y})] &= -\delta^3(\vec{x} - \vec{y})\Phi^\dagger(t, \vec{x}), \\
[J^0(t, \vec{x}), \Pi^\dagger(t, \vec{y})] &= \delta^3(\vec{x} - \vec{y})\Pi^\dagger(t, \vec{x}).
\end{aligned} \qquad (1.125)$$

In particular, performing the space integrals in Eqs. 1.125, one obtains

---

[23] The transition to normal ordering involves an (infinite) constant which does not contribute to the commutator.

$$[Q, \Phi(x)] = \Phi(x),$$
$$[Q, \Pi(x)] = -\Pi(x),$$
$$[Q, \Phi^\dagger(x)] = -\Phi^\dagger(x), \qquad (1.126)$$
$$[Q, \Pi^\dagger(x)] = \Pi^\dagger(x).$$

In order to illustrate the implications of Eqs. 1.126, let us take an eigenstate $|\alpha\rangle$ of Q with eigenvalue q_α and consider, for example, the action of $\Phi(x)$ on that state,

$$Q(\Phi(x)|\alpha\rangle) = ([Q, \Phi(x)] + \Phi(x)Q)|\alpha\rangle = (1 + q_\alpha)(\Phi(x)|\alpha\rangle).$$

We conclude that the operators $\Phi(x)$ and $\Pi^\dagger(x)$ $[\Phi^\dagger(x)$ and $\Pi(x)]$ increase (decrease) the Noether charge of a system by one unit.

We are now in the position to discuss the consequences of the U(1) symmetry of Eq. 1.114 for the Green functions of the theory. To that end, let us consider as our prototype the Green function

$$G^\mu(x, y, z) = \langle 0|T[\Phi(x)J^\mu(y)\Phi^\dagger(z)]|0\rangle, \qquad (1.127)$$

which describes the transition amplitude for the creation of a quantum of Noether charge $+1$ at x, propagation to y, interaction at y via the current operator, propagation to z with annihilation at z. In Eq. 1.127, $|0\rangle$ refers to the ground state of the quantum field theory described by the Lagrangian of Eq. 1.114 and should not be confused with the ground state of a free theory.

First of all we observe that under the global infinitesimal transformations of Eq. 1.116, $J^\mu(x) \mapsto J'^\mu(x) = J^\mu(x)$, or in other words $[Q, J^\mu(x)] = 0$. We thus obtain

$$\begin{aligned} G^\mu(x, y, z) \mapsto G'^\mu(x, y, z) &= \langle 0|T[(1 + i\varepsilon)\Phi(x)J'^\mu(y)(1 - i\varepsilon)\Phi^\dagger(z)]|0\rangle \\ &= \langle 0|T[\Phi(x)J^\mu(y)\Phi^\dagger(z)]|0\rangle \\ &= G^\mu(x, y, z), \end{aligned} \qquad (1.128)$$

the Green function remains invariant under the U(1) transformation. (In general, the transformation behavior of a Green function depends on the irreducible representations under which the fields transform. In particular, for more complicated groups such as SU(N), standard tensor methods of group theory may be applied to reduce the product representations into irreducible components. We also note that for U(1), the symmetry current is charge neutral, i.e. invariant, which for more complicated groups, in general, is not the case.)

Moreover, since $J^\mu(x)$ is the Noether current of the underlying U(1) symmetry there are further restrictions on the Green function beyond its transformation behavior under the group. In order to see this, we consider the divergence of Eq. 1.127 and apply the equal-time commutation relations of Eqs. 1.125 to obtain

$$\partial_\mu^y G^\mu(x, y, z) = [\delta^4(y - x) - \delta^4(y - z)]\langle 0|T[\Phi(x)\Phi^\dagger(z)]|0\rangle, \qquad (1.129)$$

where we made use of $\partial_\mu J^\mu = 0$. Equation 1.129 is the analogue of the Ward identity of QED [53]. In other words, the underlying symmetry not only determines the transformation behavior of Green functions under the group, but also relates n-point Green functions containing a symmetry current to $(n-1)$-point Green functions. In principle, calculations similar to those leading to Eqs. 1.128 and 1.129, can be performed for any Green function of the theory.

Exercise 1.18 Derive Eq. 1.129.
Hints: The time ordering is defined as

$$\langle 0|T[\Phi(x)J^\mu(y)\Phi^\dagger(z)]|0\rangle = \Phi(x)J^\mu(y)\Phi^\dagger(z)\Theta(x_0 - y_0)\Theta(y_0 - z_0)$$
$$+ \Phi(x)\Phi^\dagger(z)J^\mu(y)\Theta(x_0 - z_0)\Theta(z_0 - y_0) + \cdots.$$

All in all there exist $3! = 6$ distinct orderings. Make use of

$$\partial_\mu^y \Theta(x_0 - y_0) = -g_{\mu 0}\delta(x_0 - y_0),$$
$$\partial_\mu^y \Theta(y_0 - z_0) = g_{\mu 0}\delta(y_0 - z_0).$$

We will now show that the symmetry constraints imposed by the Ward identities can be compactly summarized in terms of an invariance property of a generating functional. For a discussion of functionals and partial functional derivatives, see App. B. In the present case, the generating functional depends on a set of functions denoted by j, j^*, and j_μ which are called external sources. They couple to the fields $\Phi^\dagger, \Phi$, and the U(1) current J^μ, respectively. The generating functional is defined as

$$W[j, j^*, j_\mu] = \exp(iZ[j, j^*, j_\mu])$$
$$= \langle 0|T\left(\exp\left\{i\int d^4x[j(x)\Phi^\dagger(x) + j^*(x)\Phi(x) + j_\mu(x)J^\mu(x)]\right\}\right)|0\rangle,$$

$$(1.130)$$

where $|0\rangle$ denotes the ground state of the theory described by the Lagrangian of Eq. 1.114. Moreover, $\Phi, \Phi^\dagger$, and $J^\mu(x)$ refer to the field operators and the Noether current in the Heisenberg picture. Note that the field operators and the conjugate momenta are subject to the equal-time commutation relations and, in addition, must satisfy the Heisenberg equations of motion:

$$\partial_0\Phi(x) = i[H, \Phi(x)],$$
$$\partial_0\Pi^\dagger(x) = i[H, \Pi^\dagger(x)],$$
$$\partial_0\Phi^\dagger(x) = i[H, \Phi^\dagger(x)],$$
$$\partial_0\Pi(x) = i[H, \Pi(x)],$$

$$(1.131)$$

where

$$H = \int d^3x \mathscr{H}, \tag{1.132}$$

$$\mathscr{H} = \Pi^\dagger \Pi + \vec{\nabla} \Phi^\dagger \cdot \vec{\nabla} \Phi + m^2 \Phi^\dagger \Phi + \lambda (\Phi^\dagger \Phi)^2. \tag{1.133}$$

Via the equations of motion and implicitly through the ground state, the generating functional depends on the dynamics of the system which is determined by the Lagrangian of Eq. 1.114 or the Hamiltonian of Eq. 1.132. The Green functions of the theory involving Φ, $\Phi^\dagger$, and J^μ are obtained through partial functional derivatives of Eq. 1.130. For example, the Green function of Eq. 1.127 is given by[24]

$$G^\mu(x,y,z) = (-i)^3 \frac{\delta^3 W[j,j^*,j_\mu]}{\delta j^*(x) \delta j_\mu(y) \delta j(z)}\bigg|_{j=0,j^*=0,j_\mu=0}. \tag{1.134}$$

Alternatively, the generating functional may be written as the vacuum-to-vacuum transition amplitude in the presence of external fields,

$$W[j,j^*,j_\mu] = \langle 0, \text{out} | 0, \text{in} \rangle_{j,j^*,j_\mu}. \tag{1.135}$$

In order to discuss the constraints imposed on the generating functional via the underlying symmetry of the theory, let us consider its path integral representation [14, 55],[25]

$$W[j,j^*,j_\mu] = \int [d\Phi_1][d\Phi_2] e^{iS[\Phi,\Phi^* j j^* j_\mu]}, \tag{1.136}$$

where

$$S[\Phi,\Phi^*,j,j^*,j_\mu] = S[\Phi,\Phi^*] + \int d^4x [\Phi(x)j^*(x) + \Phi^*(x)j(x) + J^\mu(x)j_\mu(x)] \tag{1.137}$$

denotes the action corresponding to the Lagrangian of Eq. 1.114 in combination with a coupling to the external sources. In the path integral formulation we deal with functional integrals instead of linear operators. In the following we will write

---

[24] In order to obtain Green functions from the generating functional, the simple rule

$$\frac{\delta f(x)}{\delta f(y)} = \delta^4(x-y)$$

is extremely useful. Furthermore, the partial functional derivative satisfies properties similar to the ordinary differentiation, namely linearity, the product and chain rules. See App. B for more details.

[25] Up to an irrelevant constant the measure $[d\Phi_1][d\Phi_2]$ is equivalent to $[d\Phi][d\Phi^*]$, with Φ and Φ^* considered as independent variables of integration.

Φ^* instead of $\Phi^\dagger$. Let us now consider a *local* infinitesimal transformation of the fields (see Eqs. 1.116) together with a *simultaneous* transformation of the external sources,

$$j'(x) = [1 + i\varepsilon(x)]j(x), \quad j'^*(x) = [1 - i\varepsilon(x)]j^*(x), \quad j'_\mu(x) = j_\mu(x) - \partial_\mu\varepsilon(x).$$
(1.138)

The action of Eq. 1.137 remains invariant under such a transformation,

$$S[\Phi', \Phi'^*, j', j'^*, j'_\mu] = S[\Phi, \Phi^*, j, j^*, j_\mu].$$
(1.139)

We stress that the transformation of the external current j_μ is necessary to cancel a term resulting from the kinetic term in the Lagrangian. Also note that the *global* symmetry of the Lagrangian determines the explicit form of the transformations of Eq. 1.138. We can now verify the invariance of the generating functional as follows,

$$
\begin{aligned}
W[j, j^*, j_\mu] &= \int [d\Phi_1][d\Phi_2] e^{iS[\Phi,\Phi^*,j,j^*,j_\mu]} \\
&= \int [d\Phi_1][d\Phi_2] e^{iS[\Phi',\Phi'^*,j',j'^*,j'_\mu]} \\
&= \int [d\Phi'_1][d\Phi'_2] \left| \left(\frac{\partial\Phi_i}{\partial\Phi'_j} \right) \right| e^{iS[\Phi',\Phi'^*,j',j'^*,j'_\mu]} \\
&= \int [d\Phi_1][d\Phi_2] e^{iS[\Phi,\Phi^*,j',j'^*,j'_\mu]} \\
&= W[j', j'^*, j'_\mu].
\end{aligned}
$$
(1.140)

We made use of the fact that the Jacobi determinant is one and renamed the integration variables. In other words, given the *global* U(1) symmetry of the Lagrangian, Eq. 1.114, the generating functional is invariant under the *local* transformations of Eq. 1.138. It is this observation which, for the more general case of the chiral group SU(N)×SU(N), was used by Gasser and Leutwyler as the starting point of chiral perturbation theory [20, 21].

We still have to discuss how this invariance allows us to collect the Ward identities in a compact formula. We start from Eq. 1.140 and perform a Taylor series expansion, keeping only terms linear in infinitesimal quantities,

$$
\begin{aligned}
0 &= \int [d\Phi_1][d\Phi_2] \left(e^{iS[\Phi,\Phi^*,j',j'^*,j'_\mu]} - e^{iS[\Phi,\Phi^*,j,j^*,j_\mu]} \right) \\
&= \int [d\Phi_1][d\Phi_2] \int d^4x \{ \varepsilon[\Phi j^* - \Phi^* j] - iJ^\mu\partial_\mu\varepsilon \} e^{iS[\Phi,\Phi^*,j,j^*,j_\mu]}.
\end{aligned}
$$

Observe that

$$\Phi(x) e^{iS[\Phi,\Phi^*,j,j^*,j_\mu]} = -i\frac{\delta}{\delta j^*(x)} e^{iS[\Phi,\Phi^*,j,j^*,j_\mu]},$$

and similarly for the other terms, resulting in

$$0 = \int [d\Phi_1][d\Phi_2] \int d^4x \left\{ \varepsilon(x) \left[-ij^*(x)\frac{\delta}{\delta j^*(x)} + ij(x)\frac{\delta}{\delta j(x)} \right] \right.$$
$$\left. - \partial_\mu \varepsilon(x)\frac{\delta}{\delta j_\mu(x)} \right\} e^{iS[\Phi,\Phi^* ,j,j^*,j_\mu]}.$$

Finally, we interchange the order of integration, make use of integration by parts, and apply the divergence theorem:

$$0 = \int d^4x \varepsilon(x) \left[ij(x)\frac{\delta}{\delta j(x)} - ij^*(x)\frac{\delta}{\delta j^*(x)} + \partial_\mu^x \frac{\delta}{\delta j_\mu(x)} \right] W[j,j^*,j_\mu]. \qquad (1.141)$$

Since Eq. 1.141 must hold for any $\varepsilon(x)$ we obtain as the master equation for deriving Ward identities,

$$\left[j(x)\frac{\delta}{\delta j(x)} - j^*(x)\frac{\delta}{\delta j^*(x)} - i\partial_\mu^x \frac{\delta}{\delta j_\mu(x)} \right] W[j,j^*,j_\mu] = 0. \qquad (1.142)$$

We note that Eqs. 1.140 and 1.142 are equivalent.

As an illustration let us re-derive the Ward identity of Eq. 1.129 using Eq. 1.142. For that purpose we start from Eq. 1.134,

$$\partial_\mu^y G^\mu(x,y,z) = (-i)^3 \partial_\mu^y \frac{\delta^3 W}{\delta j^*(x)\delta j_\mu(y)\delta j(z)} \bigg|_{j=0,j^*=0,j_\mu=0},$$

apply Eq. 1.142,

$$= (-i)^2 \left\{ \frac{\delta^2}{\delta j^*(x)\delta j(z)} \left[j^*(y)\frac{\delta}{\delta j^*(y)} - j(y)\frac{\delta}{\delta j(y)} \right] W \right\}_{j=0,j^*=0,j_\mu=0},$$

make use of $\delta j^*(y)/\delta j^*(x) = \delta^4(y-x)$ and $\delta j(y)/\delta j(z) = \delta^4(y-z)$ for the partial functional derivatives,

$$= (-i)^2 \left\{ \delta^4(y-x)\frac{\delta^2 W}{\delta j^*(y)\delta j(z)} - \delta^4(y-z)\frac{\delta^2 W}{\delta j^*(x)\delta j(y)} \right\}_{j=0,j^*=0,j_\mu=0},$$

and, finally, use the definition of Eq. 1.130,

$$\partial_\mu^y G^\mu(x,y,z) = [\delta^4(y-x) - \delta^4(y-z)]\langle 0|T[\Phi(x)\Phi^\dagger(z)]|0\rangle$$

which is the same as Eq. 1.129. In principle, any Ward identity can be obtained by taking appropriate higher partial functional derivatives of W and then using Eq. 1.142.

1.4.2 Chiral Green Functions

Let us now turn to time-ordered products of color-neutral, Hermitian quadratic forms involving the light-quark fields evaluated between the vacuum of QCD. Using the LSZ reduction formalism [32, 36] such Green functions can be related to physical processes involving mesons as well as their interactions with the electroweak gauge fields of the Standard Model. The interpretation depends on the transformation properties and quantum numbers of the quadratic forms, determining for which mesons they may serve as an interpolating field. In addition to the vector and axial-vector currents of Eqs. 1.92, 1.93, and 1.96 we want to investigate scalar and pseudoscalar densities,[26]

$$S_a(x) = \bar{q}(x)\lambda_a q(x), \quad P_a(x) = i\bar{q}(x)\gamma_5\lambda_a q(x), \quad a = 0, \ldots, 8, \qquad (1.143)$$

which enter, for example, in Eqs. 1.112 as the divergences of the vector and axial-vector currents for nonzero quark masses. Whenever it is more convenient, we will also use

$$S(x) = \bar{q}(x)q(x), \quad P(x) = i\bar{q}(x)\gamma_5 q(x), \qquad (1.144)$$

instead of S_0 and P_0.

For example, the following Green functions of the "vacuum" sector,

$$\langle 0|T[A_a^\mu(x)P_b(y)]|0\rangle,$$
$$\langle 0|T[P_a(x)J^\mu(y)P_b(z)]|0\rangle,$$
$$\langle 0|T[P_a(w)P_b(x)P_c(y)P_d(z)]|0\rangle,$$

are related to pion decay, the pion electromagnetic form factor (J^μ is the electromagnetic current), and pion-pion scattering, respectively. One may also consider similar time-ordered products evaluated between a single nucleon in the initial and final states in addition to the vacuum Green functions. This allows one to discuss properties of the nucleon as well as dynamical processes involving a single nucleon, such as

$\langle N|J^\mu(x)|N\rangle \leftrightarrow$ nucleon electromagnetic form factors,

$\langle N|A_a^\mu(x)|N\rangle \leftrightarrow$ axial form factor + induced pseudoscalar form factor,

$\langle N|T[J^\mu(x)J^\nu(y)]|N\rangle \leftrightarrow$ Compton scattering,

$\langle N|T[J^\mu(x)P_a(y)]|N\rangle \leftrightarrow$ pion photo- and electroproduction.

Generally speaking, a chiral Ward identity relates the divergence of a Green function containing at least one factor of V_a^μ or A_a^μ (see Eqs. 1.92 and 1.93) to some

---

[26] The singlet axial-vector current involves an anomaly such that the Green functions involving this current operator are related to Green functions containing the contraction of the gluon field-strength tensor with its dual.

linear combination of other Green functions. The terminology *chiral* refers to the underlying $SU(3)_L \times SU(3)_R$ group. To make this statement more precise, let us consider as a simple example the two-point Green function involving an axial-vector current and a pseudoscalar density,[27]

$$G^\mu_{APab}(x,y) = \langle 0|T[A^\mu_a(x)P_b(y)]|0\rangle$$
$$= \Theta(x_0 - y_0)\langle 0|A^\mu_a(x)P_b(y)|0\rangle + \Theta(y_0 - x_0)\langle 0|P_b(y)A^\mu_a(x)|0\rangle,$$
$$(1.145)$$

and evaluate the divergence

$$\partial^x_\mu G^\mu_{APab}(x,y) = \partial^x_\mu[\Theta(x_0 - y_0)\langle 0|A^\mu_a(x)P_b(y)|0\rangle + \Theta(y_0 - x_0)\langle 0|P_b(y)A^\mu_a(x)|0\rangle]$$
$$= \delta(x_0 - y_0)\langle 0|A^0_a(x)P_b(y)|0\rangle - \delta(x_0 - y_0)\langle 0|P_b(y)A^0_a(x)|0\rangle$$
$$+ \Theta(x_0 - y_0)\langle 0|\partial^x_\mu A^\mu_a(x)P_b(y)|0\rangle + \Theta(y_0 - x_0)\langle 0|P_b(y)\partial^x_\mu A^\mu_a(x)|0\rangle$$
$$= \delta(x_0 - y_0)\langle 0|[A^0_a(x), P_b(y)]|0\rangle + \langle 0|T[\partial^x_\mu A^\mu_a(x)P_b(y)]|0\rangle,$$

where we made use of $\partial^x_\mu \Theta(x_0 - y_0) = \delta(x_0 - y_0)g_{0\mu} = -\partial^x_\mu \Theta(y_0 - x_0)$. This simple example already shows the main features of (chiral) Ward identities. From the differentiation of the theta functions one obtains equal-time commutators between a charge density and the remaining quadratic forms. The results of such commutators are a reflection of the underlying symmetry, as will be shown below. As a second term, one obtains the divergence of the current operator in question. If the symmetry is perfect, such terms vanish identically. For example, this is always true for the electromagnetic case with its $U(1)$ symmetry. If the symmetry is only approximate, an additional term involving the symmetry breaking appears. For a soft breaking such a divergence can be treated as a perturbation.

Via induction, the generalization of the above simple example to an $(n + 1)$-point Green function is symbolically of the form

$$\partial^x_\mu \langle 0|T\{J^\mu(x)A_1(x_1)...A_n(x_n)\}|0\rangle$$
$$= \langle 0|T\{[\partial^x_\mu J^\mu(x)]A_1(x_1)...A_n(x_n)\}|0\rangle$$
$$+ \delta(x^0 - x^0_1)\langle 0|T\{[J_0(x), A_1(x_1)]A_2(x_2)...A_n(x_n)\}|0\rangle$$
$$+ \delta(x^0 - x^0_2)\langle 0|T\{A_1(x_1)[J_0(x), A_2(x_2)]...A_n(x_n)\}|0\rangle$$
$$+ \cdots + \delta(x^0 - x^0_n)\langle 0|T\{A_1(x_1)...[J_0(x), A_n(x_n)]\}|0\rangle, \qquad (1.146)$$

where J^μ now generically stands for any of the Noether currents.

---

[27] The time ordering of n points $x_1, ..., x_n$ gives rise to $n!$ distinct orderings, each involving products of $n - 1$ theta functions.

1.4.3 The Algebra of Currents

In the above example, we have seen that chiral Ward identities depend on the equal-time commutation relations of the *charge densities* of the symmetry currents with the relevant quadratic quark forms. Unfortunately, a naive application of Eq. 1.103 may lead to erroneous results. Let us illustrate this by means of a simplified example, the equal-time commutator of the time and space components of the ordinary electromagnetic current in QED. A naive use of the canonical commutation relations leads to

$$[J_0(t,\vec{x}), J_i(t,\vec{y})] = [\Psi^\dagger(t,\vec{x})\Psi(t,\vec{x}), \Psi^\dagger(t,\vec{y})\gamma_0\gamma_i\Psi(t,\vec{y})]$$
$$= \delta^3(\vec{x}-\vec{y})\Psi^\dagger(t,\vec{x})[\mathbb{1}, \gamma_0\gamma_i]\Psi(t,\vec{x}) = 0, \qquad (1.147)$$

where we made use of the delta function to evaluate the fields at $\vec{x} = \vec{y}$. It was noticed a long time ago by Schwinger that this result cannot be true [47]. In order to see this, consider the commutator

$$[J_0(t,\vec{x}), \vec{\nabla}_y \cdot \vec{J}(t,\vec{y})] = -[J_0(t,\vec{x}), \partial_t J_0(t,\vec{y})],$$

where we made use of current conservation, $\partial_\mu J^\mu = 0$. If Eq. 1.147 were true, one would necessarily also have

$$0 = [J_0(t,\vec{x}), \partial_t J_0(t,\vec{y})],$$

which we evaluate for $\vec{x} = \vec{y}$ between the ground state,

$$0 = \langle 0|[J_0(t,\vec{x}), \partial_t J_0(t,\vec{x})]|0\rangle$$
$$= \sum_n \left(\langle 0|J_0(t,\vec{x})|n\rangle\langle n|\partial_t J_0(t,\vec{x})|0\rangle - \langle 0|\partial_t J_0(t,\vec{x})|n\rangle\langle n|J_0(t,\vec{x})|0\rangle \right)$$
$$= 2i \sum_n (E_n - E_0)|\langle 0|J_0(t,\vec{x})|n\rangle|^2.$$

Here, we inserted a complete set of states and made use of

$$\partial_t J_0(t,\vec{x}) = i[H, J_0(t,\vec{x})].$$

Since every individual term in the sum is nonnegative, one would need

$$\langle 0|J_0(t,\vec{x})|n\rangle = 0$$

for any intermediate state, which is unphysical because it would imply that, for example, e^+e^- pairs cannot be created from the application of the charge density operator to the ground state. The solution is that the starting point, Eq. 1.147, is not true. The corrected version of Eq. 1.147 picks up an additional, so-called Schwinger term containing a derivative of the delta function.

Quite generally, by evaluating commutation relations with the component Θ^{00} of the energy-momentum tensor one can show that the equal-time commutation relation between a charge density and a current density can be determined up to one derivative of the δ function [33],

$$[J_a^0(0,\vec{x}), J_b^i(0,\vec{y})] = iC_{abc}J_c^i(0,\vec{x})\delta^3(\vec{x}-\vec{y}) + S_{ab}^{ij}(0,\vec{y})\partial_j\delta^3(\vec{x}-\vec{y}), \qquad (1.148)$$

where the Schwinger term possesses the symmetry

$$S_{ab}^{ij}(0,\vec{y}) = S_{ba}^{ji}(0,\vec{y}),$$

and C_{abc} denote the structure constants of the group in question.

However, in our above derivation of the chiral Ward identity, we also made use of the *naive* time-ordered product (T) as opposed to the *covariant* one (T^*) which, typically, differ by another non-covariant term which is called a seagull. Feynman's conjecture [33] states that there is a cancelation between Schwinger terms and seagull terms such that a Ward identity obtained by using the naive T product and by simultaneously omitting Schwinger terms ultimately yields the correct result to be satisfied by the Green function (involving the covariant T^* product). Although this will not be true in general, a sufficient condition for it to happen is that the time component algebra of the full theory remains the same as the one derived canonically and does not possess a Schwinger term.

Keeping the above discussion in mind, the complete list of equal-time commutation relations, omitting Schwinger terms, reads

$$[V_a^0(t,\vec{x}), V_b^\mu(t,\vec{y})] = \delta^3(\vec{x}-\vec{y})if_{abc}V_c^\mu(t,\vec{x}),$$
$$[V_a^0(t,\vec{x}), V^\mu(t,\vec{y})] = 0,$$
$$[V_a^0(t,\vec{x}), A_b^\mu(t,\vec{y})] = \delta^3(\vec{x}-\vec{y})if_{abc}A_c^\mu(t,\vec{x}),$$
$$[V_a^0(t,\vec{x}), S_b(t,\vec{y})] = \delta^3(\vec{x}-\vec{y})if_{abc}S_c(t,\vec{x}), \quad b=1,\dots,8,$$
$$[V_a^0(t,\vec{x}), S_0(t,\vec{y})] = 0,$$
$$[V_a^0(t,\vec{x}), P_b(t,\vec{y})] = \delta^3(\vec{x}-\vec{y})if_{abc}P_c(t,\vec{x}), \quad b=1,\dots,8,$$
$$[V_a^0(t,\vec{x}), P_0(t,\vec{y})] = 0,$$
$$[A_a^0(t,\vec{x}), V_b^\mu(t,\vec{y})] = \delta^3(\vec{x}-\vec{y})if_{abc}A_c^\mu(t,\vec{x}),$$
$$[A_a^0(t,\vec{x}), V^\mu(t,\vec{y})] = 0,$$
$$[A_a^0(t,\vec{x}), A_b^\mu(t,\vec{y})] = \delta^3(\vec{x}-\vec{y})if_{abc}V_c^\mu(t,\vec{x}), \qquad (1.149)$$
$$[A_a^0(t,\vec{x}), S_b(t,\vec{y})] = i\delta^3(\vec{x}-\vec{y})\left[\sqrt{\frac{2}{3}}\delta_{ab}P_0(t,\vec{x}) + d_{abc}P_c(t,\vec{x})\right],$$
$$b=1,\dots,8,$$
$$[A_a^0(t,\vec{x}), S_0(t,\vec{y})] = i\delta^3(\vec{x}-\vec{y})\sqrt{\frac{2}{3}}P_a(t,\vec{x}),$$
$$[A_a^0(t,\vec{x}), P_b(t,\vec{y})] = -i\delta^3(\vec{x}-\vec{y})\left[\sqrt{\frac{2}{3}}\delta_{ab}S_0(t,\vec{x}) + d_{abc}S_c(t,\vec{x})\right],$$
$$b=1,\dots,8,$$
$$[A_a^0(t,\vec{x}), P_0(t,\vec{y})] = -i\delta^3(\vec{x}-\vec{y})\sqrt{\frac{2}{3}}S_a(t,\vec{x}).$$

For example,

$$[V_a^0(t, \vec{x}), V_b^\mu(t, \vec{y})]$$

$$= \left[q^\dagger(t, \vec{x}) 1 \frac{\lambda_a}{2} q(t, \vec{x}), q^\dagger(t, \vec{y}) \gamma_0 \gamma^\mu \frac{\lambda_b}{2} q(t, \vec{y}) \right]$$

$$= \delta^3(\vec{x} - \vec{y}) \left[q^\dagger(t, \vec{x}) \gamma_0 \gamma^\mu \frac{\lambda_a}{2} \frac{\lambda_b}{2} q(t, \vec{y}) - q^\dagger(t, \vec{y}) \gamma_0 \gamma^\mu \frac{\lambda_b}{2} \frac{\lambda_a}{2} q(t, \vec{x}) \right]$$

$$= \delta^3(\vec{x} - \vec{y}) i f_{abc} V_c^\mu(t, \vec{x}).$$

The remaining expressions are obtained analogously.

1.4.4 QCD in the Presence of External Fields and the Generating Functional

Here, we want to consider the consequences of Eqs. 1.149 for the Green functions of QCD (in particular, at low energies). In principle, using the techniques of Sect. 1.4.2, for each Green function one can *explicitly* work out the chiral Ward identity which, however, becomes more and more tedious as the number n of quark quadratic forms increases. As seen above, there exists an elegant way of formally combining all Green functions in a generating functional. The (infinite) set of *all* chiral Ward identities is encoded as an invariance property of that functional. The rationale behind this approach is that, in the absence of anomalies, the Ward identities obeyed by the Green functions are equivalent to an invariance of the generating functional under a *local* transformation of the external fields [37]. The use of local transformations allows one to also consider divergences of Green functions. This statement has been illustrated in Sect. 1.4.1 using the U(1) invariance of the Lagrangian of Eq. 1.114.

Following the procedure of Gasser and Leutwyler [20, 21], we introduce into the Lagrangian of QCD the couplings of the nine vector currents and the eight axial-vector currents as well as the scalar and pseudoscalar quark densities to external c-number fields,

$$\mathcal{L} = \mathcal{L}_{\text{QCD}}^0 + \mathcal{L}_{\text{ext}}, \tag{1.150}$$

where

$$\mathcal{L}_{\text{ext}} = \sum_{a=1}^8 v_a^\mu \bar{q} \gamma_\mu \frac{\lambda_a}{2} q + v_{(s)}^\mu \frac{1}{3} \bar{q} \gamma_\mu q + \sum_{a=1}^8 a_a^\mu \bar{q} \gamma_\mu \gamma_5 \frac{\lambda_a}{2} q$$

$$- \sum_{a=0}^8 s_a \bar{q} \lambda_a q + \sum_{a=0}^8 p_a i \bar{q} \gamma_5 \lambda_a q$$

$$= \bar{q} \gamma_\mu \left(v^\mu + \frac{1}{3} v_{(s)}^\mu + \gamma_5 a^\mu \right) q - \bar{q}(s - i\gamma_5 p)q. \tag{1.151}$$

The 35 real functions $v_a^\mu(x), v_{(s)}^\mu(x), a_a^\mu(x), s_a(x)$, and $p_a(x)$, will collectively be denoted by $[v, a, s, p]$.[28] A precursor of this method can be found in Refs. [9, 10]. The ordinary three-flavor QCD Lagrangian is recovered by setting $v^\mu = v_{(s)}^\mu = a^\mu = p = 0$ and $s = \mathrm{diag}(m_u, m_d, m_s)$ in Eq. 1.151. The Green functions of the vacuum sector may be combined in the generating functional

$$\exp(iZ[v, a, s, p]) = \langle 0|T \exp\left[i \int d^4x \mathcal{L}_{\mathrm{ext}}(x)\right]|0\rangle_0. \qquad (1.152)$$

Note that both the quark field operators q in $\mathcal{L}_{\mathrm{ext}}$ and the ground state $|0\rangle$ refer to the chiral limit, indicated by the subscript 0 in Eq. 1.152. The quark fields are operators in the Heisenberg picture and have to satisfy the equations of motion and the canonical anticommutation relations. The generating functional is related to the vacuum-to-vacuum transition amplitude in the presence of external fields,

$$\exp(iZ[v, a, s, p]) = \langle 0, \mathrm{out}|0, \mathrm{in}\rangle_{v,a,s,p}. \qquad (1.153)$$

A particular Green function is then obtained through a partial functional derivative with respect to the external fields. As an example, suppose we are interested in the scalar u-quark condensate in the chiral limit, $\langle 0|\bar{u}u|0\rangle_0$. We express $\bar{u}u$ as

$$\bar{u}u = \frac{1}{2}\sqrt{\frac{2}{3}}\bar{q}\lambda_0 q + \frac{1}{2}\bar{q}\lambda_3 q + \frac{1}{2}\frac{1}{\sqrt{3}}\bar{q}\lambda_8 q$$

and obtain

$$\langle 0|\bar{u}(x)u(x)|0\rangle_0 = \frac{i}{2}\left[\sqrt{\frac{2}{3}}\frac{\delta}{\delta s_0(x)} + \frac{\delta}{\delta s_3(x)} + \frac{1}{\sqrt{3}}\frac{\delta}{\delta s_8(x)}\right]\exp(iZ[v, a, s, p])\Big|_{v=a=s=p=0}.$$

From the generating functional, we can even obtain Green functions of the "real world," where the quark fields and the ground state are those with finite quark masses. For example, the two-point function of two axial-vector currents of the "real world," i.e., for $s = \mathrm{diag}(m_u, m_d, m_s)$, and the "true vacuum" $|0\rangle$, is given by

$$\langle 0|T[A_a^\mu(x)A_b^\nu(0)]|0\rangle = (-i)^2\frac{\delta^2}{\delta a_{a\mu}(x)\delta a_{b\nu}(0)}\exp(iZ[v, a, s, p])\Big|_{v=a=p=0, s=\mathrm{diag}(m_u, m_d, m_s)}.$$

$$(1.154)$$

Note that the left-hand side involves the quark fields and the ground state of the "real world," whereas the right-hand side is the generating functional defined in terms of the quark fields and the ground state of the chiral limit. The actual value

---

[28] We omit the coupling to the singlet axial-vector current which has an anomaly, but include a singlet vector current $v_{(s)}^\mu$ which is of some physical relevance in the two-flavor sector.

Table 1.5 Transformation properties of the matrices Γ under parity

Γ	$\mathbb{1}$	γ^μ	$\sigma^{\mu\nu}$	γ_5	$\gamma^\mu\gamma_5$
$\gamma_0\Gamma\gamma_0$	$\mathbb{1}$	γ_μ	$\sigma_{\mu\nu}$	$-\gamma_5$	$-\gamma_\mu\gamma_5$

of the generating functional for a given configuration of external fields v, a, s, and p reflects the dynamics generated by the QCD Lagrangian. The (infinite) set of *all* chiral Ward identities resides in an invariance of the generating functional under a *local* transformation of the external fields [20, 37] (see the discussion of Sect. 1.4.1). The use of local transformations allows one to also consider divergences of Green functions. We require $\mathscr{L}$ of Eq. 1.150 to be a Hermitian Lorentz scalar, to be even under the discrete symmetries P, C, and T, and to be invariant under *local* chiral transformations. In fact, it is sufficient to consider P and C, only, because T is then automatically incorporated owing to the *CPT* theorem [38].

Under parity, the quark fields transform as

$$q_f(t, \vec{x}) \stackrel{P}{\mapsto} \gamma_0 q_f(t, -\vec{x}), \tag{1.155}$$

and the requirement of parity conservation,

$$\mathscr{L}(t, \vec{x}) \stackrel{P}{\mapsto} \mathscr{L}(t, -\vec{x}), \tag{1.156}$$

leads, using the results of Table 1.5, to the following constraints for the external fields,

$$v^\mu \stackrel{P}{\mapsto} v_\mu, \quad v^\mu_{(s)} \stackrel{P}{\mapsto} v^{(s)}_\mu, \quad a^\mu \stackrel{P}{\mapsto} -a_\mu, \quad s \stackrel{P}{\mapsto} s, \quad p \stackrel{P}{\mapsto} -p. \tag{1.157}$$

In Eq. 1.157 it is understood that the arguments change from $(t, \vec{x})$ to $(t, -\vec{x})$.

Similarly, under charge conjugation the quark fields transform as

$$q_{\alpha,f} \stackrel{C}{\mapsto} C_{\alpha\beta}\bar{q}_{\beta,f}, \quad \bar{q}_{\alpha,f} \stackrel{C}{\mapsto} -q_{\beta,f}C^{-1}_{\beta\alpha}, \tag{1.158}$$

where the subscripts α and β are Dirac-spinor indices,

$$C = i\gamma^2\gamma^0 = -C^{-1} = -C^\dagger = -C^T = \begin{pmatrix} 0 & 0 & 0 & -1 \\ 0 & 0 & 1 & 0 \\ 0 & -1 & 0 & 0 \\ 1 & 0 & 0 & 0 \end{pmatrix}$$

is the usual charge-conjugation matrix, and f refers to flavor. Taking Fermi statistics into account, one obtains

$$\bar{q}\Gamma Fq \stackrel{C}{\mapsto} -\bar{q}C\Gamma^T CF^T q,$$

where F denotes a matrix in flavor space. In combination with Table 1.6 it is straightforward to show that invariance of $\mathscr{L}_{\text{ext}}$ under charge conjugation requires the transformation properties

Table 1.6 Transformation properties of the matrices Γ under charge conjugation

Γ	$\mathbb{1}$	γ^μ	$\sigma^{\mu\nu}$	γ_5	$\gamma^\mu\gamma_5$
$-C\Gamma^T C$	$\mathbb{1}$	$-\gamma^\mu$	$-\sigma^{\mu\nu}$	γ_5	$\gamma^\mu\gamma_5$

$$v_\mu \overset{C}{\mapsto} -v_\mu^T, \quad v_\mu^{(s)} \overset{C}{\mapsto} -v_\mu^{(s)T}, \quad a_\mu \overset{C}{\mapsto} a_\mu^T, \quad s,p \overset{C}{\mapsto} s^T, p^T, \tag{1.159}$$

where the transposition refers to flavor space.

Finally, we need to discuss the requirements to be met by the external fields under local $SU(3)_L \times SU(3)_R \times U(1)_V$ transformations. In a first step, we write Eq. 1.151 in terms of the left- and right-handed quark fields.

Exercise 1.19 We first define

$$r_\mu = v_\mu + a_\mu, \quad l_\mu = v_\mu - a_\mu. \tag{1.160}$$

(a) Make use of the projection operators P_L and P_R and verify

$$\bar{q}\gamma^\mu \left(v_\mu + \frac{1}{3}v_\mu^{(s)} + \gamma_5 a_\mu \right) q = \bar{q}_R \gamma^\mu \left(r_\mu + \frac{1}{3}v_\mu^{(s)} \right) q_R + \bar{q}_L \gamma^\mu \left(l_\mu + \frac{1}{3}v_\mu^{(s)} \right) q_L.$$

(b) Also verify

$$\bar{q}(s - i\gamma_5 p)q = \bar{q}_L(s - ip)q_R + \bar{q}_R(s + ip)q_L.$$

We obtain for the Lagrangian of Eq. 1.151

$$\mathcal{L} = \mathcal{L}_{QCD}^0 + \bar{q}_L \gamma^\mu \left(l_\mu + \frac{1}{3}v_\mu^{(s)} \right) q_L + \bar{q}_R \gamma^\mu \left(r_\mu + \frac{1}{3}v_\mu^{(s)} \right) q_R$$
$$- \bar{q}_R(s + ip)q_L - \bar{q}_L(s - ip)q_R. \tag{1.161}$$

Equation 1.161 remains invariant under *local* transformations[29]

$$q_R \mapsto \exp\left(-i\frac{\Theta(x)}{3} \right) V_R(x)q_R,$$
$$q_L \mapsto \exp\left(-i\frac{\Theta(x)}{3} \right) V_L(x)q_L, \tag{1.162}$$

where $V_R(x)$ and $V_L(x)$ are independent space-time-dependent SU(3) matrices, provided the external fields are subject to the transformations

---

[29] From now on V_R and V_L will denote *local* transformations, whereas R and L will be used for *global* transformations.

$$r_\mu \mapsto V_R r_\mu V_R^\dagger + i V_R \partial_\mu V_R^\dagger,$$

$$l_\mu \mapsto V_L l_\mu V_L^\dagger + i V_L \partial_\mu V_L^\dagger,$$

$$v_\mu^{(s)} \mapsto v_\mu^{(s)} - \partial_\mu \Theta, \tag{1.163}$$

$$s + ip \mapsto V_R (s + ip) V_L^\dagger,$$

$$s - ip \mapsto V_L (s - ip) V_R^\dagger.$$

The derivative terms in Eq. 1.163 serve the same purpose as in the construction of gauge theories, i.e., they cancel analogous terms originating from the kinetic part of the quark Lagrangian.

There is another, yet, more practical aspect of the local invariance, namely: such a procedure allows one to also discuss a coupling to external gauge fields in the transition to the effective theory to be discussed later. For example, a coupling of the electromagnetic field to point-like fundamental particles results from gauging a U(1) symmetry. Here, the corresponding U(1) group is to be understood as a subgroup of a local $SU(3)_L \times SU(3)_R$. Another example deals with the interaction of the light quarks with the charged and neutral gauge bosons of the weak interactions.

Let us consider both examples explicitly. The coupling of quarks to an external electromagnetic four-vector potential $\mathcal{A}_\mu$ is given by

$$r_\mu = l_\mu = -e\mathcal{A}_\mu Q, \tag{1.164}$$

where $Q = \text{diag}(2/3, -1/3, -1/3)$ is the quark-charge matrix and $e > 0$ the elementary charge:

$$\mathcal{L}_{\text{ext}} = -e\mathcal{A}_\mu (\bar{q}_L Q \gamma^\mu q_L + \bar{q}_R Q \gamma^\mu q_R) = -e\mathcal{A}_\mu \bar{q} Q \gamma^\mu q$$

$$= -e\mathcal{A}_\mu \left(\frac{2}{3} \bar{u} \gamma^\mu u - \frac{1}{3} \bar{d} \gamma^\mu d - \frac{1}{3} \bar{s} \gamma^\mu s \right) = -e\mathcal{A}_\mu J^\mu.$$

On the other hand, if one considers only the two-flavor version of QCD one has to insert for the external fields

$$r_\mu = l_\mu = -e\mathcal{A}_\mu \frac{\tau_3}{2}, \quad v_\mu^{(s)} = -\frac{e}{2}\mathcal{A}_\mu. \tag{1.165}$$

In the description of semi-leptonic interactions such as $\pi^- \to \mu^- \bar{\nu}_\mu, \pi^- \to \pi^0 e^- \bar{\nu}_e$, or neutron decay $n \to p e^- \bar{\nu}_e$ one needs the interaction of quarks with the massive charged weak bosons $\mathcal{W}_\mu^\pm = (\mathcal{W}_{1\mu} \mp i\mathcal{W}_{2\mu})/\sqrt{2}$,

$$r_\mu = 0, \quad l_\mu = -\frac{g}{\sqrt{2}}(\mathcal{W}_\mu^+ T_+ + \text{H.c.}), \tag{1.166}$$

where H.c. refers to the Hermitian conjugate and

$$T_+ = \begin{pmatrix} 0 & V_{ud} & V_{us} \\ 0 & 0 & 0 \\ 0 & 0 & 0 \end{pmatrix}.$$

Here, V_{ij} denote the elements of the Cabibbo-Kobayashi-Maskawa quark-mixing matrix describing the transformation between the mass eigenstates of QCD and the weak eigenstates [41],

$$|V_{ud}| = 0.97425 \pm 0.00022, \quad |V_{us}| = 0.2252 \pm 0.0009.$$

At lowest order in perturbation theory, the Fermi constant is related to the gauge coupling g and the W mass by

$$G_F = \sqrt{2}\frac{g^2}{8M_W^2} = 1.16637(1) \times 10^{-5}\,\text{GeV}^{-2}. \tag{1.167}$$

Making use of

$$\bar{q}_L\gamma^\mu \mathscr{W}_\mu^+ T_+ q_L = \mathscr{W}_\mu^+ (\bar{u}\,\bar{d}\,\bar{s})P_R\gamma^\mu \begin{pmatrix} 0 & V_{ud} & V_{us} \\ 0 & 0 & 0 \\ 0 & 0 & 0 \end{pmatrix} P_L \begin{pmatrix} u \\ d \\ s \end{pmatrix}$$

$$= \mathscr{W}_\mu^+ (\bar{u}\,\bar{d}\,\bar{s})\gamma^\mu \frac{1}{2}(\mathbb{1} - \gamma_5) \begin{pmatrix} V_{ud}d + V_{us}s \\ 0 \\ 0 \end{pmatrix}$$

$$= \frac{1}{2}\mathscr{W}_\mu^+ [V_{ud}\bar{u}\gamma^\mu(\mathbb{1} - \gamma_5)d + V_{us}\bar{u}\gamma^\mu(\mathbb{1} - \gamma_5)s],$$

we see that inserting Eq. 1.166 into Eq. 1.161 leads to the standard charged-current weak interaction in the light-quark sector,

$$\mathscr{L}_{\text{ext}} = -\frac{g}{2\sqrt{2}}\Big\{\mathscr{W}_\mu^+ [V_{ud}\bar{u}\gamma^\mu(\mathbb{1} - \gamma_5)d + V_{us}\bar{u}\gamma^\mu(\mathbb{1} - \gamma_5)s] + \text{H.c.}\Big\}.$$

The situation is slightly different for the neutral weak interaction. Here, the three-flavor version requires a coupling of the Z boson to the singlet axial-vector current which, because of the anomaly of Eq. 1.112, we have dropped from our discussion. On the other hand, in the two-flavor version the axial-vector current part is traceless and we have

$$r_\mu = e\tan(\theta_W)\mathscr{Z}_\mu\frac{\tau_3}{2},$$

$$l_\mu = -\frac{g}{\cos(\theta_W)}\mathscr{Z}_\mu\frac{\tau_3}{2} + e\tan(\theta_W)\mathscr{Z}_\mu\frac{\tau_3}{2}, \tag{1.168}$$

$$v_\mu^{(s)} = \frac{e\tan(\theta_W)}{2}\mathscr{Z}_\mu,$$

where θ_W is the weak angle. With these external fields, we obtain the standard weak neutral-current interaction

$$\mathscr{L}_{ext} = -\frac{g}{2\cos(\theta_W)}\mathscr{Z}_\mu\left(\bar{u}\gamma^\mu\left\{\left[\frac{1}{2}-\frac{4}{3}\sin^2(\theta_W)\right]\mathbb{1}-\frac{1}{2}\gamma_5\right\}u\right.$$
$$\left.+\bar{d}\gamma^\mu\left\{\left[-\frac{1}{2}+\frac{2}{3}\sin^2(\theta_W)\right]\mathbb{1}+\frac{1}{2}\gamma_5\right\}d\right),$$

where we made use of $e = g\sin(\theta_W)$.

1.4.5 PCAC in the Presence of an External Electromagnetic Field

Finally, the technique of coupling the QCD Lagrangian to external fields also allows us to determine the current divergences for rigid external fields, i.e., fields which are *not* simultaneously transformed. For example, in Eqs. 1.112 we have determined the divergences of the vector and axial-vector currents due to the quark masses. The presence of an external electromagnetic field provides another example which has been used in the discussion of pion photo- and electroproduction on the nucleon. For the sake of simplicity we restrict ourselves to the two-flavor sector. (The generalization to the three-flavor case is straightforward.)

Exercise 1.20 Consider a *global* chiral transformation only and assume that the external fields are *not* simultaneously transformed. Show that the divergences of the currents read (see Eq. 1.47) [19]

$$\partial_\mu V_i^\mu = i\bar{q}\gamma^\mu\left[\frac{\tau_i}{2},v_\mu\right]q + i\bar{q}\gamma^\mu\gamma_5\left[\frac{\tau_i}{2},a_\mu\right]q - i\bar{q}\left[\frac{\tau_i}{2},s\right]q - \bar{q}\gamma_5\left[\frac{\tau_i}{2},p\right]q, \quad (1.169)$$

$$\partial_\mu A_i^\mu = i\bar{q}\gamma^\mu\gamma_5\left[\frac{\tau_i}{2},v_\mu\right]q + i\bar{q}\gamma^\mu\left[\frac{\tau_i}{2},a_\mu\right]q + i\bar{q}\gamma_5\left\{\frac{\tau_i}{2},s\right\}q + \bar{q}\left\{\frac{\tau_i}{2},p\right\}q. \quad (1.170)$$

Exercise 1.21 As an example, let us consider the QCD Lagrangian for a finite light quark mass $\hat{m} = m_u = m_d$ in combination with a coupling to an external electromagnetic four-vector potential $\mathscr{A}_\mu$ (see Eq. 1.165, $a_\mu = 0 = p$). Show that the expressions for the divergence of the vector and axial-vector currents, respectively, are given by

$$\partial_\mu V_i^\mu = -\varepsilon_{3ij}e\mathscr{A}_\mu\bar{q}\gamma^\mu\frac{\tau_j}{2}q = -\varepsilon_{3ij}e\mathscr{A}_\mu V_j^\mu, \quad (1.171)$$

$$\partial_\mu A_i^\mu = -e\mathscr{A}_\mu\varepsilon_{3ij}\bar{q}\gamma^\mu\gamma_5\frac{\tau_j}{2}q + 2\hat{m}i\bar{q}\gamma_5\frac{\tau_i}{2}q = -e\mathscr{A}_\mu\varepsilon_{3ij}A_j^\mu + \hat{m}P_i, \quad (1.172)$$

where we have introduced the isovector pseudoscalar density $P_i = i\bar{q}\gamma_5\tau_i q$. In fact, Eq. 1.172 is incomplete, because the third component of the axial-vector current, A_3^μ, has an anomaly which is related to the decay $\pi^0 \rightarrow \gamma\gamma$. The full equation reads

$$\partial_\mu A_i^\mu = \hat{m} P_i - e\mathscr{A}_\mu \varepsilon_{3ij} A_j^\mu + \delta_{i3}\frac{e^2}{32\pi^2}\varepsilon_{\mu\nu\rho\sigma}\mathscr{F}^{\mu\nu}\mathscr{F}^{\rho\sigma}, \quad \varepsilon_{0123} = 1, \qquad (1.173)$$

where $\mathscr{F}_{\mu\nu} = \partial_\mu \mathscr{A}_\nu - \partial_\nu \mathscr{A}_\mu$ is the electromagnetic field-strength tensor.

We emphasize the formal similarity of Eq. 1.172 to the (pre-QCD) PCAC (Partially Conserved Axial-Vector Current) relation obtained by Adler [2] through the inclusion of the electromagnetic interactions with minimal electromagnetic coupling. Since in QCD the quarks are taken as truly elementary, their interaction with an (external) electromagnetic field is of such a minimal type. In Adler's version, the right-hand side of Eq. 1.173 contains a renormalized field operator creating and destroying pions instead of $\hat{m} P_i$. From a modern point of view, the combination $\hat{m} P_i/(M_\pi^2 F_\pi)$ serves as an interpolating pion field. Furthermore, the anomaly term is not yet present in Ref. [2].

References

1. Abers, E.S., Lee, B.W.: Phys. Rep. **9**, 1 (1973)
2. Adler, S.L.: Phys. Rev. **139**, B1638 (1965)
3. Adler, S.L.: Phys. Rev. **177**, 2426 (1969)
4. Adler, S.L., Bardeen, W.A.: Phys. Rev. **182**, 1517 (1969)
5. Adler, S.L., Dashen, R.F.: Current Algebras and Applications to Particle Physics. Benjamin, New York (1968)
6. de Alfaro, V., Fubini, S., Furlan, G., Rossetti, C.: Currents in Hadron Physics (Chap. 2.1.1). North-Holland, Amsterdam (1973)
7. Altarelli, G.: Phys. Rep. **81**, 1 (1982)
8. Balachandran, A.P., Trahern, C.G.: Lectures on Group Theory for Physicists. Bibliopolis, Naples (1984)
9. Bell, J.S.: Nuovo Cim. A **50**, 129 (1969)
10. Bell, J.S., Jackiw, R.: Nuovo Cim. A **60**, 47 (1969)
11. Bjorken, J.D., Drell, S.D.: Relativistic Quantum Mechanics. McGraw-Hill, New York (1964)
12. Borodulin, V.I., Rogalev, R.N., Slabospitsky, S.R.: CORE: COmpendium of RElations. arXiv:hep-ph/9507456
13. Cheng, T.P., Li, L.F.: Gauge Theory of Elementary Particle Physics. Clarendon, Oxford (1984)
14. Das, A.: Field Theory: A Path Integral Approach. World Scientific, Singapore (1993)
15. Donoghue, J.F., Golowich, E., Holstein, B.R.: Dynamics of the Standard Model. Cambridge University Press, Cambridge (1992)
16. Ericson, T.E., Weise, W.: Pions and Nuclei (Apps. 3 and 8). Clarendon, Oxford (1988)
17. Fradkin, E.S.: Zh. Eksp. Teor. Fiz. **29**, 258 (1955) [Sov. Phys. JETP **2**, 361 (1955)]
18. Fritzsch, H., Gell-Mann, M., Leutwyler, H.: Phys. Lett. B **47**, 365 (1973)
19. Fuchs, T., Scherer, S.: Phys. Rev. C **68**, 055501 (2003)
20. Gasser, J., Leutwyler, H.: Ann. Phys. **158**, 142 (1984)
21. Gasser, J., Leutwyler, H.: Nucl. Phys. B **250**, 465 (1985)
22. Gell-Mann, M.: Phys. Rev. **125**, 1067 (1962)
23. Gell-Mann, M.: Physics **1**, 63 (1964)
24. Gell-Mann, M., Lévy, M.: Nuovo Cim. **16**, 705 (1960)
25. Gell-Mann, M., Ne'eman, Y.: The Eightfold Way. Benjamin, New York (1964)
26. Gell-Mann, M., Oakes, R.J., Renner, B.: Phys. Rev. **175**, 2195 (1968)

27. Glashow, S.L., Weinberg, S.: Phys. Rev. Lett. **20**, 224 (1968)
28. Greenberg, O.W.: Phys. Rev. Lett. **13**, 598 (1964)
29. Gross, D.J., Wilczek, F.: Phys. Rev. Lett. **30**, 1343 (1973)
30. Han, M.Y., Nambu, Y.: Phys. Rev. **139**, B1006 (1965)
31. Hill, E.L.: Rev. Mod. Phys. **23**, 253 (1951)
32. Itzykson, C., Zuber, J.B.: Quantum Field Theory. McGraw-Hill, New York (1980)
33. Jackiw, R.: Field theoretic investigations in current algebra. In: Treiman, S., Jackiw, R., Gross, D.J. (eds.) Lectures on Current Algebra and Its Applications. Princeton University Press, Princeton (1972)
34. Jones, H.F.: Groups, Representations and Physics. Hilger, Bristol (1990)
35. Kronfeld, A.S., Quigg, C.: Am. J. Phys. **78**, 1081 (2010)
36. Lehmann, H., Symanzik, K., Zimmermann, W.: Nuovo Cim. **1**, 205 (1955)
37. Leutwyler, H.: Ann. Phys. **235**, 165 (1994)
38. Lüders, G.: Ann. Phys. **2**, 1 (1957)
39. Manohar, A.V., Sachrajda, C.T.: Quark masses. In: Nakamura, K., et al., Particle Data Group (eds.) Review of Particle Physics. J. Phys. G **37**, 075021 (2010)
40. Marciano, W.J., Pagels, H.: Phys. Rep. **36**, 137 (1978)
41. Nakamura, K., et al., Particle Data Group: J. Phys. G **37**, 075021 (2010)
42. Noether, E.: In: Nachrichten von der Gesellschaft der Wissenschaften zu Göttingen, Mathematisch-Physikalische Klasse. Band 1918, 235–257 (1918)
43. O'Raifeartaigh, L.: Group Structure of Gauge Theories. Cambridge University Press, Cambridge (1986)
44. Ottnad, K., Kubis, B., Meißner, U.-G., Guo, F.K.: Phys. Lett. B **687**, 42 (2010)
45. Pagels, H.: Phys. Rep. **16**, 219 (1975)
46. Scheck, F.: Electroweak and Strong Interactions: An Introduction to Theoretical Particle Physics (Chap. 3.5). Springer, Berlin (1996)
47. Schwinger, J.S.: Phys. Rev. Lett. **3**, 296 (1959)
48. Takahashi, Y.: Nuovo Cim. **6**, 371 (1957)
49. 't Hooft, G.: Nucl. Phys. B **72**, 461 (1974)
50. Treiman, S., Jackiw, R., Gross, D.J.: Lectures on Current Algebra and Its Applications. Princeton University Press, Princeton (1972)
51. Weinberg, S.: Phys. Rev. Lett. **31**, 494 (1973)
52. Weinberg, S.: The Quantum Theory of Fields. Foundations, vol. 1 (Chap. 7.3). Cambridge University Press, Cambridge (1995)
53. Ward, J.C.: Phys. Rev. **78**, 182 (1950)
54. Yang, C.N., Mills, R.L.: Phys. Rev. **96**, 191 (1954)
55. Zinn-Justin, J.: Quantum Field Theory and Critical Phenomena. Clarendon, Oxford (1989)

Chapter 2
Spontaneous Symmetry Breaking and the Goldstone Theorem

2.1 Degenerate Ground States

Before discussing the case of a *continuous* symmetry, we will first have a look at a field theory with a *discrete* internal symmetry. This will allow us to distinguish between two possibilities: a dynamical system with a unique ground state or a system with a finite number of distinct degenerate ground states. In particular, we will see how, for the second case, an infinitesimal perturbation selects a particular vacuum state.

To that end we consider the Lagrangian of a real scalar field $\Phi(x)$ [8]

$$\mathscr{L}(\Phi, \partial_\mu \Phi) = \frac{1}{2} \partial_\mu \Phi \partial^\mu \Phi - \frac{m^2}{2} \Phi^2 - \frac{\lambda}{4} \Phi^4, \tag{2.1}$$

which is invariant under the discrete transformation $R : \Phi \mapsto -\Phi$. The corresponding classical energy density reads

$$\mathscr{H} = \Pi \dot{\Phi} - \mathscr{L} = \frac{1}{2} \dot{\Phi}^2 + \frac{1}{2} (\vec{\nabla}\Phi)^2 + \underbrace{\frac{m^2}{2} \Phi^2 + \frac{\lambda}{4} \Phi^4}_{\equiv \mathscr{V}(\Phi)}, \tag{2.2}$$

where one chooses $\lambda > 0$ so that $\mathscr{H}$ is bounded from below. The field Φ_0 which minimizes the Hamilton density $\mathscr{H}$ must be constant and uniform since in that case the first two terms take their minimum values of zero everywhere. It must also minimize the "potential" $\mathscr{V}$ since $\mathscr{V}(\Phi(x)) \geq \mathscr{V}(\Phi_0)$, from which we obtain the condition

$$\mathscr{V}'(\Phi) = \Phi(m^2 + \lambda \Phi^2) = 0.$$

We now distinguish two different cases:

S. Scherer and M. R. Schindler, *A Primer for Chiral Perturbation Theory*,
Lecture Notes in Physics 830, DOI: 10.1007/978-3-642-19254-8_2,
© Springer-Verlag Berlin Heidelberg 2012

Fig. 2.1
$\mathscr{V}(x) = x^2/2 + x^4/4$
(Wigner-Weyl mode)

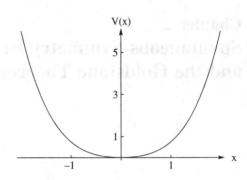

1. $m^2 > 0$ (see Fig. 2.1): In this case the potential $\mathscr{V}$ has its minimum for $\Phi = 0$. In the quantized theory we associate a unique ground state $|0\rangle$ with this minimum. Later on, in the case of a continuous symmetry, this situation will be referred to as the Wigner-Weyl realization of the symmetry.
2. $m^2 < 0$ (see Fig. 2.2): Now the potential exhibits two distinct minima. (In the continuous symmetry case this will be referred to as the Nambu-Goldstone realization of the symmetry.)

We will concentrate on the second situation, because this is the one which we would like to generalize to a continuous symmetry and which ultimately leads to the appearance of Goldstone bosons. In the present case, $\mathscr{V}(\Phi)$ has a local maximum for $\Phi = 0$ and *two* minima for

$$\Phi_\pm = \pm\sqrt{\frac{-m^2}{\lambda}} \equiv \pm\Phi_0. \tag{2.3}$$

As will be explained below, the quantized theory develops two degenerate vacua $|0, +\rangle$ and $|0, -\rangle$ which are distinguished through their vacuum expectation values of the field $\Phi(x)$:[1]

$$\langle 0, +|\Phi(x)|0, +\rangle = \langle 0, +|e^{iP\cdot x}\Phi(0)e^{-iP\cdot x}|0, +\rangle = \langle 0, +|\Phi(0)|0, +\rangle \equiv \Phi_0,$$
$$\langle 0, -|\Phi(x)|0, -\rangle = -\Phi_0. \tag{2.4}$$

We made use of translational invariance, $\Phi(x) = e^{iP\cdot x}\Phi(0)e^{-iP\cdot x}$, and the fact that the ground state is an eigenstate of energy and momentum. We associate with the transformation $R : \Phi \mapsto \Phi' = -\Phi$ a unitary operator $\mathscr{R}$ acting on the Hilbert space of our model, with the properties

---

[1] The case of a quantum field theory with an infinite volume V has to be distinguished from, say, a nonrelativistic particle in a one-dimensional potential of a shape similar to the function of Fig. 2.2. For example, in the case of a symmetric double-well potential, the solutions with positive parity always have lower energy eigenvalues than those with negative parity (see, e.g., Ref. [11]).

Fig. 2.2
$\mathscr{V}(x) = -x^2/2 + x^4/4$
(Nambu-Goldstone mode)

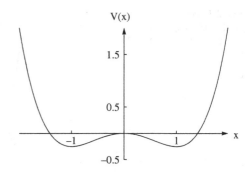

$$\mathscr{R}^2 = I, \quad \mathscr{R} = \mathscr{R}^{-1} = \mathscr{R}^\dagger.$$

In accord with Eq. 2.4 the action of the operator $\mathscr{R}$ on the ground states is given by

$$\mathscr{R}|0, \pm\rangle = |0, \mp\rangle. \tag{2.5}$$

For the moment we select one of the two expectation values and expand the Lagrangian about $\pm\Phi_0$:[2]

$$\begin{aligned}\Phi &= \pm\Phi_0 + \Phi', \\ \partial_\mu\Phi &= \partial_\mu\Phi'.\end{aligned} \tag{2.6}$$

Exercise 2.1 Show that

$$\mathscr{V}(\Phi) = \tilde{\mathscr{V}}(\Phi') = -\frac{\lambda}{4}\Phi_0^4 + \frac{1}{2}(-2m^2)\Phi'^2 \pm \lambda\Phi_0\Phi'^3 + \frac{\lambda}{4}\Phi'^4.$$

Thus, the Lagrangian in terms of the shifted dynamical variable reads

$$\mathscr{L}'(\Phi', \partial_\mu\Phi') = \frac{1}{2}\partial_\mu\Phi'\partial^\mu\Phi' - \frac{1}{2}(-2m^2)\Phi'^2 \mp \lambda\Phi_0\Phi'^3 - \frac{\lambda}{4}\Phi'^4 + \frac{\lambda}{4}\Phi_0^4. \tag{2.7}$$

In terms of the new dynamical variable Φ', the symmetry R is no longer manifest, i.e., it is hidden. Selecting one of the ground states has led to a spontaneous symmetry breaking which is always related to the existence of several degenerate vacua.

At this stage it is not clear why the ground state of the quantum system should be one or the other of $|0, \pm\rangle$ and not a superposition of both. For example, the linear combination

$$\frac{1}{\sqrt{2}}(|0, +\rangle + |0, -\rangle)$$

---

[2] The field Φ' instead of Φ is assumed to vanish at infinity.

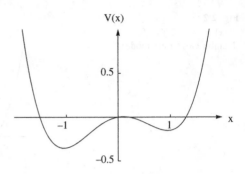

Fig. 2.3 Potential with a small odd component: $\mathscr{V}(x) = x/10 - x^2/2 + x^4/4$

is invariant under $\mathscr{R}$ as is the original Lagrangian of Eq. 2.1. However, this superposition is not stable against any infinitesimal external perturbation which is odd in Φ (see Fig. 2.3),

$$\mathscr{R}\varepsilon H' \mathscr{R}^{\dagger} = -\varepsilon H'.$$

Any such perturbation will drive the ground state into the vicinity of either $|0, +\rangle$ or $|0, -\rangle$ rather than $\frac{1}{\sqrt{2}}(|0, +\rangle \pm |0, -\rangle)$. This can easily be seen in the framework of perturbation theory for degenerate states. Consider

$$|1\rangle = \frac{1}{\sqrt{2}}(|0, +\rangle + |0, -\rangle), \quad |2\rangle = \frac{1}{\sqrt{2}}(|0, +\rangle - |0, -\rangle),$$

such that

$$\mathscr{R}|1\rangle = |1\rangle \quad \mathscr{R}|2\rangle = -|2\rangle.$$

The condition for the energy eigenvalues of the ground state, $E = E^{(0)} + \varepsilon E^{(1)} + \cdots$, to first order in ε results from

$$\det \begin{pmatrix} \langle 1|H'|1\rangle - E^{(1)} & \langle 1|H'|2\rangle \\ \langle 2|H'|1\rangle & \langle 2|H'|2\rangle - E^{(1)} \end{pmatrix} = 0.$$

Due to the symmetry properties of Eq. 2.5, we obtain

$$\langle 1|H'|1\rangle = \langle 1|\mathscr{R}^{-1}\mathscr{R}H'\mathscr{R}^{-1}\mathscr{R}|1\rangle = \langle 1| - H'|1\rangle = 0$$

and similarly $\langle 2|H'|2\rangle = 0$. Setting $\langle 1|H'|2\rangle = a > 0$, which can always be achieved by multiplication of one of the two states by an appropriate phase, one finds

$$\langle 2|H'|1\rangle \overset{H' = H'^{\dagger}}{=} \langle 1|H'|2\rangle^* = a^* = a = \langle 1|H'|2\rangle,$$

resulting in

$$\det \begin{pmatrix} -E^{(1)} & a \\ a & -E^{(1)} \end{pmatrix} = E^{(1)2} - a^2 = 0. \quad \Rightarrow E^{(1)} = \pm a.$$

Fig. 2.4 Dispersion relation $E = \sqrt{1+p_x^2}$ and asymptote $E = |p_x|$

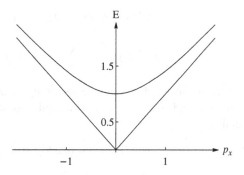

In other words, the degeneracy has been lifted and we get for the energy eigenvalues

$$E = E^{(0)} \pm \varepsilon a + \cdots. \tag{2.8}$$

The corresponding eigenstates of zeroth order in ε are $|0, +\rangle$ and $|0, -\rangle$, respectively. We thus conclude that an arbitrarily small external perturbation which is odd with respect to R will push the ground state to either $|0, +\rangle$ or $|0, -\rangle$.

In the above discussion, we have tacitly assumed that the Hamiltonian and the field $\Phi(x)$ can simultaneously be diagonalized in the vacuum sector, i.e., $\langle 0, +|0, -\rangle = 0$. Following Ref. [18], we will justify this assumption which will also be crucial for the continuous case to be discussed later.

For an infinite volume, a general vacuum state $|v\rangle$ is defined as a state with momentum eigenvalue $\vec{0}$,

$$\vec{P}|v\rangle = \vec{0},$$

where $\vec{0}$ is a *discrete* eigenvalue as opposed to an eigenvalue of single- or many-particle states for which $\vec{p} = \vec{0}$ is an element of a continuous spectrum (see Fig. 2.4). We deal with the situation of several degenerate ground states[3] which will be denoted by $|u\rangle$, $|v\rangle$, etc., and start from the identity

$$0 = \langle u|[H, \Phi(x)]|v\rangle \quad \forall \; x, \tag{2.9}$$

from which we obtain for $t = 0$

$$\int d^3y \, \langle u|\mathscr{H}(\vec{y}, 0)\Phi(\vec{x}, 0)|v\rangle = \int d^3y \, \langle u|\Phi(\vec{x}, 0)\mathscr{H}(\vec{y}, 0)|v\rangle. \tag{2.10}$$

Let us consider the left-hand side,

---

[3] For continuous symmetry groups one may have a non-countably infinite number of ground states.

$$\int d^3y \, \langle u|\mathscr{H}(\vec{y},0)\Phi(\vec{x},0)|v\rangle = \sum_w \langle u|H|w\rangle\langle w|\Phi(0)|v\rangle$$

$$+ \int d^3y \int d^3p \sum_n \langle u|\mathscr{H}(\vec{y},0)|n,\vec{p}\rangle\langle n,\vec{p}|\Phi(0)|v\rangle e^{-i\vec{p}\cdot\vec{x}},$$

where we inserted a complete set of states which we split into the vacuum contribution and the remainder, and made use of translational invariance. We now define

$$f_n(\vec{y},\vec{p}) = \langle u|\mathscr{H}(\vec{y},0)|n,\vec{p}\rangle\langle n,\vec{p}|\Phi(0)|v\rangle$$

and assume f_n to be reasonably behaved such that one can apply the lemma of Riemann and Lebesgue,

$$\lim_{|\vec{x}|\to\infty} \int d^3p f(\vec{p})e^{-i\vec{p}\cdot\vec{x}} = 0.$$

At this point the assumption of an infinite volume, $|\vec{x}| \to \infty$, is crucial. Repeating the argument for the right-hand side and taking the limit $|\vec{x}| \to \infty$, only the vacuum contributions survive in Eq. 2.10 and we obtain

$$\sum_w \langle u|H|w\rangle\langle w|\Phi(0)|v\rangle = \sum_w \langle u|\Phi(0)|w\rangle\langle w|H|v\rangle$$

for arbitrary ground states $|u\rangle$ and $|v\rangle$. In other words, the matrices $(H_{uv}) \equiv (\langle u|H|v\rangle)$ and $(\Phi_{uv}) \equiv (\langle u|\Phi(0)|v\rangle)$ commute and can be diagonalized simultaneously. Choosing an appropriate basis, one can write

$$\langle u|\Phi(0)|v\rangle = \delta_{uv}v, \quad v \in \mathbb{R},$$

where v denotes the expectation value of Φ in the state $|v\rangle$.

In the above example, the ground states $|0,+\rangle$ and $|0,-\rangle$ with vacuum expectation values $\pm\Phi_0$ are thus indeed orthogonal and satisfy

$$\langle 0,+|H|0,-\rangle = \langle 0,-|H|0,+\rangle = 0.$$

2.2 Spontaneous Breakdown of a Global, Continuous, Non-Abelian Symmetry

Using the example of the O(3) sigma model we recall a few aspects relevant to our subsequent discussion of spontaneous symmetry breaking [16].[4] To that end, we consider the Lagrangian

---

[4] The linear sigma model [6, 7, 17] is constructed in terms of the O(4) multiplet $(\sigma, \pi_1, \pi_2, \pi_3)$. Since the group O(4) is locally isomorphic to SU(2) × SU(2), the linear sigma model is a popular framework for illustrating the spontaneous symmetry breaking in two-flavor QCD.

$$\mathscr{L}(\vec{\Phi}, \partial_\mu \vec{\Phi}) = \mathscr{L}(\Phi_1, \Phi_2, \Phi_3, \partial_\mu \Phi_1, \partial_\mu \Phi_2, \partial_\mu \Phi_3)$$

$$= \frac{1}{2} \partial_\mu \Phi_i \partial^\mu \Phi_i - \frac{m^2}{2} \Phi_i \Phi_i - \frac{\lambda}{4} (\Phi_i \Phi_i)^2, \qquad (2.11)$$

where $m^2 < 0$, $\lambda > 0$, with Hermitian fields Φ_i. By choosing $m^2 < 0$, the symmetry is realized in the Nambu-Goldstone mode [9, 13].[5]

The Lagrangian of Eq. 2.11 is invariant under a global "isospin" rotation,[6]

$$g \in SO(3): \Phi_i \mapsto \Phi_i' = D_{ij}(g)\Phi_j = \left(e^{-i\alpha_k T_k}\right)_{ij} \Phi_j. \qquad (2.12)$$

For the Φ_i' to also be Hermitian, the Hermitian T_k must be purely imaginary and thus antisymmetric (see Eqs. 1.69). The iT_k provide the basis of a representation of the so(3) Lie algebra and satisfy the commutation relations $[T_i, T_j] = i\varepsilon_{ijk} T_k$. We use the representation of Eqs. 1.69, i.e., the matrix elements are given by $t_{ijk} = -i\varepsilon_{ijk}$. We now look for a minimum of the potential which does not depend on x.

Exercise 2.2 Determine the minimum of the potential

$$\mathscr{V}(\Phi_1, \Phi_2, \Phi_3) = \frac{m^2}{2} \Phi_i \Phi_i + \frac{\lambda}{4} (\Phi_i \Phi_i)^2.$$

We find

$$|\vec{\Phi}_{min}| = \sqrt{\frac{-m^2}{\lambda}} \equiv v, \quad |\vec{\Phi}| = \sqrt{\Phi_1^2 + \Phi_2^2 + \Phi_3^2}. \qquad (2.13)$$

Since $\vec{\Phi}_{min}$ can point in any direction in isospin space we have a non-countably infinite number of degenerate vacua. Any infinitesimal external perturbation that is not invariant under SO(3) will select a particular direction which, by an appropriate orientation of the internal coordinate frame, we denote as the 3 direction in our convention,

$$\vec{\Phi}_{min} = v\hat{e}_3. \qquad (2.14)$$

Clearly, $\vec{\Phi}_{min}$ of Eq. 2.14 is *not* invariant under the full group $G = SO(3)$ since rotations about the 1 and 2 axes change $\vec{\Phi}_{min}$.[7] To be specific, if

---

[5] In the beginning, the discussion of spontaneous symmetry breaking in field theories [9, 13–15] was driven by an analogy with the theory of superconductivity [1, 2, 4, 5].

[6] The Lagrangian is invariant under the full group O(3) which can be decomposed into its two components: the proper rotations connected to the identity, SO(3), and the rotation-reflections. For our purposes it is sufficient to discuss SO(3).

[7] We say, somewhat loosely, that T_1 and T_2 do not annihilate the ground state or, equivalently, finite group elements generated by T_1 and T_2 do not leave the ground state invariant. This should become clearer later on.

$$\vec{\Phi}_{\min} = v \begin{pmatrix} 0 \\ 0 \\ 1 \end{pmatrix},$$

we obtain

$$T_1 \vec{\Phi}_{\min} = v \begin{pmatrix} 0 \\ -i \\ 0 \end{pmatrix}, \quad T_2 \vec{\Phi}_{\min} = v \begin{pmatrix} i \\ 0 \\ 0 \end{pmatrix}, \quad T_3 \vec{\Phi}_{\min} = 0. \qquad (2.15)$$

Note that the set of transformations which do not leave $\vec{\Phi}_{\min}$ invariant does *not* form a group, because it does not contain the identity. On the other hand, $\vec{\Phi}_{\min}$ is invariant under a subgroup H of G, namely, the rotations about the 3 axis:

$$h \in H: \quad \vec{\Phi}' = D(h)\vec{\Phi} = e^{-i\alpha_3 T_3}\vec{\Phi}, \quad D(h)\vec{\Phi}_{\min} = \vec{\Phi}_{\min}. \qquad (2.16)$$

Exercise 2.3 Write Φ_3 as

$$\Phi_3(x) = v + \eta(x), \qquad (2.17)$$

where $\eta(x)$ is a new field replacing $\Phi_3(x)$, and express the Lagrangian in terms of the fields Φ_1, Φ_2, and η, where $v = \sqrt{-m^2/\lambda}$.

The new expression for the potential is given by

$$\tilde{\mathscr{V}} = \frac{1}{2}(-2m^2)\eta^2 + \lambda v\eta(\Phi_1^2 + \Phi_2^2 + \eta^2) + \frac{\lambda}{4}(\Phi_1^2 + \Phi_2^2 + \eta^2)^2 - \frac{\lambda}{4}v^4. \qquad (2.18)$$

Upon inspection of the terms quadratic in the fields, one finds after spontaneous symmetry breaking two massless Goldstone bosons and one massive boson:

$$\begin{aligned} m_{\Phi_1}^2 &= m_{\Phi_2}^2 = 0, \\ m_\eta^2 &= -2m^2. \end{aligned} \qquad (2.19)$$

The model-independent feature of the above example is given by the fact that for each of the two generators T_1 and T_2 which do not annihilate the ground state one obtains a *massless* Goldstone boson. By means of a two-dimensional simplification (see the "Mexican hat" potential shown in Fig. 2.5) the mechanism at hand can easily be visualized. Infinitesimal variations orthogonal to the circle of the minimum of the potential generate quadratic terms, i.e., "restoring forces" linear in the displacement, whereas tangential variations experience restoring forces only of higher orders.

Fig. 2.5 Two-dimensional rotationally invariant potential: $\mathscr{V}(x, y) = -(x^2 + y^2) + \frac{(x^2+y^2)^2}{4}$

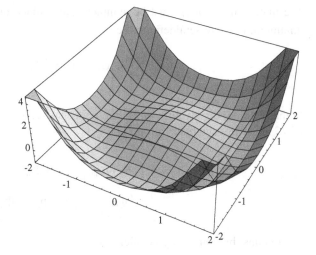

Now let us generalize the model to the case of an arbitrary compact Lie group G of order n_G resulting in n_G infinitesimal generators.[8] Once again, we start from a Lagrangian of the form [10]

$$\mathscr{L}\left(\vec{\Phi}, \partial_\mu \vec{\Phi}\right) = \frac{1}{2}\partial_\mu \vec{\Phi} \cdot \partial^\mu \vec{\Phi} - \mathscr{V}\left(\vec{\Phi}\right), \tag{2.20}$$

where $\vec{\Phi}$ is a multiplet of scalar (or pseudoscalar) Hermitian fields. The Lagrangian $\mathscr{L}$ and thus also $\mathscr{V}$ are supposed to be globally invariant under G, where the infinitesimal transformations of the fields are given by

$$g \in G: \quad \Phi_i \mapsto \Phi_i + \delta\Phi_i, \quad \delta\Phi_i = -i\varepsilon_a t_{a,ij}\Phi_j. \tag{2.21}$$

The Hermitian representation matrices $T_a = (t_{a,ij})$ are again antisymmetric and purely imaginary. We now assume that, by choosing an appropriate form of $\mathscr{V}$, the Lagrangian generates a spontaneous symmetry breaking resulting in a ground state with a vacuum expectation value $\vec{\Phi}_{\min} = \langle\vec{\Phi}\rangle$ which is invariant under a continuous subgroup H of G. We expand $\mathscr{V}$ about $\vec{\Phi}_{\min}, |\vec{\Phi}_{\min}| = v$, i.e., $\vec{\Phi} = \vec{\Phi}_{\min} + \vec{\chi}$,

$$\mathscr{V}\left(\vec{\Phi}\right) = \mathscr{V}\left(\vec{\Phi}_{\min}\right) + \underbrace{\frac{\partial\mathscr{V}\left(\vec{\Phi}_{\min}\right)}{\partial\Phi_i}}_{=\,0}\chi_i + \frac{1}{2}\underbrace{\frac{\partial^2\mathscr{V}\left(\vec{\Phi}_{\min}\right)}{\partial\Phi_i\partial\Phi_j}}_{\equiv\,m_{ij}^2}\chi_i\chi_j + \cdots. \tag{2.22}$$

---

[8] The restriction to compact groups allows for a complete decomposition into finite-dimensional irreducible unitary representations.

The matrix $M^2 = (m_{ij}^2)$ must be symmetric and, since one is expanding about a minimum, positive semidefinite, i.e.,

$$\sum_{i,j} m_{ij}^2 x_i x_j \geq 0 \quad \forall \quad \vec{x}. \tag{2.23}$$

In that case, all eigenvalues of M^2 are nonnegative. Making use of the invariance of $\mathscr{V}$ under the symmetry group G,

$$\mathscr{V}\left(\vec{\Phi}_{\min}\right) = \mathscr{V}\left(D(g)\vec{\Phi}_{\min}\right) = \mathscr{V}\left(\vec{\Phi}_{\min} + \delta\vec{\Phi}_{\min}\right)$$

$$\stackrel{(2.22)}{=} \mathscr{V}\left(\vec{\Phi}_{\min}\right) + \frac{1}{2} m_{ij}^2 \delta\Phi_{\min,i}\delta\Phi_{\min,j} + \cdots, \tag{2.24}$$

one obtains, by comparing coefficients,

$$m_{ij}^2 \delta\Phi_{\min,i}\delta\Phi_{\min,j} = 0. \tag{2.25}$$

Differentiating Eq. 2.25 with respect to $\delta\Phi_{\min,k}$ and using $m_{ij}^2 = m_{ji}^2$ results in the matrix equation

$$M^2 \delta\vec{\Phi}_{\min} = \vec{0}. \tag{2.26}$$

Inserting the variations of Eq. 2.21 for arbitrary ε_a, $\delta\vec{\Phi}_{\min} = -i\varepsilon_a T_a \vec{\Phi}_{\min}$, we conclude

$$M^2 T_a \vec{\Phi}_{\min} = \vec{0}. \tag{2.27}$$

Recall that the T_a represent generators of the symmetry transformations of the Lagrangian of Eq. 2.20. The solutions of Eq. 2.27 can be classified into two categories:

1. T_a, $a = 1, \ldots, n_H$, is a representation of an element of the Lie algebra belonging to the subgroup H of G, leaving the selected ground state invariant. Therefore, invariance under the subgroup H corresponds to

$$T_a \vec{\Phi}_{\min} = \vec{0}, \quad a = 1, \ldots, n_H,$$

   such that Eq. 2.27 is automatically satisfied without any knowledge of M^2.
2. T_a, $a = n_H + 1, \ldots, n_G$, is *not* a representation of an element of the Lie algebra belonging to the subgroup H. In that case $T_a \vec{\Phi}_{\min} \neq \vec{0}$, and $T_a \vec{\Phi}_{\min}$ is an eigenvector of M^2 with eigenvalue 0. To each such eigenvector corresponds a massless Goldstone boson. In particular, the different $T_a \vec{\Phi}_{\min} \neq \vec{0}$ are linearly independent, resulting in $n_G - n_H$ independent Goldstone bosons. (If they were not linearly independent, there would exist a nontrivial linear combination

$$\vec{0} = \sum_{a=n_H+1}^{n_G} c_a \left(T_a \vec{\Phi}_{\min} \right) = \underbrace{\left(\sum_{a=n_H+1}^{n_G} c_a T_a \right)}_{\equiv T} \vec{\Phi}_{\min},$$

such that T is an element of the Lie algebra of H in contradiction to our assumption.)

Remark It may be necessary to perform a similarity transformation on the fields in order to diagonalize the mass matrix.

Let us check these results by reconsidering the example of Eq. 2.11. In that case $n_G = 3$ and $n_H = 1$, generating two Goldstone bosons (see Eq. 2.19).

We conclude this section with two remarks.

1. The number of Goldstone bosons is determined by the structure of the symmetry groups. Let G denote the symmetry group of the Lagrangian with n_G generators, and H the subgroup with n_H generators which leaves the ground state invariant after spontaneous symmetry breaking. For each generator which does not annihilate the vacuum one obtains a massless Goldstone boson, i.e., the total number of Goldstone bosons equals $n_G - n_H$.
2. The Lagrangians used in *motivating* the phenomenon of a spontaneous symmetry breakdown are typically constructed in such a fashion that the degeneracy of the ground states is built into the potential at the classical level (the prototype being the "Mexican hat" potential of Fig. 2.5). As in the above case, it is then argued that an *elementary* Hermitian field of a multiplet transforming nontrivially under the symmetry group G acquires a vacuum expectation value signaling a spontaneous symmetry breakdown. However, there also exist theories such as QCD where one cannot infer from inspection of the Lagrangian whether the theory exhibits spontaneous symmetry breaking. Rather, the criterion for spontaneous symmetry breaking is a nonvanishing vacuum expectation value of some Hermitian operator, not an elementary field, which is generated through the dynamics of the underlying theory. In particular, we will see that the quantities developing a vacuum expectation value may also be local Hermitian operators composed of more fundamental degrees of freedom of the theory. Such a possibility was already emphasized in the derivation of Goldstone's theorem in Ref. [10].

2.3 Goldstone Theorem

By means of the above example, we motivate another approach to Goldstone's theorem without delving into all the subtleties of a quantum field-theoretical approach (for further reading, see Sect. 2 of Ref. [3]). Given a Hamilton operator with a global symmetry group $G = SO(3)$, let $\vec{\Phi}(x) = (\Phi_1(x), \Phi_2(x), \Phi_3(x))$ denote a triplet of local Hermitian operators transforming as a vector under G,

$$g \in G: \vec{\Phi}(x) \mapsto \vec{\Phi}'(x) = e^{i\sum_{k=1}^{3} \alpha_k Q_k} \vec{\Phi}(x) e^{-i\sum_{l=1}^{3} \alpha_l Q_l} = e^{-i\sum_{k=1}^{3} \alpha_k T_k} \vec{\Phi}(x), \quad (2.28)$$

where the Q_i are the generators of the $SO(3)$ transformations on the Hilbert space satisfying $[Q_i, Q_j] = i\varepsilon_{ijk}Q_k$ and the $T_i = (t_{i,jk})$ are the matrices of the three-dimensional representation satisfying $t_{i,jk} = -i\varepsilon_{ijk}$. We assume that one component of the multiplet acquires a nonvanishing vacuum expectation value:

$$\langle 0|\Phi_1(x)|0\rangle = \langle 0|\Phi_2(x)|0\rangle = 0, \quad \langle 0|\Phi_3(x)|0\rangle = v \neq 0. \quad (2.29)$$

Then the two generators Q_1 and Q_2 do not annihilate the ground state, and to each such generator corresponds a massless Goldstone boson.

In order to prove these two statements, let us expand Eq. 2.28 to first order in the α_k:

$$\vec{\Phi}' = \vec{\Phi} + i\sum_{k=1}^{3} \alpha_k[Q_k, \vec{\Phi}] = \left(1 - i\sum_{k=1}^{3} \alpha_k T_k\right)\vec{\Phi} = \vec{\Phi} + \vec{\alpha} \times \vec{\Phi}.$$

Comparing the terms linear in the α_k,

$$i[\alpha_k Q_k, \Phi_l] = \varepsilon_{lkm}\alpha_k\Phi_m,$$

and noting that all three α_k can be chosen independently, we obtain

$$i[Q_k, \Phi_l] = -\varepsilon_{klm}\Phi_m,$$

which expresses the fact that the field operators Φ_i transform as a vector.[9] Using $\varepsilon_{klm}\varepsilon_{kln} = 2\delta_{mn}$, we find

$$-\frac{i}{2}\varepsilon_{kln}[Q_k, \Phi_l] = \delta_{mn}\Phi_m = \Phi_n.$$

In particular,

$$\Phi_3 = -\frac{i}{2}([Q_1, \Phi_2] - [Q_2, \Phi_1]), \quad (2.30)$$

with cyclic permutations for the other two cases.

In order to prove that Q_1 and Q_2 do not annihilate the ground state, let us consider Eq. 2.28 for $\vec{\alpha} = (0, \pi/2, 0)$,

$$e^{-i\frac{\pi}{2}T_2}\vec{\Phi} = \begin{pmatrix} \cos\left(\frac{\pi}{2}\right) & 0 & \sin\left(\frac{\pi}{2}\right) \\ 0 & 1 & 0 \\ -\sin\left(\frac{\pi}{2}\right) & 0 & \cos\left(\frac{\pi}{2}\right) \end{pmatrix}\begin{pmatrix} \Phi_1 \\ \Phi_2 \\ \Phi_3 \end{pmatrix} = \begin{pmatrix} \Phi_3 \\ \Phi_2 \\ -\Phi_1 \end{pmatrix} = e^{i\frac{\pi}{2}Q_2}\begin{pmatrix} \Phi_1 \\ \Phi_2 \\ \Phi_3 \end{pmatrix}e^{-i\frac{\pi}{2}Q_2}.$$

---

[9] Using the replacements $Q_k \to \hat{l}_k$ and $\Phi_l \to \hat{x}_l$, note the analogy with $i[\hat{l}_k, \hat{x}_l] = -\varepsilon_{klm}\hat{x}_m$.

From the first row we obtain

$$\Phi_3 = e^{i\frac{\vec{\alpha}}{2}Q_2} \Phi_1 e^{-i\frac{\vec{\alpha}}{2}Q_2}.$$

Taking the vacuum expectation value

$$v = \langle 0|e^{i\frac{\vec{\alpha}}{2}Q_2} \Phi_1 e^{-i\frac{\vec{\alpha}}{2}Q_2}|0\rangle$$

and using Eq. 2.29, clearly $Q_2|0\rangle \neq 0$, since otherwise the exponential operator could be replaced by unity and the right-hand side would vanish. A similar argument shows $Q_1|0\rangle \neq 0$.

At this point let us make two remarks.

1. The "states" $Q_{1(2)}|0\rangle$ cannot be normalized. In a more rigorous derivation one makes use of integrals of the form

$$\int d^3x \, \langle 0|[J_k^0(t,\vec{x}), \Phi_l(0)]|0\rangle,$$

and first determines the commutator before evaluating the integral [3].
2. Some derivations of Goldstone's theorem right away start by assuming $Q_{1(2)}|0\rangle \neq 0$. However, for the discussion of spontaneous symmetry breaking in the framework of QCD it is advantageous to establish the connection between the existence of Goldstone bosons and a nonvanishing expectation value (see Sect. 3.2).

Let us now turn to the existence of Goldstone bosons, taking the vacuum expectation value of Eq. 2.30:

$$0 \neq v = \langle 0|\Phi_3(0)|0\rangle = -\frac{i}{2}\langle 0|([Q_1, \Phi_2(0)] - [Q_2, \Phi_1(0)])|0\rangle \equiv -\frac{i}{2}(A - B).$$

We will first show $A = -B$. To that end we perform a rotation of the fields as well as the generators by $\pi/2$ about the 3 axis [see Eq. 2.28 with $\vec{\alpha} = (0,0,\pi/2)$]:

$$e^{-i\frac{\pi}{2}T_3}\vec{\Phi} = \begin{pmatrix} -\Phi_2 \\ \Phi_1 \\ \Phi_3 \end{pmatrix} = e^{i\frac{\pi}{2}Q_3} \begin{pmatrix} \Phi_1 \\ \Phi_2 \\ \Phi_3 \end{pmatrix} e^{-i\frac{\pi}{2}Q_3},$$

and analogously for the charge operators

$$\begin{pmatrix} -Q_2 \\ Q_1 \\ Q_3 \end{pmatrix} = e^{i\frac{\pi}{2}Q_3} \begin{pmatrix} Q_1 \\ Q_2 \\ Q_3 \end{pmatrix} e^{-i\frac{\pi}{2}Q_3}.$$

We thus obtain

$$B = \langle 0|[Q_2, \Phi_1(0)]|0\rangle = \langle 0| \Big(e^{i\frac{\pi}{2}Q_3}(-Q_1)\underbrace{e^{-i\frac{\pi}{2}Q_3}e^{i\frac{\pi}{2}Q_3}}_{=\,1}\Phi_2(0)e^{-i\frac{\pi}{2}Q_3}$$

$$- e^{i\frac{\pi}{2}Q_3}\Phi_2(0)e^{-i\frac{\pi}{2}Q_3}e^{i\frac{\pi}{2}Q_3}(-Q_1)e^{-i\frac{\pi}{2}Q_3}\Big)|0\rangle$$

$$= -\langle 0|[Q_1, \Phi_2(0)]|0\rangle = -A,$$

where we made use of $Q_3|0\rangle = 0$, i.e., the vacuum is invariant under rotations about the 3 axis. In other words, the nonvanishing vacuum expectation value v can also be written as

$$0 \neq v = \langle 0|\Phi_3(0)|0\rangle = -i\langle 0|[Q_1, \Phi_2(0)]|0\rangle = -i\int d^3x\,\langle 0|[J_1^0(t,\vec{x}), \Phi_2(0)]|0\rangle.$$

$$(2.31)$$

We insert a complete set of states $1 = \sum_n\!\!\!\!\!\!\int |n\rangle\langle n|$ into the commutator[10]

$$v = -i\sum_n\!\!\!\!\!\!\int \int d^3x\,\big(\langle 0|J_1^0(t,\vec{x})|n\rangle\langle n|\Phi_2(0)|0\rangle - \langle 0|\Phi_2(0)|n\rangle\langle n|J_1^0(t,\vec{x})|0\rangle\big),$$

and make use of translational invariance

$$= -i\sum_n\!\!\!\!\!\!\int \int d^3x\big(e^{-iP_n\cdot x}\langle 0|J_1^0(0)|n\rangle\langle n|\Phi_2(0)|0\rangle - \cdots\big)$$

$$= -i\sum_n\!\!\!\!\!\!\int (2\pi)^3\delta^3(\vec{P}_n)\big(e^{-iE_nt}\langle 0|J_1^0(0)|n\rangle\langle n|\Phi_2(0)|0\rangle - e^{iE_nt}\langle 0|\Phi_2(0)|n\rangle\langle n|J_1^0(0)|0\rangle\big).$$

Integration with respect to the momentum of the inserted intermediate states yields an expression of the form

$$= -i(2\pi)^3\sum_n{}'\big(e^{-iE_nt}\cdots - e^{iE_nt}\cdots\big),$$

where the prime indicates that only states with $\vec{p} = \vec{0}$ need to be considered. Due to the Hermiticity of the symmetry current operators J_k^μ as well as the Φ_l, we have

$$c_n \equiv \langle 0|J_1^0(0)|n\rangle\langle n|\Phi_2(0)|0\rangle = \langle n|J_1^0(0)|0\rangle^*\langle 0|\Phi_2(0)|n\rangle^*,$$

such that

$$v = -i(2\pi)^3\sum_n{}'\big(c_ne^{-iE_nt} - c_n^*e^{iE_nt}\big). \qquad (2.32)$$

---

[10] The abbreviation $\sum_n\!\!\!\!\!\!\int |n\rangle\langle n|$ includes an integral over the total momentum $\vec{p}$ as well as all other quantum numbers necessary to fully specify the states.

From Eq. 2.32 we draw the following conclusions.

1. Due to our assumption of a nonvanishing vacuum expectation value v, there must exist states $|n\rangle$ for which both $\langle 0|J^0_{1(2)}(0)|n\rangle$ and $\langle n|\Phi_{1(2)}(0)|0\rangle$ do not vanish. The vacuum itself cannot contribute to Eq. 2.32 because $\langle 0|\Phi_{1(2)}(0)|0\rangle = 0$.

2. States with $E_n > 0$ contribute (φ_n is the phase of c_n)

$$\frac{1}{i}\left(c_n e^{-iE_n t} - c_n^* e^{iE_n t}\right) = \frac{1}{i}|c_n|\left(e^{i\varphi_n}e^{-iE_n t} - e^{-i\varphi_n}e^{iE_n t}\right) = 2|c_n|\sin(\varphi_n - E_n t)$$

to the sum. However, v is time independent and therefore the sum over states with $(E,\vec{p}) = (E_n > 0, \vec{0})$ must vanish.

3. The right-hand side of Eq. 2.32 must therefore contain the contribution from states with zero energy as well as zero momentum thus zero mass. These zero-mass states are the Goldstone bosons.

2.4 Explicit Symmetry Breaking: A First Look

Finally, let us illustrate the consequences of adding a small perturbation to our Lagrangian of Eq. 2.11 which *explicitly* breaks the symmetry. To that end, we modify the potential of Eq. 2.11 by adding a term $a\Phi_3$,

$$\mathscr{V}(\Phi_1, \Phi_2, \Phi_3) = \frac{m^2}{2}\Phi_i\Phi_i + \frac{\lambda}{4}(\Phi_i\Phi_i)^2 + a\Phi_3, \qquad (2.33)$$

where $m^2 < 0, \lambda > 0$, and $a > 0$, with Hermitian fields Φ_i. Clearly, the potential no longer has the original O(3) symmetry but is only invariant under O(2). The conditions for the new minimum, obtained from $\vec{\nabla}_\Phi \mathscr{V} = 0$, read

$$\Phi_1 = \Phi_2 = 0, \qquad \lambda\Phi_3^3 + m^2\Phi_3 + a = 0.$$

Exercise 2.4 Solve the cubic equation for Φ_3 using the perturbative ansatz

$$\langle \Phi_3 \rangle = \Phi_3^{(0)} + a\Phi_3^{(1)} + O(a^2). \qquad (2.34)$$

The solution reads

$$\Phi_3^{(0)} = \pm\sqrt{-\frac{m^2}{\lambda}}, \qquad \Phi_3^{(1)} = \frac{1}{2m^2}.$$

As expected, $\Phi_3^{(0)}$ corresponds to our result without explicit perturbation. The condition for a *minimum* (see Eq. 2.23) excludes $\Phi_3^{(0)} = +\sqrt{-\frac{m^2}{\lambda}}$. Expanding the potential with $\Phi_3 = \langle \Phi_3 \rangle + \eta$ we obtain, after a short calculation, for the masses

$$m_{\Phi_1}^2 = m_{\Phi_2}^2 = a\sqrt{\frac{\lambda}{-m^2}},$$

$$m_\eta^2 = -2m^2 + 3a\sqrt{\frac{\lambda}{-m^2}}.$$

(2.35)

The important feature here is that the original Goldstone bosons of Eq. 2.19 are now massive. The squared masses are proportional to the symmetry breaking parameter a. Calculating *quantum* corrections to observables in terms of Goldstone-boson loop diagrams will generate corrections which are nonanalytic in the symmetry breaking parameter such as $a \ln(a)$ [12]. Such so-called chiral logarithms originate from the mass terms in the Goldstone-boson propagators entering the calculation of loop integrals. We will come back to this point in the next chapter when we discuss the masses of the pseudoscalar octet in terms of the quark masses which, in QCD, represent the analogue to the parameter a in the above example.

References

1. Bardeen, J., Cooper, L.N., Schrieffer, J.R.: Phys. Rev **106**, 162 (1957)
2. Bardeen, J., Cooper, L.N., Schrieffer, J.R.: Phys. Rev. **108**, 1175 (1957)
3. Bernstein, J.: Rev. Mod. Phys **46**, 7 (1974) (Erratum, ibid. **47**, 259 (1975))
4. Bogoliubov, N.N.: Sov. Phys. JETP **7**, 41 (1958)
5. Bogoliubov, N.N.: Zh. Eksp. Teor. Fiz **34**, 58 (1958)
6. Cheng, T.P., Li, L.F.: Gauge Theory of Elementary Particle Physics (Chap. 5.3). Clarendon, Oxford (1984)
7. Gell-Mann, M., Lévy, M.: Nuovo. Cim. **16**, 705 (1960)
8. Georgi, H.: Weak Interactions and Modern Particle Theory (Chaps. 2.4–2.6). Benjamin/ Cummings, Menlo Park (1984)
9. Goldstone, J.: Nuovo. Cim. **19**, 154 (1961)
10. Goldstone, J., Salam, A., Weinberg, S.: Phys. Rev. **127**, 965 (1962)
11. Greiner, W.: Theoretical Physics. Quantum Theory, vol. 4a (in German). Deutsch, Thun (1985)
12. Li, L.F., Pagels, H.: Phys. Rev. Lett. **26**, 1204 (1971)
13. Nambu, Y.: Phys. Rev. Lett. **4**, 380 (1960)
14. Nambu, Y., Jona-Lasinio, G.: Phys. Rev. **122**, 345 (1961)
15. Nambu, Y., Jona-Lasinio, G.: Phys. Rev. **124**, 246 (1961)
16. Ryder, L.H.: Quantum Field Theory (Chap. 8). Cambridge University Press, Cambridge (1985)
17. Schwinger, J.S.: Ann. Phys. **2**, 407 (1957)
18. Weinberg, S.: The Quantum Theory of Fields. Modern Applications, vol. 2 (Chap. 19). Cambridge University Press, Cambridge (1996)

Chapter 3
Chiral Perturbation Theory for Mesons

3.1 Effective Field Theory

Before discussing chiral perturbation theory (ChPT) in detail, we want to briefly
outline some of the main features of the effective-field-theory approach, as it finds
a wide range of applications in physics. Detailed introductions and reviews are
found, e.g., in Refs. [29, 45, 48, 58, 66, 74, 78, 82]. An effective field theory (EFT)
is a low-energy approximation to some underlying, more fundamental theory. By
that we mean that the EFT is a valid approximation for energies that are small
compared to some scale Λ of the underlying theory. The specific value of Λ
depends on the theory under investigation, which may in fact contain several such
scales $\Lambda_1 < \Lambda_2 < \cdots$. The basic idea of an EFT is that one does not need to know
details of the underlying theory at energies larger than Λ in order to find a useful
description of the physics in the energy domain one is interested in.

An EFT uses the degrees of freedom suitable for the particular low-energy
domain of interest. For example, one can neglect those degrees of freedom that are
too heavy to be produced at low energies, which can simplify calculations sig-
nificantly. In fact, the degrees of freedom can be entirely different from those
appearing in the underlying theory: we will use the pseudoscalar octet (π, K, η)
and the octet of $\frac{1}{2}^+$ baryons $(p, n, \Sigma, \Xi, \Lambda)$ instead of the more fundamental quarks
and gluons as the degrees of freedom in low-energy processes in hadronic physics.

Using different degrees of freedom, we have to assure that observables calculated in the EFT are related to those of the underlying theory. This is achieved by
using the most general Lagrangian that is consistent with the symmetries of the
underlying theory, as this yields the "most general possible S-matrix consistent
with analyticity, perturbative unitarity, cluster decomposition and the assumed
symmetry principles" [100]. Since we are using the most general Lagrangian, it
actually consists of an *infinite* number of terms, each with its own coefficient, the
so-called low-energy constants (LECs). Obviously, this is not a very useful prescription without some kind of approximation to avoid having to calculate an

S. Scherer and M. R. Schindler, *A Primer for Chiral Perturbation Theory*,
Lecture Notes in Physics 830, DOI: 10.1007/978-3-642-19254-8_3,
© Springer-Verlag Berlin Heidelberg 2012

infinite number of contributions to each physical observable. The solution lies in two restrictions: first, we only demand a finite accuracy to our results, i.e., our results are allowed to differ from those of the underlying theory by a specified (small) amount; and second, we restrict ourselves to a certain energy domain, which means that the EFT has a limited range of applicability. The EFT is then used to calculate physical observables in terms of an expansion in p/Λ, where p generically stands for energy, momenta or masses that are smaller than the scale Λ related to the underlying theory. As long as $p \ll \Lambda$, only a *finite* number of terms in the expansion contribute for a specified accuracy, which correspond to a finite number of terms in the most general Lagrangian. We will see an example of a method to determine which terms to include when considering Weinberg's power counting in Sect. 3.4.9.

As explained above, the terms in the Lagrangian are constrained by the symmetries of the underlying theory. This explains our focus on the symmetries of QCD in the previous two chapters. In the following, we will study the implications of these symmetries for the interactions of the Goldstone bosons of spontaneously broken chiral symmetry. While the symmetries impose conditions on the structure of the Lagrangian, they do *not* restrict the LECs. These should in principle be calculable from the underlying theory. In the cases where this is not possible, e.g. if the underlying theory is not known or one does not (yet) know how to solve it, they can be fitted to data. Once the LECs are determined, the effective theory possesses predictive power.

On a technical note, EFTs are non-renormalizable in the traditional sense, as with increasing accuracy one needs to include more and more terms. However, as long as one considers all terms that are allowed by the symmetries, divergences that occur in calculations up to any given order of p/Λ can be renormalized by redefining fields and parameters of the EFT Lagrangian [101].

One of the best-known examples of an effective theory is Fermi's theory of beta decay. It can in fact be regarded as the leading-order piece of an EFT in which the massive gauge bosons are "integrated out." In the Standard Model, neutron beta decay $n \to pe^- \bar{\nu}_e$ is described via an intermediate W boson with mass $M_W \approx$ 80 GeV. For momentum transfers $q \ll M_W$, the W boson propagator can be replaced by the lowest-order expansion in the small quantity q/M_W, symbolically

$$\frac{1}{q^2 - M_W^2} \to -\frac{1}{M_W^2},$$

resulting in the four-fermion contact interaction of Fermi's theory. This low-energy theory includes all the ingredients of an EFT discussed above: The degrees of freedom differ from those of the underlying theory as the W boson is excluded, the underlying scale Λ is identified with the mass M_W, and the domain of applicability is restricted to momentum transfers $q \ll M_W$. It is also non-renormalizable in the traditional sense, as the Fermi constant G_F describing the four-fermion coupling has dimensions energy^{-2}.

Let us consider an analogous, but simplified, example in more detail. We choose a toy model with two massive scalar degrees of freedom,[1]

$$\mathcal{L} = \mathcal{L}_0 + \mathcal{L}_{int},$$

$$\mathcal{L}_0 = \frac{1}{2}(\partial_\mu \phi \partial^\mu \phi - M^2 \phi^2) + \frac{1}{2}(\partial_\mu \varphi \partial^\mu \varphi - m^2 \varphi^2), \tag{3.1}$$

$$\mathcal{L}_{int} = -\frac{\lambda}{2}\phi \varphi^2,$$

with $m \ll M$. The equations of motion are then derived from the Euler-Lagrange equations to be

$$\Box \phi + M^2 \phi + \frac{\lambda}{2}\varphi^2 = 0, \tag{3.2}$$

$$\Box \varphi + m^2 \varphi + \lambda \varphi \phi = 0. \tag{3.3}$$

Formally solving Eq. 3.2 for ϕ,

$$\phi = -\frac{\lambda}{2M^2}\frac{1}{1 + \frac{\Box}{M^2}}\varphi^2,$$

and inserting the solution into Eq. 3.3, we obtain

$$\Box \varphi + m^2 \varphi - \frac{\lambda^2}{2M^2}\varphi \frac{1}{1 + \frac{\Box}{M^2}}\varphi^2 = 0. \tag{3.4}$$

We see that the heavy degree of freedom ϕ has disappeared from the equation of motion of the light particle. If the momentum of φ is much less than M we can formally expand the last term in Eq. 3.4 in $1/M^2$. The leading-order expression is given by

$$\Box \varphi + m^2 \varphi - \frac{\lambda^2}{2M^2}\varphi^3 = 0. \tag{3.5}$$

The same equation of motion is generated by the effective Lagrangian

$$\mathcal{L}_{eff} = \frac{1}{2}(\partial_\mu \varphi \partial^\mu \varphi - m^2 \varphi^2) + \tilde{\lambda}\varphi^4, \tag{3.6}$$

as long as

$$\tilde{\lambda} = \frac{\lambda^2}{8M^2}.$$

---

[1] The toy model serves pedagogical purposes only. As a (quantum) field theory it is not consistent because the energy is not bounded from below [14].

Fig. 3.1 Diagrams contributing to $\varphi(p_1) + \varphi(p_2) \rightarrow \varphi(p_3) + \varphi(p_4)$ in the fundamental theory

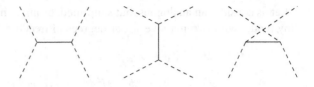

The effective Lagrangian only depends on the light field φ and contains a φ^4 interaction, which is the analogue of the Fermi interaction. Note that, instead of using the equations of motion, one can also directly "integrate out" the heavy degrees of freedom from the Lagrangian in a path-integral formalism.

Exercise 3.1 We now show that the effective Lagrangian of Eq. 3.6 produces the same low-energy scattering amplitude for $\varphi(p_1) + \varphi(p_2) \rightarrow \varphi(p_3) + \varphi(p_4)$ as the original Lagrangian of Eq. 3.1.

(a) Show that the Feynman rule for the coupling of a heavy field ϕ to two light fields φ in the original theory is given by $-i\lambda$.

(b) Calculate the amplitudes for the diagrams shown in Fig. 3.1, where the dashed and solid lines represent the fields φ and ϕ, respectively. The heavy-particle Feynman propagator is given by

$$\Delta_{F\phi}(p) = \frac{1}{p^2 - M^2 + i0^+}.$$

Show that the result can be expressed in the Mandelstam variables $s = (p_1 + p_2)^2$, $t = (p_1 - p_3)^2$, and $u = (p_1 - p_4)^2$ as

$$\mathscr{M}_{\text{fund}} = -\lambda^2 \left(\frac{i}{s - M^2 + i0^+} + \frac{i}{t - M^2 + i0^+} + \frac{i}{u - M^2 + i0^+} \right). \tag{3.7}$$

(c) We now restrict ourselves to very low energies, such that $\{s, |t|, |u|\} \ll M^2$. Show that the leading-order contribution in the low-energy expansion is given by

$$\mathscr{M}_{\text{fund}} = \frac{3i\lambda^2}{M^2} + O\left(\frac{\{s, |t|, |u|\}}{M^2} \right).$$

(d) In the effective theory, we only need to consider the diagram of Fig. 3.2. Show that the corresponding amplitude is given by

$$\mathscr{M}_{\text{eff}} = i(4!)\tilde{\lambda} = \frac{3i\lambda^2}{M^2},$$

which exactly reproduces the leading-order contribution in the fundamental theory. This calculation shows that even if we know how to calculate observables in the underlying theory, use of an EFT can simplify the necessary calculations.

Fig. 3.2 Diagram
contributing to $\varphi(p_1) +$
$\varphi(p_2) \rightarrow \varphi(p_3) + \varphi(p_4)$ in
the effective theory

In the example above we eliminated some of the heavy degrees of freedom by explicit calculation. In the case of chiral perturbation theory, the hadronic degrees of freedom are very different from even the light degrees of freedom of the underlying theory of QCD, and we therefore have to rely on the symmetries of QCD to construct the effective Lagrangian. This will be the focus of the following sections.

3.2 Spontaneous Symmetry Breaking in QCD

While the toy model of Sect. 2.2 is constructed to illustrate the concept of spontaneous symmetry breaking, it is not fully understood theoretically why QCD should exhibit this phenomenon. We will first motivate why experimental input, the hadron spectrum of the "real" world, indicates that spontaneous symmetry breaking occurs in QCD. Secondly, we will show that a nonvanishing scalar singlet quark condensate is a sufficient condition for spontaneous symmetry breaking in QCD.

3.2.1 The Hadron Spectrum

We saw in Sect. 1.3 that the QCD Lagrangian possesses an $SU(3)_L \times SU(3)_R \times U(1)_V$ symmetry in the chiral limit in which the light-quark masses vanish. From symmetry considerations involving the Hamiltonian H^0_{QCD} only, one would naively expect that hadrons are organized in approximately degenerate multiplets fitting the dimensionalities of irreducible representations of the group $SU(3)_L \times SU(3)_R \times U(1)_V$. The $U(1)_V$ symmetry results in baryon number conservation and leads to a classification of hadrons into mesons ($B = 0$) and baryons ($B = 1$). The linear combinations

$$Q_{Va} \equiv Q_{Ra} + Q_{La} \tag{3.8}$$

and

$$Q_{Aa} \equiv Q_{Ra} - Q_{La} \tag{3.9}$$

of the left- and right-handed charge operators commute with H^0_{QCD}, have opposite parity, and thus for states of positive parity one would expect the existence of degenerate states of negative parity which can be seen as follows.[2]

Let $|\alpha, +\rangle$ denote an eigenstate of H^0_{QCD} and parity with eigenvalues E_α and $+1$, respectively,

$$H^0_{QCD}|\alpha, +\rangle = E_\alpha|\alpha, +\rangle,$$
$$P|\alpha, +\rangle = |\alpha, +\rangle,$$

such as, e.g., a member of the lowest-lying baryon octet (in the chiral limit). Defining $|\psi_{a\alpha}\rangle = Q_{Aa}|\alpha, +\rangle$, because of $[H^0_{QCD}, Q_{Aa}] = 0$, we have

$$H^0_{QCD}|\psi_{a\alpha}\rangle = H^0_{QCD}Q_{Aa}|\alpha, +\rangle = Q_{Aa}H^0_{QCD}|\alpha, +\rangle = E_\alpha Q_{Aa}|\alpha, +\rangle = E_\alpha|\psi_{a\alpha}\rangle,$$
$$P|\psi_{a\alpha}\rangle = PQ_{Aa}P^{-1}P|\alpha, +\rangle = -Q_{Aa}(+|\alpha, +\rangle) = -|\psi_{a\alpha}\rangle.$$

The state $|\psi_{a\alpha}\rangle$ can be expanded in terms of the members of a multiplet with negative parity,

$$|\psi_{a\alpha}\rangle = Q_{Aa}|\alpha, +\rangle = |\beta, -\rangle\langle\beta, -|Q_{Aa}|\alpha, +\rangle = t_{a,\beta\alpha}|\beta, -\rangle.$$

However, the low-energy spectrum of baryons does not contain a degenerate baryon octet of negative parity. Naturally the question arises whether the above chain of arguments is incomplete. Indeed, we have tacitly assumed that the ground state of QCD is annihilated by the generators Q_{Aa}. Let $b^\dagger_{\alpha+}$ denote an operator creating quanta with the quantum numbers of the state $|\alpha, +\rangle$. Similarly, let $b^\dagger_{\alpha-}$ create degenerate quanta of opposite parity. Expanding

$$\left[Q_{Aa}, b^\dagger_{\alpha+}\right] = b^\dagger_{\beta-}t_{a,\beta\alpha},$$

the usual chain of arguments then works as

$$Q_{Aa}|\alpha, +\rangle = Q_{Aa}b^\dagger_{\alpha+}|0\rangle = ([Q_{Aa}, b^\dagger_{\alpha+}] + b^\dagger_{\alpha+}\underbrace{Q_{Aa}}_{\hookrightarrow 0})|0\rangle = t_{a,\beta\alpha}b^\dagger_{\beta-}|0\rangle. \qquad (3.10)$$

However, if the ground state is *not* annihilated by Q_{Aa}, the reasoning of Eq. 3.10 does no longer apply. In that case the ground state is not invariant under the full symmetry group of the Lagrangian, resulting in a spontaneous symmetry breaking. In other words, the non-existence of degenerate multiplets of opposite parity points to the fact that $SU(3)_V$ instead of $SU(3)_L \times SU(3)_R$ is approximately realized as a symmetry of the hadrons. Furthermore, the octet of the pseudoscalar mesons is special in the sense that the masses of its members are small in comparison with

---

[2] The existence of mass-degenerate states of opposite parity is referred to as parity doubling.

Fig. 3.3 Pseudoscalar meson octet in an (I_3, S) diagram. Baryon number $B = 0$. Masses in MeV

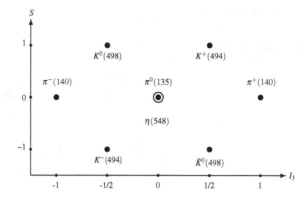

the corresponding vector mesons $(J^P = 1^-)$. They are the candidates for the Goldstone bosons of spontaneous symmetry breaking in QCD.

In order to understand the origin of the $SU(3)_V$ symmetry, let us consider the vector charges $Q_{Va} = Q_{Ra} + Q_{La}$.[3]

Exercise 3.2 Using Eqs. 1.104–1.106, show that the vector charges satisfy the commutation relations of an su(3) Lie algebra,

$$[Q_{Va}, Q_{Vb}] = i f_{abc} Q_{Vc}. \tag{3.11}$$

It was shown by Vafa and Witten [94] that, in the chiral limit, the ground state is necessarily invariant under $SU(3)_V \times U(1)_V$, i.e., the eight vector charges Q_{Va} as well as the baryon number operator[4] $Q_V/3$ annihilate the ground state,

$$Q_{Va}|0\rangle = Q_V|0\rangle = 0. \tag{3.12}$$

According to the Coleman theorem [41], the symmetry of the ground state determines the symmetry of the spectrum, i.e., Eq. 3.12 implies $SU(3)_V$ multiplets which can be classified according to their baryon number. In the reverse conclusion, the symmetry of the ground state can be inferred from the symmetry pattern of the spectrum. Figures 3.3 and 3.4 show the octets of the lowest-lying pseudoscalar-meson states and the lowest-lying baryon states of spin-parity $\frac{1}{2}^+$, respectively.

Let us now turn to the linear combinations $Q_{Aa} = Q_{Ra} - Q_{La}$.

Exercise 3.3 Using Eqs. 1.104–1.106, verify the commutation relations

$$[Q_{Aa}, Q_{Ab}] = i f_{abc} Q_{Vc}, \tag{3.13}$$

---

[3] The subscript V (for vector) indicates that the generators result from integrals of the zeroth component of vector-current operators and thus transform with a positive sign under parity.

[4] Recall that each quark is assigned a baryon number 1/3.

Fig. 3.4 Baryon octet
$\left(J^P = \frac{1}{2}^+\right)$ in an (I_3, S)
diagram. Baryon number
$B = 1$. Masses in MeV

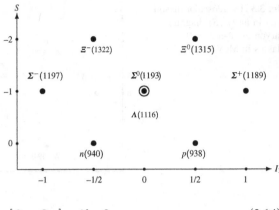

$$[Q_{Va}, Q_{Ab}] = if_{abc}Q_{Ac}. \tag{3.14}$$

Note that these charge operators do *not* form a closed algebra, i.e., the commutator of two axial-charge operators is not again an axial-charge operator. Since the parity doubling is not observed for the low-lying states, one assumes that the Q_{Aa} do *not* annihilate the ground state,

$$Q_{Aa}|0\rangle \neq 0, \tag{3.15}$$

i.e., the ground state of QCD is not invariant under "axial" transformations. In the present case, $G = \mathrm{SU}(3)_L \times \mathrm{SU}(3)_R$ with $n_G = 16$ and $H = \mathrm{SU}(3)_V$ with $n_H = 8$ and we expect eight Goldstone bosons. According to the Goldstone theorem [60], each axial generator Q_{Aa}, which does not annihilate the ground state, corresponds to a massless Goldstone-boson field ϕ_a with spin 0, whose symmetry properties are tightly connected to the generator in question. The Goldstone bosons have the same transformation behavior under parity as the axial generators,

$$\phi_a(t, \vec{x}) \overset{P}{\mapsto} -\phi_a(t, -\vec{x}), \tag{3.16}$$

i.e., they are pseudoscalars, and transform under the subgroup $H = \mathrm{SU}(3)_V$, which leaves the vacuum invariant, as an octet (see Eq. 3.14):

$$[Q_{Va}, \phi_b(x)] = if_{abc}\phi_c(x). \tag{3.17}$$

3.2.2 The Scalar Singlet Quark Condensate

In this section all quantities such as the ground state, the quark operators, etc. are considered in the chiral limit. We will show that a nonvanishing vacuum expectation value of the operator $\bar{q}q$ in the chiral limit is a sufficient (but not a necessary) condition for spontaneous symmetry breaking in zero-temperature QCD. An increase in temperature will ultimately lead to a phase transition into the regime where chiral symmetry is restored, i.e., no longer spontaneously broken (see Ref. [59] for a discussion of the low-temperature behavior of the quark condensate). The terminology scalar singlet quark condensate originates from the fact that $\bar{q}q$ transforms as a scalar under the full Lorentz group and as singlet under $SU(3)_V$, respectively. The "condensation" is a non-perturbative phenomenon of the QCD ground state that is driven by the formation of quark-antiquark pairs. The subsequent discussion will parallel that of the toy model in Sect. 2.3 after replacement of the elementary fields Φ_i by appropriate composite Hermitian operators of QCD.

Let us first recall the definition of the nine scalar and pseudoscalar quark densities:

$$S_a(y) = \bar{q}(y)\lambda_a q(y), \quad a = 0, \ldots, 8, \tag{3.18}$$

$$P_a(y) = i\bar{q}(y)\gamma_5\lambda_a q(y), \quad a = 0, \ldots, 8. \tag{3.19}$$

Exercise 3.4 Show that S_a and P_a transform under $SU(3)_L \times SU(3)_R$, i.e., $q_L \mapsto q'_L = U_L q_L$ and $q_R \mapsto q'_R = U_R q_R$, as

$$S_a \mapsto S'_a = \bar{q}_L U_L^\dagger \lambda_a U_R q_R + \bar{q}_R U_R^\dagger \lambda_a U_L q_L,$$
$$P_a \mapsto P'_a = i\bar{q}_L U_L^\dagger \lambda_a U_R q_R - i\bar{q}_R U_R^\dagger \lambda_a U_L q_L.$$

Hint: Express S_a and P_a in terms of left- and right-handed quark fields.

In technical terms: The components S_a and P_a transform as members of $(3^*, 3) \oplus (3, 3^*)$ representations.

The equal-time commutation relation of two quark operators of the form $A_i(x) = q^\dagger(x)\hat{A}_i q(x)$, where $\hat{A}_i$ symbolically denotes Γ and flavor matrices and a summation over color indices is implied, can be compactly written as (see Eq. 1.103)

$$[A_1(t, \vec{x}), A_2(t, \vec{y})] = \delta^3(\vec{x} - \vec{y})q^\dagger(x)[\hat{A}_1, \hat{A}_2]q(x). \tag{3.20}$$

In the following, let y denote $(t, \vec{y})$. With the definition

$$Q_{Va}(t) = \int d^3x\, q^\dagger(t, \vec{x})\frac{\lambda_a}{2}q(t, \vec{x}),$$

and using[5]

$$\left[1\frac{\lambda_a}{2}, \gamma_0\lambda_0\right] = 0,$$

$$\left[1\frac{\lambda_a}{2}, \gamma_0\lambda_b\right] = \gamma_0\sum_{c=1}^{8} if_{abc}\lambda_c,$$

---

[5] In this section, we explicitly write out sums over flavor indices, because a summation over repeated indices is *not* implied in the final results of Eqs. 3.27 and 3.28.

we see, after integration of Eq. 3.20 over $\vec{x}$, that the scalar quark densities of Eq. 3.18 transform under $SU(3)_V$ as a singlet and as an octet, respectively,

$$[Q_{Va}(t), S_0(y)] = 0, \quad a = 1,\ldots,8, \tag{3.21}$$

$$[Q_{Va}(t), S_b(y)] = i \sum_{c=1}^{8} f_{abc} S_c(y), \quad a, b = 1,\ldots,8, \tag{3.22}$$

with analogous results for the pseudoscalar quark densities. In the $SU(3)_V$ limit and, of course, also in the even more restrictive chiral limit, the charge operators in Eqs. (3.21) and (3.22) are actually time independent.[6] Using the relation

$$\sum_{a,b=1}^{8} f_{abc} f_{abd} = 3\delta_{cd} \tag{3.23}$$

for the structure constants of su(3), we re-express the octet components of the scalar quark densities as

$$S_a(y) = -\frac{i}{3} \sum_{b,c=1}^{8} f_{abc}[Q_{Vb}(t), S_c(y)], \tag{3.24}$$

which represents the analogue of Eq. 2.30 in the discussion of the Goldstone theorem.

In the chiral limit the ground state is necessarily invariant under $SU(3)_V$ [94], i.e., $Q_{Va}|0\rangle = 0$, and we obtain from Eq. 3.24

$$\langle 0|S_a(y)|0\rangle = \langle 0|S_a(0)|0\rangle \equiv \langle S_a\rangle = 0, \quad a = 1,\ldots,8, \tag{3.25}$$

where we made use of translational invariance of the ground state. In other words, the octet components of the scalar quark condensate *must* vanish in the chiral limit. From Eq. 3.25, we obtain for $a = 3$

$$\langle \bar{u}u\rangle - \langle \bar{d}d\rangle = 0,$$

i.e., $\langle \bar{u}u\rangle = \langle \bar{d}d\rangle$ and for $a = 8$

$$\langle \bar{u}u\rangle + \langle \bar{d}d\rangle - 2\langle \bar{s}s\rangle = 0,$$

i.e., $\langle \bar{u}u\rangle = \langle \bar{d}d\rangle = \langle \bar{s}s\rangle$.

Because of Eq. 3.21 a similar argument cannot be used for the singlet condensate, and if we assume a nonvanishing scalar singlet quark condensate in the chiral limit, we find using $\langle \bar{u}u\rangle = \langle \bar{d}d\rangle = \langle \bar{s}s\rangle$:

---

[6] The commutation relations also remain valid for *equal* times if the symmetry is explicitly broken.

$$0 \neq \langle \bar{q}q \rangle = \langle \bar{u}u + \bar{d}d + \bar{s}s \rangle = 3\langle \bar{u}u \rangle = 3\langle \bar{d}d \rangle = 3\langle \bar{s}s \rangle. \tag{3.26}$$

Finally, we make use of (no summation implied!)

$$i^2 \left[\gamma_5 \frac{\lambda_a}{2}, \gamma_0 \gamma_5 \lambda_a \right] = \gamma_0 \lambda_a^2$$

in combination with

$$\lambda_1^2 = \lambda_2^2 = \lambda_3^2 = \begin{pmatrix} 1 & 0 & 0 \\ 0 & 1 & 0 \\ 0 & 0 & 0 \end{pmatrix},$$

$$\lambda_4^2 = \lambda_5^2 = \begin{pmatrix} 1 & 0 & 0 \\ 0 & 0 & 0 \\ 0 & 0 & 1 \end{pmatrix},$$

$$\lambda_6^2 = \lambda_7^2 = \begin{pmatrix} 0 & 0 & 0 \\ 0 & 1 & 0 \\ 0 & 0 & 1 \end{pmatrix},$$

$$\lambda_8^2 = \frac{1}{3} \begin{pmatrix} 1 & 0 & 0 \\ 0 & 1 & 0 \\ 0 & 0 & 4 \end{pmatrix},$$

to obtain

$$i[Q_{Aa}(t), P_a(y)] = \begin{cases} \bar{u}u + \bar{d}d, & a = 1, 2, 3 \\ \bar{u}u + \bar{s}s, & a = 4, 5 \\ \bar{d}d + \bar{s}s, & a = 6, 7 \\ \frac{1}{3}(\bar{u}u + \bar{d}d + 4\bar{s}s), & a = 8 \end{cases} \tag{3.27}$$

where we have suppressed the y dependence on the right-hand side. We evaluate Eq. 3.27 for a ground state which is invariant under $SU(3)_V$, assuming a non-vanishing singlet scalar quark condensate,

$$\langle 0|i[Q_{Aa}(t), P_a(y)]|0 \rangle = \frac{2}{3}\langle \bar{q}q \rangle, \quad a = 1, \ldots, 8, \tag{3.28}$$

where, because of translational invariance, the right-hand side is independent of y. Inserting a complete set of states into the commutator of Eq. 3.28 yields, in complete analogy to Sect. 2.3 (see the discussion following Eq. 2.31) that both the pseudoscalar densities $P_a(y)$ as well as the axial charge operators Q_{Aa} must have a nonvanishing matrix element between the vacuum and massless one-particle states $|\phi_b\rangle$. In particular, because of Lorentz covariance, the matrix element of the axial-vector current operator between the vacuum and these massless states, appropriately normalized, can be written as

Table 3.1 Comparison of spontaneous symmetry breaking patterns

	Section 2.3	$O(N)$ sigma model	QCD
Symmetry group G of Lagrangian	$O(3)$	$O(N)$	$SU(3)_L \times SU(3)_R$
Number of generators n_G	3	$N(N-1)/2$	16
Symmetry group H of ground state	$O(2)$	$O(N-1)$	$SU(3)_V$
Number of generators n_H	1	$(N-1)(N-2)/2$	8
Number of Goldstone bosons $n_G - n_H$	2	$N-1$	8
Multiplet of Goldstone-boson fields	$(\Phi_1(x), \Phi_2(x))$	$(\Phi_1(x), \ldots, \Phi_{N-1}(x))$	$i\bar{q}(x)\gamma_5\lambda_a q(x)$
Vacuum expectation value	$v = \langle\Phi_3\rangle$	$v = \langle\Phi_N\rangle$	$v = \langle\bar{q}q\rangle$

$$\langle 0|A_a^\mu(0)|\phi_b(p)\rangle = ip^\mu F_0 \delta_{ab}, \tag{3.29}$$

where F_0 denotes the "decay" constant of the Goldstone bosons in the three-flavor chiral limit ($m_u = m_d = m_s = 0$). From Eq. 3.29 we see that a nonzero value of F_0 is a necessary and sufficient criterion for spontaneous chiral symmetry breaking. On the other hand, because of Eq. 3.28 a nonvanishing scalar singlet quark condensate $\langle\bar{q}q\rangle$ is a sufficient (but not a necessary) condition for a spontaneous symmetry breakdown in QCD. Table 3.1 contains a summary of the patterns of spontaneous symmetry breaking as discussed in Sect. 2.3, the generalization of Sect. 2.2 to the so-called $O(N)$ linear sigma model, and QCD.

3.3 Transformation Properties of the Goldstone Bosons

The purpose of this section is to discuss the transformation properties of the field variables describing the Goldstone bosons [10, 33, 42, 70, 99]. We will need the concept of a *nonlinear realization* of a group in addition to a *representation* of a group which one usually encounters in physics. We will first discuss a few general group-theoretical properties before specializing to QCD.

3.3.1 General Considerations

Let us consider a physical system described by a Lagrangian which is invariant under a compact Lie group G. We assume the ground state of the system to be invariant under only a subgroup H of G, giving rise to $n = n_G - n_H$ Goldstone bosons. Each of these Goldstone bosons will be described by an independent field ϕ_i which is a smooth real function on Minkowski space M^4. These fields are collected in an n-component vector $\Phi = (\phi_1, \ldots, \phi_n)$, defining the real vector space

$$M_1 \equiv \{\Phi : M^4 \to \mathbb{R}^n | \phi_i : M^4 \to \mathbb{R} \text{ smooth}\}. \tag{3.30}$$

Our aim is to find a mapping φ which uniquely associates with each pair $(g, \Phi) \in G \times M_1$ an element $\varphi(g, \Phi) \in M_1$ with the following properties:

$$\varphi(e, \Phi) = \Phi \quad \forall \quad \Phi \in M_1, \quad e \text{ identity of } G, \qquad (3.31)$$

$$\varphi(g_1, \varphi(g_2, \Phi)) = \varphi(g_1 g_2, \Phi) \quad \forall \quad g_1, g_2 \in G, \quad \forall \quad \Phi \in M_1. \qquad (3.32)$$

Such a mapping defines an *operation* of the group G on M_1. The second condition is the so-called group-homomorphism property [9, 64, 77]. The mapping will, in general, *not* define a *representation* of the group G, because we do not require the mapping to be linear, i.e., $\varphi(g, \lambda \Phi) \neq \lambda \varphi(g, \Phi)$. The construction proceeds as follows [70]. Let $\Phi = 0$ denote the "origin" of M_1 which, in a theory containing Goldstone bosons only, loosely speaking corresponds to the ground state config-uration. Since the ground state is supposed to be invariant under the subgroup H we require the mapping φ to be such that all elements $h \in H$ map the origin onto itself. In this context the subgroup H is also known as the little group of $\Phi = 0$.

We will establish a connection between the Goldstone-boson fields and the set of all left cosets $\{gH | g \in G\}$ which is also referred to as the quotient G/H. For a subgroup H of G the set $gH = \{gh | h \in H\}$ defines the left coset of g which is one element of G/H.[7] For our purposes we need the property that cosets either completely overlap or are completely disjoint, i.e., the quotient is a set whose elements themselves are sets of group elements, and these sets are completely disjoint.

As an illustration of these properties, consider the symmetry group C_4 of a square with directed sides:

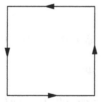

The corresponding abstract group G consists of four elements, $G = C_4 = \{e, a, a^2, a^3\}$ with the defining relation $a^4 = e$. Geometrically, a may be repre-sented by a rotation through $90°$ about an axis through the center and normal to the plane of the square. Using $a^4 = e$, the left cosets of the (nontrivial) subgroup $H = \{e, a^2\}$ are given by

$$eH = \{e, a^2\} = a^2 H, \quad aH = \{a, a^3\} = a^3 H.$$

---

[7] Accordingly, the right coset of g is defined as $Hg = \{hg | h \in H\}$. An *invariant* subgroup has the additional property that the left and right cosets coincide for each g which allows for a definition of the factor group G/H in terms of the complex product. However, here we do not need this property.

The quotient $G/H = \{gH | g \in G\}$ therefore consists of the two elements $\{e, a^2\}$ and $\{a, a^3\}$. Since the elements of the quotient are completely disjoint, any element of a given coset uniquely represents the coset in question.

Returning to the discussion of the mapping φ, Eqs. 3.31 and 3.32 result in two important properties when considering the quotient G/H. Let us first show that the origin is mapped onto the same vector in $\mathbb{R}^n$ under all elements of a given coset gH:

$$\varphi(gh, 0) = \varphi(g, \varphi(h, 0)) = \varphi(g, 0) \quad \forall \quad g \in G \quad \text{and} \quad h \in H.$$

Secondly, the mapping is injective with respect to the elements of G/H. Consider two elements g and g' of G where $g' \notin gH$. Let us assume $\varphi(g, 0) = \varphi(g', 0)$:

$$0 = \varphi(e, 0) = \varphi(g^{-1}g, 0) = \varphi(g^{-1}, \varphi(g, 0)) = \varphi(g^{-1}, \varphi(g', 0)) = \varphi(g^{-1}g', 0).$$

However, this implies $g^{-1}g' \in H$ or $g' \in gH$ in contradiction to the assumption $g' \notin gH$ and therefore $\varphi(g, 0) = \varphi(g', 0)$ cannot be true. In other words, the mapping can be inverted on the image of $\varphi(g, 0)$. The conclusion is that there exists an *isomorphic mapping* between the quotient G/H and the Goldstone-boson fields. Of course, the Goldstone-boson fields are not constant vectors in $\mathbb{R}^n$ but functions on Minkowski space. This is accomplished by allowing the cosets gH to also depend on x.

Now let us discuss the transformation behavior of the Goldstone-boson fields under an arbitrary $g \in G$ in terms of the isomorphism established above. To each Φ corresponds a coset $\tilde{g}H$ with appropriate $\tilde{g}$. Let $f = \tilde{g}h \in \tilde{g}H$ denote a representative of this coset such that

$$\Phi = \varphi(f, 0) = \varphi(\tilde{g}h, 0).$$

Now apply the mapping $\varphi(g)$ to Φ:

$$\varphi(g, \Phi) = \varphi(g, \varphi(\tilde{g}h, 0)) = \varphi(g\tilde{g}h, 0) = \varphi(f', 0) = \Phi', \quad f' \in g(\tilde{g}H).$$

In order to obtain the transformed Φ' from a given Φ, we simply need to multiply the left coset $\tilde{g}H$ representing Φ by g in order to obtain the new left coset representing Φ':

$$
\begin{array}{ccc}
\Phi & \xrightarrow{g} & \Phi' \\
\downarrow & & \uparrow \\
\tilde{g}H & \xrightarrow{g} & g\tilde{g}H
\end{array}
\tag{3.33}
$$

This procedure uniquely determines the transformation behavior of the Goldstone bosons up to an appropriate choice of variables parameterizing the elements of the quotient G/H.

3.3.2 Application to QCD

The symmetry groups relevant to the application in QCD are

$$G = \mathrm{SU}(N) \times \mathrm{SU}(N) = \{(L,R)|L \in \mathrm{SU}(N), R \in \mathrm{SU}(N)\}$$

and

$$H = \{(V,V)|V \in \mathrm{SU}(N)\} \cong \mathrm{SU}(N),$$

with $N = 2$ for massless u and d quarks and $N = 3$ for massless u, d, and s quarks. Let $\tilde{g} = (\tilde{L}, \tilde{R}) \in G$. We characterize the left coset $\tilde{g}H$ through the $\mathrm{SU}(N)$ matrix $U = \tilde{R}\tilde{L}^\dagger$ [10] such that $\tilde{g}H = (\mathbb{1}, \tilde{R}\tilde{L}^\dagger)H$. This corresponds to the convention of choosing as the representative of the coset the element which has the unit matrix in its first argument. The transformation behavior of U under $g = (L,R) \in G$ is obtained by multiplying the left coset $\tilde{g}H$ from the left with g (see Eq. 3.33):

$$g\tilde{g}H = (L, R\tilde{R}\tilde{L}^\dagger)H = (\mathbb{1}, R\tilde{R}\tilde{L}^\dagger L^\dagger)(L,L)H = (\mathbb{1}, R(\tilde{R}\tilde{L}^\dagger)L^\dagger)H,$$

where we made use of the fact that a multiplication of H with any element $(L,L) \in H$ simply reproduces H. According to our convention, the representative of the transformed left coset is $(\mathbb{1}, R\tilde{R}\tilde{L}^\dagger L^\dagger)$. The transformation behavior of U is therefore given by

$$U = \tilde{R}\tilde{L}^\dagger \mapsto U' = R(\tilde{R}\tilde{L}^\dagger)L^\dagger = RUL^\dagger. \tag{3.34}$$

As mentioned above, we need to introduce an x dependence to account for the fact that we are dealing with fields:

$$U(x) \mapsto RU(x)L^\dagger. \tag{3.35}$$

Let us now restrict ourselves to the physically relevant cases of $N = 2$ and $N = 3$ and define

$$M_1 \equiv \begin{cases} \{\Phi : M^4 \to \mathbb{R}^3 | \phi_i : M^4 \to \mathbb{R} \text{ smooth}\} & \text{for } N = 2, \\ \{\Phi : M^4 \to \mathbb{R}^8 | \phi_i : M^4 \to \mathbb{R} \text{ smooth}\} & \text{for } N = 3. \end{cases}$$

Furthermore let $\tilde{\mathscr{H}}(N)$ denote the set of all Hermitian and traceless $N \times N$ matrices,

$$\tilde{\mathscr{H}}(N) \equiv \{A \in \mathrm{gl}(N,\mathbb{C})|A^\dagger = A \wedge \mathrm{Tr}(A) = 0\},$$

which under addition of matrices defines a real vector space. We define a second set $M_2 \equiv \{\phi : M^4 \to \tilde{\mathscr{H}}(N)|\phi \text{ smooth}\}$, where the entries are smooth functions. For $N = 2$, the elements of M_1 and M_2 are related to each other according to

$$\phi = \sum_{i=1}^{3} \phi_i \tau_i = \begin{pmatrix} \phi_3 & \phi_1 - i\phi_2 \\ \phi_1 + i\phi_2 & -\phi_3 \end{pmatrix} \equiv \begin{pmatrix} \pi^0 & \sqrt{2}\pi^+ \\ \sqrt{2}\pi^- & -\pi^0 \end{pmatrix}, \qquad (3.36)$$

where the τ_i are the usual Pauli matrices and $\phi_i = \frac{1}{2}\text{Tr}[\tau_i \phi]$. Analogously for $N = 3$,

$$\phi = \sum_{a=1}^{8} \phi_a \lambda_a \equiv \begin{pmatrix} \pi^0 + \frac{1}{\sqrt{3}}\eta & \sqrt{2}\pi^+ & \sqrt{2}K^+ \\ \sqrt{2}\pi^- & -\pi^0 + \frac{1}{\sqrt{3}}\eta & \sqrt{2}K^0 \\ \sqrt{2}K^- & \sqrt{2}\bar{K}^0 & -\frac{2}{\sqrt{3}}\eta \end{pmatrix}, \qquad (3.37)$$

with the Gell-Mann matrices λ_a and $\phi_a = \frac{1}{2}\text{Tr}[\lambda_a \phi]$. Again, M_2 forms a real vector space.

Exercise 3.5 Make use of the Gell-Mann matrices of Eq. 1.6 and express the physical fields in terms of the Cartesian components, e.g.,

$$\pi^+ = \frac{1}{\sqrt{2}}(\phi_1 - i\phi_2).$$

Let us finally define

$$M_3 \equiv \left\{ U : M^4 \to \text{SU}(N) | U = \exp\left(i\frac{\phi}{F_0}\right), \phi \in M_2 \right\}.$$

At this stage, the constant F_0 is introduced to make the argument of the exponential function dimensionless. Since a bosonic field has the dimension of energy, F_0 also has the dimension of energy. Later on, F_0 will be identified with the "decay" constant of the Goldstone bosons in the chiral limit.[8] At this point it is important to note that M_3 does not define a vector space because the sum of two SU(N) matrices is not an SU(N) matrix.

We are now in the position to discuss a realization of SU(N) $\times$ SU(N) on M_3. The homomorphism

$$\varphi : G \times M_3 \to M_3 \quad \text{with} \quad \varphi[(L, R), U] \equiv RUL^\dagger,$$

defines an operation of G on M_3, because

---

[8] There is a subtlety here, because F_0 is traditionally reserved for the three-flavor chiral limit, whereas the two-flavor chiral limit (at fixed m_s) is denoted by F. In this section, we will use F_0 for both cases.

1. $RUL^\dagger \in M_3$, since $U \in M_3$ and $R, L^\dagger \in SU(N)$, i.e., $RUL^\dagger$ is a smooth $SU(N)$-valued function.
2. $\varphi[(\mathbb{1}, \mathbb{1}), U] = \mathbb{1} U \mathbb{1} = U$.
3. Let $g_i = (L_i, R_i) \in G$ and thus $g_1 g_2 = (L_1 L_2, R_1 R_2) \in G$.

$$\varphi[g_1, \varphi[g_2, U]] = \varphi[g_1, (R_2 U L_2^\dagger)] = R_1 R_2 U L_2^\dagger L_1^\dagger,$$
$$\varphi[g_1 g_2, U] = R_1 R_2 U (L_1 L_2)^\dagger = R_1 R_2 U L_2^\dagger L_1^\dagger.$$

The mapping φ is homogeneous of degree one but it is not a representation, because M_3 is *not* a vector space.

The origin $\phi = 0$ (for all x), i.e., $U_0 = \mathbb{1}$, denotes the ground state of the system. Under transformations of the subgroup $H = \{(V, V) | V \in SU(N)\}$ corresponding to rotating both left- and right-handed quark fields in QCD by the same V, the ground state remains invariant,

$$\varphi[g = (V, V), U_0] = V U_0 V^\dagger = V V^\dagger = \mathbb{1} = U_0.$$

On the other hand, under "axial transformations," i.e., rotating the left-handed quarks by a nontrivial A and the right-handed quarks by $A^\dagger$, the ground state does *not* remain invariant,

$$\varphi[g = (A, A^\dagger), U_0] = A^\dagger U_0 A^\dagger = A^\dagger A^\dagger \neq U_0,$$

which is consistent with the assumed spontaneous symmetry breakdown.

The traceless and Hermitian matrices of Eqs. 3.36 and 3.37 contain the Goldstone-boson fields. We want to discuss their transformation behavior under the subgroup $H = \{(V, V) | V \in SU(N)\}$. Expanding

$$U = \mathbb{1} + i\frac{\phi}{F_0} - \frac{\phi^2}{2F_0^2} + \cdots,$$

we immediately see that the realization φ restricted to the subgroup H,

$$\mathbb{1} + i\frac{\phi}{F_0} - \frac{\phi^2}{2F_0^2} + \cdots \mapsto V\left(\mathbb{1} + i\frac{\phi}{F_0} - \frac{\phi^2}{2F_0^2} + \cdots\right)V^\dagger$$
$$= \mathbb{1} + i\frac{V\phi V^\dagger}{F_0} - \frac{V\phi V^\dagger V\phi V^\dagger}{2F_0^2} + \cdots, \tag{3.38}$$

defines a representation on $M_2 \ni \phi \mapsto V\phi V^\dagger \in M_2$, because

$$(V\phi V^\dagger)^\dagger = V\phi V^\dagger, \quad \text{Tr}(V\phi V^\dagger) = \text{Tr}(\phi) = 0,$$
$$V_1(V_2\phi V_2^\dagger)V_1^\dagger = (V_1 V_2)\phi(V_1 V_2)^\dagger.$$

Let us consider the SU(3) case and parameterize

$$V = \exp\left(-i\Theta_{Vb}\frac{\lambda_b}{2}\right),$$

from which we obtain, by comparing both sides of Eq. 3.38,

$$\phi = \phi_a\lambda_a \overset{h\in SU(3)_V}{\longmapsto} V\phi V^\dagger = \phi - i\Theta_{Vb}\phi_a\underbrace{\left[\frac{\lambda_b}{2},\lambda_a\right]}_{=if_{bac}\lambda_c} +\cdots = \phi + f_{abc}\Theta_{Va}\phi_b\lambda_c +\cdots.$$

$$(3.39)$$

However, this corresponds exactly to the adjoint representation, i.e., in SU(3) the fields ϕ_a transform as an octet which is also consistent with the transformation behavior we discussed in Eq. 3.17:

$$e^{i\Theta_{Va}Q_{Va}}\phi_b\lambda_b e^{-i\Theta_{Vc}Q_{Vc}} = \phi_b\lambda_b + i\Theta_{Va}\underbrace{[Q_{Va},\phi_b]}_{=if_{abc}\phi_c}\lambda_b +\cdots$$

$$= \phi + f_{abc}\Theta_{Va}\phi_b\lambda_c +\cdots. \qquad (3.40)$$

For group elements of G of the form $(A,A^\dagger)$ one may proceed in a completely analogous fashion. However, one finds that the fields ϕ_a do *not* have a simple transformation behavior under these group elements. In other words, the commutation relations of the fields with the *axial* charges are complicated nonlinear functions of the fields. This is the origin for the terminology *nonlinear realization of chiral symmetry* [33, 42, 99].

3.4 Effective Lagrangian and Power-Counting Scheme

Having discussed the transformation behavior of the Goldstone-boson fields, we now turn to describing their interactions with each other and with external fields at energies far below 1 GeV.

3.4.1 The Lowest-Order Effective Lagrangian

Our goal is the construction of the most general theory describing the dynamics of the Goldstone bosons associated with the spontaneous symmetry breakdown in QCD. In the chiral limit, we want the effective Lagrangian to be invariant under $SU(3)_L \times SU(3)_R \times U(1)_V$. It should contain exactly eight pseudoscalar degrees of freedom transforming as an octet under the subgroup $H = SU(3)_V$. Moreover, taking account of spontaneous symmetry breaking, the ground state should only be invariant under $SU(3)_V \times U(1)_V$.

In terms of the SU(3) matrix

$$U(x) = \exp\left(i\frac{\phi(x)}{F_0}\right), \tag{3.41}$$

where ϕ is given in Eq. 3.37, the most general, chirally invariant, effective Lagrangian with the minimal number of derivatives reads

$$\mathscr{L}_{\text{eff}} = \frac{F_0^2}{4}\text{Tr}\left(\partial_\mu U \partial^\mu U^\dagger\right). \tag{3.42}$$

The parameter F_0 will be related to the pion decay $\pi^+ \to \mu^+ \nu_\mu$ later on (see Sect. 3.4.4).

First of all, the Lagrangian is invariant under the *global* $\text{SU}(3)_L \times \text{SU}(3)_R$ transformations of Eq. 3.35:

$$U \mapsto RUL^\dagger,$$

$$\partial_\mu U \mapsto \partial_\mu(RUL^\dagger) = \underbrace{\partial_\mu R}_{=0} UL^\dagger + R\partial_\mu U L^\dagger + RU\underbrace{\partial_\mu L^\dagger}_{=0} = R\partial_\mu U L^\dagger,$$

$$U^\dagger \mapsto LU^\dagger R^\dagger,$$

$$\partial_\mu U^\dagger \mapsto L\partial_\mu U^\dagger R^\dagger,$$

because

$$\mathscr{L}_{\text{eff}} \mapsto \frac{F_0^2}{4}\text{Tr}\left(R\partial_\mu U \underbrace{L^\dagger L}_{=\mathbb{1}} \partial^\mu U^\dagger R^\dagger\right) = \frac{F_0^2}{4}\text{Tr}\left(\underbrace{R^\dagger R}_{=\mathbb{1}} \partial_\mu U \partial^\mu U^\dagger\right) = \mathscr{L}_{\text{eff}},$$

where we made use of the trace property $\text{Tr}(AB) = \text{Tr}(BA)$. The global $\text{U}(1)_V$ invariance is trivially satisfied, because the Goldstone bosons have baryon number zero, thus transforming as $\phi \mapsto \phi$ under $\text{U}(1)_V$ which also implies $U \mapsto U$.

The substitution $\phi_a(t,\vec{x}) \mapsto -\phi_a(t,\vec{x})$ or, equivalently, $U(t,\vec{x}) \mapsto U^\dagger(t,\vec{x})$ provides a simple method of testing whether an expression is of so-called even or odd *intrinsic* parity,[9] i.e., even or odd in the number of Goldstone-boson fields. For example, it is easy to show, using the trace property, that the Lagrangian of Eq. 3.42 is even.

The purpose of the multiplicative constant $F_0^2/4$ in Eq. 3.42 is to generate the standard form of the kinetic term $\frac{1}{2}\partial_\mu\phi_a\partial^\mu\phi_a$, which can be seen by expanding the exponential $U = \mathbb{1} + i\phi/F_0 + \cdots$, $\partial_\mu U = i\partial_\mu\phi/F_0 + \cdots$, resulting in

---

[9] Since the Goldstone bosons are pseudoscalars, a true parity transformation is given by $\phi_a(t,\vec{x}) \mapsto -\phi_a(t,-\vec{x})$ or, equivalently, $U(t,\vec{x}) \mapsto U^\dagger(t,-\vec{x})$.

$$\mathscr{L}_{\text{eff}} = \frac{F_0^2}{4}\text{Tr}\left[\frac{i\partial_\mu\phi}{F_0}\left(-\frac{i\partial^\mu\phi}{F_0}\right)\right] + \cdots = \frac{1}{4}\text{Tr}(\partial_\mu\phi_a\lambda_a\partial^\mu\phi_b\lambda_b) + \cdots$$

$$= \frac{1}{4}\partial_\mu\phi_a\partial^\mu\phi_b\underbrace{\text{Tr}(\lambda_a\lambda_b)}_{= 2\delta_{ab}} + \cdots \equiv \frac{1}{2}\partial_\mu\phi_a\partial^\mu\phi_a + \mathscr{L}_{\text{int}}.$$

In particular, since there are no other terms containing only two fields ($\mathscr{L}_{\text{int}}$ starts with interaction terms containing at least four Goldstone bosons) the eight fields ϕ_a describe eight independent *massless* particles. At this stage this is only a tree-level argument. We will see in Sect. 3.5.2 that the Goldstone bosons remain massless in the chiral limit even when loop corrections have been included.

A term proportional to $\text{Tr}(UU^\dagger) = 3$ produces a constant which is irrelevant for the dynamics of the Goldstone bosons and will therefore be omitted. A term of the type $\text{Tr}[(\partial_\mu\partial^\mu U)U^\dagger]$ may be re-expressed as

$$\text{Tr}[(\partial_\mu\partial^\mu U)U^\dagger] = \partial_\mu[\text{Tr}(\partial^\mu UU^\dagger)] - \text{Tr}(\partial^\mu U\partial_\mu U^\dagger),$$

i.e., up to a total derivative it is proportional to the Lagrangian of Eq. 3.42. However, in the present context, total derivatives do not have a dynamical significance, i.e., they leave the equations of motion unchanged and can thus be dropped. The product of two invariant traces is excluded at lowest order, because $\text{Tr}(\partial_\mu UU^\dagger) = 0$.

Exercise 3.6 Prove

$$\text{Tr}(\partial_\mu UU^\dagger) = 0 \tag{3.43}$$

for the general SU(N) case by considering an SU(N)-valued field

$$U(x) = \exp\left(i\frac{\phi_a(x)\Lambda_a}{F_0}\right),$$

with $N^2 - 1$ Hermitian, traceless matrices Λ_a and real fields ϕ_a. Defining $\phi = \phi_a\Lambda_a$, expand the exponential

$$U = \mathbb{1} + i\frac{\phi}{F_0} + \frac{1}{2F_0^2}(i\phi)^2 + \frac{1}{3!F_0^3}(i\phi)^3 + \cdots$$

and consider the derivative

$$\partial_\mu U = i\frac{\partial_\mu\phi}{F_0} + \frac{i\partial_\mu\phi i\phi + i\phi i\partial_\mu\phi}{2F_0^2} + \frac{i\partial_\mu\phi(i\phi)^2 + i\phi i\partial_\mu\phi i\phi + (i\phi)^2 i\partial_\mu\phi}{3!F_0^3} + \cdots.$$

Remark ϕ and $\partial_\mu\phi$ are matrices which, in general, do not commute!
Hint: $[\phi, U^\dagger] = 0$.

Let us turn to the vector and axial-vector currents associated with the global $SU(3)_L \times SU(3)_R$ symmetry of the effective Lagrangian of Eq. 3.42. To that end, we consider the infinitesimal transformations

$$L = 1 - i\varepsilon_{La}\frac{\lambda_a}{2}, \tag{3.44}$$

$$R = 1 - i\varepsilon_{Ra}\frac{\lambda_a}{2}. \tag{3.45}$$

In order to construct J_{La}^{μ}, set $\varepsilon_{Ra} = 0$ and choose $\varepsilon_{La} = \varepsilon_{La}(x)$ (see Sect. 1.3.3):

$$U \mapsto U' = RUL^{\dagger} = U\left(1 + i\varepsilon_{La}\frac{\lambda_a}{2}\right),$$

$$U^{\dagger} \mapsto U'^{\dagger} = \left(1 - i\varepsilon_{La}\frac{\lambda_a}{2}\right)U^{\dagger},$$

$$\partial_{\mu}U \mapsto \partial_{\mu}U' = \partial_{\mu}U\left(1 + i\varepsilon_{La}\frac{\lambda_a}{2}\right) + Ui\partial_{\mu}\varepsilon_{La}\frac{\lambda_a}{2}, \tag{3.46}$$

$$\partial_{\mu}U^{\dagger} \mapsto \partial_{\mu}U'^{\dagger} = \left(1 - i\varepsilon_{La}\frac{\lambda_a}{2}\right)\partial_{\mu}U^{\dagger} - i\partial_{\mu}\varepsilon_{La}\frac{\lambda_a}{2}U^{\dagger},$$

from which we obtain for $\delta\mathscr{L}_{\text{eff}}$:

$$\begin{aligned}
\delta\mathscr{L}_{\text{eff}} &= \frac{F_0^2}{4}\text{Tr}\left[Ui\partial_{\mu}\varepsilon_{La}\frac{\lambda_a}{2}\partial^{\mu}U^{\dagger} + \partial_{\mu}U\left(-i\partial^{\mu}\varepsilon_{La}\frac{\lambda_a}{2}U^{\dagger}\right)\right] \\
&= \frac{F_0^2}{4}i\partial_{\mu}\varepsilon_{La}\text{Tr}\left[\frac{\lambda_a}{2}(\partial^{\mu}U^{\dagger}U - U^{\dagger}\partial^{\mu}U)\right] \\
&= \frac{F_0^2}{4}i\partial_{\mu}\varepsilon_{La}\text{Tr}\left(\lambda_a\partial^{\mu}U^{\dagger}U\right). \tag{3.47}
\end{aligned}$$

In the last step we made use of

$$\partial^{\mu}U^{\dagger}U = -U^{\dagger}\partial^{\mu}U,$$

which follows from differentiating $U^{\dagger}U = 1$. We thus obtain for the left currents

$$J_{La}^{\mu} = \frac{\partial\delta\mathscr{L}_{\text{eff}}}{\partial\partial_{\mu}\varepsilon_{La}} = i\frac{F_0^2}{4}\text{Tr}\left(\lambda_a\partial^{\mu}U^{\dagger}U\right), \tag{3.48}$$

and, completely analogously, choosing $\varepsilon_{La} = 0$ and $\varepsilon_{Ra} = \varepsilon_{Ra}(x)$,

$$J_{Ra}^{\mu} = \frac{\partial\delta\mathscr{L}_{\text{eff}}}{\partial\partial_{\mu}\varepsilon_{Ra}} = -i\frac{F_0^2}{4}\text{Tr}\left(\lambda_a U\partial^{\mu}U^{\dagger}\right) \tag{3.49}$$

for the right currents. Combining Eqs. 3.48 and 3.49, the vector and axial-vector currents read

$$J^\mu_{Va} = J^\mu_{Ra} + J^\mu_{La} = -i\frac{F_0^2}{4}\text{Tr}\big(\lambda_a[U, \partial^\mu U^\dagger]\big), \tag{3.50}$$

$$J^\mu_{Aa} = J^\mu_{Ra} - J^\mu_{La} = -i\frac{F_0^2}{4}\text{Tr}\big(\lambda_a\{U, \partial^\mu U^\dagger\}\big). \tag{3.51}$$

Furthermore, because of the symmetry of $\mathscr{L}_{\text{eff}}$ under $SU(3)_L \times SU(3)_R$, both vector and axial-vector currents are conserved. The vector currents J^μ_{Va} of Eq. 3.50 contain only terms with an even number of Goldstone bosons,

$$J^\mu_{Va} \overset{\phi \mapsto -\phi}{\longmapsto} -i\frac{F_0^2}{4}\text{Tr}[\lambda_a(U^\dagger\partial^\mu U - \partial^\mu U U^\dagger)]$$
$$= -i\frac{F_0^2}{4}\text{Tr}[\lambda_a(-\partial^\mu U^\dagger U + U\partial^\mu U^\dagger)] = J^\mu_{Va}.$$

On the other hand, the expression for the axial-vector currents is *odd* in the number of Goldstone bosons,

$$J^\mu_{Aa} \overset{\phi \mapsto -\phi}{\longmapsto} -i\frac{F_0^2}{4}\text{Tr}[\lambda_a(U^\dagger\partial^\mu U + \partial^\mu U U^\dagger)]$$
$$= i\frac{F_0^2}{4}\text{Tr}[\lambda_a(\partial^\mu U^\dagger U + U\partial^\mu U^\dagger)] = -J^\mu_{Aa}.$$

To find the leading term, let us expand Eq. 3.51 in the fields,

$$J^\mu_{Aa} = -i\frac{F_0^2}{4}\text{Tr}\left(\lambda_a\left\{\mathbb{1} + \cdots, -i\frac{\partial^\mu \phi_b \lambda_b}{F_0} + \cdots\right\}\right) = -F_0\partial^\mu\phi_a + \cdots,$$

from which we conclude that the axial-vector current has a nonvanishing matrix element when evaluated between the vacuum and a one-Goldstone-boson state:

$$\langle 0|J^\mu_{Aa}(x)|\phi_b(p)\rangle = \langle 0| - F_0\partial^\mu\phi_a(x)|\phi_b(p)\rangle = -F_0\partial^\mu\exp(-ip\cdot x)\delta_{ab}$$
$$= ip^\mu F_0\exp(-ip\cdot x)\delta_{ab}. \tag{3.52}$$

Equation 3.52 is the manifestation of Eq. 3.29 at lowest order in the effective field theory.

3.4.2 Symmetry Breaking by the Quark Masses

So far we have assumed a perfect $SU(3)_L \times SU(3)_R$ symmetry. However, in Sect. 2.4 we saw, by means of a simple example, how an explicit symmetry breaking may lead to finite masses of the Goldstone bosons. As has been discussed in Sect. 1.3.6, the quark-mass term of QCD results in such an explicit symmetry breaking,[10]

$$\mathscr{L}_{\mathscr{M}} = -\bar{q}_R \mathscr{M} q_L - \bar{q}_L \mathscr{M}^\dagger q_R, \quad \mathscr{M} = \begin{pmatrix} m_u & 0 & 0 \\ 0 & m_d & 0 \\ 0 & 0 & m_s \end{pmatrix}. \tag{3.53}$$

In order to incorporate the consequences of Eq. 3.53 into the effective-Lagrangian framework, one makes use of the following argument [57]: Although $\mathscr{M}$ is in reality just a constant matrix and does not transform along with the quark fields, $\mathscr{L}_{\mathscr{M}}$ of Eq. 3.53 *would be* invariant *if* $\mathscr{M}$ transformed as

$$\mathscr{M} \mapsto R\mathscr{M}L^\dagger. \tag{3.54}$$

One then constructs the most general Lagrangian $\mathscr{L}(U, \mathscr{M})$ which is invariant under Eqs. 3.35 and 3.54 and expands this function in powers of $\mathscr{M}$. At lowest order in $\mathscr{M}$ one obtains

$$\mathscr{L}_{\text{s.b.}} = \frac{F_0^2 B_0}{2} \text{Tr}(\mathscr{M} U^\dagger + U\mathscr{M}^\dagger), \tag{3.55}$$

where the subscript s.b. refers to symmetry breaking. In order to interpret the new parameter B_0, let us consider the Hamiltonian density corresponding to the sum of the Lagrangians of Eq. 3.42 and 3.55:

$$\mathscr{H}_{\text{eff}} = \frac{F_0^2}{4} \text{Tr}(\dot{U}\dot{U}^\dagger) + \frac{F_0^2}{4} \text{Tr}(\vec{\nabla}U \cdot \vec{\nabla}U^\dagger) - \mathscr{L}_{\text{s.b.}}.$$

Since the first two terms are always larger than or equal to zero, $\mathscr{H}_{\text{eff}}$ is minimized by constant and uniform fields. Using the ansatz

$$\phi = \phi_0 + \frac{1}{F_0^2}\phi_2 + \frac{1}{F_0^4}\phi_4 + \cdots$$

for the minimizing field values and organizing the individual terms in powers of $1/F_0^2$, one finds $\phi = 0$ as the classical solution even in the presence of quark-mass terms.

Exercise 3.7 We prove the statement above to order $1/F_0^2$ in the ansatz for ϕ. Since we are considering constant and uniform fields, we only have to take into account the symmetry-breaking Lagrangian $\mathscr{L}_{\text{s.b.}}$ of Eq. 3.55.

(a) Calculate the derivative of $\mathscr{L}_{\text{s.b.}}$ with respect to ϕ_a, where $\phi = \phi_a \lambda_a$. Using $\mathscr{M} = \mathscr{M}^\dagger$, show that

---

[10] In view of the coupling to the external fields $s + ip$ and $s - ip$ (see Eq. 1.161) to be discussed in Sect. 3.4.3, we distinguish between $\mathscr{M}$ and $\mathscr{M}^\dagger$ even though for a real, diagonal matrix they are the same.

$$\frac{\partial \mathscr{L}_{\text{s.b.}}}{\partial \phi_a} = F_0^2 B_0 \text{Tr}\left[\mathscr{M} \left(-\frac{\lambda_a \phi + \phi \lambda_a}{2F_0^2} + \frac{\lambda_a \phi^3 + \phi \lambda_a \phi^2 + \phi^2 \lambda_a \phi + \phi^3 \lambda_a}{24F_0^4} \right) \right]$$
$$+ O\left(\frac{1}{F_0^4}\right).$$

(b) Insert the ansatz $\phi = \phi_0 + \frac{1}{F_0^2}\phi_2 + \cdots$ and show that the trace is given by

$$-\frac{1}{2F_0^2}\text{Tr}[\mathscr{M}(\lambda_a \phi_0 + \phi_0 \lambda_a)]$$
$$-\frac{1}{2F_0^4}\text{Tr}\left[\mathscr{M}\left(\lambda_a \phi_2 + \phi_2 \lambda_a - \frac{\lambda_a \phi_0^3 + \phi_0 \lambda_a \phi_0^2 + \phi_0^2 \lambda_a \phi_0 + \phi_0^3 \lambda_a}{12}\right)\right].$$

(c) The quark-mass matrix can be parameterized as $\mathscr{M} = m_0 \lambda_0 + m_3 \lambda_3 + m_8 \lambda_8$ (see Exercise 1.17), with

$$m_0 = \frac{m_u + m_d + m_s}{\sqrt{6}}, \quad m_3 = \frac{m_u - m_d}{2}, \quad m_8 = \frac{m_u + m_d - 2m_s}{2\sqrt{3}},$$

while $\phi_0 = \phi_{0b}\lambda_b$. Considering $a = 1$, show that $\phi_{01} = 0$ for ϕ to minimize $\mathscr{H}_{\text{eff}}$. Analogous calculations hold for $a = 2, \ldots, 8$.
Hint: $\{\lambda_a, \lambda_b\} = \frac{4}{3}\delta_{ab}\mathbb{1} + 2d_{abc}\lambda_c$. The values for d_{abc} are given in Table 1.2.

(d) Using the result for ϕ_0, show that also $\phi_2 = 0$. Analogous calculations apply for higher orders in $1/F_0^2$.

Now consider the energy density of the ground state ($U_{\text{min}} = U_0 = \mathbb{1}$),

$$\langle \mathscr{H}_{\text{eff}} \rangle_{\text{min}} = -F_0^2 B_0 (m_u + m_d + m_s), \tag{3.56}$$

and compare its derivative with respect to (any of) the light-quark masses m_q with the corresponding quantity in QCD,

$$\frac{\partial \langle 0 | \mathscr{H}_{\text{QCD}} | 0 \rangle}{\partial m_q}\bigg|_{m_u=m_d=m_s=0} = \frac{1}{3}\langle 0|\bar{q}q|0\rangle_0 = \frac{1}{3}\langle \bar{q}q \rangle_0,$$

where $\langle \bar{q}q \rangle_0$ is the scalar singlet quark condensate of Eq. 3.26. Within the framework of the lowest-order effective Lagrangian, the constant B_0 is thus related to the scalar singlet quark condensate by

$$3F_0^2 B_0 = -\langle \bar{q}q \rangle_0. \tag{3.57}$$

Let us add a few remarks.

1. A term $\text{Tr}(\mathscr{M})$ by itself is not invariant.
2. The combination $\text{Tr}(\mathscr{M}U^\dagger - U\mathscr{M}^\dagger)$ has the wrong behavior under parity $\phi(t, \vec{x}) \mapsto -\phi(t, -\vec{x})$, because

$$\text{Tr}[\mathcal{M}U^\dagger(t,\vec{x}) - U(t,\vec{x})\mathcal{M}^\dagger] \overset{P}{\mapsto} \text{Tr}[\mathcal{M}U(t,-\vec{x}) - U^\dagger(t,-\vec{x})\mathcal{M}^\dagger]$$
$$\overset{\mathcal{M}=\mathcal{M}^\dagger}{=} -\text{Tr}[\mathcal{M}U^\dagger(t,-\vec{x}) - U(t,-\vec{x})\mathcal{M}^\dagger].$$

3. Because $\mathcal{M} = \mathcal{M}^\dagger$, $\mathcal{L}_{\text{s.b.}}$ contains only terms even in ϕ.

In order to determine the masses of the Goldstone bosons, we identify the terms of second order in the fields in $\mathcal{L}_{\text{s.b.}}$,

$$\mathcal{L}_{\text{s.b}} = -\frac{B_0}{2}\text{Tr}(\phi^2\mathcal{M}) + \cdots. \tag{3.58}$$

Exercise 3.8 Expand the mass term to second order in the physical fields and determine the squared masses of the Goldstone bosons.

Using Eq. 3.37 we find

$$\text{Tr}(\phi^2\mathcal{M}) = 2(m_u+m_d)\pi^+\pi^- + 2(m_u+m_s)K^+K^- + 2(m_d+m_s)K^0\bar{K}^0$$
$$+ (m_u+m_d)\pi^0\pi^0 + \frac{2}{\sqrt{3}}(m_u-m_d)\pi^0\eta + \frac{m_u+m_d+4m_s}{3}\eta^2.$$

For the sake of simplicity we consider the isospin-symmetric limit $m_u = m_d = \hat{m}$ so that the $\pi^0\eta$ term vanishes and there is no π^0-η mixing. We then obtain for the masses of the Goldstone bosons, to lowest order in the quark masses,

$$M_\pi^2 = 2B_0\hat{m}, \tag{3.59}$$

$$M_K^2 = B_0(\hat{m}+m_s), \tag{3.60}$$

$$M_\eta^2 = \frac{2}{3}B_0(\hat{m}+2m_s). \tag{3.61}$$

These results, in combination with Eq. 3.57, $B_0 = -\langle\bar{q}q\rangle/(3F_0^2)$, correspond to relations obtained in Ref. [56] and are referred to as the Gell-Mann, Oakes, and Renner relations. Furthermore, the squared masses of Eqs. 3.59–3.61 satisfy the Gell-Mann-Okubo relation

$$4M_K^2 = 4B_0(\hat{m}+m_s) = 2B_0(\hat{m}+2m_s) + 2B_0\hat{m} = 3M_\eta^2 + M_\pi^2, \tag{3.62}$$

independent of the value of B_0. Without additional input regarding the numerical value of B_0, Eqs. 3.59–3.61 do not allow for an extraction of the absolute values of the quark masses $\hat{m}$ and m_s, because rescaling $B_0 \to \lambda B_0$ in combination with $m_q \to m_q/\lambda$ leaves the relations invariant. For the ratio of the quark masses one obtains, using the empirical values $M_\pi = 135\,\text{MeV}$, $M_K = 496\,\text{MeV}$, and $M_\eta = 548\,\text{MeV}$,

Table 3.2 Comparison between the symmetry-breaking patterns of a Heisenberg ferromagnet and QCD

	Heisenberg ferromagnet	QCD
Symmetry of Hamiltonian	$O(3)$	$SU(3)_L \times SU(3)_R$
Symmetry of $|0\rangle$	$O(2)$	$SU(3)_V$
Vacuum expectation value	$\langle \vec{M} \rangle$	$\langle \bar{q}q \rangle_0$
Explicit symmetry breaking	External magnetic field	Quark masses
Interaction	$-\langle \vec{M} \rangle \cdot \vec{H}$	$\langle \mathcal{H}_{\text{eff}} \rangle$ of Eq. 3.56

$$\frac{M_K^2}{M_\pi^2} = \frac{\hat{m} + m_s}{2\hat{m}} \Rightarrow \frac{m_s}{\hat{m}} = 25.9,$$

$$\frac{M_\eta^2}{M_\pi^2} = \frac{2m_s + \hat{m}}{3\hat{m}} \Rightarrow \frac{m_s}{\hat{m}} = 24.3. \tag{3.63}$$

Let us conclude this section with a remark on $\langle \bar{q}q \rangle_0$. A nonvanishing quark condensate in the chiral limit is a sufficient but not a necessary condition for spontaneous chiral symmetry breaking. The effective-Lagrangian term of Eq. 3.55 not only results in a shift of the vacuum energy but also in finite Goldstone-boson masses, and both effects are proportional to the parameter B_0. We recall that it was a symmetry argument which excluded a term $\text{Tr}(\mathcal{M})$ which, at leading order in $\mathcal{M}$, would decouple the vacuum energy shift from the Goldstone-boson masses. The scenario underlying $\mathscr{L}_{\text{s.b.}}$ of Eq. 3.55 is similar to that of a Heisenberg ferromagnet which exhibits a spontaneous magnetization $\langle \vec{M} \rangle$, breaking the $O(3)$ symmetry of the Heisenberg Hamiltonian down to $O(2)$. In the present case, the analogue of the order parameter $\langle \vec{M} \rangle$ is the quark condensate $\langle \bar{q}q \rangle_0$. In the case of the ferromagnet, the interaction with an external magnetic field $\vec{H}$ is given by $-\langle \vec{M} \rangle \cdot \vec{H}$, which corresponds to Eq. 3.56, with the quark masses playing the role of the external field $\vec{H}$ (see Table 3.2). However, in principle, it is also possible that B_0 vanishes or is rather small. In such a case the quadratic masses of the Goldstone bosons might be dominated by terms which are nonlinear in the quark masses, i.e., by higher-order terms in the expansion of $\mathscr{L}(U, \mathcal{M})$. Such a scenario is the origin of the so-called generalized chiral perturbation theory [68]. The analogue would be an antiferromagnet which shows a spontaneous symmetry breaking but with $\langle \vec{M} \rangle = 0$. The analysis of the s-wave $\pi\pi$-scattering lengths [35, 36] supports the conjecture that the quark condensate is indeed the leading order parameter of the spontaneously broken chiral symmetry (see also Sect. 3.5.4).

3.4.3 Construction of the Effective Lagrangian

In Sect. 3.4.1 we have derived the lowest-order effective Lagrangian for a *global* $SU(3)_L \times SU(3)_R$ symmetry. On the other hand, the Ward identities originating in the global $SU(3)_L \times SU(3)_R$ symmetry of QCD are obtained from a *locally* invariant

Table 3.3 Transformation properties under the group (G), charge conjugation (C), and parity (P). The expressions for adjoint matrices are trivially obtained by taking the Hermitian conjugate of each entry. In the parity-transformed expression it is understood that the argument is $(t, -\vec{x})$ and that partial derivatives ∂_μ act with respect to x and not with respect to the argument of the corresponding function

Element	G	C	P
U	$V_R U V_L^\dagger$	U^T	$U^\dagger$
$D_{\lambda_1} \ldots D_{\lambda_n} U$	$V_R D_{\lambda_1} \ldots D_{\lambda_n} U V_L^\dagger$	$(D_{\lambda_1} \ldots D_{\lambda_n} U)^T$	$(D^{\lambda_1} \ldots D^{\lambda_n} U)^\dagger$
χ	$V_R \chi V_L^\dagger$	χ^T	$\chi^\dagger$
$D_{\lambda_1} \ldots D_{\lambda_n} \chi$	$V_R D_{\lambda_1} \ldots D_{\lambda_n} \chi V_L^\dagger$	$(D_{\lambda_1} \ldots D_{\lambda_n} \chi)^T$	$(D^{\lambda_1} \ldots D^{\lambda_n} \chi)^\dagger$
r_μ	$V_R r_\mu V_R^\dagger + i V_R \partial_\mu V_R^\dagger$	$-l_\mu^T$	l^μ
l_μ	$V_L l_\mu V_L^\dagger + i V_L \partial_\mu V_L^\dagger$	$-r_\mu^T$	r^μ
$f_{R\mu\nu}$	$V_R f_{R\mu\nu} V_R^\dagger$	$-(f_{L\mu\nu})^T$	$f_L^{\mu\nu}$
$f_{L\mu\nu}$	$V_L f_{L\mu\nu} V_L^\dagger$	$-(f_{R\mu\nu})^T$	$f_R^{\mu\nu}$

generating functional involving a coupling to external fields (see Sects. 1.4.1 and 1.4.4). Our goal is to approximate the "true" generating functional $Z_{\text{QCD}}[v, a, s, p]$ of Eq. 1.153 by a sequence $Z_{\text{eff}}^{(2)}[v, a, s, p] + Z_{\text{eff}}^{(4)}[v, a, s, p] + \cdots$, where the effective generating functionals are obtained using the effective field theory.[11] Therefore, we need to promote the global symmetry of the effective Lagrangian to a local one and introduce a coupling to the *same* external fields v, a, s, and p as in QCD [52, 53, 71].

In the following we will outline the principles entering the construction of the effective Lagrangian for a local $G = \text{SU}(3)_L \times \text{SU}(3)_R$ symmetry.[12] The matrix U transforms as

$$U(x) \mapsto U'(x) = V_R(x) U(x) V_L^\dagger(x), \tag{3.64}$$

where $V_L(x)$ and $V_R(x)$ are independent space-time-dependent SU(3) matrices. As in the case of gauge theories, we need external fields $l_a^\mu(x)$ and $r_a^\mu(x)$ (see Eqs. 1.151, 1.160, and 1.163 and Table 3.3) corresponding to the parameters $\Theta_{La}(x)$ and $\Theta_{Ra}(x)$ of $V_L(x)$ and $V_R(x)$, respectively. For any object A transforming as $V_R A V_L^\dagger$ such as, e.g., U we define the covariant derivative $D_\mu A$ as

$$D_\mu A \equiv \partial_\mu A - i r_\mu A + i A l_\mu. \tag{3.65}$$

---

[11] Including all of the infinite number of effective functionals $Z_{\text{eff}}^{(2n)}[v, a, s, p]$ will generate a result which is equivalent to that obtained from $Z_{\text{QCD}}[v, a, s, p]$.

[12] In principle, we could also "gauge" the $\text{U}(1)_V$ symmetry. However, this is primarily of relevance to the two-flavor sector in order to fully incorporate the coupling to the electromagnetic four-vector potential (see Eq. 1.165). Since in the three-flavor sector the quark-charge matrix is traceless, this important case is included in our considerations.

Exercise 3.9 Verify the transformation behavior

$$D_\mu A \mapsto V_R(D_\mu A)V_L^\dagger.$$

Hint: Make use of $V_R \partial_\mu V_R^\dagger = -\partial_\mu V_R V_R^\dagger$.

Again, the defining property of the covariant derivative is that it should transform in the same way as the object it acts on.[13] Since the effective Lagrangian will ultimately contain arbitrarily high powers of derivatives we also need the field-strength tensors $f_{L\mu\nu}$ and $f_{R\mu\nu}$ corresponding to the external fields r_μ and l_μ,

$$f_{R\mu\nu} \equiv \partial_\mu r_\nu - \partial_\nu r_\mu - i[r_\mu, r_\nu], \tag{3.66}$$

$$f_{L\mu\nu} \equiv \partial_\mu l_\nu - \partial_\nu l_\mu - i[l_\mu, l_\nu]. \tag{3.67}$$

The field-strength tensors are traceless,

$$\text{Tr}(f_{L\mu\nu}) = \text{Tr}(f_{R\mu\nu}) = 0, \tag{3.68}$$

because $\text{Tr}(l_\mu) = \text{Tr}(r_\mu) = 0$ and the trace of any commutator vanishes. Finally, following the convention of Gasser and Leutwyler [53] we introduce the linear combination $\chi \equiv 2B_0(s + ip)$ with the scalar and pseudoscalar external fields of Eq. 1.151, where B_0 is defined in Eq. 3.57. Table 3.3 contains the transformation properties of all building blocks under the group (G) and the discrete symmetries C and P.

In the counting scheme of chiral perturbation theory the elements count as[14]

$$U = \mathcal{O}(q^0),\ D_\mu U = \mathcal{O}(q),\ r_\mu, l_\mu = \mathcal{O}(q),\ f_{L/R\mu\nu} = \mathcal{O}(q^2),\ \chi = \mathcal{O}(q^2). \tag{3.69}$$

The external fields r_μ and l_μ count as $\mathcal{O}(q)$ to match $\partial_\mu A$, and χ is of $\mathcal{O}(q^2)$ because of Eqs. 3.59–3.61. Any additional covariant derivative counts as $\mathcal{O}(q)$.

The construction of the effective Lagrangian in terms of the building blocks of Eq. 3.69 proceeds as follows.[15] Given objects $A, B, \ldots$, all of which transform as $A' = V_R A V_L^\dagger$, $B' = V_R B V_L^\dagger$, $\ldots$, one can form invariants by taking the trace of products of the type $AB^\dagger$:

---

[13] Under certain circumstances it is advantageous to introduce for each object with a well-defined transformation behavior a separate covariant derivative. One may then use a product rule similar to the one of ordinary differentiation.

[14] Throughout this monograph we will reserve the notation $\mathcal{O}(q^n)$ for power counting in chiral perturbation theory, whereas $O(x^n)$ denotes terms of order x^n in the usual mathematical sense.

[15] There is a certain freedom in the choice of the elementary building blocks. For example, by a suitable multiplication with U or $U^\dagger$ any building block can be made to transform as $V_R \ldots V_L^\dagger$ without changing its chiral order. The present approach most naturally leads to the Lagrangian of Gasser and Leutwyler [53].

$$\mathrm{Tr}(AB^\dagger) \mapsto \mathrm{Tr}[V_R A V_L^\dagger (V_R B V_L^\dagger)^\dagger] = \mathrm{Tr}(V_R A \underbrace{V_L^\dagger V_L}_{=1} B^\dagger V_R^\dagger) = \mathrm{Tr}(AB^\dagger \underbrace{V_R^\dagger V_R}_{=1})$$

$$= \mathrm{Tr}(AB^\dagger).$$

The generalization to more terms is obvious and, of course, the product of invariant traces is invariant:

$$\mathrm{Tr}(AB^\dagger CD^\dagger), \quad \mathrm{Tr}(AB^\dagger)\mathrm{Tr}(CD^\dagger), \quad \ldots \tag{3.70}$$

The complete list of relevant elements up to and including $\mathcal{O}(q^2)$ transforming as $V_R \ldots V_L^\dagger$ reads

$$U, D_\mu U, D_\mu D_\nu U, \chi, U f_{L\mu\nu}, f_{R\mu\nu} U. \tag{3.71}$$

For the invariants up to $\mathcal{O}(q^2)$ we then obtain

$$\mathcal{O}(q^0) : \mathrm{Tr}(UU^\dagger) = 3,$$
$$\mathcal{O}(q) : \mathrm{Tr}(D_\mu UU^\dagger) \overset{*}{=} -\mathrm{Tr}[U(D_\mu U)^\dagger] \overset{*}{=} 0,$$
$$\mathcal{O}(q^2) : \mathrm{Tr}(D_\mu D_\nu UU^\dagger) \overset{**}{=} -\mathrm{Tr}[D_\nu U(D_\mu U)^\dagger] \overset{**}{=} \mathrm{Tr}[U(D_\nu D_\mu U)^\dagger],$$
$$\mathrm{Tr}(\chi U^\dagger), \tag{3.72}$$
$$\mathrm{Tr}(U\chi^\dagger),$$
$$\mathrm{Tr}(U f_{L\mu\nu} U^\dagger) = \mathrm{Tr}(f_{L\mu\nu}) = 0,$$
$$\mathrm{Tr}(f_{R\mu\nu}) = 0.$$

In $*$ we made use of two important properties of the covariant derivative $D_\mu U$:

$$D_\mu UU^\dagger = -U(D_\mu U)^\dagger, \tag{3.73}$$

$$\mathrm{Tr}(D_\mu UU^\dagger) = 0. \tag{3.74}$$

The first relation results from the unitarity of U in combination with the definition of the covariant derivative, Eq. 3.65:

$$D_\mu UU^\dagger = \underbrace{\partial_\mu UU^\dagger}_{=-U\partial_\mu U^\dagger} -ir_\mu \underbrace{UU^\dagger}_{=1} +iUl_\mu U^\dagger,$$

$$-U(D_\mu U)^\dagger = -U\partial_\mu U^\dagger - \underbrace{UU^\dagger}_{=1} ir_\mu - U(-il_\mu U^\dagger).$$

Equation 3.74 is shown using $\mathrm{Tr}(r_\mu) = \mathrm{Tr}(l_\mu) = 0$ together with Eq. 3.43, $\mathrm{Tr}(\partial_\mu UU^\dagger) = 0$:

$$\mathrm{Tr}(D_\mu UU^\dagger) = \mathrm{Tr}(\partial_\mu UU^\dagger - ir_\mu UU^\dagger + iUl_\mu U^\dagger) = 0.$$

Exercise 3.10 Verify **,

$$\text{Tr}(D_\mu D_\nu U U^\dagger) = -\text{Tr}[D_\nu U (D_\mu U)^\dagger] = \text{Tr}[U(D_\nu D_\mu U)^\dagger],$$

by explicit calculation.

Finally, we impose Lorentz invariance, i.e., Lorentz indices have to be contracted, resulting in three candidate terms:

$$\text{Tr}[D_\mu U (D^\mu U)^\dagger], \tag{3.75}$$

$$\text{Tr}(\chi U^\dagger \pm U \chi^\dagger). \tag{3.76}$$

The term in Eq. 3.76 with the minus sign is excluded because it has the wrong sign under parity (see Table 3.3), and we end up with the most general, *locally* invariant, effective Lagrangian at lowest chiral order [53],[16]

$$\mathcal{L}_2 = \frac{F_0^2}{4}\text{Tr}[D_\mu U (D^\mu U)^\dagger] + \frac{F_0^2}{4}\text{Tr}(\chi U^\dagger + U \chi^\dagger). \tag{3.77}$$

At $\mathcal{O}(q^2)$ it contains two low-energy constants: the SU(3) chiral limit of the Goldstone-boson decay constant F_0, and $B_0 = -\langle 0|\bar{q}q|0\rangle_0/(3F_0^2)$ (hidden in the definition of χ).

Exercise 3.11 Under charge conjugation fields describing particles are mapped onto fields describing antiparticles, i.e., $\pi^0 \mapsto \pi^0$, $\eta \mapsto \eta$, $\pi^+ \leftrightarrow \pi^-$, $K^+ \leftrightarrow K^-$, $K^0 \leftrightarrow \bar{K}^0$.

(a) What does that mean for the matrix

$$\phi = \begin{pmatrix} \pi^0 + \frac{1}{\sqrt{3}}\eta & \sqrt{2}\pi^+ & \sqrt{2}K^+ \\ \sqrt{2}\pi^- & -\pi^0 + \frac{1}{\sqrt{3}}\eta & \sqrt{2}K^0 \\ \sqrt{2}K^- & \sqrt{2}\bar{K}^0 & -\frac{2}{\sqrt{3}}\eta \end{pmatrix} ?$$

(b) Using $A^T B^T = (BA)^T$, show by induction $(A^T)^n = (A^n)^T$. In combination with (a) show that $U = \exp(i\phi/F_0) \overset{C}{\mapsto} U^T$.

(c) Under charge conjugation the external fields transform as

$$v_\mu \mapsto -v_\mu^T, \quad a_\mu \mapsto a_\mu^T, \quad s \mapsto s^T, \quad p \mapsto p^T.$$

Derive the transformation behavior of $r_\mu = v_\mu + a_\mu$, $l_\mu = v_\mu - a_\mu$, $\chi = 2B_0(s + ip)$, and $\chi^\dagger$.

(d) Using (b) and (c), show that the covariant derivative of U under charge conjugation transforms as

---

[16] At $\mathcal{O}(q^2)$ invariance under C does not provide any additional constraints.

$$D_\mu U \mapsto (D_\mu U)^T.$$

(e) Show that

$$\mathcal{L}_2 = \frac{F_0^2}{4}\mathrm{Tr}[D_\mu U(D^\mu U)^\dagger] + \frac{F_0^2}{4}\mathrm{Tr}(\chi U^\dagger + U\chi^\dagger)$$

is invariant under charge conjugation. Note that $(A^T)^\dagger = (A^\dagger)^T$ and $\mathrm{Tr}(A^T) = \mathrm{Tr}(A)$.

(f) As an example, show the invariance of the L_3 term of $\mathcal{L}_4$ (see Sect. 3.5.1) under charge conjugation:

$$L_3 \mathrm{Tr}\left[D_\mu U(D^\mu U)^\dagger D_\nu U(D^\nu U)^\dagger\right].$$

Hint: At the end you will need $(D_\mu U)^\dagger = -U^\dagger D_\mu U U^\dagger$ and $U^\dagger D_\mu U U^\dagger = -(D_\mu U)^\dagger$.

The lowest-order equation of motion corresponding to Eq. 3.77 is obtained by considering small variations of the SU(3) matrix,

$$U'(x) = U(x) + \delta U(x) = \left(\mathbb{1} + i\sum_{a=1}^{8}\Delta_a(x)\lambda_a\right)U(x), \qquad (3.78)$$

where the $\Delta_a(x)$ are real functions. The matrix U' satisfies both conditions $U'U'^\dagger = \mathbb{1}$ and $\det(U') = 1$ up to and including terms linear in Δ_a. Applying the principle of stationary action, the variation of the action reads

$$\delta S = i\frac{F_0^2}{4}\int_{t_1}^{t_2} dt \int d^3x \sum_{a=1}^{8}\Delta_a(x)\mathrm{Tr}\left\{\lambda_a[D_\mu D^\mu U U^\dagger - U(D_\mu D^\mu U)^\dagger - \chi U^\dagger + U\chi^\dagger]\right\},$$

where we made use of integration by parts, the standard boundary conditions $\Delta_a(t_1, \vec{x}) = \Delta_a(t_2, \vec{x}) = 0$, the divergence theorem, and the definition of the covariant derivative of Eq. 3.65. Since the test functions $\Delta_a(x)$ may be chosen arbitrarily, we obtain eight Euler-Lagrange equations

$$\mathrm{Tr}\left\{\lambda_a[D^2 U U^\dagger - U(D^2 U)^\dagger - \chi U^\dagger + U\chi^\dagger]\right\} = 0, \qquad a = 1, \ldots, 8, \qquad (3.79)$$

which may be combined into a compact matrix form

$$\mathcal{O}_{\mathrm{EOM}}^{(2)}(U) \equiv D^2 U U^\dagger - U(D^2 U)^\dagger - \chi U^\dagger + U\chi^\dagger + \frac{1}{3}\mathrm{Tr}(\chi U^\dagger - U\chi^\dagger) = 0. \quad (3.80)$$

The trace term in Eq. 3.80 appears, because Eq. 3.79 contains eight and not nine independent equations.

Fig. 3.5 Pion decay
$\pi^+ \to \mu^+ \nu_\mu$

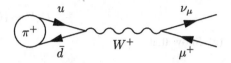

3.4.4 Application at Lowest Order: Pion Decay

The Lagrangian $\mathscr{L}_2$ of Eq. 3.77 has predictive power once the low-energy constant F_0 is identified. This LEC may be obtained from the weak decay of the pion, $\pi^+ \to \mu^+ \nu_\mu$.

At the level of the Standard Model degrees of freedom, pion decay is described by the annihilation of a u quark and a $\bar{d}$ antiquark, forming the π^+, into a W^+ boson, propagation of the intermediate W^+, and creation of the lepton ν_μ and the antilepton μ^+ in the final state (see Fig. 3.5). The coupling of the W bosons to the leptons is given by

$$\mathscr{L} = -\frac{g}{2\sqrt{2}} \left[\mathscr{W}_\rho^+ \bar{\nu}_\mu \gamma^\rho (\mathbb{1} - \gamma_5)\mu + \mathscr{W}_\rho^- \bar{\mu}\gamma^\rho (\mathbb{1} - \gamma_5)\nu_\mu \right], \tag{3.81}$$

whereas their interaction with the quarks forming the Goldstone bosons is effectively taken into account by inserting Eq. 1.166 into the Lagrangian of Eq. 3.77. Let us consider the first term of Eq. 3.77 and set $r_\mu = 0$ with, at this point, still arbitrary l_μ.

Exercise 3.12 Using $D_\mu U = \partial_\mu U + iUl_\mu$, derive

$$\frac{F_0^2}{4}\mathrm{Tr}[D_\mu U (D^\mu U)^\dagger] = i\frac{F_0^2}{2}\mathrm{Tr}(l_\mu \partial^\mu U^\dagger U) + \cdots,$$

where only the term linear in l_μ is shown.

If we parameterize

$$l_\mu = \sum_{a=1}^{8} l_{\mu a}\frac{\lambda_a}{2},$$

the interaction term linear in l_μ reads

$$\mathscr{L}_{\mathrm{int}} = \sum_{a=1}^{8} l_{\mu a} \left[i\frac{F_0^2}{4}\mathrm{Tr}(\lambda_a \partial^\mu U^\dagger U) \right] = \sum_{a=1}^{8} l_{\mu a} J_{La}^\mu, \tag{3.82}$$

where we made use of Eq. 3.48 defining J_{La}^μ. Again, by using Eq. 3.41, we expand J_{La}^μ to first order in ϕ,

$$J_{La}^\mu = \frac{F_0}{2}\partial^\mu \phi_a + O(\phi^2), \tag{3.83}$$

from which we obtain the matrix element

$$\langle 0|J_{La}^\mu(0)|\phi_b(p)\rangle = \frac{F_0}{2}\langle 0|\partial^\mu\phi_a(0)|\phi_b(p)\rangle = -ip^\mu\frac{F_0}{2}\delta_{ab}. \quad (3.84)$$

Inserting l_μ of Eq. 1.166, we find for the interaction term of a single Goldstone boson with a W

$$\mathcal{L}_{W\phi} = \frac{F_0}{2}\mathrm{Tr}(l_\mu\partial^\mu\phi) = -\frac{g}{\sqrt{2}}\frac{F_0}{2}\mathrm{Tr}[(\mathcal{W}_\mu^+T_+ + \mathcal{W}_\mu^-T_-)\partial^\mu\phi].$$

Thus, we need to calculate[17]

$$\mathrm{Tr}(T_+\partial^\mu\phi) = \mathrm{Tr}\left[\begin{pmatrix} 0 & V_{ud} & V_{us} \\ 0 & 0 & 0 \\ 0 & 0 & 0 \end{pmatrix}\partial^\mu\begin{pmatrix} \pi^0+\frac{1}{\sqrt{3}}\eta & \sqrt{2}\pi^+ & \sqrt{2}K^+ \\ \sqrt{2}\pi^- & -\pi^0+\frac{1}{\sqrt{3}}\eta & \sqrt{2}K^0 \\ \sqrt{2}K^- & \sqrt{2}\bar{K}^0 & -\frac{2}{\sqrt{3}}\eta \end{pmatrix}\right]$$

$$= V_{ud}\sqrt{2}\partial^\mu\pi^- + V_{us}\sqrt{2}\partial^\mu K^-,$$

$$\mathrm{Tr}(T_-\partial^\mu\phi) = \mathrm{Tr}\left[\begin{pmatrix} 0 & 0 & 0 \\ V_{ud} & 0 & 0 \\ V_{us} & 0 & 0 \end{pmatrix}\partial^\mu\begin{pmatrix} \pi^0+\frac{1}{\sqrt{3}}\eta & \sqrt{2}\pi^+ & \sqrt{2}K^+ \\ \sqrt{2}\pi^- & -\pi^0+\frac{1}{\sqrt{3}}\eta & \sqrt{2}K^0 \\ \sqrt{2}K^- & \sqrt{2}\bar{K}^0 & -\frac{2}{\sqrt{3}}\eta \end{pmatrix}\right]$$

$$= V_{ud}\sqrt{2}\partial^\mu\pi^+ + V_{us}\sqrt{2}\partial^\mu K^+.$$

We then obtain for the interaction term

$$\mathcal{L}_{W\phi} = -g\frac{F_0}{2}[\mathcal{W}_\mu^+(V_{ud}\partial^\mu\pi^- + V_{us}\partial^\mu K^-) + \mathcal{W}_\mu^-(V_{ud}\partial^\mu\pi^+ + V_{us}\partial^\mu K^+)]. \quad (3.85)$$

Expanding the Feynman propagator for W bosons in Landau gauge,

$$\frac{-g_{\mu\nu}+\frac{k_\mu k_\nu}{M_W^2}}{k^2-M_W^2} = \frac{g_{\mu\nu}}{M_W^2} + O\left(\frac{k_\rho k_\sigma}{M_W^4}\right), \quad (3.86)$$

and neglecting terms which are of higher order in $(\mathrm{momentum}/M_W)^2$, the Feynman rule for the invariant amplitude for weak pion decay has the form "leptonic vertex $\times$ W propagator $\times$ hadronic vertex,"

$$\mathcal{M} = i\left[-\frac{g}{2\sqrt{2}}\bar{u}_{v_\mu}\gamma^\rho(1-\gamma_5)v_{\mu^+}\right]\frac{ig_{\rho\sigma}}{M_W^2}i\left[-g\frac{F_0}{2}V_{ud}(-ip^\sigma)\right]$$

$$= -G_F V_{ud}F_0\bar{u}_{v_\mu}\not{p}(1-\gamma_5)v_{\mu^+}, \quad (3.87)$$

where p denotes the four-momentum of the pion and G_F is the Fermi constant of Eq. 1.167. The corresponding decay rate is

$$\Gamma = \frac{1}{\tau} = \frac{G_F^2 V_{ud}^2}{4\pi}F_0^2 M_\pi m_\mu^2\left(1-\frac{m_\mu^2}{M_\pi^2}\right)^2. \quad (3.88)$$

---

[17] Recall that the entries V_{ud} and V_{us} of the Cabibbo-Kobayashi-Maskawa matrix are real.

The constant F_0 is referred to as the pion-decay constant in the (three-flavor) chiral limit.[18] It measures the strength of the matrix element of the axial-vector-current operator between a one-Goldstone-boson state and the vacuum (see Eq. 3.29). Since the interaction of the W boson with the quarks is of $V - A$ type and the vector-current operator does not contribute to the matrix element between a single pion and the vacuum, pion decay is completely determined by the axial-vector current. The degeneracy of a single coupling constant F_0 is removed at next-to-leading order, $\mathcal{O}(q^4)$ [53], once SU(3) symmetry breaking is taken into account. The empirical numbers for F_π and F_K are 92.4 MeV and 113 MeV, respectively [75].

Exercise 3.13 The differential decay rate for $\pi^+(p_\pi) \rightarrow v_\mu(p_v) + \mu^+(p_\mu)$ in the pion rest frame is given by

$$d\Gamma = \frac{1}{2M_\pi} \overline{|\mathcal{M}|^2} \frac{d^3 p_v}{2E_v(2\pi)^3} \frac{d^3 p_\mu}{2E_\mu(2\pi)^3} (2\pi)^4 \delta^4(p_\pi - p_v - p_\mu).$$

Here, we assume the neutrino to be massless and make use of the normalization $u^\dagger u = 2E = v^\dagger v$. The invariant amplitude is given by Eq. 3.87. Neutrinos in the Standard Model are left-handed and their spinors therefore satisfy

$$\frac{\mathbb{1} - \gamma_5}{2} u_{v_\mu}(p_v) = u_{v_\mu}(p_v),$$

$$\frac{\mathbb{1} + \gamma_5}{2} u_{v_\mu}(p_v) = 0.$$

(a) Make use of the Dirac equation

$$\bar{u}_{v_\mu}(p_v) \not{p}_v = 0,$$

$$\not{p}_\mu v_{\mu^+}(p_\mu, s_\mu) = -m_\mu v_{\mu^+}(p_\mu, s_\mu),$$

and show

$$\bar{u}_{v_\mu}(p_v)(p_v + p_\mu)_\rho \gamma^\rho (\mathbb{1} - \gamma_5) v_{\mu^+}(p_\mu, s_\mu) = -m_\mu \bar{u}_{v_\mu}(p_v)(\mathbb{1} + \gamma_5) v_{\mu^+}(p_\mu, s_\mu).$$

Hint: $\{\gamma^\rho, \gamma_5\} = 0$.

---

[18] Of course, in the chiral limit, the pion is massless and, in such a world, the massive leptons would decay into Goldstone bosons, e.g., $e^- \rightarrow \pi^- v_e$. However, at $\mathcal{O}(q^2)$, the symmetry-breaking term of Eq. 3.55 gives rise to Goldstone-boson masses, whereas the decay constant is not modified at $\mathcal{O}(q^2)$.

(b) Verify, using trace techniques,

$$\left[\bar{u}_{v_{\mu}}(p_v)(p_v + p_{\mu})_{\rho}\gamma^{\rho}(1 - \gamma_5)v_{\mu^+}(p_{\mu}, s_{\mu})\right]$$

$$\times \left[\bar{u}_{v_{\mu}}(p_v)(p_v + p_{\mu})_{\sigma}\gamma^{\sigma}(1 - \gamma_5)v_{\mu^+}(p_{\mu}, s_{\mu})\right]^*$$

$$= m_{\mu}^2 \bar{u}_{v_{\mu}}(p_v)(1 + \gamma_5)v_{\mu^+}(p_{\mu}, s_{\mu})\bar{v}_{\mu^+}(p_{\mu}, s_{\mu})(1 - \gamma_5)u_{v_{\mu}}(p_v)$$

$$= m_{\mu}^2 \text{Tr}[u_{v_{\mu}}(p_v)\bar{u}_{v_{\mu}}(p_v)(1 + \gamma_5)v_{\mu^+}(p_{\mu}, s_{\mu})\bar{v}_{\mu^+}(p_{\mu}, s_{\mu})(1 - \gamma_5)]$$

$$= \ldots$$

$$= 4m_{\mu}^2 M_{\pi}^2 \left[\frac{1}{2}\left(1 - \frac{m_{\mu}^2}{M_{\pi}^2}\right) - \frac{m_{\mu}p_v \cdot s_{\mu}}{M_{\pi}^2}\right].$$

Hints:

$$(1 - \gamma_5)u_{v_{\mu}}(p_v)\bar{u}_{v_{\mu}}(p_v)(1 + \gamma_5) = (1 - \gamma_5)\not{p}_v(1 + \gamma_5),$$

$$v_{\mu^+}(p_{\mu}, s_{\mu})\bar{v}_{\mu^+}(p_{\mu}, s_{\mu}) = (\not{p}_{\mu} - m_{\mu})\frac{1 + \gamma_5 \not{s}_{\mu}}{2},$$

Tr(odd # of gamma matrices) = 0,

$$\gamma_5 = \text{product of 4 gamma matrices},$$

$$\gamma_5^2 = 1,$$

$$\text{Tr}(\not{a}\,\not{b}) = 4a \cdot b,$$

$$\text{Tr}(\gamma_5\,\not{a}\,\not{b}) = 0.$$

(c) Sum over the spin projections of the muon and integrate with respect to the (unobserved) neutrino

$$d\Gamma = \frac{1}{8\pi^2}G_F^2 V_{ud}^2 F_0^2 m_{\mu}^2 M_{\pi}\left(1 - \frac{m_{\mu}^2}{M_{\pi}^2}\right)\int \frac{d^3p_{\mu}}{E_{\mu}E_v}\delta(M_{\pi} - E_{\mu} - E_v).$$

Make use of

$$d^3p_{\mu} = p_{\mu}^2 dp_{\mu}d\Omega_{\mu}$$

and note that the argument of the delta function implicitly depends on $p_{\mu} = |\vec{p}_{\mu}|$. Moreover,

$$E_v + E_{\mu} = M_{\pi},$$
$$E_v = |\vec{p}_v| = |\vec{p}_{\mu}|.$$

The final result reads

$$\Gamma = \frac{1}{\tau} = \underbrace{G_F^2 V_{ud}^2 F_0^2 4 m_\mu^2 M_\pi^2 \left(1 - \frac{m_\mu^2}{M_\pi^2}\right) \frac{1}{16\pi M_\pi} \left(1 - \frac{m_\mu^2}{M_\pi^2}\right)}_{= |\mathcal{M}|^2}$$

$$= \frac{1}{4\pi} G_F^2 V_{ud}^2 F_0^2 m_\mu^2 M_\pi \left(1 - \frac{m_\mu^2}{M_\pi^2}\right)^2. \tag{3.89}$$

3.4.5 Application at Lowest Order: Pion-Pion Scattering

Now that the LEC F_0 has been identified with the pion-decay constant in the chiral limit, we will show how the lowest-order Lagrangian predicts the prototype of a Goldstone-boson reaction, namely, $\pi\pi$ scattering.

Exercise 3.14 Consider the Lagrangian $\mathcal{L}_2$ in the $SU(2)_L \times SU(2)_R$ sector with $r_\mu = l_\mu = 0$,

$$\mathcal{L}_2 = \frac{F^2}{4} \text{Tr}(\partial_\mu U \partial^\mu U^\dagger) + \frac{F^2}{4} \text{Tr}(\chi U^\dagger + U \chi^\dagger),$$

where

$$\chi = 2B\mathcal{M} = 2B \begin{pmatrix} \hat{m} & 0 \\ 0 & \hat{m} \end{pmatrix},$$

$$U = \exp\left(i\frac{\phi}{F}\right), \quad \phi = \sum_{i=1}^{3} \phi_i \tau_i \equiv \begin{pmatrix} \pi^0 & \sqrt{2}\pi^+ \\ \sqrt{2}\pi^- & -\pi^0 \end{pmatrix}.$$

In the $SU(2)_L \times SU(2)_R$ sector it is common to express quantities in the chiral limit without subscript 0, e.g., F and B. By this one means the $SU(2)_L \times SU(2)_R$ chiral limit, i.e., $m_u = m_d = 0$ but m_s at its physical value. In the $SU(3)_L \times SU(3)_R$ sector the quantities F_0 and B_0 denote the chiral limit for all three light quarks: $m_u = m_d = m_s = 0$.

(a) Using the substitution $U \leftrightarrow U^\dagger$, show that $\mathcal{L}_2$ contains only even powers of ϕ,

$$\mathcal{L}_2 = \mathcal{L}_2^{2\phi} + \mathcal{L}_2^{4\phi} + \cdots.$$

(b) Since $\mathcal{L}_2$ does not produce a three-Goldstone-boson vertex, the scattering of two Goldstone bosons is described by a four-Goldstone-boson contact interaction. Verify

$$\mathcal{L}_2^{4\phi} = \frac{1}{48F^2} \left[\text{Tr}([\phi, \partial_\mu \phi][\phi, \partial^\mu \phi]) + 2B\text{Tr}(\mathcal{M}\phi^4)\right]$$

Fig. 3.6 Lowest-order
Feynman diagram for $\pi\pi$
scattering. The vertex is
derived from $\mathscr{L}_2$, denoted by
2 in the interaction blob

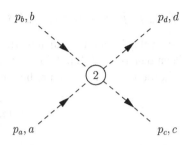

by using the expansion

$$U = 1 + i\frac{\phi}{F} - \frac{1}{2}\frac{\phi^2}{F^2} - \frac{i}{6}\frac{\phi^3}{F^3} + \frac{1}{24}\frac{\phi^4}{F^4} + \cdots.$$

Remark Substituting $F \to F_0$, $B \to B_0$, and the relevant expressions for ϕ and
the quark-mass matrix $\mathscr{M}$, the corresponding formula for $SU(3)_L \times SU(3)_R$
looks identical.

(c) Inserting $\phi = \phi_i \tau_i$ and working out the traces, show that the interaction
Lagrangian can be written as

$$\mathscr{L}_2^{4\phi} = \frac{1}{6F^2}(\phi_i \partial^\mu \phi_i \partial_\mu \phi_j \phi_j - \phi_i \phi_i \partial_\mu \phi_j \partial^\mu \phi_j) + \frac{M^2}{24F^2}\phi_i \phi_i \phi_j \phi_j,$$

where $M^2 = 2B\hat{m}$.

(d) From $\mathscr{L}_2^{4\phi}$, derive the Feynman rule for incoming pions with Cartesian isospin
indices a and b, and outgoing pions with isospin indices c, d (see Fig. 3.6):

$$\mathscr{M} = i\left[\delta_{ab}\delta_{cd}\frac{s - M^2}{F^2} + \delta_{ac}\delta_{bd}\frac{t - M^2}{F^2} + \delta_{ad}\delta_{bc}\frac{u - M^2}{F^2}\right]$$
$$- \frac{i}{3F^2}(\delta_{ab}\delta_{cd} + \delta_{ac}\delta_{bd} + \delta_{ad}\delta_{bc})(\Lambda_a + \Lambda_b + \Lambda_c + \Lambda_d), \quad (3.90)$$

where $\Lambda_k = p_k^2 - M^2$ and s, t, and u are the usual Mandelstam variables,

$$s = (p_a + p_b)^2, \quad t = (p_a - p_c)^2, \quad u = (p_a - p_d)^2.$$

(e) Using four-momentum conservation, show that the Mandelstam variables
satisfy the relation

$$s + t + u = p_a^2 + p_b^2 + p_c^2 + p_d^2.$$

The T-matrix element $(\mathscr{M} = iT)$ of the scattering process $\pi_a(p_a) + \pi_b(p_b) \to \pi_c(p_c) + \pi_d(p_d)$ can be parameterized as

$$T_{ab;cd}(p_a, p_b; p_c, p_d) = \delta_{ab}\delta_{cd}A(s, t, u) + \delta_{ac}\delta_{bd}A(t, s, u) + \delta_{ad}\delta_{bc}A(u, t, s), \quad (3.91)$$

where the function A satisfies $A(s, t, u) = A(s, u, t)$ [97]. Since the last line of the Feynman rule of Eq. 3.90 vanishes for external lines satisfying on-mass-shell conditions, at $\mathcal{O}(q^2)$ the prediction for the function A is given by

$$A(s, t, u) = \frac{s - M_\pi^2}{F_\pi^2}. \tag{3.92}$$

In Eq. 3.92 we substituted F_π for F and M_π for M, because the difference is of $\mathcal{O}(q^4)$ in T. Equation 3.92 illustrates an important general property of Goldstone-boson interactions. If we consider the (theoretical) limit $M_\pi^2, s, t, u \to 0$, the T matrix vanishes, $T \to 0$. In other words, the strength of Goldstone-boson interactions vanishes in the zero-energy and zero-mass limit.

Usually, $\pi\pi$ scattering is discussed in terms of its isospin decomposition. Since the pions form an isospin triplet, the two isovectors of both the initial and final states may be coupled to $I = 0, 1, 2$. For $m_u = m_d = \hat{m}$ the strong interactions are invariant under isospin transformations, implying that scattering-matrix elements can be decomposed as

$$\langle I', I_3' | T | I, I_3 \rangle = T^I \delta_{II'} \delta_{I_3 I_3'}. \tag{3.93}$$

For the case of $\pi\pi$ scattering the three isospin amplitudes are given in terms of the invariant amplitude A of Eq. 3.91 by [52]

$$\begin{aligned} T^{I=0} &= 3A(s, t, u) + A(t, u, s) + A(u, s, t), \\ T^{I=1} &= A(t, u, s) - A(u, s, t), \\ T^{I=2} &= A(t, u, s) + A(u, s, t). \end{aligned} \tag{3.94}$$

For example, the physical $\pi^+\pi^+$ scattering process is described by $T^{I=2}$. Other physical processes are obtained using the appropriate Clebsch-Gordan coefficients.

Evaluating the T matrices at threshold, one obtains the s-wave $\pi\pi$-scattering lengths

$$T^{I=0}\big|_{\text{thr}} = 32\pi a_0^0, \quad T^{I=2}\big|_{\text{thr}} = 32\pi a_0^2, \tag{3.95}$$

where the subscript 0 refers to s-wave scattering and the superscript to the isospin. ($T^{I=1}\big|_{\text{thr}}$ vanishes because of Bose symmetry.) The convention in ChPT differs from the usual definition of a scattering length in the effective-range expansion by a factor $(-M_\pi)$ [83]. The current-algebra prediction of Ref. [97] is identical with the lowest-order result obtained from Eq. 3.92,

$$a_0^0 = \frac{7M_\pi^2}{32\pi F_\pi^2} = 0.159, \quad a_0^2 = -\frac{M_\pi^2}{16\pi F_\pi^2} = -0.0454, \tag{3.96}$$

where we made use of the numerical values $F_\pi = 92.4\,\text{MeV}$ and $M_\pi = M_{\pi^+} = 139.57\,\text{MeV}$.

Exercise 3.15 Verify Eq. 3.96.
Hint: Make use of $s_{\text{thr}} = 4M_\pi^2$ and $s + t + u = 4M_\pi^2$.

Equations 3.96 represent an absolute prediction of chiral symmetry. Once F_π is known (from pion decay), the scattering lengths are predicted. We will come back to $\pi\pi$ scattering in Sect. 3.5.4 when we also discuss corrections of higher order [20, 21, 52].

Exercise 3.16 Sometimes it is more convenient to use a different parameterization of U which is very popular in two-flavor calculations:[19]

$$U(x) = \frac{1}{F}[\sigma(x)\mathbb{1} + i\vec{\pi}(x) \cdot \vec{\tau}], \quad \sigma(x) = \sqrt{F^2 - \vec{\pi}^2(x)}.$$

The fields of the two parameterizations are nonlinearly related by a field transformation,

$$\frac{\vec{\pi}}{F} = \hat{\phi}\sin\left(\frac{|\vec{\phi}|}{F}\right) = \frac{\vec{\phi}}{F}\left(1 - \frac{1}{6}\frac{\vec{\phi}^2}{F^2} + \cdots\right). \tag{3.97}$$

Repeat the above steps with the new parameterization. Because of the equivalence theorem of field theory [34, 42, 65], the results for observables (such as, e.g., S-matrix elements) do not depend on the parameterization. On the other hand, intermediate building blocks such as Feynman rules with off-mass-shell momenta depend on the parameterization chosen.

Exercise 3.17 The three-flavor calculation proceeds analogously to Exercise 3.14. Using the properties of the Gell-Mann matrices and the results of Exercise 1.4, show that in the isospin-symmetric case

$$\mathscr{L}_2^{4\phi} = -\frac{1}{6F_0^2}\phi_a\partial_\mu\phi_b\phi_c\partial^\mu\phi_d f_{abe}f_{cde} + \frac{(2\hat{m} + m_s)B_0}{36F_0^2}\phi_a\phi_a\phi_b\phi_b$$

$$+ \frac{(\hat{m} - m_s)B_0}{12\sqrt{3}F_0^2}\left(\frac{2}{3}\phi_8\phi_a\phi_b\phi_c d_{abc} + \phi_a\phi_a\phi_b\phi_c d_{bc8}\right).$$

Hint:

$$d_{abe}f_{ecd} + d_{bce}f_{ead} + d_{cae}f_{ebd} = 0,$$

$$d_{abe}d_{cde} = \frac{1}{3}(\delta_{ac}\delta_{bd} + \delta_{ad}\delta_{bc} - \delta_{ab}\delta_{cd} + f_{ace}f_{bde} + f_{ade}f_{bce}).$$

---

[19] We will refer to this parameterization as the square-root parameterization because of the square root multiplying the unit matrix.

3.4.6 Application at Lowest Order: Compton Scattering

Exercise 3.18 We will investigate the reaction $\gamma(q) + \pi^+(p) \to \gamma(q') + \pi^+(p')$ at lowest order in the momentum expansion $[\mathcal{O}(q^2)]$.

(a) Consider the first term of $\mathcal{L}_2$ of Eq. 3.77 and substitute

$$r_\mu = l_\mu = -e\mathcal{A}_\mu Q, \quad Q = \begin{pmatrix} \frac{2}{3} & 0 & 0 \\ 0 & -\frac{1}{3} & 0 \\ 0 & 0 & -\frac{1}{3} \end{pmatrix}, \quad e > 0, \quad \frac{e^2}{4\pi} \approx \frac{1}{137},$$

where $\mathcal{A}_\mu$ is a Hermitian (external) electromagnetic four-vector potential (see Eq. 1.164). Show that

$$D_\mu U = \partial_\mu U + ie\mathcal{A}_\mu[Q, U],$$

$$(D^\mu U)^\dagger = \partial^\mu U^\dagger + ie\mathcal{A}^\mu[Q, U^\dagger].$$

Using the substitution $U \leftrightarrow U^\dagger$, show that the resulting Lagrangian consists of terms involving only even numbers of Goldstone-boson fields.

(b) Insert the result of (a) into $\mathcal{L}_2$ and verify

$$\frac{F_0^2}{4}\text{Tr}[D_\mu U(D^\mu U)^\dagger] = \frac{F_0^2}{4}\text{Tr}[\partial_\mu U\partial^\mu U^\dagger] - ie\mathcal{A}_\mu \frac{F_0^2}{2}\text{Tr}(Q[\partial^\mu U, U^\dagger])$$

$$- e^2\mathcal{A}_\mu\mathcal{A}^\mu\frac{F_0^2}{4}\text{Tr}([Q, U][Q, U^\dagger]).$$

Hint: $U\partial^\mu U^\dagger = -\partial^\mu UU^\dagger$ and $\partial^\mu U^\dagger U = -U^\dagger\partial^\mu U$.
The second term describes interactions with a single photon and the third term with two photons.

(c) Using $U = \exp(i\phi/F_0) = \mathbb{1} + i\phi/F_0 - \phi^2/(2F_0^2) + \cdots$, identify those interaction terms which contain exactly two Goldstone bosons:

$$\mathcal{L}_2^{A-2\phi} = -e\mathcal{A}_\mu\frac{i}{2}\text{Tr}(Q[\partial^\mu\phi, \phi]),$$

$$\mathcal{L}_2^{2A-2\phi} = -\frac{1}{4}e^2\mathcal{A}_\mu\mathcal{A}^\mu\text{Tr}([Q, \phi][Q, \phi]).$$

(d) Insert ϕ expressed in terms of physical fields (see Eq. 3.37). Verify the intermediate steps

$$([\partial^\mu \phi, \phi])_{11} = 2(\partial^\mu \pi^+ \pi^- - \pi^+ \partial^\mu \pi^- + \partial^\mu K^+ K^- - K^+ \partial^\mu K^-),$$
$$([\partial^\mu \phi, \phi])_{22} = 2(\partial^\mu \pi^- \pi^+ - \pi^- \partial^\mu \pi^+ + \partial^\mu K^0 \bar{K}^0 - K^0 \partial^\mu \bar{K}^0),$$
$$([\partial^\mu \phi, \phi])_{33} = 2(\partial^\mu K^- K^+ - K^- \partial^\mu K^+ + \partial^\mu \bar{K}^0 K^0 - \bar{K}^0 \partial^\mu K^0),$$

$$[Q, \phi] = \sqrt{2} \begin{pmatrix} 0 & \pi^+ & K^+ \\ -\pi^- & 0 & 0 \\ -K^- & 0 & 0 \end{pmatrix},$$

$$[Q, \phi][Q, \phi] = -2 \begin{pmatrix} \pi^+ \pi^- + K^+ K^- & 0 & 0 \\ 0 & \pi^- \pi^+ & \pi^- K^+ \\ 0 & K^- \pi^+ & K^- K^+ \end{pmatrix}.$$

Now show

$$\mathcal{L}_2^{A-2\phi} = -\mathcal{A}_\mu ie(\partial^\mu \pi^+ \pi^- - \pi^+ \partial^\mu \pi^- + \partial^\mu K^+ K^- - K^+ \partial^\mu K^-),$$
$$\mathcal{L}_2^{2A-2\phi} = e^2 \mathcal{A}_\mu \mathcal{A}^\mu (\pi^+ \pi^- + K^+ K^-).$$

(e) The corresponding Feynman rules read

$$\mathcal{L}_2^{A-2\phi} \Rightarrow \text{vertex for } \gamma(q, \varepsilon) + \pi^\pm(p) \to \pi^\pm(p') : \mp ie\varepsilon \cdot (p + p'),$$
$$\mathcal{L}_2^{2A-2\phi} \Rightarrow \text{vertex for } \gamma(q, \varepsilon) + \pi^\pm(p) \to \gamma(q', \varepsilon') + \pi^\pm(p') : 2ie^2 \varepsilon'^* \cdot \varepsilon,$$

and analogously for charged kaons. An internal line of momentum p is described by the propagator $i/(p^2 - M^2 + i0^+)$. Determine the Compton scattering amplitude for $\gamma(q, \varepsilon) + \pi^+(p) \to \gamma(q', \varepsilon') + \pi^+(p')$:

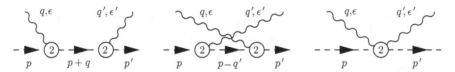

What is the scattering amplitude for $\gamma(q, \varepsilon) + \pi^-(p) \to \gamma(q', \varepsilon') + \pi^-(p')$?
(f) Verify gauge invariance in terms of the substitution $\varepsilon \to q$.
(g) Verify the invariance of the matrix element under the substitution $(q, \varepsilon) \leftrightarrow (-q', \varepsilon'^*)$ (photon crossing).

A discussion of the scattering amplitude beyond leading order may be found in Refs. [16, 93].

3.4.7 Dimensional Regularization

In the 1960s, when phenomenological Lagrangians were developed as an alternative to current-algebra techniques, it was the common understanding that such Lagrangians should only be used at tree level [33, 42, 88, 98, 99]. For example, in Ref. [88] Schwinger made the point that "it is not meaningful to question the use

of the coupling terms "in lowest order". That is the nature of a numerical effective
Lagrange function, which gives a direct description of the phenomena." However,
with the pioneering work of Weinberg [100] it became clear that one can even
calculate quantum corrections to phenomenological Lagrangians. To quote from
Ref. [101]: "... the cancellation of ultraviolet divergences does not really depend
on renormalizability; as long as we include every one of the infinite number of
interactions allowed by symmetries, the so-called non-renormalizable theories are
actually just as renormalizable as renormalizable theories." If we use the
Lagrangian of Eq. 3.77 beyond tree level, we will encounter ultraviolet divergences from loop integrals. For the regularization of the loop diagrams we will
make use of dimensional regularization [69, 91, 92, 95], because it preserves
algebraic relations among the Green functions (Ward identities). As discussed in
Sect. 3.5.1, the infinities will be absorbed in a renormalization of the coupling
constants of the most general Lagrangian.

For the sake of completeness we provide a simple illustration of the method of
dimensional regularization. Let us consider the integral

$$I = \int \frac{d^4k}{(2\pi)^4} \frac{i}{k^2 - M^2 + i0^+}, \tag{3.98}$$

which appears in the generic diagram of Fig. 3.7. We introduce

$$a \equiv \sqrt{\vec{k}^2 + M^2} > 0,$$

so that

$$k^2 - M^2 + i0^+ = k_0^2 - \vec{k}^2 - M^2 + i0^+ = k_0^2 - a^2 + i0^+ = k_0^2 - (a - i0^+)^2$$
$$= [k_0 + (a - i0^+)][k_0 - (a - i0^+)],$$

and define

$$f(k_0) = \frac{1}{[k_0 + (a - i0^+)][k_0 - (a - i0^+)]}.$$

In order to determine $\int_{-\infty}^{\infty} dk_0 f(k_0)$ as part of the calculation of I, we consider f in
the complex k_0 plane and make use of Cauchy's theorem

$$\oint_C dz f(z) = 0 \tag{3.99}$$

for functions which are differentiable in every point inside the closed contour C.
We choose the path as shown in Fig. 3.8,

$$0 = \sum_{i=1}^{4} \int_{\gamma_i} dz f(z),$$

and make use of

Fig. 3.7 Generic one-loop diagram: The black box denotes some unspecified vertex structure which is irrelevant for the discussion

$$\int_{\gamma} dz f(z) = \int_{a}^{b} dt f[\gamma(t)]\gamma'(t)$$

to obtain for the individual integrals:

1. $\gamma_1(t) = t$, $\gamma_1'(t) = 1$, $a = -\infty$, $b = \infty$:

$$\int_{\gamma_1} dz f(z) = \int_{-\infty}^{\infty} dt f(t),$$

2. $\gamma_2(t) = Re^{it}$, $\gamma_2'(t) = iRe^{it}$, $a = 0$, $b = \frac{\pi}{2}$:

$$\int_{\gamma_2} dz f(z) = \lim_{R\to\infty} \int_{0}^{\frac{\pi}{2}} dt f(Re^{it})iRe^{it} = 0, \text{ because } \lim_{R\to\infty} \underbrace{Rf(Re^{it})}_{\sim \frac{1}{R}} = 0,$$

3. $\gamma_3(t) = it$, $\gamma_3'(t) = i$, $a = \infty$, $b = -\infty$:

$$\int_{\gamma_3} dz f(z) = i \int_{\infty}^{-\infty} dt f(it),$$

4. $\gamma_4(t) = Re^{it}$, $\gamma_4'(t) = iRe^{it}$, $a = \frac{3}{2}\pi$, $b = \pi$:

$$\int_{\gamma_4} dz f(z) = 0 \text{ analogous to } \gamma_2.$$

The quarter circles at infinity do not contribute, because the function $f(z)$ vanishes sufficiently fast as $|z| \to \infty$. In combination with Eq. 3.99 we obtain the so-called Wick rotation

$$\int_{-\infty}^{\infty} dt f(t) = -i \int_{\infty}^{-\infty} dt f(it) = i \int_{-\infty}^{\infty} dt f(it). \tag{3.100}$$

As an intermediate result the integral of Eq. 3.98 reads

$$I = \frac{1}{(2\pi)^4} i \int_{-\infty}^{\infty} dk_0 \int d^3k \frac{i}{(ik_0)^2 - \vec{k}^2 - M^2 + i0^+} = \int \frac{d^4l}{(2\pi)^4} \frac{1}{l^2 + M^2 - i0^+},$$

Fig. 3.8 Path of integration
in the complex k_0 plane

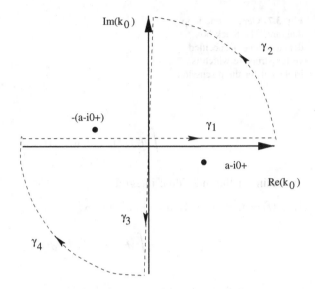

where $l^2 = l_1^2 + l_2^2 + l_3^2 + l_4^2$ denotes a Euclidian scalar product. In this *special* case, the integrand does not have a pole and we can thus omit the $-i0^+$ which gave the positions of the poles in the original integral consistent with the boundary conditions. Performing the angular integration in four dimensions (see Exercise 3.20) and introducing a cutoff Λ for the radial integration, the integral I diverges quadratically for large values of l (ultraviolet divergence):

$$I(\Lambda) = \frac{1}{8\pi^2} \int_0^\Lambda dl \, \frac{l^3}{l^2 + M^2} = \frac{\Lambda^2}{(4\pi)^2} + \frac{M^2}{(4\pi)^2} \ln\left(\frac{M^2}{\Lambda^2 + M^2}\right)$$

$$= \frac{M^2}{(4\pi)^2} \left[\frac{1}{x^2} + \ln(x^2) - \ln(1 + x^2)\right], \qquad (3.101)$$

where $x^2 = M^2/\Lambda^2 \to 0$ as $\Lambda \to \infty$. The degree of divergence can be estimated by simply counting powers of momenta. If the integral behaves asymptotically as $\int d^4l/l^2$, $\int d^4l/l^3$, $\int d^4l/l^4$, the integral is said to diverge quadratically, linearly, and logarithmically, respectively.

Various methods have been devised to regularize divergent integrals. Unlike in Eq. 3.101 where we used a cutoff Λ, we will make use of *dimensional* regularization. Note that the degree of divergence of the integral depends on the number of dimensions. The method of dimensional regularization relies on the fact that the ultraviolet degree of divergence decreases with a decreasing number of dimensions. Here we will make use of dimensional regularization because it also preserves algebraic relations among Green functions (Ward identities) if the underlying symmetries do not depend on the number of space-time dimensions.

In dimensional regularization, we generalize the integral from four to n dimensions and introduce polar coordinates

$$l_1 = l\cos(\theta_1),$$
$$l_2 = l\sin(\theta_1)\cos(\theta_2),$$
$$l_3 = l\sin(\theta_1)\sin(\theta_2)\cos(\theta_3),$$
$$\vdots \qquad\qquad\qquad (3.102)$$
$$l_{n-1} = l\sin(\theta_1)\sin(\theta_2)\dots\cos(\theta_{n-1}),$$
$$l_n = l\sin(\theta_1)\sin(\theta_2)\dots\sin(\theta_{n-1}),$$

where $0 \le l$, $\theta_i \in [0,\pi]$ $(i = 1,\dots,n-2)$, and $\theta_{n-1} \in [0, 2\pi]$. A general integral is then symbolically of the form

$$\int d^n l\dots = \int\limits_0^\infty dl\, l^{n-1} \int\limits_0^{2\pi} d\theta_{n-1} \int\limits_0^\pi d\theta_{n-2}\sin(\theta_{n-2})\dots \int\limits_0^\pi d\theta_1 \sin^{n-2}(\theta_1)\dots.$$
$$(3.103)$$

If the integrand does not depend on the angles, the angular integration can be carried out explicitly. To that end one makes use of

$$\int\limits_0^\pi d\theta\,\sin^m(\theta) = \frac{\sqrt{\pi}\,\Gamma\left(\frac{m+1}{2}\right)}{\Gamma\left(\frac{m+2}{2}\right)},$$

which can be shown by induction (see Exercise 3.21). We then obtain for the angular integration

$$\int\limits_0^{2\pi} d\theta_{n-1}\dots\int\limits_0^\pi d\theta_1 \sin^{n-2}(\theta_1) = 2\pi\underbrace{\frac{\sqrt{\pi}\,\Gamma(1)}{\Gamma\left(\frac{3}{2}\right)}\frac{\sqrt{\pi}\,\Gamma\left(\frac{3}{2}\right)}{\Gamma(2)}\dots\frac{\sqrt{\pi}\,\Gamma\left(\frac{n-1}{2}\right)}{\Gamma\left(\frac{n}{2}\right)}}_{(n-2)\text{ factors}}$$
$$(3.104)$$
$$= 2\pi\frac{\pi^{\frac{n-2}{2}}}{\Gamma\left(\frac{n}{2}\right)} = 2\frac{\pi^{\frac{n}{2}}}{\Gamma\left(\frac{n}{2}\right)}.$$

We define the integral for n dimensions (n integer) as

$$I_n(M^2,\mu^2) = \mu^{4-n}\int\frac{d^n k}{(2\pi)^n}\frac{i}{k^2 - M^2 + i0^+},$$
$$(3.105)$$

where the scale μ ('t Hooft parameter, renormalization scale) has been introduced so that the integral has the same dimension for arbitrary n. (The integral of Eq. 3.105 is convergent only for $n = 1$.) After the Wick rotation of Eq. 3.100 and the angular integration of Eq. 3.104 the integral formally reads

$$I_n(M^2,\mu^2) = \mu^{4-n}2\frac{\pi^{\frac{n}{2}}}{\Gamma\left(\frac{n}{2}\right)}\frac{1}{(2\pi)^n}\int\limits_0^\infty dl\,\frac{l^{n-1}}{l^2 + M^2}.$$

For later use, we investigate the (more general) integral

$$\int_0^\infty dl \, \frac{l^{n-1}}{(l^2+M^2)^\alpha} = \frac{1}{(M^2)^\alpha} \int_0^\infty dl \, \frac{l^{n-1}}{\left(\frac{l^2}{M^2}+1\right)^\alpha} = \frac{1}{2}(M^2)^{\frac{n}{2}-\alpha} \int_0^\infty dt \, \frac{t^{\frac{n}{2}-1}}{(t+1)^\alpha}, \quad (3.106)$$

where we substituted $t = l^2/M^2$. We then make use of the Beta function

$$B(x,y) = \int_0^\infty dt \, \frac{t^{x-1}}{(1+t)^{x+y}} = \frac{\Gamma(x)\Gamma(y)}{\Gamma(x+y)}, \quad (3.107)$$

where the *integral* converges for $x > 0$, $y > 0$ and diverges if $x \leq 0$ or $y \leq 0$. For nonpositive values of x or y we analytically continue in terms of the Gamma function to define the Beta function and thus the integral of Eq. 3.106.[20] Setting $x = n/2$, $x + y = \alpha$, and $y = \alpha - n/2$ our (intermediate) integral reads

$$\int_0^\infty dl \, \frac{l^{n-1}}{(l^2+M^2)^\alpha} = \frac{1}{2}(M^2)^{\frac{n}{2}-\alpha} \frac{\Gamma\left(\frac{n}{2}\right)\Gamma\left(\alpha-\frac{n}{2}\right)}{\Gamma(\alpha)}, \quad (3.108)$$

which, for $\alpha = 1$, yields for our original integral

$$I_n(M^2, \mu^2) = \mu^{4-n} \underbrace{2 \frac{\pi^{\frac{n}{2}}}{\Gamma\left(\frac{n}{2}\right)}}_{\text{angular integration}} \frac{1}{(2\pi)^n} \frac{1}{2}(M^2)^{\frac{n}{2}-1} \underbrace{\frac{\Gamma\left(\frac{n}{2}\right)\Gamma\left(1-\frac{n}{2}\right)}{\Gamma(1)}}_{=1} \quad (3.109)$$

$$= \frac{\mu^{4-n}}{(4\pi)^{\frac{n}{2}}}(M^2)^{\frac{n}{2}-1}\Gamma\left(1-\frac{n}{2}\right).$$

Since $\Gamma(z)$ is an analytic function in the complex plane except for poles of first order in $0, -1, -2, \ldots$, and $a^z = \exp[\ln(a)z]$, $a \in \mathbb{R}^+$, is an analytic function in $\mathbb{C}$, the right-hand side of Eq. 3.109 can be thought of as a function of a *complex* variable n which is analytic in $\mathbb{C}$ except for poles of first order for $n = 2, 4, 6, \ldots$. Making use of

$$\mu^{4-n} = (\mu^2)^{2-\frac{n}{2}}, \quad (M^2)^{\frac{n}{2}-1} = M^2(M^2)^{\frac{n}{2}-2}, \quad (4\pi)^{\frac{n}{2}} = (4\pi)^2(4\pi)^{\frac{n}{2}-2},$$

we define (for complex n)

$$I(M^2, \mu^2, n) = \frac{M^2}{(4\pi)^2}\left(\frac{4\pi\mu^2}{M^2}\right)^{2-\frac{n}{2}}\Gamma\left(1-\frac{n}{2}\right).$$

For $n \rightarrow 4$ the Gamma function has a pole and we want to investigate how this pole is approached. The property $\Gamma(z+1) = z\Gamma(z)$ allows one to rewrite

---

[20] Recall that $\Gamma(z)$ is single-valued and analytic over the entire complex plane, save for the points $z = -n$, $n = 0, 1, 2, \ldots$, where it possesses simple poles with residue $(-1)^n/n!$.

$$\Gamma\left(1 - \frac{n}{2}\right) = \frac{\Gamma\left(1 - \frac{n}{2} + 1\right)}{1 - \frac{n}{2}} = \frac{\Gamma\left(2 - \frac{n}{2} + 1\right)}{\left(1 - \frac{n}{2}\right)\left(2 - \frac{n}{2}\right)} = \frac{\Gamma\left(1 + \frac{\varepsilon}{2}\right)}{(-1)\left(1 - \frac{\varepsilon}{2}\right)\frac{\varepsilon}{2}},$$

where we defined $\varepsilon \equiv 4 - n$.[21] Making use of $a^x = \exp[\ln(a)x] = 1 + \ln(a)x + O(x^2)$, we expand the integral for small ε,

$$
\begin{aligned}
I(M^2, \mu^2, n) &= \frac{M^2}{16\pi^2} \left[1 + \frac{\varepsilon}{2}\ln\left(\frac{4\pi\mu^2}{M^2}\right) + O(\varepsilon^2)\right] \\
&\quad \times \left(-\frac{2}{\varepsilon}\right)\left[1 + \frac{\varepsilon}{2} + O(\varepsilon^2)\right]\Big[\underbrace{\Gamma(1)}_{=1} + \frac{\varepsilon}{2}\Gamma'(1) + O(\varepsilon^2)\Big] \\
&= \frac{M^2}{16\pi^2}\left[-\frac{2}{\varepsilon} - \Gamma'(1) - 1 - \ln(4\pi) + \ln\left(\frac{M^2}{\mu^2}\right) + O(\varepsilon)\right],
\end{aligned}
$$

where $-\Gamma'(1) = \gamma_E = 0.5772\ldots$ is Euler's constant. We finally obtain

$$I(M^2, \mu^2, n) = \frac{M^2}{16\pi^2}\left[R + \ln\left(\frac{M^2}{\mu^2}\right)\right] + O(n - 4), \tag{3.110}$$

where

$$R = \frac{2}{n-4} - [\ln(4\pi) + \Gamma'(1) + 1]. \tag{3.111}$$

The comparison between Eqs. 3.110 and 3.101 illustrates the following general observations: in dimensional regularization power-law divergences are analytically continued to zero and logarithmic ultraviolet divergences of one-loop integrals show up as single poles in $\varepsilon = 4 - n$.

Using the same techniques, one can easily derive a very useful expression for the more general integral (see Exercise 3.22)

$$\int \frac{d^n k}{(2\pi)^n} \frac{(k^2)^p}{(k^2 - M^2 + i0^+)^q} = i(-)^{p-q}\frac{1}{(4\pi)^{\frac{n}{2}}}(M^2)^{p + \frac{n}{2} - q}\frac{\Gamma\left(p + \frac{n}{2}\right)\Gamma\left(q - p - \frac{n}{2}\right)}{\Gamma\left(\frac{n}{2}\right)\Gamma(q)}. \tag{3.112}$$

In the case of integrals containing more than one propagator, one can combine these to obtain integrals of the type of Eq. 3.112 with M^2 replaced by $A - i0^+$, where A is a real number. In this context it is important to consistently deal with the boundary condition $-i0^+$ [95]. To that end, one expresses a complex number z in its polar form $z = |z| \exp(i\varphi)$, where the argument φ of z is uniquely determined if, in addition, we demand $-\pi \leq \varphi < \pi$. For example, let us consider a term of the type $\ln(A - i0^+)$. For $A > 0$ one simply has $\ln(A - i0^+) = \ln(A)$. For $A < 0$ the infinitesimal imaginary part indicates that $-|A|$ is reached in the third quadrant

---

[21] Note that the convention $\varepsilon = 2 - \frac{n}{2}$ is also commonly used in the literature.

from below the real axis so that we have to use $\varphi = -\pi$. We then make use of $\ln(ab) = \ln(a) + \ln(b)$ and obtain

$$\ln(A - i0^+) = \ln(|A|) + \ln(e^{-i\pi}) = \ln(|A|) - i\pi, \quad A < 0.$$

Both cases can be summarized in a single expression

$$\ln(A - i0^+) = \ln(|A|) - i\pi\Theta(-A) \quad \text{for} \quad A \in \mathbb{R}. \tag{3.113}$$

The preceding discussion is of importance for consistently determining imaginary parts of loop integrals.

Let us conclude with the general observation that (ultraviolet) divergences of one-loop integrals in dimensional regularization always show up as single poles in $\varepsilon = 4 - n$.

The following five exercises are related to dimensional regularization.

Exercise 3.19 We consider the integral

$$I = \int \frac{d^4 k}{(2\pi)^4} \frac{i}{k^2 - M^2 + i0^+}.$$

Introduce $a \equiv \sqrt{\vec{k}^2 + M^2}$ and define

$$f(k_0) = \frac{1}{[k_0 + (a - i0^+)][k_0 - (a - i0^+)]}.$$

In order to determine $\int_{-\infty}^{\infty} dk_0 f(k_0)$ as part of the calculation of I, we consider f in the complex k_0 plane and choose the paths

$$\gamma_1(t) = t, \quad t_1 = -\infty, \quad t_2 = \infty \quad \text{and} \quad \gamma_2(t) = R e^{it}, \quad t_1 = 0, \quad t_2 = \pi.$$

(a) Using the residue theorem, determine

$$\oint_C dz f(z) = \int_{\gamma_1} dz f(z) + \lim_{R \to \infty} \int_{\gamma_2} dz f(z) = 2\pi i \text{Res}[f(z), -(a - i0^+)].$$

Verify

$$\int_{-\infty}^{\infty} dk_0 f(k_0) = \frac{-i\pi}{\sqrt{\vec{k}^2 + M^2} - i0^+}.$$

(b) Using (a), show

$$\int \frac{d^4 k}{(2\pi)^4} \frac{i}{k^2 - M^2 + i0^+} = \frac{1}{2} \int \frac{d^3 k}{(2\pi)^3} \frac{1}{\sqrt{\vec{k}^2 + M^2} - i0^+}.$$

(c) Now consider the generalization from $4 \to n$ dimensions:

$$\int \frac{d^{n-1}k}{(2\pi)^{n-1}} \frac{1}{\sqrt{\vec{k}^2 + M^2}}, \qquad \vec{k}^2 = k_1^2 + k_2^2 + \cdots + k_{n-1}^2.$$

We can omit the $-i0^+$, because the integrand no longer has a pole. Introduce polar coordinates in $n-1$ dimensions and perform the angular integration to obtain

$$\int \frac{d^{n-1}k}{(2\pi)^{n-1}} \frac{1}{\sqrt{\vec{k}^2 + M^2}} = \frac{1}{2^{n-2}} \pi^{-\frac{n-1}{2}} \frac{1}{\Gamma\left(\frac{n-1}{2}\right)} \int_0^\infty dr \, \frac{r^{n-2}}{\sqrt{r^2 + M^2}}.$$

(d) Using the substitutions $t = r/M$ and $y = t^2$, show

$$\int_0^\infty dr \, \frac{r^{n-2}}{\sqrt{r^2 + M^2}} = \frac{1}{2} M^{n-2} \underbrace{\frac{\Gamma\left(\frac{n-1}{2}\right)\Gamma\left(1 - \frac{n}{2}\right)}{\Gamma\left(\frac{1}{2}\right)}}_{= \sqrt{\pi}}.$$

Hint: Make use of the Beta function

$$B(x,y) = \int_0^\infty dt \, \frac{t^{x-1}}{(1+t)^{x+y}} = \frac{\Gamma(x)\Gamma(y)}{\Gamma(x+y)}.$$

(e) Now put the results together to obtain

$$\int \frac{d^n k}{(2\pi)^n} \frac{i}{k^2 - M^2 + i0^+} = \frac{1}{(4\pi)^{\frac{n}{2}}} M^{n-2} \Gamma\left(1 - \frac{n}{2}\right),$$

which agrees with the result of Eq. 3.109 once the factor μ^{4-n} is taken into account.

Exercise 3.20 Consider polar coordinates in four dimensions:

$$l_1 = l\cos(\theta_1),$$
$$l_2 = l\sin(\theta_1)\cos(\theta_2),$$
$$l_3 = l\sin(\theta_1)\sin(\theta_2)\cos(\theta_3),$$
$$l_4 = l\sin(\theta_1)\sin(\theta_2)\sin(\theta_3),$$

where $l = \sqrt{l_1^2 + l_2^2 + l_3^2 + l_4^2}$, $\theta_1 \in [0, \pi]$, $\theta_2 \in [0, \pi]$, and $\theta_3 \in [0, 2\pi]$. The transition from four-dimensional Cartesian coordinates to polar coordinates introduces the determinant of the Jacobi or functional matrix

$$J = \begin{pmatrix} \frac{\partial l_1}{\partial l} & \cdots & \frac{\partial l_1}{\partial \theta_3} \\ \vdots & & \vdots \\ \frac{\partial l_4}{\partial l} & \cdots & \frac{\partial l_4}{\partial \theta_3} \end{pmatrix}.$$

Show that

$$\det(J) = l^3 \sin^2(\theta_1) \sin(\theta_2),$$

and thus

$$dl_1 dl_2 dl_3 dl_4 = \underbrace{l^3 dl \sin^2(\theta_1) \sin(\theta_2) d\theta_1 d\theta_2 d\theta_3}_{= d\Omega},$$

with

$$\int d\Omega = 2\pi^2.$$

Exercise 3.21 Show by induction

$$\int\limits_0^\pi d\theta \sin^m(\theta) = \frac{\sqrt{\pi}\,\Gamma\!\left(\frac{m+1}{2}\right)}{\Gamma\!\left(\frac{m+2}{2}\right)}$$

for $m \geq 1$.

Hints: Make use of integration by parts. $\Gamma(1) = 1$, $\Gamma(1/2) = \sqrt{\pi}$, $x\Gamma(x) = \Gamma(x+1)$.

Exercise 3.22 Show that in dimensional regularization

$$\int \frac{d^n k}{(2\pi)^n} \frac{(k^2)^p}{(k^2 - M^2 + i0^+)^q} = i(-)^{p-q} \frac{1}{(4\pi)^{\frac{n}{2}}} (M^2)^{p+\frac{n}{2}-q} \frac{\Gamma\!\left(p+\frac{n}{2}\right)\Gamma\!\left(q-p-\frac{n}{2}\right)}{\Gamma\!\left(\frac{n}{2}\right)\Gamma(q)}.$$

We first assume $M^2 > 0$, $p = 0, 1, \ldots$, $q = 1, 2, \ldots$, and $p < q$. The last condition is used in the Wick rotation to guarantee that the quarter circles at infinity do not contribute to the integral.

(a) Show that the transition to the Euclidian metric produces the factor $i(-)^{p-q}$.
(b) Perform the angular integration in n dimensions. Perform the remaining radial integration using

$$\int\limits_0^\infty dl \frac{l^{n-1}}{(l^2 + M^2)^\alpha} = \frac{1}{2}(M^2)^{\frac{n}{2}-\alpha} \frac{\Gamma\!\left(\frac{n}{2}\right)\Gamma\!\left(\alpha - \frac{n}{2}\right)}{\Gamma(\alpha)}.$$

What do you have to substitute for $n - 1$ and α, respectively?

Now put the results together. The analytic continuation of the right-hand side is used to also define expressions with (integer) $q \leq p$ in dimensional regularization.

Exercise 3.23 Consider the complex function

$$f(z) = a^z = \exp(\ln(a)z) \equiv u(x,y) + iv(x,y), \quad a \in \mathbb{R}, \quad z = x + iy.$$

(a) Determine $u(x,y)$ and $v(x,y)$. Note that $u,v \in C^\infty(\mathbb{R}^2)$.
(b) Determine $\partial u/\partial x$, $\partial u/\partial y$, $\partial v/\partial x$, and $\partial v/\partial y$. Show that the Cauchy-Riemann differential equations $\partial u/\partial x = \partial v/\partial y$ and $\partial u/\partial y = -\partial v/\partial x$ are satisfied. The complex function f is thus holomorphic in $\mathbb{C}$. We made use of this fact when discussing $I(M^2, \mu^2, n)$ as a function of the complex variable n in the context of dimensional regularization.

3.4.8 The Generation of Counter Terms

Regularization, such as dimensional regularization discussed in the previous section, is a method to systematically separate divergences that appear in loop diagrams from finite contributions. We now briefly discuss renormalization, i.e., how to absorb the divergences in the parameters of the Lagrangian (see Ref. [43] for details).[22] For simplicity, we consider a toy-model Lagrangian for two massive scalar degrees of freedom,

$$\mathscr{L} = \frac{1}{2}\left(\partial_\mu \phi_B \partial^\mu \phi_B - M_B^2 \phi_B^2\right) + \frac{1}{2}\left(\partial_\mu \varphi_B \partial^\mu \varphi_B - m_B^2 \varphi_B^2\right) - \frac{\lambda_B}{4}\phi_B^2 \varphi_B^2, \quad (3.114)$$

where the subscripts B indicate bare quantities. By expressing the bare fields and parameters in terms of renormalized quantities, one generates counter terms which are responsible for the absorption of all divergences occurring in the calculation of loop diagrams. We first introduce renormalized fields ϕ and φ,

$$\phi_B = \sqrt{Z_\phi}\,\phi, \quad \varphi_B = \sqrt{Z_\varphi}\,\varphi,$$

and then rewrite the field renormalization constants $\sqrt{Z_\phi}$ and $\sqrt{Z_\varphi}$ as well as the remaining bare quantities in terms of renormalized parameters:

---

[22] More generally, renormalization is simply the process of expressing the parameters of the Lagrangian in terms of physical observables, independent of the presence of divergences [51].

$$Z_\phi = 1 + \delta Z_\phi(M, m, \lambda, v),$$

$$Z_\varphi = 1 + \delta Z_\varphi(M, m, \lambda, v),$$

$$M_B^2 = M^2(v) + \delta M^2(M, m, \lambda, v),$$

$$m_B^2 = m^2(v) + \delta m^2(M, m, \lambda, v),$$

$$\lambda_B = \lambda(v) + \delta\lambda(M, m, \lambda, v).$$

The parameter v indicates the dependence on the choice of a renormalization prescription. For example, we could require the masses M and m to be the physical masses of ϕ and φ, respectively. The freedom of choosing the renormalization condition will play a crucial role in baryonic ChPT. With these substitutions the Lagrangian takes the form

$$\mathscr{L} = \mathscr{L}_{\text{basic}} + \mathscr{L}_{\text{ct}}, \tag{3.115}$$

with the so-called basic and counter-term Lagrangians, respectively,

$$\mathscr{L}_{\text{basic}} = \frac{1}{2}\left(\partial_\mu\phi\partial^\mu\phi - M^2\phi^2\right) + \frac{1}{2}\left(\partial_\mu\varphi\partial^\mu\varphi - m^2\varphi^2\right) - \frac{\lambda}{4}\phi^2\varphi^2, \tag{3.116}$$

$$\mathscr{L}_{\text{ct}} = \frac{1}{2}\delta Z_\phi\partial_\mu\phi\partial^\mu\phi - \frac{1}{2}\delta\{M^2\}\phi^2 + \frac{1}{2}\delta Z_\varphi\partial_\mu\varphi\partial^\mu\varphi - \frac{1}{2}\delta\{m^2\}\varphi^2$$
$$- \frac{\delta\{\lambda\}}{4}\phi^2\varphi^2, \tag{3.117}$$

where we have introduced the abbreviations[23]

$$\delta\{M^2\} = \delta Z_\phi M^2 + Z_\phi\delta M^2,$$

$$\delta\{m^2\} = \delta Z_\varphi m^2 + Z_\varphi\delta m^2,$$

$$\delta\{\lambda\} = \delta\lambda Z_\phi Z_\varphi + \lambda\left(\delta Z_\phi + \delta Z_\varphi + \delta Z_\phi\delta Z_\varphi\right).$$

Expanding the counter-term Lagrangian of Eq. 3.117 in powers of the renormalized couplings generates an infinite series. By suitably adjusting the expansion coefficients, the individual terms are responsible for the subtraction of divergences appearing in loop diagrams. In the following, whenever we speak of renormalized diagrams, we refer to diagrams which have been calculated with a basic Lagrangian and to which the contributions of the counter-term Lagrangian have been added.

---

[23] Note that Ref. [43] uses a slightly different convention which corresponds to the replacement $(\delta Z_\phi M^2 + Z_\phi\delta M^2) \to \delta M^2$; analogously for δm^2.

3.4.9 Power-Counting Scheme

An essential prerequisite for the construction of effective field theories is a "theorem" of Weinberg stating that a perturbative description in terms of the most general effective Lagrangian containing all possible terms compatible with assumed symmetry principles yields the most general S-matrix consistent with the fundamental principles of quantum field theory and the assumed symmetry principles [100]. The corresponding effective Lagrangian contains an infinite number of terms with an infinite number of free parameters. Turning Weinberg's theorem into a practical tool requires two steps: one needs some scheme to organize the effective Lagrangian and a systematic method of assessing the importance of diagrams generated by the interaction terms of this Lagrangian when calculating a physical matrix element.

In the framework of mesonic chiral perturbation theory, the most general chiral Lagrangian describing the dynamics of the Goldstone bosons is organized as a string of terms with an increasing number of derivatives and quark-mass terms,

$$\mathscr{L}_{\text{eff}} = \mathscr{L}_2 + \mathscr{L}_4 + \mathscr{L}_6 + \cdots, \tag{3.118}$$

where the subscripts refer to the order in the momentum and quark-mass expansion. The subscript 2, for example, denotes either two derivatives or one quark-mass term (see Eq. 3.77). In terms of Feynman rules, derivatives generate four-momenta. A quark-mass term counts as two derivatives because of Eqs. 3.59–3.61 $(M^2 \sim m_q)$ in combination with the on-shell condition $p^2 = M^2$. We will generically count a small four-momentum—or the corresponding derivative—and a Goldstone-boson mass as of $\mathcal{O}(q)$. The chiral orders in Eq. 3.118 are all even $[\mathcal{O}(q^{2k}), k \geq 1]$, because Lorentz indices of derivatives always have to be contracted and quark-mass terms count as $\mathcal{O}(q^2)$.

Besides the knowledge of the most general Lagrangian, we need a method to assess the importance of different renormalized diagrams contributing to a given process. For that purpose we analyze a given diagram under a simultaneous rescaling of all *external* momenta, $p_i \mapsto tp_i$, and the light-quark masses, $m_q \mapsto t^2 m_q$ (corresponding to $M^2 \mapsto t^2 M^2$). As we will show below, this results in an overall rescaling of the amplitude $\mathscr{M}$ of a given diagram,

$$\mathscr{M}(tp_i, t^2 m_q) = t^D \mathscr{M}(p_i, m_q). \tag{3.119}$$

Equation 3.119 defines the chiral dimension D of the diagram. The chiral dimension is given by

$$D = nN_L - 2N_I + \sum_{k=1}^{\infty} 2kN_{2k} \tag{3.120}$$

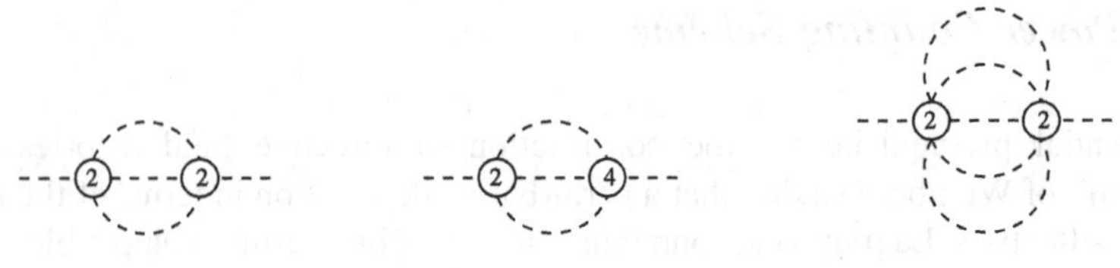

$$D = 4\cdot2 - 2\cdot3 + 2\cdot2 = 6. \qquad D = 4\cdot2 - 2\cdot3 + 2\cdot1 + 4\cdot1 = 8. \qquad D = 4\cdot4 - 2\cdot5 + 2\cdot2 = 10.$$

Fig. 3.9 Application of the power-counting formula of Eq. 3.120 in $n = 4$ dimensions

$$D = 2 + (n - 2)N_L + \sum_{k=1}^{\infty} 2(k - 1)N_{2k} \qquad (3.121)$$

$$\geq 2 \text{ in four dimensions,}$$

where n is the number of space-time dimensions, N_L the number of independent loops, N_l the number of internal Goldstone-boson lines, and N_{2k} the number of vertices from $\mathscr{L}_{2k}$.[24] Going to smaller momenta and masses corresponds to a rescaling with $0 < t < 1$. Clearly, for small enough momenta and masses contributions with increasing D become less important and diagrams with small D, such as $D = 2$ or $D = 4$, should dominate. Of course, the rescaling of Eq. 3.119 must be viewed as a mathematical tool. While external three-momenta can, to a certain extent, be made arbitrarily small, the rescaling of the quark masses is a theoretical instrument only. Note that loop diagrams are always suppressed due to the term $(n - 2)N_L$ in Eq. 3.121. It may happen, though, that the leading-order tree diagrams vanish and therefore the lowest-order contribution to a certain process is a one-loop diagram. An example is the reaction $\gamma\gamma \to \pi^0\pi^0$ [16].

Equation 3.121 establishes a relation between the momentum and loop expansions, because at each chiral order the maximum number of loops is bounded from above. In other words, we have a perturbative scheme in terms of external momenta and masses which are small compared to some scale Λ. With the aid of Eq. 3.110, we can estimate the so-called chiral-symmetry-breaking scale Λ_χ to be $\Lambda_\chi = 4\pi F_0 = O(1\,\text{GeV})$ [73]. In a loop correction every endpoint of an internal Goldstone-boson line is multiplied by a factor $1/F_0$, since the SU(N) matrix of Eq. 3.41 contains the Goldstone-boson fields in the combination ϕ/F_0. On the other hand, external momenta q or Goldstone-boson masses produce factors of q^2 or M^2 as, e.g., in Eq. 3.110. Together with the factor $1/(16\pi^2)$ of Eq. 3.110 remaining after integration in four dimensions they combine to corrections of the order of $[q/(4\pi F_0)]^2$ for each independent loop. Examples of the application of the power-counting formula are shown in Fig. 3.9.

---

[24] Note that the number of independent momenta is *not* the number of faces or closed circuits that may be drawn on the internal lines of a diagram. This may, for example, be seen using a diagram with the topology of a tetrahedron which has four faces but $N_L = 6 - (4 - 1) = 3$ (see, e.g., Chap. 6-2 of Ref. [63]).

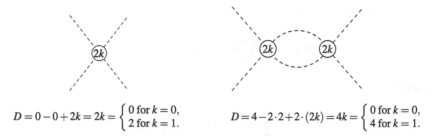

$$D = 0 - 0 + 2k = 2k = \begin{cases} 0 \text{ for } k = 0, \\ 2 \text{ for } k = 1. \end{cases} \qquad D = 4 - 2 \cdot 2 + 2 \cdot (2k) = 4k = \begin{cases} 0 \text{ for } k = 0, \\ 4 \text{ for } k = 1. \end{cases}$$

Fig. 3.10 The loop diagram is only suppressed if $k_{\min} > 0$

In order to prove the power-counting formula, we start from the Feynman rules for evaluating an S-matrix element and investigate the behavior of the individual building blocks. Internal lines are described by a propagator in n dimensions which under rescaling behaves as

$$\int \frac{d^n k}{(2\pi)^n} \frac{i}{k^2 - M^2 + i0^+} \mapsto \int \frac{d^n k}{(2\pi)^n} \frac{i}{t^2 (k^2/t^2 - M^2 + i0^+)}$$

$$\overset{k = tk'}{=} t^{n-2} \int \frac{d^n k'}{(2\pi)^n} \frac{i}{k'^2 - M^2 + i0^+}. \qquad (3.122)$$

Vertices with $2k$ derivatives or k quark-mass terms rescale as

$$\delta^n(q) q^{2k} \mapsto t^{2k-n} \delta^n(q) q^{2k},$$

since $p \mapsto tp$ if q is an external momentum, and $k = tk'$ if q is an internal momentum (see above). These are the rules to calculate $S \sim \delta^n(P) \mathcal{M}$. We need to add n to compensate for the overall momentum-conserving delta function. Applying these rules, the scaling behavior of the contribution to $\mathcal{M}$ of a given diagram reads

$$D = n + (n-2)N_I + \sum_{k=1}^{\infty} N_{2k}(2k - n).$$

The relation between the number of independent loops, the number of internal lines, and the total number of vertices $N_V = \sum_{k=1}^{\infty} N_{2k}$ is given by $N_L = N_I - (N_V - 1)$. The product of N_V momentum-conserving delta functions contains overall momentum conservation. Therefore, one has $N_V - 1$ rather than N_V restrictions on the internal momenta. Applying

$$-n \sum_{k=1}^{\infty} N_{2k} = -n N_V = n(N_L - N_I - 1)$$

results in Eq. 3.120:

$$D = n N_L - 2N_I + \sum_{k=1}^{\infty} 2k N_{2k}.$$

On the other hand, applying

$$-n \sum_{k=1}^{\infty} N_{2k} = -2 \sum_{k=1}^{\infty} N_{2k} + (n-2)(N_L - N_I - 1),$$

results in Eq. 3.121:

$$D = 2 + \sum_{k=1}^{\infty} 2(k-1)N_{2k} + (n-2)N_L.$$

In the discussion of loop integrals we need to address the question of convergence, since applying the substitution $tk' = k$ in Eq. 3.122 is well-defined only for convergent integrals. As discussed above, we regularize loop integrals by use of dimensional regularization. We therefore need to introduce a renormalization scale μ which also has to be rescaled linearly. However, at a given chiral order, the sum of all diagrams does not, by construction, depend on the renormalization scale.

Finally, note that a minimal $k > 0$ is important. Otherwise, an infinite number of diagrams containing vertices from $\mathcal{L}_0$ would have to be summed (see Fig. 3.10). This is for example the case when dealing with the nucleon-nucleon interaction.

3.5 Beyond Leading Order

Already in 1967 it was shown by Weinberg [98] that an effective Lagrangian is a convenient tool for reproducing the results of current algebra in terms of tree-level calculations. In the purely mesonic sector, $\mathcal{L}_2$ of Eq. 3.77 represents the corresponding Lagrangian. It was noted by Li and Pagels [72] that a perturbation theory around a symmetry which is realized in the Nambu-Goldstone mode, in general, leads to observables which are nonanalytic functions of the symmetry-breaking parameters, here the quark masses. In 1979 Weinberg initiated the application of an effective-field-theory program beyond tree level allowing for a systematic calculation of corrections to the chiral limit [100]. When calculating one-loop graphs, using vertices from $\mathcal{L}_2$, one generates ultraviolet divergences which in the framework of dimensional regularization appear as poles at space-time dimension $n = 4$. The loop diagrams are renormalized by absorbing the infinite parts into the redefinition of the fields and the parameters of the most general Lagrangian. Since $\mathcal{L}_2$ is not renormalizable in the traditional sense, the infinities cannot be absorbed by a renormalization of the coefficients F_0 and B_0. According to Weinberg's power counting of Eq. 3.121, one-loop graphs with vertices from $\mathcal{L}_2$ are of $\mathcal{O}(q^4)$. Therefore, one needs to construct the most general Lagrangian $\mathcal{L}_4$ and adjust (renormalize) its parameters to cancel the one-loop infinities originating from $\mathcal{L}_2$.

Beyond the quantum corrections to processes already described by $\mathcal{L}_2$, at next-to-leading order we encounter another important feature, namely, the effective

Wess-Zumino-Witten (WZW) action [102, 103]. The WZW action provides an effective description of the constraints due to the anomalous Ward identities. In general, anomalies arise if the symmetries of the Lagrangian at the classical level are not supported by the quantized theory after renormalization.

3.5.1 The $\mathcal{O}(q^4)$ Lagrangian of Gasser and Leutwyler

Applying the ideas outlined in Sect. 3.4.3, it is possible to construct the most general $SU(3)_L \times SU(3)_R$-invariant Lagrangian at $\mathcal{O}(q^4)$. Here we only quote the result of Gasser and Leutwyler [53]:

$$
\begin{aligned}
\mathcal{L}_4 = {}& L_1 \left\{ \mathrm{Tr}[D_\mu U (D^\mu U)^\dagger] \right\}^2 + L_2 \mathrm{Tr}\left[D_\mu U (D_\nu U)^\dagger\right] \mathrm{Tr}\left[D^\mu U (D^\nu U)^\dagger\right] \\
& + L_3 \mathrm{Tr}\left[D_\mu U (D^\mu U)^\dagger D_\nu U (D^\nu U)^\dagger\right] + L_4 \mathrm{Tr}\left[D_\mu U (D^\mu U)^\dagger\right] \mathrm{Tr}\left(\chi U^\dagger + U\chi^\dagger\right) \\
& + L_5 \mathrm{Tr}\left[D_\mu U (D^\mu U)^\dagger (\chi U^\dagger + U\chi^\dagger)\right] + L_6 \left[\mathrm{Tr}\left(\chi U^\dagger + U\chi^\dagger\right)\right]^2 \\
& + L_7 \left[\mathrm{Tr}\left(\chi U^\dagger - U\chi^\dagger\right)\right]^2 + L_8 \mathrm{Tr}\left(U\chi^\dagger U\chi^\dagger + \chi U^\dagger \chi U^\dagger\right) \\
& - iL_9 \mathrm{Tr}\left[f^R_{\mu\nu} D^\mu U (D^\nu U)^\dagger + f^L_{\mu\nu} (D^\mu U)^\dagger D^\nu U\right] + L_{10} \mathrm{Tr}\left(U f^L_{\mu\nu} U^\dagger f^{\mu\nu}_R\right) \\
& + H_1 \mathrm{Tr}\left(f^R_{\mu\nu} f^{\mu\nu}_R + f^L_{\mu\nu} f^{\mu\nu}_L\right) + H_2 \mathrm{Tr}\left(\chi \chi^\dagger\right).
\end{aligned}
\tag{3.123}
$$

The numerical values of the low-energy constants L_i are not determined by chiral symmetry. In analogy to F_0 and B_0 of $\mathcal{L}_2$ they are parameters containing information on the underlying dynamics and should, in principle, be calculable in terms of the (remaining) parameters of QCD, namely, the heavy-quark masses and the QCD scale Λ_{QCD}. In practice, they parameterize our inability to solve the dynamics of QCD in the non-perturbative regime. So far they have either been fixed using empirical input (see, e.g., Refs. [19, 25, 53]) or theoretically using QCD-inspired models, meson-resonance saturation [47, 79], and lattice QCD (see Ref. [76] for a recent overview and the report of the Flavianet Lattice Averaging Group (FLAG) on the lattice determination of LECs for a detailed review [40]).

By construction, Eq. 3.123 represents the most general Lagrangian at $\mathcal{O}(q^4)$, and it is thus possible to absorb all one-loop divergences originating from $\mathcal{L}_2$ by an appropriate renormalization of the coefficients L_i and H_i:

$$
L_i = L_i^r + \frac{\Gamma_i}{32\pi^2} R \quad (i = 1, \ldots, 10), \quad H_i = H_i^r + \frac{\Delta_i}{32\pi^2} R \quad (i = 1, 2), \tag{3.124}
$$

where R has already been defined in Eq. 3.111:

$$
R = \frac{2}{n-4} - [\ln(4\pi) + \Gamma'(1) + 1],
$$

Table 3.4 Renormalized low-energy constants L_i^r in units of 10^{-3} at the scale $\mu = M_\rho$, see Ref. [19]. $\Delta_1 = -1/8$, $\Delta_2 = 5/24$. Recent preliminary results of a global fit of the renormalized LECs L_i^r including $\mathcal{O}(q^6)$ corrections are discussed in Ref. [25]

Coefficient	Empirical Value	Γ_i
L_1^r	0.4 ± 0.3	$\frac{3}{32}$
L_2^r	1.35 ± 0.3	$\frac{3}{16}$
L_3^r	-3.5 ± 1.1	0
L_4^r	-0.3 ± 0.5	$\frac{1}{8}$
L_5^r	1.4 ± 0.5	$\frac{3}{8}$
L_6^r	-0.2 ± 0.3	$\frac{11}{144}$
L_7^r	-0.4 ± 0.2	0
L_8^r	0.9 ± 0.3	$\frac{5}{48}$
L_9^r	6.9 ± 0.7	$\frac{1}{4}$
L_{10}^r	-5.5 ± 0.7	$-\frac{1}{4}$

with n denoting the number of space-time dimensions and $\gamma_E = -\Gamma'(1)$ being Euler's constant. The constants Γ_i and Δ_i are given in Table 3.4. Except for L_3 and L_7, the low-energy constants L_i and the "contact terms"—i.e., pure external-field terms—H_1 and H_2 are required in the renormalization of the one-loop graphs. Since H_1 and H_2 contain only external fields, they are of no physical relevance. The idea of renormalization consists of adjusting the parameters of the counter terms of the most general effective Lagrangian so that they cancel the divergences of (multi-) loop diagrams. In doing so, one still has the freedom of choosing a suitable renormalization condition. For example, in the minimal subtraction scheme (MS) one would fix the parameters of the counter-term Lagrangian such that they would precisely absorb the contributions proportional to $2/(n-4)$ in R, while the modified minimal subtraction scheme of ChPT ($\widetilde{\text{MS}}$) would, in addition, cancel the term in the square brackets.[25]

The renormalized coefficients L_i^r depend on the scale μ introduced by dimensional regularization (see Eq. 3.105) and their values at two different scales μ_1 and μ_2 are related by

$$L_i^r(\mu_2) = L_i^r(\mu_1) + \frac{\Gamma_i}{16\pi^2} \ln\left(\frac{\mu_1}{\mu_2}\right). \tag{3.125}$$

We will see that the scale dependence of the coefficients and the finite part of the loop diagrams compensate each other in such a way that physical observables are scale independent.

---

[25] In distinction to the $\overline{\text{MS}}$ scheme commonly used in Standard Model calculations, the $\widetilde{\text{MS}}$ scheme contains an additional finite subtraction term. To be specific, in $\widetilde{\text{MS}}$ one uses multiples of $2/(n-4) - [\ln(4\pi) + \Gamma'(1) + 1]$ instead of $2/(n-4) - [\ln(4\pi) + \Gamma'(1)]$ in $\overline{\text{MS}}$.

Fig. 3.11 Self-energy
diagrams at $\mathcal{O}(q^4)$. Vertices
derived from $\mathscr{L}_{2n}$ are
denoted by $2n$ in the
interaction blobs

3.5.2 Masses of the Goldstone Bosons at $\mathcal{O}(q^4)$

A discussion of the masses at $\mathcal{O}(q^4)$ is one of the simplest applications of chiral
perturbation theory beyond tree level and will allow us to illustrate various
characteristic properties:

1. The relation between the bare low-energy constants L_i and the renormalized
   coefficients L_i^r in Eq. 3.124 is such that the divergences of one-loop diagrams
   are canceled.
2. Similarly, the scale dependence of the coefficients $L_i^r(\mu)$ on the one hand and of
   the finite contributions of the one-loop diagrams on the other hand lead to scale-
   independent predictions for physical observables.
3. A perturbative expansion in the explicit symmetry-breaking parameter with
   respect to a symmetry that is realized in the Nambu-Goldstone mode generates
   corrections which are nonanalytic in the symmetry-breaking parameter, here
   the quark masses.

Let us consider $\mathscr{L}_2 + \mathscr{L}_4$ for QCD with finite quark masses, but in the absence
of external fields. We restrict ourselves to the limit of isospin symmetry, i.e.,
$m_u = m_d = \hat{m}$. In order to determine the masses, we calculate the self energies
$\Sigma(p^2)$ of the Goldstone bosons.

Let

$$\Delta_{F\phi}(p) = \frac{1}{p^2 - M_{\phi,2}^2 + i0^+}, \quad \phi = \pi, K, \eta, \tag{3.126}$$

denote the Feynman propagator containing the lowest-order masses of
Eqs. 3.59–3.61,

$$M_{\pi,2}^2 = 2B_0\hat{m}, \quad M_{K,2}^2 = B_0(\hat{m} + m_s), \quad M_{\eta,2}^2 = \frac{2}{3}B_0(\hat{m} + 2m_s).$$

The subscript 2 refers to chiral order 2. The proper self-energy insertions,
$-i\Sigma_\phi(p^2)$, consist of one-particle-irreducible diagrams only, i.e., diagrams which
do not fall apart into two separate pieces when cutting an arbitrary internal line. At
chiral order $D = 4$, the contributions to $-i\Sigma_{\phi,4}(p^2)$ are those shown in Fig. 3.11.
In general, the full (unrenormalized) propagator may be summed using a geo-
metric series (see Fig. 3.12):

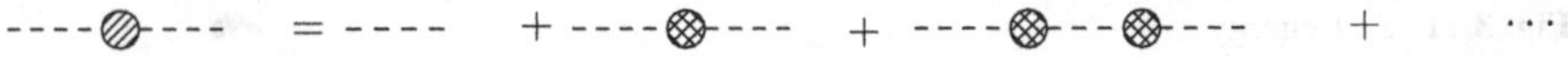

Fig. 3.12 Unrenormalized propagator as the sum of irreducible self-energy diagrams. Hatched and cross-hatched "vertices" denote one-particle-reducible and one-particle-irreducible contributions, respectively

$$
i\Delta_\phi(p) = \frac{i}{p^2 - M_{\phi,2}^2 + i0^+} + \frac{i}{p^2 - M_{\phi,2}^2 + i0^+}[-i\Sigma_\phi(p^2)]\frac{i}{p^2 - M_{\phi,2}^2 + i0^+} + \cdots
$$

$$
= \frac{i}{p^2 - M_{\phi,2}^2 - \Sigma_\phi(p^2) + i0^+}.
$$

$$(3.127)$$

The physical mass, including the interaction, is defined as the pole of Eq. 3.127,

$$
M_\phi^2 - M_{\phi,2}^2 - \Sigma_\phi(M_\phi^2) = 0, \tag{3.128}
$$

where the accuracy of the determination of M_ϕ^2 depends on the accuracy of the calculation of Σ_ϕ.

For our particular application with exactly two external meson lines, the relevant interaction Lagrangians can be written as

$$
\mathcal{L}_{\text{int}} = \mathcal{L}_2^{4\phi} + \mathcal{L}_4^{2\phi}, \tag{3.129}
$$

where $\mathcal{L}_2^{4\phi}$ is given by (see Exercise 3.14)

$$
\mathcal{L}_2^{4\phi} = \frac{1}{48F_0^2}\left\{\text{Tr}([\phi,\partial_\mu\phi][\phi,\partial^\mu\phi]) + 2B_0\text{Tr}(\mathcal{M}\phi^4)\right\}. \tag{3.130}
$$

The terms of $\mathcal{L}_4$ proportional to L_9, L_{10}, H_1, and H_2 do not contribute, because they either contain field-strength tensors or external fields only. Since $\partial_\mu U = O(\phi)$, the L_1, L_2, and L_3 terms of Eq. 3.123 are $O(\phi^4)$ and need not be considered. The only candidates are the $L_4 - L_8$ terms, of which we consider the L_4 term as an explicit example,[26]

$$
L_4\text{Tr}(\partial_\mu U\partial^\mu U^\dagger)\text{Tr}(\chi U^\dagger + U\chi^\dagger) = L_4\frac{2}{F_0^2}[\partial_\mu\eta\partial^\mu\eta + \partial_\mu\pi^0\partial^\mu\pi^0 + 2\partial_\mu\pi^+\partial^\mu\pi^-
$$

$$
+ 2\partial_\mu K^+\partial^\mu K^- + 2\partial_\mu K^0\partial^\mu\bar{K}^0 + O(\phi^4)]
$$

$$
\times [4B_0(2\hat{m} + m_s) + O(\phi^2)].
$$

The remaining terms are treated analogously and we obtain for $\mathcal{L}_4^{2\phi}$

---

[26] For pedagogical reasons, we make use of the physical fields. From a technical point of view, it is often advantageous to work with the Cartesian fields and, at the end of the calculation, express physical processes in terms of the Cartesian components.

$$\mathcal{L}_4^{2\phi} = -\frac{1}{2}\left(a_\pi \pi^0 \pi^0 + b_\pi \partial_\mu \pi^0 \partial^\mu \pi^0\right) - a_\pi \pi^+ \pi^- - b_\pi \partial_\mu \pi^+ \partial^\mu \pi^-$$
$$- a_K K^+ K^- - b_K \partial_\mu K^+ \partial^\mu K^- - a_K K^0 \bar{K}^0 - b_K \partial_\mu K^0 \partial^\mu \bar{K}^0$$
$$- \frac{1}{2}\left(a_\eta \eta^2 + b_\eta \partial_\mu \eta \partial^\mu \eta\right), \tag{3.131}$$

where the constants a_ϕ and b_ϕ are given by

$$a_\pi = \frac{64 B_0^2}{F_0^2}\left[(2\hat{m} + m_s)\hat{m} L_6 + \hat{m}^2 L_8\right],$$

$$b_\pi = -\frac{16 B_0}{F_0^2}\left[(2\hat{m} + m_s) L_4 + \hat{m} L_5\right],$$

$$a_K = \frac{32 B_0^2}{F_0^2}\left[(2\hat{m} + m_s)(\hat{m} + m_s) L_6 + \frac{1}{2}(\hat{m} + m_s)^2 L_8\right],$$

$$b_K = -\frac{16 B_0}{F_0^2}\left[(2\hat{m} + m_s) L_4 + \frac{1}{2}(\hat{m} + m_s) L_5\right],$$

$$a_\eta = \frac{64 B_0^2}{3 F_0^2}\left[(2\hat{m} + m_s)(\hat{m} + 2m_s) L_6 + 2(\hat{m} - m_s)^2 L_7 + (\hat{m}^2 + 2m_s^2) L_8\right],$$

$$b_\eta = -\frac{16 B_0}{F_0^2}\left[(2\hat{m} + m_s) L_4 + \frac{1}{3}(\hat{m} + 2m_s) L_5\right]. \tag{3.132}$$

At $\mathcal{O}(q^4)$ the self energies are of the form

$$\Sigma_{\phi,4}(p^2) = A_\phi + B_\phi p^2, \tag{3.133}$$

where the constants A_ϕ and B_ϕ receive a tree-level contribution from $\mathcal{L}_4$ and a one-loop contribution with a vertex from $\mathcal{L}_2$ (see Fig. 3.11). For the tree-level contribution of $\mathcal{L}_4$ this is easily seen, because the Lagrangians of Eq. 3.131 contain either exactly two derivatives of the fields or no derivatives at all. For example, the tree contribution for the η reads

$$-i\Sigma_{\eta,4}^{\text{tree}}(p^2) = 2i\left[-\frac{1}{2}a_\eta - b_\eta \frac{1}{2}(ip_\mu)(-ip^\mu)\right] = -i(a_\eta + b_\eta p^2),$$

where, as usual, $\partial_\mu \phi$ generates $-ip_\mu$ and ip_μ for initial and final lines, respectively, and the factor two takes account of two combinations of contracting the fields with external lines.

For the one-loop contribution the argument is as follows. The Lagrangian $\mathcal{L}_2^{4\phi}$ contains either two derivatives or no derivatives at all which, symbolically, can be written as $\phi\phi\partial\phi\partial\phi$ and ϕ^4, respectively. The first term results in M^2 (see below) or p^2, depending on whether the ϕ or the $\partial\phi$ are contracted with the external fields.

The "mixed" situation vanishes upon integration. The second term, ϕ^4, does not generate a momentum dependence.

As a specific example, we evaluate the pion-loop contribution to the π^0 self energy (see Fig. 3.13) by applying the Feynman rule of Eq. 3.90 for $a = c = 3$, $p_a = p_c = p$, $b = d = j$, and $p_b = p_d = k$:[27]

$$\frac{1}{2}\int \frac{d^4k}{(2\pi)^4} i \left[\underbrace{\delta_{3j}\delta_{3j}}_{=1} \frac{(p+k)^2 - M_{\pi,2}^2}{F_0^2} + \underbrace{\delta_{33}\delta_{jj}}_{=3} \frac{-M_{\pi,2}^2}{F_0^2} + \underbrace{\delta_{3j}\delta_{j3}}_{=1} \frac{(p-k)^2 - M_{\pi,2}^2}{F_0^2} \right.$$
$$\left. - \frac{1}{3F_0^2}\underbrace{(\delta_{3j}\delta_{3j} + \delta_{33}\delta_{jj} + \delta_{3j}\delta_{j3})}_{=5}\left(2p^2 + 2k^2 - 4M_{\pi,2}^2\right)\right] \frac{i}{k^2 - M_{\pi,2}^2 + i0^+}$$
$$= \frac{1}{2}\int \frac{d^4k}{(2\pi)^4} \frac{i}{3F_0^2}\left(-4p^2 - 4k^2 + 5M_{\pi,2}^2\right)\frac{i}{k^2 - M_{\pi,2}^2 + i0^+},$$

(3.134)

where $1/2$ is a symmetry factor.[28] The integral of Eq. 3.134 diverges and we thus consider its extension to n dimensions in order to make use of the dimensional-regularization technique described in Sect. 3.4.7. In addition to the loop integral of Eq. 3.110,

$$I(M^2, \mu^2, n) = \mu^{4-n}\int \frac{d^nk}{(2\pi)^n}\frac{i}{k^2 - M^2 + i0^+} = \frac{M^2}{16\pi^2}\left[R + \ln\left(\frac{M^2}{\mu^2}\right)\right] + O(n-4),$$

(3.135)

where R is given in Eq. 3.111, we need

$$\mu^{4-n}i\int \frac{d^nk}{(2\pi)^n}\frac{k^2}{k^2 - M^2 + i0^+} = \mu^{4-n}i\int \frac{d^nk}{(2\pi)^n}\frac{k^2 - M^2 + M^2}{k^2 - M^2 + i0^+},$$

where we have added $0 = -M^2 + M^2$ in the numerator. We make use of

$$\mu^{4-n}i\int \frac{d^nk}{(2\pi)^n} = 0$$

in dimensional regularization which is "shown" as follows. Consider the (more general) integral

---

[27] Note that we work in the three-flavor sector and thus with the exponential parameterization of U.

[28] When deriving the Feynman rule of Exercise 3.14, we took account of 24 distinct combinations of contracting four field operators with four external lines. However, the Feynman diagram of Eq. 3.134 involves only 12 possibilities to contract two fields with each other and the remaining two fields with two external lines.

Fig. 3.13 Contribution of the pion loops ($j = 1, 2, 3$) to the π^0 self energy

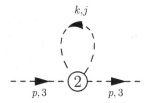

$$\int d^n k \, (k^2)^p,$$

substitute $k = \lambda k' \, (\lambda > 0)$, and relabel $k' = k$,

$$\int d^n k \, (k^2)^p = \lambda^{n+2p} \int d^n k \, (k^2)^p. \tag{3.136}$$

Since $\lambda > 0$ is arbitrary and, for fixed p, the result is to hold for arbitrary n, Eq. 3.136 is set to zero in dimensional regularization. We emphasize that the vanishing of Eq. 3.136 has the character of a prescription. The integral does not depend on any scale and its analytic continuation is ill defined in the sense that there is no dimension n where it is meaningful. It is ultraviolet divergent for $n + 2p \geq 0$ and infrared divergent for $n + 2p \leq 0$.

We then obtain

$$\mu^{4-n} i \int \frac{d^n k}{(2\pi)^n} \frac{k^2}{k^2 - M^2 + i0^+} = M^2 I(M^2, \mu^2, n),$$

with $I(M^2, \mu^2, n)$ of Eq. 3.135. The pion-loop contribution to the π^0 self energy is thus

$$\frac{i}{6F_0^2} (-4p^2 + M_{\pi,2}^2) I(M_{\pi,2}^2, \mu^2, n),$$

which is indeed of the type discussed in Eq. 3.133 and diverges as $n \to 4$.

After analyzing all loop contributions and combining them with the tree-level contributions of Eqs. 3.132, the constants A_ϕ and B_ϕ of Eq. 3.133 are given by

$$A_\pi = \frac{M_{\pi,2}^2}{F_0^2} \left\{ \underbrace{-\frac{1}{6} I(M_{\pi,2}^2) - \frac{1}{6} I(M_{\eta,2}^2) - \frac{1}{3} I(M_{K,2}^2)}_{\text{loop contribution}} \underbrace{+ 32[(2\hat{m} + m_s)B_0 L_6 + \hat{m} B_0 L_8]}_{\text{tree-level contribution}} \right\},$$

$$B_\pi = \frac{2}{3} \frac{I(M_{\pi,2}^2)}{F_0^2} + \frac{1}{3} \frac{I(M_{K,2}^2)}{F_0^2} - \frac{16 B_0}{F_0^2} [(2\hat{m} + m_s)L_4 + \hat{m} L_5],$$

$$A_K = \frac{M_{K,2}^2}{F_0^2} \left\{ \frac{1}{12} I(M_{\eta,2}^2) - \frac{1}{4} I(M_{\pi,2}^2) - \frac{1}{2} I(M_{K,2}^2) \right.$$

$$\left. + 32\left[(2\hat{m} + m_s)B_0 L_6 + \frac{1}{2}(\hat{m} + m_s)B_0 L_8 \right] \right\},$$

$$B_K = \frac{1}{4}\frac{I(M_{\eta,2}^2)}{F_0^2} + \frac{1}{4}\frac{I(M_{\pi,2}^2)}{F_0^2} + \frac{1}{2}\frac{I(M_{K,2}^2)}{F_0^2} - 16\frac{B_0}{F_0^2}\left[(2\hat{m}+m_s)L_4 + \frac{1}{2}(\hat{m}+m_s)L_5\right],$$

$$A_\eta = \frac{M_{\eta,2}^2}{F_0^2}\left[-\frac{2}{3}I(M_{\eta,2}^2)\right] + \frac{M_{\pi,2}^2}{F_0^2}\left[\frac{1}{6}I(M_{\eta,2}^2) - \frac{1}{2}I(M_{\pi,2}^2) + \frac{1}{3}I(M_{K,2}^2)\right]$$
$$+ \frac{M_{\eta,2}^2}{F_0^2}[16M_{\eta,2}^2 L_8 + 32(2\hat{m}+m_s)B_0 L_6] + \frac{128}{9}\frac{B_0^2(\hat{m}-m_s)^2}{F_0^2}(3L_7 + L_8),$$

$$B_\eta = \frac{I(M_{K,2}^2)}{F_0^2} - \frac{16}{F_0^2}(2\hat{m}+m_s)B_0 L_4 - 8\frac{M_{\eta,2}^2}{F_0^2}L_5, \tag{3.137}$$

where, for simplicity, we have suppressed the dependence on the scale μ and the number of dimensions n in the integrals $I(M^2, \mu^2, n)$ (see Eq. 3.135). Both the integrals I and the bare coefficients L_i (with the exception of L_7) have $1/(n-4)$ poles and finite pieces. In particular, the coefficients A_ϕ and B_ϕ are *not* finite as $n \to 4$, showing that they do not correspond to observables.

The masses at $\mathcal{O}(q^4)$ are determined by solving Eq. 3.128 with the predictions of Eq. 3.133 for the self energies,

$$M_\phi^2 = M_{\phi,2}^2 + A_\phi + B_\phi M_\phi^2,$$

from which we obtain

$$M_\phi^2 = \frac{M_{\phi,2}^2 + A_\phi}{1 - B_\phi} = M_{\phi,2}^2(1 + B_\phi) + A_\phi + \mathcal{O}(q^6),$$

because $A_\phi = \mathcal{O}(q^4)$ and $\{B_\phi, M_{\phi,2}^2\} = \mathcal{O}(q^2)$. Expressing the bare coefficients L_i in Eq. 3.137 in terms of the renormalized coefficients by using Eq. 3.124, the results for the masses of the Goldstone bosons at $\mathcal{O}(q^4)$ read [53]

$$M_{\pi,4}^2 = M_{\pi,2}^2\left\{1 + \frac{M_{\pi,2}^2}{32\pi^2 F_0^2}\ln\left(\frac{M_{\pi,2}^2}{\mu^2}\right) - \frac{M_{\eta,2}^2}{96\pi^2 F_0^2}\ln\left(\frac{M_{\eta,2}^2}{\mu^2}\right)\right.$$
$$\left. + \frac{16}{F_0^2}\left[(2\hat{m}+m_s)B_0(2L_6^r - L_4^r) + \hat{m}B_0(2L_8^r - L_5^r)\right]\right\}, \tag{3.138}$$

$$M_{K,4}^2 = M_{K,2}^2\left\{1 + \frac{M_{\eta,2}^2}{48\pi^2 F_0^2}\ln\left(\frac{M_{\eta,2}^2}{\mu^2}\right)\right.$$
$$\left. + \frac{16}{F_0^2}\left[(2\hat{m}+m_s)B_0(2L_6^r - L_4^r) + \frac{1}{2}(\hat{m}+m_s)B_0(2L_8^r - L_5^r)\right]\right\}, \tag{3.139}$$

$$M_{\eta,4}^2 = M_{\eta,2}^2 \left[1 + \frac{M_{K,2}^2}{16\pi^2 F_0^2} \ln\left(\frac{M_{K,2}^2}{\mu^2}\right) - \frac{M_{\eta,2}^2}{24\pi^2 F_0^2} \ln\left(\frac{M_{\eta,2}^2}{\mu^2}\right) \right.$$

$$\left. + \frac{16}{F_0^2}(2\hat{m} + m_s)B_0(2L_6^r - L_4^r) + 8\frac{M_{\eta,2}^2}{F_0^2}(2L_8^r - L_5^r) \right]$$

$$+ M_{\pi,2}^2 \left[\frac{M_{\eta,2}^2}{96\pi^2 F_0^2} \ln\left(\frac{M_{\eta,2}^2}{\mu^2}\right) - \frac{M_{\pi,2}^2}{32\pi^2 F_0^2} \ln\left(\frac{M_{\pi,2}^2}{\mu^2}\right) + \frac{M_{K,2}^2}{48\pi^2 F_0^2} \ln\left(\frac{M_{K,2}^2}{\mu^2}\right) \right]$$

$$+ \frac{128\, B_0^2(\hat{m} - m_s)^2}{9} \frac{}{F_0^2}(3L_7^r + L_8^r). \tag{3.140}$$

First of all, we note that the expressions for the masses are finite. The infinite parts of the coefficients L_i of the Lagrangian of Gasser and Leutwyler exactly cancel the divergent terms resulting from the integrals. This is the reason why the bare coefficients L_i must be infinite. Furthermore, at $\mathcal{O}(q^4)$ the masses of the Goldstone bosons vanish if the quark masses are sent to zero. This is, of course, what we expected from QCD in the chiral limit but it is comforting to see that the self interaction in $\mathscr{L}_2$ (in the absence of quark masses) does not generate Goldstone-boson masses at higher order. At $\mathcal{O}(q^4)$, the squared Goldstone-boson masses contain terms which are analytic in the quark masses, namely, of the form m_q^2 multiplied by the renormalized low-energy constants L_i^r. However, there are also nonanalytic terms of the type $m_q^2 \ln(m_q)$—so-called chiral logarithms—which do not involve new parameters. Such a behavior is an illustration of the mechanism found by Li and Pagels [72], who noticed that a perturbation theory around a symmetry which is realized in the Nambu-Goldstone mode results in both analytic as well as nonanalytic expressions in the perturbation. Finally, the scale dependence of the renormalized coefficients L_i^r of Eq. 3.124 is by construction such that it cancels the scale dependence of the chiral logarithms. Thus, physical observables do not depend on the scale μ.

Exercise 3.24 We want to verify this statement by differentiating Eq. 3.138 with respect to μ.

(a) Using Eq. 3.125, show

$$\frac{dL_i^r(\mu)}{d\mu} = -\frac{\Gamma_i}{16\pi^2 \mu}.$$

(b) Verify

$$\frac{dM_{\pi,4}^2}{d\mu} = 0.$$

Hints: Make use of Eqs. 3.59–3.61 and the values of the coefficients Γ_i of Table 3.4.

Exercise 3.25 In this exercise we want to familiarize ourselves with the conventions of the two-flavor sector of ChPT. Moreover, it will serve as an illustration of the equivalence theorem of field theory [34, 42, 65] beyond tree level: results for observables (such as, e.g., S-matrix elements) do not depend on the parameterization of the fields. In the discussion of $\pi\pi$ scattering we have already seen an example at tree level.

In the two-flavor sector two different parameterizations of the SU(2) matrix $U(x)$ are popular,

$$U(x) = \exp\left[i\frac{\vec{\phi}(x) \cdot \vec{\tau}}{F}\right], \tag{3.141}$$

$$U(x) = \frac{1}{F}[\sigma(x)\mathbb{1} + i\vec{\pi}(x) \cdot \vec{\tau}], \quad \sigma(x) = \sqrt{F^2 - \vec{\pi}^2(x)}, \tag{3.142}$$

where the pion fields of the two parameterizations are nonlinearly related (see Eq. 3.97). Furthermore, independent of the parameterizations of Eqs. 3.141 and 3.142, at $\mathcal{O}(q^4)$ two Lagrangians are commonly used, namely, those of Gasser and Leutwyler [52] and of Gasser, Sainio, and Švarc [54], respectively:

$$\begin{aligned}
\mathscr{L}_4^{\text{GL}} &= \frac{l_1}{4}\left\{\text{Tr}[D_\mu U(D^\mu U)^\dagger]\right\}^2 + \frac{l_2}{4}\text{Tr}[D_\mu U(D_\nu U)^\dagger]\text{Tr}[D^\mu U(D^\nu U)^\dagger] \\
&\quad + \frac{l_3}{16}\left[\text{Tr}(\chi U^\dagger + U\chi^\dagger)\right]^2 + \frac{l_4}{4}\text{Tr}[D_\mu U(D^\mu \chi)^\dagger + D_\mu\chi(D^\mu U)^\dagger] \\
&\quad + l_5\left[\text{Tr}(f_{R\mu\nu}Uf_L^{\mu\nu}U^\dagger) - \frac{1}{2}\text{Tr}(f_{L\mu\nu}f_L^{\mu\nu} + f_{R\mu\nu}f_R^{\mu\nu})\right] \\
&\quad + i\frac{l_6}{2}\text{Tr}[f_{R\mu\nu}D^\mu U(D^\nu U)^\dagger + f_{L\mu\nu}(D^\mu U)^\dagger D^\nu U] \\
&\quad - \frac{l_7}{16}\left[\text{Tr}(\chi U^\dagger - U\chi^\dagger)\right]^2 \\
&\quad + \frac{h_1 + h_3}{4}\text{Tr}(\chi\chi^\dagger) + \frac{h_1 - h_3}{16}\left\{\left[\text{Tr}(\chi U^\dagger + U\chi^\dagger)\right]^2\right. \\
&\quad \left. + \left[\text{Tr}(\chi U^\dagger - U\chi^\dagger)\right]^2 - 2\text{Tr}(\chi U^\dagger \chi U^\dagger + U\chi^\dagger U\chi^\dagger)\right\} \\
&\quad - 2h_2\text{Tr}(f_{L\mu\nu}f_L^{\mu\nu} + f_{R\mu\nu}f_R^{\mu\nu}), \tag{3.143}
\end{aligned}$$

$$
\begin{aligned}
\mathscr{L}_4^{\mathrm{GSS}} = {} & \frac{l_1}{4}\left\{\mathrm{Tr}[D_\mu U(D^\mu U)^\dagger]\right\}^2 + \frac{l_2}{4}\mathrm{Tr}[D_\mu U(D_\nu U)^\dagger]\mathrm{Tr}[D^\mu U(D^\nu U)^\dagger] \\
& + \frac{l_3 + l_4}{16}\left[\mathrm{Tr}(\chi U^\dagger + U\chi^\dagger)\right]^2 + \frac{l_4}{8}\mathrm{Tr}[D_\mu U(D^\mu U)^\dagger]\mathrm{Tr}(\chi U^\dagger + U\chi^\dagger) \\
& + l_5\mathrm{Tr}(f_{R\mu\nu}Uf_L^{\mu\nu}U^\dagger) + i\frac{l_6}{2}\mathrm{Tr}[f_{R\mu\nu}D^\mu U(D^\nu U)^\dagger + f_{L\mu\nu}(D^\mu U)^\dagger D^\nu U] \\
& - \frac{l_7}{16}\left[\mathrm{Tr}(\chi U^\dagger - U\chi^\dagger)\right]^2 + \frac{h_1 + h_3 - l_4}{4}\mathrm{Tr}(\chi\chi^\dagger) \\
& + \frac{h_1 - h_3 - l_4}{16}\left\{\left[\mathrm{Tr}(\chi U^\dagger + U\chi^\dagger)\right]^2 + \left[\mathrm{Tr}(\chi U^\dagger - U\chi^\dagger)\right]^2\right. \\
& \left. - 2\mathrm{Tr}(\chi U^\dagger \chi U^\dagger + U\chi^\dagger U\chi^\dagger)\right\} - \frac{4h_2 + l_5}{2}\mathrm{Tr}(f_{L\mu\nu}f_L^{\mu\nu} + f_{R\mu\nu}f_R^{\mu\nu}).
\end{aligned}
$$

$$\tag{3.144}$$

When comparing with the three-flavor version of Eq. 3.123 we first note that Eqs. 3.143 and 3.144 contain fewer independent terms. This follows from the application of certain trace relations which reduce the number of independent structures for 2×2 matrices in comparison with 3×3 matrices. The expressions proportional to $(h_1 - h_3)$ and $(h_1 - h_3 - l_4)$ in $\mathscr{L}_4^{\mathrm{GL}}$ and $\mathscr{L}_4^{\mathrm{GSS}}$, respectively, can be rewritten so that the U's completely drop out, i.e., they contain only external fields. The trick is to use

$$
\begin{aligned}
2\mathrm{Tr}(\chi U^\dagger \chi U^\dagger + U\chi^\dagger U\chi^\dagger) = {} & [\mathrm{Tr}(\chi U^\dagger + U\chi^\dagger)]^2 + [\mathrm{Tr}(\chi U^\dagger - U\chi^\dagger)]^2 + \mathrm{Tr}(\tau_i\chi)\mathrm{Tr}(\tau_i\chi) \\
& + \mathrm{Tr}(\tau_i\chi^\dagger)\mathrm{Tr}(\tau_i\chi^\dagger) - [\mathrm{Tr}(\chi)]^2 - [\mathrm{Tr}(\chi^\dagger)]^2.
\end{aligned}
$$

In terms of a field transformation [86] the two Lagrangians $\mathscr{L}_4^{\mathrm{GL}}$ and $\mathscr{L}_4^{\mathrm{GSS}}$ can be shown to be equivalent (see App. D.1 of Ref. [87] for details). In principle, we are free to combine any of the two parameterizations for U with any of the two Lagrangians $\mathscr{L}_4$. The outcome for physical observables should not depend on the specific choice.

Remark Like in Eq. 3.124, the bare and the renormalized low-energy constants l_i and l_i^r are related by

$$
l_i = l_i^r + \gamma_i\frac{R}{32\pi^2},
$$

where $R = 2/(n - 4) - [\ln(4\pi) + \Gamma'(1) + 1]$ and

$$
\gamma_1 = \frac{1}{3}, \quad \gamma_2 = \frac{2}{3}, \quad \gamma_3 = -\frac{1}{2}, \quad \gamma_4 = 2, \quad \gamma_5 = -\frac{1}{6}, \quad \gamma_6 = -\frac{1}{3}, \quad \gamma_7 = 0.
$$

In the two-flavor sector one often uses the scale-independent parameters $\bar{l}_i$ which are defined by

Fig. 3.14 Self-energy diagrams at $\mathcal{O}(q^4)$. Vertices derived from $\mathscr{L}_{2n}$ are denoted by $2n$ in the interaction blobs

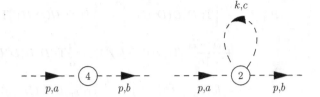

$$l_i^r = \frac{\gamma_i}{32\pi^2}\left[\bar{l}_i + \ln\left(\frac{M^2}{\mu^2}\right)\right], \quad i = 1,\ldots,6, \tag{3.145}$$

where $M^2 = 2B\hat{m}$. Since $\ln(1) = 0$, the $\bar{l}_i$ are proportional to the renormalized low-energy constants at the scale $\mu = M$.

We will now turn to the calculation of the squared pion mass at $\mathcal{O}(q^4)$. For the two-flavor calculation of the Goldstone-boson self energies at $\mathcal{O}(q^4)$ we need the interaction Lagrangian

$$\mathscr{L}_{\text{int}} = \mathscr{L}_2^{4\phi} + \mathscr{L}_4^{2\phi}.$$

Setting the external fields to zero and inserting $\chi = 2B\hat{m}$, derive $\mathscr{L}_4^{2\phi}$ for Eqs. 3.143 and 3.144 for both parameterizations of U.

Exercise 3.26 Using isospin symmetry, at $\mathcal{O}(q^4)$ the pion self energy is of the form

$$\Sigma_{ba}(p^2) = \delta_{ab}(A + Bp^2).$$

The constants A and B (not to be confused with the low-energy constant related to the quark condensate) receive a tree-level contribution from $\mathscr{L}_4$ and a one-loop contribution from $\mathscr{L}_2$ (see Fig. 3.14). Their numerical values depend on the parameterization of U and the version of $\mathscr{L}_4$.

(a) Using the results of Exercises 3.14 and 3.25, derive the expressions of Table 3.5 for the self-energy coefficients.
(b) Using

$$M_{\pi,4}^2 = \frac{M_{\pi,2}^2 + A}{1 - B} = M_{\pi,2}^2(1 + B) + A + \mathcal{O}(q^6),$$

derive the squared pion mass at $\mathcal{O}(q^4)$:

$$M_{\pi,4}^2 = M^2 - \frac{\bar{l}_3}{32\pi^2 F^2}M^4 + \mathcal{O}(M^6),$$

where $M^2 = 2B\hat{m}$. Note that the result for the pion mass is, as expected, independent of the Lagrangian and parameterization used. On the other hand, the constants A and B are auxiliary mathematical quantities and thus depend on both Lagrangian and parameterization.

Table 3.5 Self-energy coefficients and wave function renormalization constants. I denotes the dimensionally regularized integral $I = I(M^2, \mu^2, n) = \frac{M^2}{16\pi^2}\left[R + \ln\left(\frac{M^2}{\mu^2}\right)\right] + O(n-4)$, $R = \frac{2}{n-4} - [\ln(4\pi) + \Gamma'(1) + 1]$, $M^2 = 2B\hat{m}$. The abbreviations GL and GSS refer to the Lagrangians of Eqs. 3.143 and 3.144, respectively, exponential and square-root to the parameterizations of U of Eqs. 3.141 and 3.142, respectively

Lagrangian and parameterization	A	B
GL, exponential	$-\frac{1}{6}\frac{M^2}{F^2}I + 2l_3\frac{M^4}{F^2}$	$\frac{2}{3}\frac{I}{F^2}$
GL, square-root	$\frac{3}{2}\frac{M^2}{F^2}I + 2l_3\frac{M^4}{F^2}$	$-\frac{I}{F^2}$
GSS, exponential	$-\frac{1}{6}\frac{M^2}{F^2}I + 2(l_3+l_4)\frac{M^4}{F^2}$	$\frac{2}{3}\frac{I}{F^2} - 2l_4\frac{M^2}{F^2}$
GSS, square-root	$\frac{3}{2}\frac{M^2}{F^2}I + 2(l_3+l_4)\frac{M^4}{F^2}$	$-\frac{I}{F^2} - 2l_4\frac{M^2}{F^2}$

3.5.3 The Effective Wess-Zumino-Witten Action

The Lagrangians discussed so far have a larger symmetry than QCD [103]. For example, if we consider the case of "pure" QCD, i.e., no external fields except for $\chi = 2B_0\mathcal{M}$ with the quark-mass matrix of Eq. 3.53, the two Lagrangians $\mathcal{L}_2$ and $\mathcal{L}_4$ are invariant under the substitution $\phi(x) \mapsto -\phi(x)$. As discussed in Sect. 3.4.1, they contain interaction terms with an even number of Goldstone bosons only, i.e., they are of even intrinsic parity. In other words, they cannot describe, e.g, $K^+K^- \to \pi^+\pi^-\pi^0$. Analogously, $\mathcal{L}_2$ and $\mathcal{L}_4$ including a coupling to electromagnetic fields cannot describe the decay $\pi^0 \to \gamma\gamma$.

These observations lead us to a discussion of the effective Wess-Zumino-Witten (WZW) action [102, 103]. Whereas normal Ward identities are related to the *invariance* of the generating functional under local transformations of the external fields (see Sects. 1.4.1 and 1.4.4), the anomalous Ward identities [2–4, 11, 15], which were first obtained in the framework of renormalized perturbation theory, give a particular form to the *variation* of the generating functional [52, 102]. Wess and Zumino derived consistency or integrability relations which are satisfied by the anomalous Ward identities and then explicitly constructed a functional involving the pseudoscalar octet which satisfies the anomalous Ward identities [102]. In particular, Wess and Zumino emphasized that their interaction Lagrangians cannot be obtained as part of a chirally invariant Lagrangian.

Witten suggested to add to the lowest-order equation of motion the simplest term possible which breaks the symmetry of having only an even number of Goldstone bosons at the Lagrangian level [103]. For the case of massless Goldstone bosons without any external fields the modified equation of motion reads[29]

---

[29] In order to conform with our previous convention of Eq. 3.35, we need to replace U of Ref. [103] by $U^\dagger$. Furthermore, F_π of Ref. [103] corresponds to $2F_0$. Finally, $\partial^2 U U^\dagger - U\partial^2 U^\dagger = 2\partial_\mu(\partial^\mu U U^\dagger)$.

$$\partial_\mu \left(\frac{F_0^2}{2} U \partial^\mu U^\dagger \right) + \lambda \varepsilon^{\mu\nu\rho\sigma} U \partial_\mu U^\dagger U \partial_\nu U^\dagger U \partial_\rho U^\dagger U \partial_\sigma U^\dagger = 0, \qquad (3.146)$$

where λ is a (purely imaginary) constant and $\varepsilon_{0123} = 1$. Substituting $U \leftrightarrow U^\dagger$ in Eq. 3.146 and subsequently multiplying from the left by U and from the right by $U^\dagger$, we verify that the two terms transform with opposite relative signs. Recall that a term which is even (odd) in the Lagrangian leads to a term which is odd (even) in the equation of motion. For the purpose of writing down an action corresponding to Eq. 3.146, we extend the domain of definition of U to a hypothetical fifth dimension,

$$U(y) = \exp\left(i\alpha \frac{\phi(x)}{F_0} \right), \quad y^i = (x^\mu, \alpha), \ i = 0, \ldots, 4, \ 0 \le \alpha \le 1, \qquad (3.147)$$

where Minkowski space is defined as the surface of the five-dimensional space for $\alpha = 1$. Let us first quote the result of the effective Wess-Zumino-Witten (WZW) action in the absence of external fields (denoted by a superscript 0) [103]:

$$S_{WZW}^0 = -\frac{i}{240\pi^2} \int\limits_0^1 d\alpha \int d^4x \, \varepsilon^{ijklm} \text{Tr}\left(\mathcal{U}_{Li} \mathcal{U}_{Lj} \mathcal{U}_{Lk} \mathcal{U}_{Ll} \mathcal{U}_{Lm} \right), \qquad (3.148)$$

where the indices $i, \ldots, m$ run from 0 to 4, $y_4 = y^4 = \alpha$, ε_{ijklm} is the completely antisymmetric tensor with $\varepsilon_{01234} = -\varepsilon^{01234} = 1$, and $\mathcal{U}_{Li} \equiv U^\dagger \partial U / \partial y^i$.

Exercise 3.27 Consider the action[30]

$$S = S_2 + S_{ano}^0,$$

where S_2 is the action corresponding to the Lagrangian of Eq. 3.42 and $S_{ano}^0 = nS_{WZW}^0$ is the anomalous action, with n an integer still to be determined.

(a) Using the ansatz

$$U'(y) = [\mathbb{1} + i\alpha\Delta(x)]U(y), \quad \Delta(x) = \sum_{a=1}^8 \Delta_a(x)\lambda_a,$$

verify

$$\delta S_{ano}^0 = \frac{n}{48\pi^2} \int\limits_0^1 d\alpha \int d^4x \, \varepsilon^{ijklm} \text{Tr}\left(U^\dagger \partial_i(\alpha\Delta) U \mathcal{U}_{Lj} \mathcal{U}_{Lk} \mathcal{U}_{Ll} \mathcal{U}_{Lm} \right).$$

Hint: Make use of permutation symmetries.

(b) Make use of integration by parts, the boundary conditions $\Delta(\vec{x}, t_1) = \Delta(\vec{x}, t_2) = 0$ for the test functions, and the permutation symmetries to obtain

---

[30] The subscript ano refers to anomalous.

$$\delta S^0_{\text{ano}} = \frac{n}{48\pi^2} \int d^4x\, \varepsilon^{\mu\nu\rho\sigma} \text{Tr}(\Delta \mathcal{U}_{R\mu} \mathcal{U}_{R\nu} \mathcal{U}_{R\rho} \mathcal{U}_{R\sigma}),$$

where $\mathcal{U}_{R\mu} \equiv U\partial_\mu U^\dagger$. In combination with the result for δS_2 (see Sect. 3.4.3), this yields

$$\int d^4x\, \Delta_a \text{Tr}\left\{ \lambda_a \left[i\frac{F_0^2}{4}(\Box U U^\dagger - U\Box U^\dagger) + \frac{n}{48\pi^2} \varepsilon^{\mu\nu\rho\sigma} \mathcal{U}_{R\mu} \mathcal{U}_{R\nu} \mathcal{U}_{R\rho} \mathcal{U}_{R\sigma} \right] \right\} = 0$$

$$(3.149)$$

for arbitrary test functions Δ_a. Using $\Box U U^\dagger - U \Box U^\dagger = -2\partial_\mu(U\partial^\mu U^\dagger)$ and the fact that the expression inside the square brackets is traceless, Eqs. 3.146 and 3.149 are equivalent provided that the constants λ and n are related by $\lambda = in/(48\pi^2)$.

A rather unusual and surprising feature of Eq. 3.148 is that the action functional corresponding to the new term cannot be written as the four-dimensional integral of a Lagrangian expressed in terms of U and its derivatives. Expanding the SU(3) matrix $U(y)$ in terms of the Goldstone-boson fields, $U(y) = 1 + i\alpha\phi(x)/F_0 + O(\phi^2)$, one obtains an infinite series of terms, each involving an odd number of Goldstone bosons, i.e., the WZW action S^0_{WZW} is of odd intrinsic parity. For each individual term the α integration can be performed explicitly resulting in an ordinary action in terms of a four-dimensional integral of a local Lagrangian. For example, the term with the smallest number of Goldstone bosons reads

$$S^{5\phi}_{\text{WZW}} = \frac{1}{240\pi^2 F_0^5} \int_0^1 d\alpha \int d^4x\, \varepsilon^{ijklm} \text{Tr}[\partial_i(\alpha\phi)\partial_j(\alpha\phi)\partial_k(\alpha\phi)\partial_l(\alpha\phi)\partial_m(\alpha\phi)]$$

$$= \frac{1}{240\pi^2 F_0^5} \int_0^1 d\alpha \int d^4x\, \varepsilon^{ijklm} \partial_i \text{Tr}[\alpha\phi\partial_j(\alpha\phi)\partial_k(\alpha\phi)\partial_l(\alpha\phi)\partial_m(\alpha\phi)]$$

$$= \frac{1}{240\pi^2 F_0^5} \int d^4x\, \varepsilon^{\mu\nu\rho\sigma} \text{Tr}(\phi\partial_\mu\phi\partial_\nu\phi\partial_\rho\phi\partial_\sigma\phi). \qquad (3.150)$$

In the last step we made use of the fact that exactly one index can take the value 4. The term involving $i = 4$ has been integrated with respect to α, whereas the other four possibilities cancel each other because the ε tensor in four dimensions is antisymmetric under a cyclic permutation of the indices, whereas the trace is symmetric under a cyclic permutation. In particular, the WZW action without external fields involves at least five Goldstone bosons [102]. Once the constant n is known, it allows, e.g., for the description of the process $K^+K^- \rightarrow \pi^+\pi^-\pi^0$.

Using topological arguments, Witten showed that the constant n appearing in Eq. 3.148 must be an integer. However, it was pointed out in Ref. [12] that the traditional argument relating n with the number of colors N_c is incomplete. The connection to N_c is established by introducing the coupling to electromagnetism

[102, 103]. In the presence of external fields there will be an additional term in the anomalous action,

$$S_{\mathrm{ano}} = S_{\mathrm{ano}}^0 + S_{\mathrm{ano}}^{\mathrm{ext}} = n(S_{\mathrm{WZW}}^0 + S_{\mathrm{WZW}}^{\mathrm{ext}}), \tag{3.151}$$

given by (see, e.g., Ref. [18])

$$S_{\mathrm{WZW}}^{\mathrm{ext}} = -\frac{i}{48\pi^2} \int d^4x \, \varepsilon^{\mu\nu\rho\sigma} \mathrm{Tr}(Z_{\mu\nu\rho\sigma}) \tag{3.152}$$

with

$$
\begin{aligned}
Z_{\mu\nu\rho\sigma} &= \frac{1}{2} U l_\mu U^\dagger r_\nu U l_\rho U^\dagger r_\sigma + U l_\mu l_\nu l_\rho U^\dagger r_\sigma - U^\dagger r_\mu r_\nu r_\rho U l_\sigma \\
&+ i U \partial_\mu l_\nu l_\rho U^\dagger r_\sigma - i U^\dagger \partial_\mu r_\nu r_\rho U l_\sigma + i \partial_\mu r_\nu U l_\rho U^\dagger r_\sigma - i \partial_\mu l_\nu U^\dagger r_\rho U l_\sigma \\
&- i \mathscr{U}_{L\mu} l_\nu U^\dagger r_\rho U l_\sigma + i \mathscr{U}_{R\mu} r_\nu U l_\rho U^\dagger r_\sigma - i \mathscr{U}_{L\mu} l_\nu l_\rho l_\sigma + i \mathscr{U}_{R\mu} r_\nu r_\rho r_\sigma \\
&+ \frac{1}{2}\left(\mathscr{U}_{L\mu} U^\dagger \partial_\nu r_\rho U l_\sigma - \mathscr{U}_{R\mu} U \partial_\nu l_\rho U^\dagger r_\sigma + \mathscr{U}_{L\mu} U^\dagger r_\nu U \partial_\rho l_\sigma - \mathscr{U}_{R\mu} U l_\nu U^\dagger \partial_\rho r_\sigma \right) \\
&- \mathscr{U}_{L\mu} \mathscr{U}_{L\nu} U^\dagger r_\rho U l_\sigma + \mathscr{U}_{R\mu} \mathscr{U}_{R\nu} U l_\rho U^\dagger r_\sigma + \frac{1}{2} \mathscr{U}_{L\mu} l_\nu \mathscr{U}_{L\rho} l_\sigma - \frac{1}{2} \mathscr{U}_{R\mu} r_\nu \mathscr{U}_{R\rho} r_\sigma \\
&+ \mathscr{U}_{L\mu} l_\nu \partial_\rho l_\sigma - \mathscr{U}_{R\mu} r_\nu \partial_\rho r_\sigma + \mathscr{U}_{L\mu} \partial_\nu l_\rho l_\sigma - \mathscr{U}_{R\mu} \partial_\nu r_\rho r_\sigma \\
&- i \mathscr{U}_{L\mu} \mathscr{U}_{L\nu} \mathscr{U}_{L\rho} l_\sigma + i \mathscr{U}_{R\mu} \mathscr{U}_{R\nu} \mathscr{U}_{R\rho} r_\sigma,
\end{aligned}
\tag{3.153}
$$

with the abbreviations $\mathscr{U}_{L\mu} \equiv U^\dagger \partial_\mu U$ and $\mathscr{U}_{R\mu} \equiv U \partial_\mu U^\dagger$.

As a special case, let us consider the coupling to external electromagnetic four-vector potentials by inserting

$$r_\mu = l_\mu = -e\mathscr{A}_\mu Q,$$

where Q is the quark-charge matrix. The terms involving three and four electromagnetic four-vector potentials vanish upon contraction with the totally antisymmetric tensor $\varepsilon^{\mu\nu\rho\sigma}$, because their contributions to $Z_{\mu\nu\rho\sigma}$ are symmetric in at least two indices, and we obtain

$$
\begin{aligned}
n\mathscr{L}_{\mathrm{WZW}}^{\mathrm{ext}} &= -en\mathscr{A}_\mu J^\mu + i\frac{ne^2}{48\pi^2} \varepsilon^{\mu\nu\rho\sigma} \partial_\nu \mathscr{A}_\rho \mathscr{A}_\sigma \\
&\quad \times \mathrm{Tr}[2Q^2(U\partial_\mu U^\dagger - U^\dagger \partial_\mu U) - QU^\dagger Q\partial_\mu U + QUQ\partial_\mu U^\dagger].
\end{aligned}
\tag{3.154}
$$

We note that the current

$$J^\mu = \frac{\varepsilon^{\mu\nu\rho\sigma}}{48\pi^2} \mathrm{Tr}(Q\partial_\nu UU^\dagger \partial_\rho UU^\dagger \partial_\sigma UU^\dagger + QU^\dagger \partial_\nu UU^\dagger \partial_\rho UU^\dagger \partial_\sigma U), \quad \varepsilon_{0123} = 1, \tag{3.155}$$

by itself is not gauge invariant and the additional terms of Eq. 3.154 are required to obtain a gauge-invariant action. The standard procedure of determining n is to investigate the interaction Lagrangian which is relevant to the decay $\pi^0 \to \gamma\gamma$ by expanding $U = 1 + i\,\mathrm{diag}(\pi^0, -\pi^0, 0)/F_0 + \cdots$. However, as pointed out by Bär and Wiese [12], when considering the electromagnetic interaction for an arbitrary number of colors one should replace the ordinary quark-charge matrix in the Standard Model by

$$Q = \begin{pmatrix} \frac{2}{3} & 0 & 0 \\ 0 & -\frac{1}{3} & 0 \\ 0 & 0 & -\frac{1}{3} \end{pmatrix} \to \begin{pmatrix} \frac{1}{2N_c} + \frac{1}{2} & 0 & 0 \\ 0 & \frac{1}{2N_c} - \frac{1}{2} & 0 \\ 0 & 0 & \frac{1}{2N_c} - \frac{1}{2} \end{pmatrix}.$$

Exercise 3.28 From Eq. 3.154, derive the corresponding effective Lagrangian for $\pi^0 \to \gamma\gamma$ decay,

$$\mathscr{L}_{\pi^0\gamma\gamma} = -\frac{n}{N_c}\frac{e^2}{32\pi^2}\varepsilon^{\mu\nu\rho\sigma}\mathscr{F}_{\mu\nu}\mathscr{F}_{\rho\sigma}\frac{\pi^0}{F_0}.$$

Hint: Make use of integration by parts to shift the derivative from the pion field onto the electromagnetic four-vector potential.

The corresponding invariant amplitude at tree level reads

$$\mathscr{M} = i\frac{n}{N_c}\frac{e^2}{4\pi^2 F_0}\varepsilon^{\mu\nu\rho\sigma}q_{1\mu}\varepsilon^*_{1\nu}q_{2\rho}\varepsilon^*_{2\sigma}. \tag{3.156}$$

Exercise 3.29 Sum over the final photon polarizations and integrate over phase space to obtain the decay rate (see Exercise 3.13)

$$\Gamma_{\pi^0\to\gamma\gamma} = \frac{\alpha^2 M^3_{\pi^0}}{64\pi^3 F_0^2}\frac{n^2}{N_c^2} = 7.6\,\mathrm{eV}\times\left(\frac{n}{N_c}\right)^2, \tag{3.157}$$

where $\alpha = e^2/(4\pi)$ denotes the fine-structure constant.
Hints: Let $\varepsilon_{1\mu}(\lambda_1)\varepsilon_{2\nu}(\lambda_2)M^{\mu\nu}$ denote the invariant amplitude of a general process involving two real photons. As a consequence of electromagnetic current conservation, $q_{1\mu}M^{\mu\nu} = 0$ and $q_{2\nu}M^{\mu\nu} = 0$, the sum over photon polarizations is given by

$$\sum_{\lambda_1,\lambda_2=1}^{2}|\varepsilon_{1\mu}(\lambda_1)\varepsilon_{2\nu}(\lambda_2)M^{\mu\nu}|^2 = M_{\mu\nu}M^{\mu\nu*}.$$

Finally, make use of

$$\varepsilon_{\mu\nu\alpha\beta}\varepsilon^{\mu\nu\rho\sigma} = -2(g_\alpha{}^\rho g_\beta{}^\sigma - g_\alpha{}^\sigma g_\beta{}^\rho).$$

Equation 3.157 is in good agreement with the experimental value $(7.7 \pm 0.6)\,\mathrm{eV}$ for $n = N_c$. However, the result is no indication for $N_c = 3$ [12]. Bär and Wiese conclude from their analysis that one should rather consider three-flavor processes such as $\eta \to \pi^+\pi^-\gamma$ or $K\gamma \to K\pi$ to test the expected N_c dependence in a low-energy reaction. For example, the Lagrangian relevant to the decay $\eta \to \pi^+\pi^-\gamma$ is given by

$$\mathscr{L}_{\eta\pi^+\pi^-\gamma} = \frac{ien}{12\sqrt{3}\pi^2 F_0^3}(Q_u - Q_d)\varepsilon^{\mu\nu\rho\sigma}\mathscr{A}_\mu \partial_\nu \eta \partial_\rho \pi^+ \partial_\sigma \pi^-,$$

where the quark-charge difference $Q_u - Q_d = 1$ is independent of N_c. However, by investigating the corresponding η and η' decays up to next-to-leading order in the framework of the combined $1/N_c$ and chiral expansions, Borasoy and Lipartia have concluded that the number of colors cannot be determined from these decays due to the importance of sub-leading terms which are needed to account for the experimental decay widths and photon spectra [28].

3.5.4 Chiral Perturbation Theory at $\mathcal{O}(q^6)$

Mesonic chiral perturbation theory at $\mathcal{O}(q^4)$ has led to a host of successful applications and may be considered a full-grown and mature area of low-energy particle physics. For the time being, calculations at $\mathcal{O}(q^6)$ are state of the art (see Ref. [24] for an overview). Calculations in the even-intrinsic-parity sector start at $\mathcal{O}(q^2)$, and two-loop calculations at $\mathcal{O}(q^6)$ are thus of next-to-next-to-leading order (NNLO). The corresponding effective Lagrangian $\mathscr{L}_6$ was constructed in Refs. [22, 49] and contains, in its final form, 90 terms in the three-flavor sector (plus four contact terms analogous to the H_i terms of $\mathscr{L}_4$). The odd-intrinsic-parity sector starts at $\mathcal{O}(q^4)$ with the anomalous WZW action, as discussed in Sect. 3.5.3. In this sector next-to-leading-order (NLO), i.e. one-loop, calculations are of $\mathcal{O}(q^6)$. It has been known for some time that quantum corrections to the WZW classical action do not renormalize the coefficient of the WZW term [6, 17, 44] (D. Issler, 1990, SLAC-PUB-4943-REV, unpublished). The counter terms needed to renormalize the one-loop singularities at $\mathcal{O}(q^6)$ are of a conventional chirally invariant structure. In the three-flavor sector, the most general odd-intrinsic-parity Lagrangian at $\mathcal{O}(q^6)$ contains 23 independent terms [23, 46]. For an overview of applications in the odd-intrinsic-parity sector, we refer to Ref. [18].

Although an explicit calculation at the two-loop level is beyond the scope of these lecture notes, we want to discuss the results for the s-wave $\pi\pi$-scattering lengths a_0^0 and a_0^2 of Eq. 3.95. The s-wave $\pi\pi$-scattering lengths have been calculated at next-to-leading order [52] and at next-to-next-to-leading order [20, 21]. Let us have a closer look at the individual contributions to a_0^0 as reported in Ref. [20]:

$$
a_0^0 = \overbrace{0.156}^{\mathcal{O}(q^2)} + \overbrace{\underbrace{0.039}_{L} + \underbrace{0.005}_{anal.}}^{\mathcal{O}(q^4)\,:\,+28\%} + \overbrace{\underbrace{0.013}_{k_i} + \underbrace{0.003}_{L} + \underbrace{0.001}_{anal.}}^{\mathcal{O}(q^6)\,:\,+8.5\%} = \overbrace{0.217}^{total}. \tag{3.158}
$$

The corrections at $\mathcal{O}(q^4)$ consist of a dominant part from the chiral logarithms (L) of the one-loop diagrams and a less important analytical contribution (anal.) resulting from the one-loop diagrams as well as the tree graphs of $\mathscr{L}_4$. The total corrections at $\mathcal{O}(q^4)$ amount to 28% of the $\mathcal{O}(q^2)$ prediction. At $\mathcal{O}(q^6)$, one obtains two-loop corrections, one-loop corrections, and $\mathscr{L}_6$-tree-level contributions. Once again, the loop corrections (k_i, involving double chiral logarithms, and L) are more important than the analytical contributions. The influence of $\mathscr{L}_6$ was estimated via scalar- and vector-meson exchange and found to be very small. The result of Eq. 3.158 reveals a nice convergence and is in excellent agreement with the empirical data to be discussed below. Due to the relatively large strange-quark mass, the convergence in three-flavor calculations is usually slower.

By matching the chiral representation of the scattering amplitude with a dispersive representation [7, 85], the predictions for the s-wave $\pi\pi$-scattering lengths are [35, 37]

$$
a_0^0 = 0.220 \pm 0.005, \quad a_0^2 = -0.0444 \pm 0.0010. \tag{3.159}
$$

The empirical results for the s-wave $\pi\pi$-scattering lengths have been obtained from various sources. In the K_{e4} decay $K^+ \to \pi^+\pi^- e^+ \nu_e$, the connection with low-energy $\pi\pi$ scattering stems from a partial-wave analysis of the form factors relevant for the K_{e4} decay in terms of $\pi\pi$ angular momentum eigenstates. In the low-energy regime, the phases of these form factors are related by (a generalization of) Watson's theorem [96] to the corresponding phases of $I = 0$ s-wave and $I = 1$ p-wave elastic scattering [36]. Using effective-field-theory techniques, isospin-symmetry-breaking effects, generated by real and virtual photons and by the mass difference of the up and down quarks, were discussed in Ref. [39]. Performing a combined analysis of the Geneva-Saclay data [84], the BNL-E865 data [80, 81], and the NA48/2 data [13] results in [39]

$$
a_0^0 = 0.217 \pm 0.008_{\text{exp}} \pm 0.006_{\text{th}}, \tag{3.160}
$$

which is in excellent agreement with the prediction of Eq. 3.159. The $\pi^{\pm}p \to \pi^{\pm}\pi^+ n$ reactions require an extrapolation to the pion pole to extract the $\pi\pi$ amplitude and are thus regarded as containing more model dependence, $a_0^0 = 0.204 \pm 0.014\,(\text{stat}) \pm 0.008\,(\text{syst})$ [67]. The DIRAC Collaboration [1] makes use of a lifetime measurement of pionium to extract $|a_0^0 - a_0^2| = 0.264^{+0.033}_{-0.020}$. Finally, in the $K^{\pm} \to \pi^{\pm}\pi^0\pi^0$ decay, isospin-symmetry breaking leads to a cusp structure $\sim a_0 - a_2$ in the $\pi^0\pi^0$ invariant mass distribution near $s_{\pi^0\pi^0} \approx 4M_{\pi^+}^2$ [31, 32]. Based on the model of Ref. [32], the NA48/2 Collaboration extracts $a_0^0 - a_0^2 = 0.268 \pm 0.010\,(\text{stat}) \pm 0.004\,(\text{syst}) \pm 0.013\,(\text{ext})$. A more sophisticated analysis of the cusps in $K \rightarrow 3\pi$ within an effective-field-theory framework can be found in Refs. [26, 27, 38].

In particular, when analyzing the data of Ref. [80] in combination with the Roy equations, an upper limit $|\bar{l}_3| \leq 16$ was obtained in Ref. [36] for the scale-independent low-energy constant of the two-flavor Lagrangian $\mathscr{L}_4$ (see Eq. 3.145). The great interest generated by this result is to be understood in the context of the pion mass at $\mathcal{O}(q^4)$ (see Exercise 3.26),

$$M_\pi^2 = M^2 - \frac{\bar{l}_3}{32\pi^2 F^2} M^4 + \mathcal{O}(M^6), \tag{3.161}$$

where $M^2 = 2\hat{m}B$. Recall that the constant B is related to the scalar quark condensate in the chiral limit and that a nonvanishing quark condensate is a sufficient criterion for spontaneous chiral symmetry breakdown in QCD. If the expansion of M_π^2 in powers of the quark masses is dominated by the linear term in Eq. 3.161, the result is often referred to as the Gell-Mann-Oakes-Renner relation [56]. If the terms of order $\hat{m}^2$ were comparable or even larger than the linear terms, a different power counting or bookkeeping in ChPT would be required [68]. The estimate $|\bar{l}_3| \leq 16$ implies that the Gell-Mann-Oakes-Renner relation is indeed a decent starting point, because the contribution of the second term of Eq. 3.161 to the pion mass is approximately given by

$$-\frac{\bar{l}_3 M_\pi^2}{64\pi^2 F_\pi^2} M_\pi = -0.054 M_\pi \quad \text{for} \quad \bar{l}_3 = 16,$$

i.e., more than 94 % of the pion mass must stem from the quark condensate [36].

As our final example, let us discuss a constraint provided by chiral symmetry, relating the electromagnetic polarizabilities of the charged pion and radiative pion beta decay. In the framework of classical electrodynamics, the electric and magnetic polarizabilities α and β describe the response of a system to a static, uniform, external electric and magnetic field in terms of induced electric and magnetic dipole moments [62]. In principle, empirical information on the pion polarizabilities can be obtained from the differential cross section of low-energy Compton scattering on a charged pion, $\gamma(q) + \pi^+(p) \rightarrow \gamma(q') + \pi^+(p')$ (see Exercise 3.18),

$$\frac{d\sigma}{d\Omega_{\text{lab}}} = \left(\frac{\omega'}{\omega}\right)^2 \frac{e^2}{4\pi M_\pi} \left\{ \frac{e^2}{4\pi M_\pi} \frac{1+z^2}{2} \right.$$
$$\left. - \frac{\omega\omega'}{2} \left[(\alpha + \beta)_{\pi^+} (1+z)^2 + (\alpha - \beta)_{\pi^+} (1-z)^2 \right] \right\} + \cdots,$$

where $z = \hat{q} \cdot \hat{q}'$ and $\omega'/\omega = [1 + \omega(1-z)/M_\pi]$. The forward and backward differential cross sections are sensitive to $(\alpha + \beta)_{\pi^+}$ and $(\alpha - \beta)_{\pi^+}$, respectively.

Within the framework of the partially conserved axial-vector current (PCAC) hypothesis and current algebra the electromagnetic polarizabilities of the charged pion are related to the radiative charged-pion beta decay $\pi^+ \rightarrow e^+ \nu_e \gamma$ [89, 90].

The result obtained using ChPT at leading nontrivial order $[\mathcal{O}(q^4)]$ [16] is equivalent to the original PCAC result,

$$\alpha_{\pi^+} = -\beta_{\pi^+} = 2\frac{e^2}{4\pi}\frac{1}{(4\pi F_\pi)^2 M_\pi}\frac{\bar{l}_\Delta}{6},$$

where $\bar{l}_\Delta \equiv (\bar{l}_6 - \bar{l}_5)$ is a linear combination of scale-independent parameters of the two-flavor Lagrangian $\mathcal{L}_4$ (see Eq. 3.145). At $\mathcal{O}(q^4)$, this difference is related to the ratio $\gamma = F_A/F_V$ of the pion axial-vector form factor F_A and the vector form factor F_V of radiative pion beta decay [52], $\gamma = \bar{l}_\Delta/6$. Once this ratio is known, chiral symmetry makes an *absolute* prediction for the polarizabilities. This situation is similar to the s-wave $\pi\pi$-scattering lengths of Eq. 3.96 which are predicted once F_π is known. Using the most recent determination $\gamma = 0.443 \pm 0.015$ by the PIBETA Collaboration [50] (assuming $F_V = 0.0259$ obtained from the conserved vector current hypothesis) results in the $\mathcal{O}(q^4)$ prediction $\alpha_{\pi^+} = (2.64 \pm 0.09) \times 10^{-4}\,\mathrm{fm}^3$, where the estimate of the error is only the one due to the error of γ and does not include effects from higher orders in the quark-mass expansion.

Corrections to the leading-order PCAC result have been calculated at $\mathcal{O}(q^6)$ and turn out to be rather small [30, 55]. Using updated values for the LECs, the predictions of Ref. [55] are

$$(\alpha + \beta)_{\pi^+} = 0.16 \times 10^{-4}\,\mathrm{fm}^3, \tag{3.162}$$

$$(\alpha - \beta)_{\pi^+} = (5.7 \pm 1.0) \times 10^{-4}\,\mathrm{fm}^3. \tag{3.163}$$

The corresponding corrections to the $\mathcal{O}(q^4)$ result indicate a similar rate of convergence as for the $\pi\pi$-scattering lengths [20, 52]. The error for $(\alpha + \beta)_{\pi^+}$ is of the order $0.1 \times 10^{-4}\,\mathrm{fm}^3$, mostly from the dependence on the scale at which the $\mathcal{O}(q^6)$ low-energy coupling constants are estimated by resonance saturation.

As there is no stable pion target, empirical information about the pion polarizabilities is not easy to obtain. For that purpose, one has to consider reactions which contain the Compton scattering amplitude as a building block, such as, e.g., the Primakoff effect in high-energy pion-nucleus bremsstrahlung, $\pi^- Z \to \pi^- Z\gamma$, radiative pion photoproduction on the nucleon, $\gamma p \to \gamma\pi^+ n$, and pion pair production in $e^+ e^-$ scattering, $e^+ e^- \to e^+ e^- \pi^+ \pi^-$. Unfortunately, at present, the experimental situation looks rather contradictory (see Refs. [5, 55] for recent reviews of the data and further references to the experiments).

In terms of Feynman diagrams, the reaction $\gamma p \to \gamma\pi^+ n$ contains real Compton scattering on a charged pion as a pion pole diagram (see Fig. 3.15). This reaction was recently investigated at the Mainz Microtron MAMI with the result [5]

$$(\alpha - \beta)_{\pi^+} = (11.6 \pm 1.5_{\mathrm{stat}} \pm 3.0_{\mathrm{syst}} \pm 0.5_{\mathrm{mod}}) \times 10^{-4}\,\mathrm{fm}^3. \tag{3.164}$$

Fig. 3.15 The reaction $\gamma p \to \gamma \pi^+ n$ contains Compton scattering on a pion as a sub-diagram in the t channel, where $t = (p_n - p_p)^2$

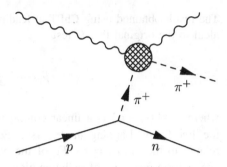

A similar result was obtained at Serpukhov using the Primakoff method [8],

$$(\alpha - \beta)_{\pi^+} = (13.6 \pm 2.8_{\text{stat}} \pm 2.4_{\text{syst}}) \times 10^{-4}\,\text{fm}^3, \qquad (3.165)$$

in agreement with the value from MAMI. Recently, also the COMPASS Collaboration at CERN has investigated this reaction [61] but a final result is not yet available. Unfortunately, the third method based on the reactions $e^+e^- \to \gamma\gamma \to \pi^+\pi^-$, has led to even more contradictory results (see Ref. [55]).

Comparing the empirical results of Eqs. 3.164 and 3.165 with the ChPT result of $(5.7 \pm 1.0) \times 10^{-4}\,\text{fm}^3$, we conclude that the electromagnetic polarizabilities of the charged pion remain one of the challenging topics of hadronic physics in the low-energy domain. Chiral symmetry provides a strong constraint in terms of radiative pion beta decay and mesonic chiral perturbation theory makes a firm prediction beyond the current-algebra result at the two-loop level. Both the experimental determination as well as the theoretical extraction from experiment require further efforts.

References

1. Adeva, B., et al., DIRAC Collaboration: Phys. Lett. B **619**, 50 (2005)
2. Adler, S.L.: Phys. Rev. **177**, 2426 (1969)
3. Adler, S.L., Bardeen, W.A.: Phys. Rev. **182**, 1517 (1969)
4. Adler, S.L.: In: Deser, S., Grisaru, M., Pendleton, H. (eds.) Lectures on Elementary Particles and Quantum Field Theory, 1970 Brandeis University Summer Institute in Theoretical Physics, vol. 1. M.I.T. Press, Cambridge (1970)
5. Ahrens, J., et al.: Eur. Phys. J. A **23**, 113 (2005)
6. Akhoury, R., Alfakih, A.: Ann. Phys. **210**, 81 (1991)
7. Ananthanarayan, B., Colangelo, G., Gasser, J., Leutwyler, H.: Phys. Rep. **353**, 207 (2001)
8. Antipov, Y.M., et al.: Phys. Lett. B **121**, 445 (1983)
9. Balachandran, A.P., Trahern, C.G.: Lectures on Group Theory for Physicists. Bibliopolis, Naples (1984)
10. Balachandran, A.P., Marmo, G., Skagerstam, B.S., Stern, A.: Classical Topology and Quantum States (Chap. 12.2). World Scientific, Singapore (1991)
11. Bardeen, W.A.: Phys. Rev. **184**, 1848 (1969)

12. Bär, O., Wiese, U.J.: Nucl. Phys. B **609**, 225 (2001)
13. Batley, J.R., et al., NA48/2 Collaboration: Eur. Phys. J. C **54**, 411 (2008)
14. Baym, G.: Phys. Rev. **117**, 886 (1960)
15. Bell, J.S., Jackiw, R.: Nuovo Cim. A **60**, 47 (1969)
16. Bijnens, J., Cornet, F.: Nucl. Phys. B **296**, 557 (1988)
17. Bijnens, J., Bramon, A., Cornet, F.: Z. Phys. C **46**, 599 (1990)
18. Bijnens, J.: Int. J. Mod. Phys. A **8**, 3045 (1993)
19. Bijnens, J., Ecker, G., Gasser, J.: In: Maiani, L., Pancheri, G., Paver, N. (eds.) The Second DAΦNE Physics Handbook, vol. 2, pp. 125–143. SIS, Frascati (1995)
20. Bijnens, J., Colangelo, G., Ecker, G., Gasser, J., Sainio, M.E.: Phys. Lett. B **374**, 210 (1996)
21. Bijnens, J., Colangelo, G., Ecker, G., Gasser, J., Sainio, M.E.: Nucl. Phys. B **508**, 263 (1997)
22. Bijnens, J., Colangelo, G., Ecker, G.: JHEP **9902**, 020 (1999)
23. Bijnens, J., Girlanda, L., Talavera, P.: Eur. Phys. J. C **23**, 539 (2002)
24. Bijnens, J.: Prog. Part. Nucl. Phys. **58**, 521 (2007)
25. Bijnens, J., Jemos, I.: In: PoS (CD09), 087 (2009)
26. Bissegger, M., Fuhrer, A., Gasser, J., Kubis, B., Rusetsky, A.: Phys. Lett. B **659**, 576 (2008)
27. Bissegger, M., Fuhrer, A., Gasser, J., Kubis, B., Rusetsky, A.: Nucl. Phys. B **806**, 178 (2009)
28. Borasoy, B., Lipartia, E.: Phys. Rev. D **71**, 014027 (2005)
29. Burgess, C.P.: Phys. Rep. **330**, 193 (2000)
30. Bürgi, U.: Nucl. Phys. B **479**, 392 (1996)
31. Cabibbo, N.: Phys. Rev. Lett. **93**, 121801 (2004)
32. Cabibbo, N., Isidori, G.: JHEP **0503**, 021 (2005)
33. Callan, C.G., Coleman, S.R., Wess, J., Zumino, B.: Phys. Rev. **177**, 2247 (1969)
34. Chisholm, J.: Nucl. Phys. **26**, 469 (1961)
35. Colangelo, G., Gasser, J., Leutwyler, H.: Phys. Lett. B **488**, 261 (2000)
36. Colangelo, G., Gasser, J., Leutwyler, H.: Phys. Rev. Lett. **86**, 5008 (2001)
37. Colangelo, G., Gasser, J., Leutwyler, H.: Nucl. Phys. B **603**, 125 (2001)
38. Colangelo, G., Gasser, J., Kubis, B., Rusetsky, A.: Phys. Lett. B **638**, 187 (2006)
39. Colangelo, G., Gasser, J., Rusetsky, A.: Eur. Phys. J. C **59**, 777 (2009)
40. Colangelo, G., et al.: arXiv:1011.4408 [hep-lat]
41. Coleman, S.: J. Math. Phys. **7**, 787 (1966)
42. Coleman, S.R., Wess, J., Zumino, B.: Phys. Rev. **177**, 2239 (1969)
43. Collins, J.C.: Renormalization. Cambridge University Press, Cambridge (1984)
44. Donoghue, J.F., Wyler, D.: Nucl. Phys. B **316**, 289 (1989)
45. Donoghue, J.F.: In: PoS (EFT09), 001 (2009)
46. Ebertshäuser, T., Fearing, H.W., Scherer, S.: Phys. Rev. D **65**, 054033 (2002)
47. Ecker, G., Gasser, J., Pich, A., de Rafael, E.: Nucl. Phys. B **321**, 311 (1989)
48. Ecker, G.: arXiv:hep-ph/0507056
49. Fearing, H.W., Scherer, S.: Phys. Rev. D **53**, 315 (1996)
50. Frlež, E., et al.: Phys. Rev. Lett. **93**, 181804 (2004)
51. Gasser, J., Leutwyler, H.: Phys. Rep. **87**, 77 (1982)
52. Gasser, J., Leutwyler, H.: Ann. Phys. **158**, 142 (1984)
53. Gasser, J., Leutwyler, H.: Nucl. Phys. B **250**, 465 (1985)
54. Gasser, J., Sainio, M.E., Švarc, A.: Nucl. Phys. B **307**, 779 (1988)
55. Gasser, J., Ivanov, M.A., Sainio, M.E.: Nucl. Phys. B **745**, 84 (2006)
56. Gell-Mann, M., Oakes, R.J., Renner, B.: Phys. Rev. **175**, 2195 (1968)
57. Georgi, H.: Weak Interactions and Modern Particle Theory. Benjamin/Cummings, Menlo Park (1984)
58. Georgi, H.: Ann. Rev. Nucl. Part. Sci. **43**, 207 (1993)
59. Gerber, P., Leutwyler, H.: Nucl. Phys. B **321**, 387 (1989)
60. Goldstone, J., Salam, A., Weinberg, S.: Phys. Rev. **127**, 965 (1962)

61. Guskov, A., On behalf of the COMPASS collaboration: J. Phys. Conf. Ser. **110**, 022016 (2008)
62. Holstein, B.R.: Comm. Nucl. Part. Phys. A **19**, 221 (1990)
63. Itzykson, C., Zuber, J.B.: Quantum Field Theory. McGraw-Hill, New York (1980)
64. Jones, H.F.: Groups, Representations and Physics. Hilger, Bristol (1990)
65. Kamefuchi, S., O'Raifeartaigh, L., Salam, A.: Nucl. Phys. **28**, 529 (1961)
66. Kaplan, D.B.: arXiv:nucl-th/9506035
67. Kermani, M., et al., CHAOS Collaboration: Phys. Rev. C **58**, 3431 (1998)
68. Knecht, M., Moussallam, B., Stern, J., Fuchs, N.H.: Nucl. Phys. B **457**, 513 (1995)
69. Leibbrandt, G.: Rev. Mod. Phys. **47**, 849 (1975)
70. Leutwyler, H.: In: Ellis, R.K., Hill, C.T., Lykken, J.D. (eds.) Perspectives in the Standard Model. Proceedings of the 1991 Advanced Theoretical Study Institute in Elementary Particle Physics, Boulder, CO, 2–28 June 1991. World Scientific, Singapore (1992)
71. Leutwyler, H.: Ann. Phys. **235**, 165 (1994)
72. Li, L.F., Pagels, H.: Phys. Rev. Lett. **26**, 1204 (1971)
73. Manohar, A.V., Georgi, H.: Nucl. Phys. B **234**, 189 (1984)
74. Manohar, A.V.: arXiv:hep-ph/9606222
75. Nakamura, K., et al., Particle Data Group: J. Phys. G **37**, 075021 (2010)
76. Necco, S.: In: PoS (Confinement8), 024 (2008)
77. O'Raifeartaigh, L.: Group Structure of Gauge Theories. Cambridge University Press, Cambridge (1986)
78. Pich, A.: arXiv:hep-ph/9806303
79. Pich, A.: In: PoS (Confinement8), 026 (2008)
80. Pislak, S., et al., BNL-E865 Collaboration: Phys. Rev. Lett. **87**, 221801 (2001)
81. Pislak, S., et al.: Phys. Rev. D **67**, 072004 (2003) [Erratum, ibid. D **81**, 119903 (2010)]
82. Polchinski, J.: arXiv:hep-th/9210046
83. Preston, M.A.: Physics of the Nucleus. Addison-Wesley, Reading (1962)
84. Rosselet, L., et al.: Phys. Rev. D **15**, 574 (1977)
85. Roy, S.M.: Phys. Lett. B **36**, 353 (1971)
86. Scherer, S., Fearing, H.W.: Phys. Rev. D **52**, 6445 (1995)
87. Scherer, S.: Adv. Nucl. Phys. **27**, 277 (2003)
88. Schwinger, J.S.: Phys. Lett. B **24**, 473 (1967)
89. Terent'ev, M.V.: Yad. Fiz. **16**, 162 (1972)
90. Terent'ev, M.V.: Sov. J. Nucl. Phys. **16**, 87 (1973)
91. 't Hooft, G., Veltman, M.J.: Nucl. Phys. B **44**, 189 (1972)
92. 't Hooft, G., Veltman, M.J.: Nucl. Phys. B **153**, 365 (1979)
93. Unkmeir, C., Scherer, S., L'vov, A.I., Drechsel, D.: Phys. Rev. D **61**, 034002 (2000)
94. Vafa, C., Witten, E.: Nucl. Phys. B **234**, 173 (1984)
95. Veltman, M.J.: Diagrammatica. The Path to Feynman Rules. Cambridge University Press, Cambridge (1994)
96. Watson, K.M.: Phys. Rev. **95**, 228 (1954)
97. Weinberg, S.: Phys. Rev. Lett. **17**, 616 (1966)
98. Weinberg, S.: Phys. Rev. Lett. **18**, 188 (1967)
99. Weinberg, S.: Phys. Rev. **166**, 1568 (1968)
100. Weinberg, S.: Physica A **96**, 327 (1979)
101. Weinberg, S.: The Quantum Theory of Fields. Foundations, vol. I (Chap. 12). Cambridge University Press, Cambridge (1995)
102. Wess, J., Zumino, B.: Phys. Lett. B **37**, 95 (1971)
103. Witten, E.: Nucl. Phys. B **223**, 422 (1983)

Chapter 4
Chiral Perturbation Theory for Baryons

In this chapter we will discuss matrix elements with a single baryon in the initial and final states. With such matrix elements we can, e.g., describe static properties such as masses or magnetic moments, form factors, or, finally, more complicated processes, such as pion-nucleon scattering, Compton scattering, pion photoproduction etc. Technically speaking, we are interested in the baryon-to-baryon transition amplitude in the presence of external fields (as opposed to the vacuum-to-vacuum transition amplitude of Sect. 1.4.4),

$$\mathcal{F}(\vec{p}\,', \vec{p}; v, a, s, p) = \langle \vec{p}\,', \text{out} | \vec{p}, \text{in} \rangle^c_{v,a,s,p}, \quad \vec{p} \neq \vec{p}\,',$$

determined by the Lagrangian of Eq. 1.151,

$$\mathcal{L} = \mathcal{L}^0_{\text{QCD}} + \mathcal{L}_{\text{ext}} = \mathcal{L}^0_{\text{QCD}} + \bar{q}\gamma_\mu \left(v^\mu + \frac{1}{3} v^\mu_{(s)} + \gamma_5 a^\mu \right) q - \bar{q}(s - i\gamma_5 p)q.$$

In the above equation, $|\vec{p}, \text{in}\rangle$ and $|\vec{p}\,', \text{out}\rangle$ denote asymptotic one-baryon in- and out-states, i.e., states which in the remote past and distant future behave as free one-particle states of momentum $\vec{p}$ and $\vec{p}\,'$, respectively. The functional $\mathcal{F}$ consists of connected diagrams only (superscript c). For example, the matrix elements of the vector and axial-vector currents between one-baryon states are given by

$$\langle \vec{p}\,' | V^\mu_a(x) | \vec{p} \rangle = \frac{\delta}{i\delta v_{a\mu}(x)} \mathcal{F}(\vec{p}\,', \vec{p}; v, a, s, p) \bigg|_{v=0, a=0, s=\mathcal{M}, p=0},$$

$$\langle \vec{p}\,' | A^\mu_a(x) | \vec{p} \rangle = \frac{\delta}{i\delta a_{a\mu}(x)} \mathcal{F}(\vec{p}\,', \vec{p}; v, a, s, p) \bigg|_{v=0, a=0, s=\mathcal{M}, p=0},$$

where $\mathcal{M} = \text{diag}(m_u, m_d, m_s)$ denotes the quark-mass matrix and

$$V^\mu_a(x) = \bar{q}(x)\gamma^\mu \frac{\lambda_a}{2} q(x), \quad A^\mu_a(x) = \bar{q}(x)\gamma^\mu \gamma_5 \frac{\lambda_a}{2} q(x).$$

S. Scherer and M. R. Schindler, *A Primer for Chiral Perturbation Theory*,
Lecture Notes in Physics 830, DOI: 10.1007/978-3-642-19254-8_4,
© Springer-Verlag Berlin Heidelberg 2012

As in the mesonic sector the method of calculating the Green functions associated
with the above functional consists of an effective-Lagrangian approach in combination with an appropriate power counting. Specific matrix elements will be
calculated applying the Gell-Mann and Low formula of perturbation theory [29].

4.1 Transformation Properties of the Fields

The group-theoretical foundations of constructing phenomenological Lagrangians
in the presence of spontaneous symmetry breaking were developed in Refs.
[13, 15, 63]. The fields entering the Lagrangian are assumed to transform under
irreducible representations of the subgroup H which leaves the vacuum invariant,
whereas the symmetry group G of the Hamiltonian or Lagrangian is nonlinearly
realized (for the transformation behavior of the Goldstone bosons, see Sect. 3.3).

Our aim is a description of the interaction of baryons with the Goldstone bosons
as well as the external fields at low energies. To that end we need to specify the
transformation properties of the dynamical fields entering the Lagrangian. Our
discussion follows Refs. [26, 30].

To be specific, we consider the octet of the $\frac{1}{2}^+$ baryons (see Fig. 3.4). With each
member of the octet we associate a complex, four-component Dirac field which we
arrange in a traceless 3×3 matrix B,

$$B = \sum_{a=1}^{8} \frac{B_a \lambda_a}{\sqrt{2}} = \begin{pmatrix} \frac{1}{\sqrt{2}}\Sigma^0 + \frac{1}{\sqrt{6}}\Lambda & \Sigma^+ & p \\ \Sigma^- & -\frac{1}{\sqrt{2}}\Sigma^0 + \frac{1}{\sqrt{6}}\Lambda & n \\ \Xi^- & \Xi^0 & -\frac{2}{\sqrt{6}}\Lambda \end{pmatrix}, \qquad (4.1)$$

where we have suppressed the dependence on x. For later use, we have to keep in
mind that each entry of Eq. 4.1 is a Dirac field, but for the purpose of discussing
the transformation properties under global flavor SU(3) this can be ignored,
because these transformations act on each of the four components in the same way.
In contrast to the mesonic case of Eq. 3.37, where we collected the fields of the
Goldstone-boson octet in a Hermitian traceless matrix ϕ, the B_a of the spin-1/2
case are not real (Hermitian), i.e., $B \neq B^\dagger$.

Exercise 4.1 Using Eq. 4.1, express the physical fields in terms of Cartesian
fields.

Now let us define the set

$$M \equiv \{B(x)|B(x) \text{ complex, traceless } 3 \times 3 \text{ matrix}\}, \qquad (4.2)$$

which under the addition of matrices is a complex vector space. The following
homomorphism is a representation of the abstract group $H = \text{SU}(3)_V$ on the vector
space M (see also Eq. 3.38):

$$\varphi : H \to \varphi(H), \quad V \mapsto \varphi(V) \quad \text{where } \varphi(V) : M \to M,$$
$$B(x) \mapsto B'(x) = \varphi(V)B(x) \equiv VB(x)V^\dagger. \tag{4.3}$$

$B'(x)$ is again an element of M, because $\text{Tr}[B'(x)] = \text{Tr}[VB(x)V^\dagger] = \text{Tr}[B(x)] = 0$. Equation 4.3 satisfies the homomorphism property,

$$\varphi(V_1)\varphi(V_2)B(x) = \varphi(V_1)V_2 B(x)V_2^\dagger = V_1 V_2 B(x)V_2^\dagger V_1^\dagger = (V_1 V_2)B(x)(V_1 V_2)^\dagger$$
$$= \varphi(V_1 V_2)B(x),$$

and is indeed a *representation* of SU(3), because

$$\varphi(V)[\lambda_1 B_1(x) + \lambda_2 B_2(x)] = V[\lambda_1 B_1(x) + \lambda_2 B_2(x)]V^\dagger = \lambda_1 VB_1(x)V^\dagger + \lambda_2 VB_2(x)V^\dagger$$
$$= \lambda_1 \varphi(V)B_1(x) + \lambda_2 \varphi(V)B_2(x).$$

Equation 4.3 is just the familiar statement that B transforms as an octet under (the adjoint representation of) SU(3)$_V$.[1]

Let us now turn to various representations and realizations of the group SU(3)$_L \times$ SU(3)$_R$. We consider two explicit examples and refer the interested reader to the textbook by Georgi [30] for more details. In analogy to the discussion of the quark fields in QCD, we may introduce left- and right-handed components of the baryon fields (see Eq. 1.37):

$$B_1 = P_L B_1 + P_R B_1 = B_L + B_R. \tag{4.4}$$

We define the set $M_1 \equiv \{(B_L(x), B_R(x))\}$ which under the addition of matrices is a complex vector space. The following homomorphism is a representation of the abstract group $G = $ SU(3)$_L \times$ SU(3)$_R$ on M_1:

$$(B_L, B_R) \mapsto (B_L', B_R') \equiv (LB_L L^\dagger, RB_R R^\dagger), \tag{4.5}$$

where we have suppressed the x dependence. The proof proceeds in complete analogy to that of Eq. 4.3.

As a second example, consider the set $M_2 \equiv \{B_2(x)\}$ with the homomorphism

$$B_2 \mapsto B_2' \equiv LB_2 L^\dagger, \tag{4.6}$$

i.e., the transformation behavior is independent of R. The mapping defines a representation of the group $G = $ SU(3)$_L \times$ SU(3)$_R$, although the transformation behavior is drastically different from the first example. However, the important feature which both mappings have in common is that under the subgroup $H = \{(V, V)|V \in$ SU(3)$\}$ of G both fields B_i transform as an octet:

---

[1] Technically speaking the adjoint representation is faithful (one-to-one) modulo the center Z of SU(3), which is defined as the set of all elements commuting with all elements of SU(3) and is given by $Z = \{\mathbb{1}, \exp(2\pi i/3)\mathbb{1}, \exp(4\pi i/3)\mathbb{1}\}$.

$$B_1 = B_L + B_R \overset{H}{\mapsto} VB_L V^\dagger + VB_R V^\dagger = VB_1 V^\dagger,$$

$$B_2 \overset{H}{\mapsto} VB_2 V^\dagger.$$

We will now show how in a theory also containing Goldstone bosons the various realizations may be connected to each other using field redefinitions. Here we consider the second example, with the fields B_2 of Eq. 4.6 and U of Eq. 3.41 transforming as

$$B_2 \mapsto LB_2 L^\dagger, \quad U \mapsto RUL^\dagger,$$

and define new baryon fields by

$$\tilde{B} \equiv UB_2 - \frac{1}{3}\mathrm{Tr}(UB_2),$$

so that the new pair $(\tilde{B}, U)$ transforms as

$$\tilde{B} \mapsto RUB_2 L^\dagger - \frac{1}{3}\mathrm{Tr}(RUB_2 L^\dagger), \quad U \mapsto RUL^\dagger.$$

Note in particular that $\tilde{B}$ still transforms as an octet under the subgroup $H = \mathrm{SU}(3)_V$.

Given that physical observables are invariant under field transformations we may choose a description of baryons that is maximally convenient for the construction of the effective Lagrangian [30] and which is commonly used in chiral perturbation theory. We start with $G = \mathrm{SU}(2)_L \times \mathrm{SU}(2)_R$ and consider the case of $G = \mathrm{SU}(3)_L \times \mathrm{SU}(3)_R$ later. Let

$$\Psi = \begin{pmatrix} p \\ n \end{pmatrix} \tag{4.7}$$

denote the nucleon field with two four-component Dirac fields for the proton and the neutron, and U the $\mathrm{SU}(2)$ matrix containing the pion fields. We have already seen in Sect. 3.3.2 that the mapping $U \mapsto RUL^\dagger$ defines a realization of G. We denote the unitary square root of U by u, $u^2(x) = U(x)$, and define the $\mathrm{SU}(2)$-valued function $K(L, R, U)$ by

$$u(x) \mapsto u'(x) = \sqrt{RUL^\dagger} \equiv RuK^{-1}(L, R, U), \tag{4.8}$$

i.e.,

$$K(L, R, U) = u'^{-1}Ru = \sqrt{RUL^\dagger}^{-1} R\sqrt{U}.$$

The following homomorphism defines an operation of G on the set $\{(U, \Psi)\}$:

$$\varphi(g) : \begin{pmatrix} U \\ \Psi \end{pmatrix} \mapsto \begin{pmatrix} U' \\ \Psi' \end{pmatrix} = \begin{pmatrix} RUL^\dagger \\ K(L, R, U)\Psi \end{pmatrix}, \tag{4.9}$$

because the identity leaves (U, Ψ) invariant and

$$\varphi(g_1)\varphi(g_2)\begin{pmatrix} U \\ \Psi \end{pmatrix} = \varphi(g_1)\begin{pmatrix} R_2 U L_2^\dagger \\ K(L_2, R_2, U)\Psi \end{pmatrix}$$

$$= \begin{pmatrix} R_1 R_2 U L_2^\dagger L_1^\dagger \\ K(L_1, R_1, R_2 U L_2^\dagger) K(L_2, R_2, U)\Psi \end{pmatrix}$$

$$= \begin{pmatrix} R_1 R_2 U (L_1 L_2)^\dagger \\ K(L_1 L_2, R_1 R_2, U)\Psi \end{pmatrix}$$

$$= \varphi(g_1 g_2)\begin{pmatrix} U \\ \Psi \end{pmatrix}.$$

Exercise 4.2 Consider the SU(N)-valued function ($N = 2, 3$)

$$K(L, R, U) = \sqrt{RUL^\dagger}^{-1} R\sqrt{U}.$$

Verify the homomorphism property

$$K(L_1, R_1, R_2 U L_2^\dagger) K(L_2, R_2, U) = K[(L_1 L_2), (R_1 R_2), U].$$

Note that for a general group element $g = (L, R)$ the transformation behavior of Ψ depends on U. For the special case of an isospin transformation, $R = L = V$, one obtains $u' = VuV^\dagger$, because

$$U' = u'^2 = VuV^\dagger VuV^\dagger = Vu^2V^\dagger = VUV^\dagger.$$

Comparison with Eq. 4.8 yields $K^{-1}(V, V, U) = V^\dagger$ or $K(V, V, U) = V$, i.e., Ψ transforms linearly as an isospin doublet under the isospin subgroup $H = \text{SU}(2)_V$ of $\text{SU}(2)_L \times \text{SU}(2)_R$. A general feature here is that the transformation behavior under the subgroup which leaves the ground state invariant is independent of U. Moreover, as already discussed in Sect. 3.3.2, the Goldstone bosons ϕ transform according to the adjoint representation of $\text{SU}(2)_V$, i.e., as an isospin triplet.

For the case $G = \text{SU}(3)_L \times \text{SU}(3)_R$ one uses the realization

$$\varphi(g) : \begin{pmatrix} U \\ B \end{pmatrix} \mapsto \begin{pmatrix} U' \\ B' \end{pmatrix} = \begin{pmatrix} RUL^\dagger \\ K(L, R, U)BK^\dagger(L, R, U) \end{pmatrix}, \tag{4.10}$$

where K is defined in complete analogy to Eq. 4.8 after inserting the corresponding SU(3) matrices.

4.2 Baryonic Effective Lagrangian at Lowest Order

Given the dynamical fields of Eqs. 4.9 and 4.10 and their transformation properties, we will now discuss the most general effective baryonic Lagrangian at lowest order. As in the vacuum sector, chiral symmetry provides constraints among the single-baryon Green functions. Analogous to the mesonic sector, these Ward identities will be satisfied if the Green functions are calculated from the most general effective Lagrangian coupled to external fields with a *local* invariance under the chiral group (see Sect. 1.4).

Let us start with the construction of the πN effective Lagrangian $\mathscr{L}_{\pi N}^{(1)}$ which we demand to have a *local* $SU(2)_L \times SU(2)_R \times U(1)_V$ symmetry. The transformation behavior of the external fields is given in Eq. 1.163, whereas the nucleon doublet Ψ and U transform as

$$\begin{pmatrix} U(x) \\ \Psi(x) \end{pmatrix} \mapsto \begin{pmatrix} V_R(x)U(x)V_L^\dagger(x) \\ \exp[-i\Theta(x)]K[V_L(x), V_R(x), U(x)]\Psi(x) \end{pmatrix}. \qquad (4.11)$$

The phase factor $\exp[-i\Theta(x)]$ is responsible for the $U(1)_V$ transformation of the nucleon field (see Eq. 1.162 for the corresponding transformation behavior of the quark fields). The local character of the transformation implies that we need to introduce a covariant derivative $D_\mu \Psi$ with the usual property that it transforms in the same way as Ψ:

$$D_\mu\Psi(x) \mapsto [D_\mu\Psi(x)]' = \exp[-i\Theta(x)]K[V_L(x), V_R(x), U(x)]D_\mu\Psi(x). \qquad (4.12)$$

Since K not only depends on V_L and V_R but also on U, we may expect the covariant derivative to contain u and $u^\dagger$ as well as their derivatives.

The so-called chiral connection (recall $\partial_\mu u u^\dagger = -u\partial_\mu u^\dagger$),

$$\Gamma_\mu = \frac{1}{2}[u^\dagger(\partial_\mu - ir_\mu)u + u(\partial_\mu - il_\mu)u^\dagger], \qquad (4.13)$$

is an integral part of the covariant derivative of the nucleon doublet:

$$D_\mu\Psi = \left(\partial_\mu + \Gamma_\mu - iv_\mu^{(s)}\right)\Psi. \qquad (4.14)$$

What needs to be shown is

$$\begin{aligned} D_\mu'\Psi' &= \left[\partial_\mu + \Gamma_\mu' - i\left(v_\mu^{(s)} - \partial_\mu\Theta\right)\right]\exp(-i\Theta)K\Psi \\ &= \exp(-i\Theta)K\left(\partial_\mu + \Gamma_\mu - iv_\mu^{(s)}\right)\Psi. \end{aligned} \qquad (4.15)$$

To that end, we make use of the product rule,

$$\partial_\mu[\exp(-i\Theta)K\Psi] = -i\partial_\mu\Theta\,\exp(-i\Theta)K\Psi + \exp(-i\Theta)\partial_\mu K\Psi + \exp(-i\Theta)K\partial_\mu\Psi,$$

in Eq. 4.15 and multiply by $\exp(i\Theta)$, reducing it to

$$\partial_\mu K = K\Gamma_\mu - \Gamma_\mu'K.$$

Using Eq. 4.8,

$$K = u'^\dagger V_R u = \underbrace{u' u'^\dagger}_{= \mathbb{1}} u'^\dagger V_R u = u' U'^\dagger V_R u = u' V_L \underbrace{U^\dagger}_{= u^\dagger u^\dagger} \underbrace{V_R^\dagger V_R}_{= \mathbb{1}} u = u' V_L u^\dagger,$$

we find

$$2(K\Gamma_\mu - \Gamma'_\mu K)$$
$$= K\left[u^\dagger(\partial_\mu - ir_\mu)u\right] - \left[u'^\dagger(\partial_\mu - iV_R r_\mu V_R^\dagger + V_R \partial_\mu V_R^\dagger)u'\right]K$$
$$+ (R \to L, r_\mu \to l_\mu, u \leftrightarrow u^\dagger, u' \leftrightarrow u'^\dagger)$$
$$= u'^\dagger V_R(\partial_\mu u - ir_\mu u) - u'^\dagger \partial_\mu u' \underbrace{K}_{= u'^\dagger V_R u} + iu'^\dagger V_R r_\mu \underbrace{V_R^\dagger u' K}_{= u} - u'^\dagger V_R \partial_\mu V_R^\dagger \underbrace{u' K}_{= V_R u}$$
$$+ (R \to L, r_\mu \to l_\mu, u \leftrightarrow u^\dagger, u' \leftrightarrow u'^\dagger)$$
$$= u'^\dagger V_R \partial_\mu u - iu'^\dagger V_R r_\mu u - \underbrace{u'^\dagger \partial_\mu u' u'^\dagger}_{= -\partial_\mu u'^\dagger} V_R u + iu'^\dagger V_R r_\mu u - u'^\dagger \underbrace{V_R \partial_\mu V_R^\dagger V_R}_{= -\partial_\mu V_R} u$$
$$+ (R \to L, r_\mu \to l_\mu, u \leftrightarrow u^\dagger, u' \leftrightarrow u'^\dagger)$$
$$= u'^\dagger V_R \partial_\mu u + \partial_\mu u'^\dagger V_R u + u'^\dagger \partial_\mu V_R u + (R \to L, u \leftrightarrow u^\dagger, u' \leftrightarrow u'^\dagger)$$
$$= \partial_\mu(u'^\dagger V_R u + u' V_L u^\dagger) = 2\partial_\mu K,$$

i.e., the covariant derivative defined in Eq. 4.14 indeed satisfies the condition of Eq. 4.12. At $\mathcal{O}(q)$, another Hermitian building block exists,[2] the so-called chiral vielbein,

$$u_\mu \equiv i\left[u^\dagger(\partial_\mu - ir_\mu)u - u(\partial_\mu - il_\mu)u^\dagger\right], \tag{4.16}$$

which under parity transforms as an axial vector:

$$u_\mu \overset{P}{\mapsto} i\left[u(\partial^\mu - il^\mu)u^\dagger - u^\dagger(\partial^\mu - ir^\mu)u\right] = -u^\mu.$$

Exercise 4.3   Using

$$u' = V_R u K^\dagger = KuV_L^\dagger,$$

show that, under $SU(2)_L \times SU(2)_R \times U(1)_V$, u_μ transforms as

$$u_\mu \mapsto K u_\mu K^\dagger.$$

The most general effective πN Lagrangian describing processes with a single nucleon in the initial and final states is then of the type $\bar{\Psi}\hat{O}\Psi$, where $\hat{O}$ is an operator acting in Dirac and isospin space, transforming under $SU(2)_L \times SU(2)_R \times U(1)_V$

---

[2] The power counting will be discussed below.

as $K\widehat{O}K^\dagger$. As in the mesonic sector, the Lagrangian must be a Hermitian Lorentz scalar which is even under the discrete symmetries C, P, and T.

The most general such Lagrangian with the smallest number of derivatives is given by [26]

$$\mathscr{L}_{\pi N}^{(1)} = \bar{\Psi}\left(i\slashed{D} - m + \frac{g_A}{2}\gamma^\mu\gamma_5 u_\mu\right)\Psi. \tag{4.17}$$

It contains two parameters (LECs) not determined by chiral symmetry: m, the nucleon mass in the chiral limit, and g_A, the axial-vector coupling constant in the chiral limit. We denote the physical values of these two quantities by m_N and g_A, respectively. The physical value of g_A is determined from neutron beta decay and is given by $g_A = 1.2694 \pm 0.0028$ [47]. The overall normalization of the Lagrangian is chosen such that in the case of no external fields and no pion fields it reduces to that of a free nucleon of mass m.

Exercise 4.4 Consider the lowest-order πN Lagrangian of Eq. 4.17. Assume that there are no external fields, $l_\mu = r_\mu = v_\mu^{(s)} = 0$, so that

$$\Gamma_\mu = \frac{1}{2}(u^\dagger\partial_\mu u + u\partial_\mu u^\dagger), \quad u_\mu = i(u^\dagger\partial_\mu u - u\partial_\mu u^\dagger).$$

By expanding

$$u = \exp\left(i\frac{\vec{\phi}\cdot\vec{\tau}}{2F}\right) = \mathbb{1} + i\frac{\vec{\phi}\cdot\vec{\tau}}{2F} - \frac{\vec{\phi}^2}{8F^2} + \cdots,$$

derive the interaction Lagrangians containing one and two pion fields, respectively.

Exercise 4.5 Consider the two-flavor Lagrangian

$$\mathscr{L}_{\text{eff}} = \mathscr{L}_{\pi N}^{(1)} + \mathscr{L}_2,$$

where

$$\mathscr{L}_{\pi N}^{(1)} = \bar{\Psi}\left(i\slashed{D} - m + \frac{g_A}{2}\gamma^\mu\gamma_5 u_\mu\right)\Psi,$$

$$\mathscr{L}_2 = \frac{F^2}{4}\text{Tr}[D_\mu U(D^\mu U)^\dagger] + \frac{F^2}{4}\text{Tr}(\chi U^\dagger + U\chi^\dagger).$$

(a) We would like to study this Lagrangian in the presence of an (external) electromagnetic four-vector potential $\mathscr{A}_\mu$. For that purpose we need to insert for the external fields (see Eq. 1.165)

$$r_\mu = l_\mu = -e\mathscr{A}_\mu\frac{\tau_3}{2}, \quad v_\mu^{(s)} = -\frac{e}{2}\mathscr{A}_\mu, \quad e > 0, \quad \frac{e^2}{4\pi} \approx \frac{1}{137}.$$

Derive the interaction Lagrangians $\mathscr{L}_{\gamma NN}$, $\mathscr{L}_{\pi NN}$, $\mathscr{L}_{\gamma\pi NN}$, and $\mathscr{L}_{\gamma\pi\pi}$. Here, the nomenclature is such that $\mathscr{L}_{\gamma NN}$ denotes the interaction Lagrangian describing the interaction of an external electromagnetic four-vector potential with a single nucleon in the initial and final states, respectively. For example, $\mathscr{L}_{\gamma\pi NN}$ must be symbolically of the type $\bar{\Psi}\phi\mathscr{A}\Psi$. Using Feynman rules, these four interaction Lagrangians would be sufficient to describe pion photoproduction on the nucleon, $\gamma N \rightarrow \pi N$, at lowest order in ChPT.

(b) Now we would like to describe the interaction with a massive charged weak boson $\mathscr{W}_\mu^\pm = (\mathscr{W}_{1\mu} \mp i\mathscr{W}_{2\mu})/\sqrt{2}$ (see Eq. 1.166),

$$r_\mu = 0, \quad l_\mu = -\frac{g}{\sqrt{2}}(\mathscr{W}_\mu^+ T_+ + \text{H.c.}),$$

where H.c. refers to the Hermitian conjugate and

$$T_+ = \begin{pmatrix} 0 & V_{ud} \\ 0 & 0 \end{pmatrix}.$$

Here, V_{ud} denotes an element of the Cabibbo-Kobayashi-Maskawa quark-mixing matrix,

$$|V_{ud}| = 0.97425 \pm 0.00022.$$

At lowest order in perturbation theory, the Fermi constant is related to the gauge coupling g and the W mass by

$$G_F = \sqrt{2}\frac{g^2}{8M_W^2} = 1.16637(1) \times 10^{-5}\,\text{GeV}^{-2}.$$

Derive the interaction Lagrangians $\mathscr{L}_{WNN}$ and $\mathscr{L}_{W\pi}$.

(c) Finally, we consider the neutral weak interaction (see Eq. 1.168),

$$r_\mu = e\tan(\theta_W)\mathscr{Z}_\mu\frac{\tau_3}{2},$$
$$l_\mu = -\frac{g}{\cos(\theta_W)}\mathscr{Z}_\mu\frac{\tau_3}{2} + e\tan(\theta_W)\mathscr{Z}_\mu\frac{\tau_3}{2},$$
$$v_\mu^{(s)} = \frac{e\tan(\theta_W)}{2}\mathscr{Z}_\mu,$$

where θ_W is the weak angle, $e = g\sin(\theta_W)$. Derive the interaction Lagrangians $\mathscr{L}_{ZNN}$ and $\mathscr{L}_{Z\pi}$.

Since the nucleon mass m_N does not vanish in the chiral limit, the zeroth component ∂^0 of the partial derivative acting on the nucleon field does not produce a "small" quantity. We thus have to address the new features of chiral power counting in the baryonic sector. The counting of the external fields as well as of covariant derivatives acting on the mesonic fields remains the same as in mesonic chiral perturbation theory (see Eq. 3.69). On the other hand, the counting of bilinears $\bar{\Psi}\Gamma\Psi$ is probably easiest understood by investigating the matrix elements of positive-energy plane-wave solutions to the free Dirac equation in the Dirac representation:

$$\psi^{(+)}(\vec{x}, t) = \exp(-ip \cdot x)\sqrt{E + m_N}\left(\begin{array}{c} \chi \\ \frac{\vec{\sigma} \cdot \vec{p}}{E + m_N}\chi \end{array}\right), \qquad (4.18)$$

where χ denotes a two-component Pauli spinor and $p = (E, \vec{p})$ with $E = \sqrt{\vec{p}^2 + m_N^2}$. In the low-energy limit, i.e., for nonrelativistic kinematics, the lower (small) component is suppressed as $|\vec{p}|/m_N$ in comparison with the upper (large) component. For the analysis of the bilinears it is convenient to divide the 16 Γ matrices into even and odd ones, $\mathscr{E} = \{\mathbb{1}, \gamma_0, \gamma_5\gamma_i, \sigma_{ij}\}$ and $\mathscr{O} = \{\gamma_5, \gamma_5\gamma_0, \gamma_i, \sigma_{i0}\}$ [19, 23], respectively, where odd matrices couple large and small components but not large with large, whereas even matrices do the opposite. Finally, $i\partial^\mu$ acting on the nucleon solution produces p^μ which we write symbolically as $p = (m_N, \vec{0}) + (E - m_N, \vec{p})$, where we count the second term as $\mathscr{O}(q)$, i.e., as a small quantity. Therefore, $\not{p} - m$ counts as $\mathscr{O}(q)$.[3] We are now in the position to summarize the chiral counting scheme for the (new) elements of baryon chiral perturbation theory [40]:

$$\begin{aligned} \Psi, \bar{\Psi} &= \mathscr{O}(q^0), D_\mu\Psi = \mathscr{O}(q^0), (i\not{D} - m)\Psi = \mathscr{O}(q), \\ \mathbb{1}, \gamma_\mu, \gamma_5\gamma_\mu, \sigma_{\mu\nu} &= \mathscr{O}(q^0), \gamma_5 = \mathscr{O}(q), \end{aligned} \qquad (4.19)$$

where the order given is the minimal one. For example, γ_μ has both an $\mathscr{O}(q^0)$ piece, γ_0, as well as an $\mathscr{O}(q)$ piece, γ_i. A rigorous nonrelativistic reduction may be achieved in the framework of the Foldy-Wouthuysen method [19, 23] or the heavy-baryon approach [5, 37] (see Sect. 4.6.1).

The construction of the $SU(3)_L \times SU(3)_R$ Lagrangian proceeds similarly except for the fact that the baryon fields are contained in the 3×3 matrix of Eq. 4.1 transforming as $KBK^\dagger$. Analogously to the mesonic sector, the building blocks are written as products transforming as $K...K^\dagger$ with a trace taken at the end. The lowest-order Lagrangian reads [30, 40]

$$\mathscr{L}_{MB}^{(1)} = \text{Tr}[\bar{B}(i\not{D} - M_0)B] - \frac{D}{2}\text{Tr}(\bar{B}\gamma^\mu\gamma_5\{u_\mu, B\}) - \frac{F}{2}\text{Tr}(\bar{B}\gamma^\mu\gamma_5[u_\mu, B]), \qquad (4.20)$$

where M_0 denotes the mass of the baryon octet in the chiral limit. The covariant derivative of B is defined as

$$D_\mu B = \partial_\mu B + [\Gamma_\mu, B], \qquad (4.21)$$

with Γ_μ of Eq. 4.13 (for $SU(3)_L \times SU(3)_R$). The constants D and F may be determined by fitting the semi-leptonic decays $B \to B' + e^- + \bar{\nu}_e$ at tree level [9]:

---

[3] The quantity $m_N - m$ is of $\mathscr{O}(q^2)$ as we will see in Sect. 4.5.3.

$$D = 0.80, \quad F = 0.50. \tag{4.22}$$

Exercise 4.6 Consider the three-flavor Lagrangian of Eq. 4.20 in the absence of external fields:

$$D_\mu B = \partial_\mu B + \frac{1}{2}[u^\dagger \partial_\mu u + u \partial_\mu u^\dagger, B], \quad u_\mu = i(u^\dagger \partial_\mu u - u \partial_\mu u^\dagger).$$

Using

$$B = \frac{B_a \lambda_a}{\sqrt{2}}, \quad \bar{B} = \frac{\bar{B}_b \lambda_b}{\sqrt{2}},$$

show that the interaction Lagrangians with one and two mesons can be written as

$$\mathcal{L}_{\phi BB}^{(1)} = \frac{1}{F_0}(d_{abc}D + if_{abc}F)\bar{B}_b \gamma^\mu \gamma_5 B_a \partial_\mu \phi_c,$$

$$\mathcal{L}_{\phi\phi BB}^{(1)} = -\frac{i}{2F_0^2} f_{abe} f_{cde} \bar{B}_b \gamma^\mu B_a \phi_c \partial_\mu \phi_d.$$

Hint: $u^\dagger \partial_\mu u + u \partial_\mu u^\dagger = u^\dagger \partial_\mu u - \partial_\mu u u^\dagger = [u^\dagger, \partial_\mu u]$. Make use of Eqs. 1.10 and 1.12.

4.3 Applications at Lowest Order

4.3.1 Goldberger-Treiman Relation and the Axial-Vector Current Matrix Element

We have seen in Sect. 1.3.6 that the quark masses in QCD give rise to a non-vanishing divergence of the axial-vector current operator (see Eq. 1.112). Here we will discuss the implications for the matrix elements of the pseudoscalar density and of the axial-vector current evaluated between single-nucleon states in terms of the lowest-order Lagrangians of Eqs. 3.77 and 4.17. In particular, we will see that the divergence equation

$$\langle N(p')|\partial_\mu A_i^\mu(0)|N(p)\rangle = \langle N(p')|\hat{m}P_i(0)|N(p)\rangle, \tag{4.23}$$

where $\hat{m} = m_u = m_d$, is satisfied in ChPT.

The nucleon matrix element of the pseudoscalar density can be parameterized as[4]

$$\hat{m}\langle N(p')|P_i(0)|N(p)\rangle = \frac{M_\pi^2 F_\pi}{M_\pi^2 - t} G_{\pi N}(t) i\bar{u}(p')\gamma_5 \tau_i u(p), \tag{4.24}$$

---

[4] In the following, spin and isospin quantum numbers as well as isospinors are suppressed.

where $t = (p' - p)^2$. Equation 4.24 *defines* the form factor $G_{\pi N}(t)$ in terms of the QCD operator $\hat{m}P_i(x)$. The operator $\hat{m}P_i(x)/(M_\pi^2 F_\pi)$ serves as an interpolating pion field and thus $G_{\pi N}(t)$ is also referred to as the pion-nucleon form factor (for this specific choice of the interpolating pion field). The pion-nucleon coupling constant $g_{\pi N}$ is defined as $g_{\pi N} \equiv G_{\pi N}(t = M_\pi^2)$.

The Lagrangian $\mathcal{L}_{\pi N}^{(1)}$ of Eq. 4.17 does not generate a direct coupling of an external pseudoscalar field $p_i(x)$ to the nucleon, i.e., it does not contain any terms involving χ or $\chi^\dagger$. At lowest order in the chiral expansion, the matrix element of the pseudoscalar density is therefore given in terms of the diagram of Fig. 4.1, i.e., the pseudoscalar source produces a pion which propagates and is then absorbed by the nucleon. The coupling of a pseudoscalar field to the pion in the framework of $\mathcal{L}_2$ is given by

$$\mathcal{L}_{\text{ext}} = i\frac{F^2 B}{2}\text{Tr}(pU^\dagger - Up) = 2BFp_i\phi_i + \cdots. \tag{4.25}$$

When working with the realization of Eq. 4.9 it is convenient to use the exponential parameterization

$$U(x) = \exp\left[i\frac{\vec{\phi}(x)\cdot\vec{\tau}}{F}\right],$$

because in that case the square root is simply given by

$$u(x) = \exp\left[i\frac{\vec{\phi}(x)\cdot\vec{\tau}}{2F}\right].$$

According to Fig. 4.1, we need to identify the interaction term of a nucleon with a single pion. In the absence of external fields the chiral vielbein of Eq. 4.16 is odd in the pion fields,

$$u_\mu = i\left[u^\dagger\partial_\mu u - u\partial_\mu u^\dagger\right] \overset{\phi_i \mapsto -\phi_i}{\longmapsto} i\left[u\partial_\mu u^\dagger - u^\dagger\partial_\mu u\right] = -u_\mu. \tag{4.26}$$

Expanding u and $u^\dagger$ as

$$u = 1 + i\frac{\vec{\phi}\cdot\vec{\tau}}{2F} + O(\phi^2), \quad u^\dagger = 1 - i\frac{\vec{\phi}\cdot\vec{\tau}}{2F} + O(\phi^2), \tag{4.27}$$

we obtain

$$u_\mu = -\frac{\partial_\mu\vec{\phi}\cdot\vec{\tau}}{F} + O(\phi^3), \tag{4.28}$$

which, when inserted into $\mathcal{L}_{\pi N}^{(1)}$ of Eq. 4.17, generates the following interaction Lagrangian (see Exercise 4.4):

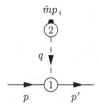

Fig. 4.1 Lowest-order contribution to the single-nucleon matrix element of the pseudoscalar density. The mesonic and baryonic vertices derived from $\mathscr{L}_2$ and $\mathscr{L}_{\pi N}^{(1)}$, respectively, are denoted by the numbers 2 and 1 in the interaction blobs

$$\mathscr{L}_{\text{int}} = -\frac{1}{2}\frac{g_A}{F}\bar{\Psi}\gamma^\mu\gamma_5 \underbrace{\partial_\mu\vec{\phi}\cdot\vec{\tau}}_{= \partial_\mu\phi_j\tau_j} \Psi. \tag{4.29}$$

(Note that the sign is opposite to the conventionally-used pseudovector pion-nucleon coupling.[5]) The Feynman rule for the vertex of an incoming pion with four-momentum q and Cartesian isospin index i is given by

$$i\left(-\frac{1}{2}\frac{g_A}{F}\right)\gamma^\mu\gamma_5\tau_j\delta_{ji}(-iq_\mu) = -\frac{1}{2}\frac{g_A}{F}\,\not{q}\gamma_5\tau_i. \tag{4.30}$$

On the other hand, the chiral connection of Eq. 4.13 with the external fields set to zero is even in the pion fields,

$$\Gamma_\mu = \frac{1}{2}\left[u^\dagger\partial_\mu u + u\partial_\mu u^\dagger\right] \overset{\phi_i \mapsto -\phi_i}{\longmapsto} \frac{1}{2}\left[u\partial_\mu u^\dagger + u^\dagger\partial_\mu u\right] = \Gamma_\mu, \tag{4.31}$$

i.e., it does not contribute to the single-pion vertex.

We now put the individual pieces together and obtain for the diagram of Fig. 4.1

$$\hat{m}2BF\frac{i}{t - M^2}\bar{u}(p')\left(-\frac{1}{2}\frac{g_A}{F}\,\not{q}\gamma_5\tau_i\right)u(p) = M^2F\frac{mg_A}{F}\frac{1}{M^2 - t}\bar{u}(p')\gamma_5 i\tau_i u(p),$$

where we used $M^2 = 2B\hat{m}$ and the Dirac equation to show $\bar{u}\,\not{q}\gamma_5 u = 2m\bar{u}\gamma_5 u$. At $\mathcal{O}(q^2)$, $F_\pi = F$ and $M_\pi^2 = M^2$ so that, by comparison with Eq. 4.24, we can read off the lowest-order result

$$G_{\pi N}(t) = \frac{m}{F}g_A, \tag{4.32}$$

i.e., at this order the form factor does not depend on t. In general, the pion-nucleon coupling constant is defined at $t = M_\pi^2$ which, in the present case, simply yields

---

[5] In fact, also the definition of the pion-nucleon form factor of Eq. 4.24 contains a sign opposite to the standard convention so that, in the end, the Goldberger-Treiman relation emerges with the conventional sign.

$$g_{\pi N} = G_{\pi N}(M_\pi^2) = \frac{m}{F} g_A.$$ (4.33)

Equation 4.33 represents the famous Goldberger-Treiman relation [32, 48] which establishes a connection between quantities entering weak processes, F_π and g_A (to be discussed below), and a typical strong-interaction quantity, namely the pion-nucleon coupling constant $g_{\pi N}$. The numerical violation of the Goldberger-Treiman relation, as expressed in the so-called Goldberger-Treiman discrepancy [50],

$$\Delta \equiv 1 - \frac{m_N g_A}{F_\pi g_{\pi N}},$$ (4.34)

is at the percent level,[6] although one has to keep in mind that *all four* physical quantities move from their chiral-limit values g_A etc. to the empirical ones g_A etc.

Using Lorentz covariance and isospin symmetry, the matrix element of the axial-vector current between initial and final nucleon states—excluding second-class currents [61]—can be parameterized as[7]

$$\langle N(p')|A_i^\mu(0)|N(p)\rangle = \bar{u}(p')\left[\gamma^\mu G_A(t) + \frac{(p'-p)^\mu}{2m_N}G_P(t)\right]\gamma_5\frac{\tau_i}{2}u(p),$$ (4.35)

where $t = (p'-p)^2$, and $G_A(t)$ and $G_P(t)$ are the axial and induced pseudoscalar form factors, respectively.

At lowest order, an external axial-vector field $a_{i\mu}$ couples directly to the nucleon as

$$\mathcal{L}_{\text{ext}} = g_A\bar{\Psi}\gamma^\mu\gamma_5\frac{\tau_i}{2}\Psi a_{i\mu}+\cdots,$$ (4.36)

which is obtained from $\mathcal{L}_{\pi N}^{(1)}$ with $u_\mu = (r_\mu - l_\mu)+\cdots = 2a_\mu+\cdots$. A second contribution results from the coupling of an external axial-vector field to the pion, propagation of the pion, and subsequent absorption of the pion by the nucleon. The coupling of an external axial-vector field to pions is obtained from $\mathcal{L}_2$ with $r_\mu = -l_\mu = a_\mu$,

$$\mathcal{L}_{\text{ext}} = -F\partial^\mu\phi_i a_{i\mu}+\cdots,$$ (4.37)

which gives rise to a diagram similar to Fig. 4.1, with $\hat{m}p_i$ replaced by $a_{i\mu}$.

---

[6] Using $m_N = (m_p + m_n)/2 = 938.92$ MeV, $g_A = 1.2695(29)$, $F_\pi = 92.42(26)$ MeV, and $g_{\pi N} = 13.21^{+0.11}_{-0.05}$ [56], one obtains $\Delta_{\pi N} = (2.37^{+0.89}_{-0.51})$ %.

[7] The terminology "first and second classes" refers to the transformation property of strangeness-conserving semi-leptonic weak interactions under $\mathcal{G}$ conjugation [61] which is the product of charge symmetry and charge conjugation $\mathcal{G} = \mathcal{C}\exp(i\pi Q_{V2})$. A second-class contribution would appear in terms of a third form factor G_T contributing as

$$G_T(t)\bar{u}(p')i\frac{\sigma^{\mu\nu}q_\nu}{2m_N}\gamma_5\frac{\tau_i}{2}u(p).$$

Assuming perfect $\mathcal{G}$-conjugation symmetry, the form factor G_T vanishes.

The matrix element is thus given by

$$\bar{u}(p')\left\{g_A\gamma^\mu\gamma_5\frac{\tau_i}{2} + \left[-\frac{1}{2}\frac{g_A}{F}(\not{p}'-\not{p})\gamma_5\tau_i\right]\frac{i}{q^2-M^2}(-iFq^\mu)\right\}u(p),$$

from which we obtain, by applying the Dirac equation,

$$G_A(t) = g_A, \tag{4.38}$$

$$G_P(t) = -\frac{4m^2g_A}{t-M^2}. \tag{4.39}$$

At this order, the axial form factor does not yet show a t dependence. The axial-vector coupling constant is defined as $G_A(0)$, which is simply given by g_A. We have thus identified the second new parameter of $\mathscr{L}_{\pi N}^{(1)}$ besides the nucleon mass m. Both quantities refer to the chiral limit. The induced pseudoscalar form factor is determined by pion exchange, which is the simplest version of the so-called pion-pole dominance. The $1/(t-M^2)$ behavior of G_P is not in conflict with the book-keeping of a calculation at $\mathcal{O}(q)$, because, according to Eq. 3.69, the external axial-vector field a_μ counts as $\mathcal{O}(q)$, and the definition of the matrix element contains a momentum $(p'-p)^\mu$ and the chirality matrix γ_5 (see Eq. 4.19) so that the combined order of all elements is indeed $\mathcal{O}(q)$.

It is straightforward to verify that the form factors of Eqs. 4.32, 4.38, and 4.39 satisfy the relation

$$2m_N G_A(t) + \frac{t}{2m_N}G_P(t) = 2\frac{M_\pi^2 F_\pi}{M_\pi^2-t}G_{\pi N}(t), \tag{4.40}$$

which is required by the divergence equation of Eq. 4.23 with the parameterizations of Eqs. 4.24 and 4.35 for the matrix elements. In other words, only two of the three form factors G_A, G_P, and $G_{\pi N}$ are independent. Note that this relation is not restricted to small values of t but holds for any t.

Exercise 4.7 According to Eq. 1.112, the divergence of the axial-vector current in the two-flavor sector is given by

$$\partial_\mu A_i^\mu(x) = \hat{m}P_i(x), \quad i = 1,2,3,$$

where we have assumed $\hat{m} = m_u = m_d$. Let $|A\rangle$ and $|B\rangle$ denote some (arbitrary) hadronic states which are eigenstates of the four-momentum operator P with eigenvalues p_A and p_B, respectively. Evaluating the above operator equation between $|A\rangle$ and $\langle B|$ and using translational invariance, one obtains

$$\langle B|\partial_\mu A_i^\mu(x)|A\rangle = \partial_\mu\langle B|A_i^\mu(x)|A\rangle = \partial_\mu\left[\langle B|e^{iP\cdot x}A_i^\mu(0)e^{-iP\cdot x}|A\rangle\right]$$
$$= \partial_\mu\left[e^{i(p_B-p_A)\cdot x}\langle B|A_i^\mu(0)|A\rangle\right] = iq_\mu e^{iq\cdot x}\langle B|A_i^\mu(0)|A\rangle$$
$$= e^{iq\cdot x}\hat{m}\langle B|P_i(0)|A\rangle,$$

where we introduced $q = p_B - p_A$. Dividing both sides by $e^{iq \cdot x} \neq 0$, we obtain

$$iq_\mu \langle B|A_i^\mu(0)|A \rangle = \hat{m} \langle B|P_i(0)|A \rangle.$$

(a) Make use of the parameterizations of Eqs. 4.24 and 4.35 for the nucleon matrix elements and derive Eq. 4.40.
   Hint: Make use of the Dirac equation.
(b) Verify that the lowest-order predictions

$$G_A(t) = g_A, \quad G_P(t) = -\frac{4m^2 g_A}{t - M^2}, \quad G_{\pi N}(t) = \frac{m}{F} g_A,$$

indeed satisfy this constraint. Note that $m_N = m + \mathcal{O}(q^2)$.

4.3.2 Pion-Nucleon Scattering

As another example, we will consider pion-nucleon scattering and show how the effective Lagrangian of Eq. 4.17 reproduces the Weinberg-Tomozawa predictions for the s-wave scattering lengths [60, 62]. We will contrast the results with those of a tree-level calculation within pseudoscalar (PS) and pseudovector (PV) pion-nucleon couplings.

Before calculating the πN scattering amplitude within ChPT, we introduce a general parameterization of the invariant amplitude $\mathcal{M} = iT$ for the process $\pi_a(q) + N(p) \rightarrow \pi_b(q') + N(p')$ [14]:[8]

$$T_{ab}(p,q;p',q') = \frac{1}{2}\{\tau_b, \tau_a\}T^+(p,q;p',q') + \frac{1}{2}[\tau_b, \tau_a]T^-(p,q;p',q')$$
$$= \delta_{ab}T^+(p,q;p',q') - i\varepsilon_{abc}\tau_c T^-(p,q;p',q'), \qquad (4.41)$$

where

$$T^\pm(p,q;p',q') = \bar{u}(p')\left[A^\pm(\nu,\nu_B) + \frac{1}{2}(\slashed{q} + \slashed{q}')B^\pm(\nu,\nu_B)\right]u(p). \qquad (4.42)$$

The amplitudes $A^\pm$ and $B^\pm$ are functions of the two independent scalar kinematical variables

---

[8] One also finds the parameterization

$$T = \bar{u}(p')\left(D - \frac{1}{4m_N}[\slashed{q}', \slashed{q}]B\right)u(p)$$

with $D = A + \nu B$, where, for simplicity, we have omitted the isospin labels.

$$v = \frac{s - u}{4m_N} = \frac{(p + p') \cdot q}{2m_N} = \frac{(p + p') \cdot q'}{2m_N}, \tag{4.43}$$

$$v_B = -\frac{q \cdot q'}{2m_N} = \frac{t - 2M_\pi^2}{4m_N}, \tag{4.44}$$

where $s = (p + q)^2$, $t = (p' - p)^2$, and $u = (p' - q)^2$ are the usual Mandelstam variables satisfying $s + t + u = 2m_N^2 + 2M_\pi^2$. From pion-crossing symmetry,

$$T_{ab}(p, q; p', q') = T_{ba}(p, -q'; p', -q),$$

we obtain for the crossing behavior of the amplitudes:

$$\begin{aligned} A^+(-v, v_B) &= A^+(v, v_B), \quad A^-(-v, v_B) = -A^-(v, v_B), \\ B^+(-v, v_B) &= -B^+(v, v_B), \quad B^-(-v, v_B) = B^-(v, v_B). \end{aligned} \tag{4.45}$$

As in $\pi\pi$ scattering, one often also finds the isospin decomposition as in Exercise 3.14,

$$\langle I', I_3' | T | I, I_3 \rangle = T^I \delta_{II'} \delta_{I_3 I_3'}.$$

In this context we would like to point out that our convention for the physical pion fields (and states) (see Exercise 3.5) differs by a factor (-1) for the π^+ from the spherical convention which is commonly used in the context of applying the Wigner-Eckart theorem. Taking into account a factor of (-1) for each π^+ in the initial and final states, the relation between the two sets is given by [18]

$$\begin{aligned} T^{\frac{1}{2}} &= T^+ + 2T^-, \\ T^{\frac{3}{2}} &= T^+ - T^-. \end{aligned} \tag{4.46}$$

To verify Eqs. 4.46, we consider

$$\begin{aligned} T^{\pi^+\pi^+} &= \frac{1}{2}(T_{11} - iT_{12} + iT_{21} + T_{22}) = T^+ - \tau_3 T^-, \\ T^{\pi^+\pi^0} &= \frac{1}{\sqrt{2}}(T_{13} + iT_{23}) = \tau_+ T^-, \end{aligned}$$

where $\tau_\pm = (\tau_1 \pm i\tau_2)/\sqrt{2}$. We then evaluate the matrix elements

$$\begin{aligned} \langle p\pi^+ | T | p\pi^+ \rangle &= T^+ - T^-, \\ \langle p\pi^0 | T | n\pi^+ \rangle &= \sqrt{2}T^-. \end{aligned}$$

A comparison with the results of Exercise 4.8 below,

$$\text{sph.}\langle p\pi^+|T|p\pi^+\rangle_{\text{sph.}} = T^{\frac{3}{2}} = (-1)^2\langle p\pi^+|T|p\pi^+\rangle = T^+ - T^-,$$

$$\text{sph.}\langle p\pi^0|T|n\pi^+\rangle_{\text{sph.}} = \frac{\sqrt{2}}{3}(T^{\frac{3}{2}} - T^{\frac{1}{2}}) = (-1)\langle p\pi^0|T|n\pi^+\rangle = -\sqrt{2}T^-,$$

results in Eq. 4.46. (The subscript sph. serves to distinguish the spherical convention from our convention.)

Exercise 4.8 Consider the general parameterization of the invariant amplitude $\mathcal{M} = iT$ for the process $\pi_a(q) + N(p) \to \pi_b(q') + N(p')$ of Eqs. 4.41 and 4.42 with the kinematical variables of Eqs. 4.43 and 4.44.

(a) Show that

$$s - m_N^2 = 2m_N(v - v_B), \quad u - m_N^2 = -2m_N(v + v_B).$$

Hint: Make use of four-momentum conservation, $p + q = p' + q'$, and of the mass-shell conditions, $p^2 = p'^2 = m_N^2$, $q^2 = q'^2 = M_\pi^2$.
Derive the threshold values

$$v|_{\text{thr}} = M_\pi, \quad v_B|_{\text{thr}} = -\frac{M_\pi^2}{2m_N}.$$

(b) Show that from pion-crossing symmetry,

$$T_{ab}(p, q; p', q') = T_{ba}(p, -q'; p', -q),$$

we obtain the crossing behavior of Eq. 4.45.
(c) The physical πN channels may be expressed in terms of the isospin eigenstates as (a spherical convention is understood)

$$|p\pi^+\rangle = \left|\frac{3}{2}, \frac{3}{2}\right\rangle,$$

$$|p\pi^0\rangle = \sqrt{\frac{2}{3}}\left|\frac{3}{2}, \frac{1}{2}\right\rangle + \frac{1}{\sqrt{3}}\left|\frac{1}{2}, \frac{1}{2}\right\rangle,$$

$$|n\pi^+\rangle = \frac{1}{\sqrt{3}}\left|\frac{3}{2}, \frac{1}{2}\right\rangle - \sqrt{\frac{2}{3}}\left|\frac{1}{2}, \frac{1}{2}\right\rangle,$$

$$|p\pi^-\rangle = \frac{1}{\sqrt{3}}\left|\frac{3}{2}, -\frac{1}{2}\right\rangle + \sqrt{\frac{2}{3}}\left|\frac{1}{2}, -\frac{1}{2}\right\rangle,$$

$$|n\pi^0\rangle = \sqrt{\frac{2}{3}}\left|\frac{3}{2}, -\frac{1}{2}\right\rangle - \frac{1}{\sqrt{3}}\left|\frac{1}{2}, -\frac{1}{2}\right\rangle,$$

$$|n\pi^-\rangle = \left|\frac{3}{2}, -\frac{3}{2}\right\rangle.$$

Using

$$\langle I', I_3'|T|I, I_3\rangle = T^I \delta_{II'}\delta_{I_3 I_3'},$$

derive the expressions for $\langle p\pi^0|T|n\pi^+\rangle$, $\langle p\pi^0|T|p\pi^0\rangle$, and $\langle n\pi^+|T|n\pi^+\rangle$. Verify that

$$\langle p\pi^0|T|p\pi^0\rangle - \langle n\pi^+|T|n\pi^+\rangle = \frac{1}{\sqrt{2}}\langle p\pi^0|T|n\pi^+\rangle.$$

Exercise 4.9 Consider the so-called pseudoscalar pion-nucleon interaction[9]

$$\mathscr{L}_{\pi NN}^{PS} = ig_{\pi N}\bar{\Psi}\gamma_5\vec{\phi}\cdot\vec{\tau}\Psi.$$

The Feynman rule for both the absorption and the emission of a pion with Cartesian isospin index a is given by

$$-g_{\pi N}\gamma_5\tau_a.$$

Derive the s- and u-channel pole contributions to the invariant amplitude of pion-nucleon scattering (see Fig. 4.3). A t-channel pion-pole contribution does not exist, because a triple-pion vertex does not exist.

Remark The Feynman propagator of a nucleon with mass m (in the chiral limit) reads

$$iS_F(p) = \frac{i}{\not{p} - m + i0^+},$$

where the unit matrix in isospin space has been omitted.

Let us turn to the tree-level approximation to the πN scattering amplitude as obtained from $\mathscr{L}_{\pi N}^{(1)}$ of Eq. 4.17. In order to derive the relevant interaction Lagrangians from Eq. 4.17, we reconsider the chiral connection of Eq. 4.13 with the external fields set to zero and obtain

$$\Gamma_\mu = \frac{i}{4F^2}\vec{\phi}\times\partial_\mu\vec{\phi}\cdot\vec{\tau} + O(\phi^4). \tag{4.47}$$

The linear pion-nucleon interaction term was already derived in Eq. 4.29 so that we obtain the following interaction Lagrangian:

$$\mathscr{L}_{\text{int}} = -\frac{1}{2}\frac{g_A}{F}\bar{\Psi}\gamma^\mu\gamma_5\partial_\mu\phi_b\tau_b\Psi - \frac{1}{4F^2}\bar{\Psi}\gamma^\mu\underbrace{\vec{\phi}\times\partial_\mu\vec{\phi}\cdot\vec{\tau}}_{=\,\varepsilon_{cde}\phi_d\partial_\mu\phi_e\tau_c}\Psi. \tag{4.48}$$

The first term is the pseudovector pion-nucleon coupling and the second the contact interaction with two factors of the pion field interacting with the nucleon at a single point. The Feynman rules for the vertices derived from Eq. 4.48 read

1. for an incoming pion with four-momentum q and Cartesian isospin index a:

Fig. 4.2 Contact
contribution to the
pion-nucleon scattering
amplitude

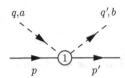

Fig. 4.3 s- and u-channel
pole contributions to the
pion-nucleon scattering
amplitude

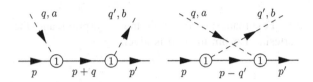

$$-\frac{1}{2}\frac{g_A}{F}\,\slashed{q}\gamma_5\tau_a,\tag{4.49}$$

2. for an incoming pion with q, a and an outgoing pion with q', b:

$$i\left(-\frac{1}{4F^2}\right)\gamma^\mu\varepsilon_{cde}\left[\delta_{da}\delta_{eb}iq'_\mu+\delta_{db}\delta_{ea}(-iq_\mu)\right]\tau_c=\frac{\slashed{q}+\slashed{q}'}{4F^2}\varepsilon_{abc}\tau_c.\tag{4.50}$$

The latter gives a contact contribution to $\mathscr{M}$ (see Fig. 4.2),

$$\mathscr{M}_{\text{cont}}=\bar{u}(p')\frac{\slashed{q}+\slashed{q}'}{4F^2}\underbrace{\varepsilon_{abc}\tau_c}_{=i\frac{1}{2}[\tau_b,\tau_a]}u(p)=i\frac{1}{2F^2}\bar{u}(p')\frac{1}{2}(\slashed{q}+\slashed{q}')\frac{1}{2}[\tau_b,\tau_a]u(p).$$

$$\tag{4.51}$$

We emphasize that such a term is not present in a conventional calculation with either a pseudoscalar or a pseudovector pion-nucleon interaction.

For the s- and u-channel nucleon-pole diagrams the pseudovector vertex appears twice (see Fig. 4.3) and we obtain

$$\mathscr{M}_{s+u}=i\frac{g_A^2}{4F^2}\bar{u}(p')(-\slashed{q}')\gamma_5\frac{1}{\slashed{p}'+\slashed{q}'-m}\,\slashed{q}\gamma_5\tau_b\tau_au(p)$$

$$+i\frac{g_A^2}{4F^2}\bar{u}(p')\slashed{q}\gamma_5\frac{1}{\slashed{p}'-\slashed{q}-m}(-\slashed{q}')\gamma_5\tau_a\tau_bu(p).\tag{4.52}$$

The s- and u-channel pole contributions are related to each other through pion crossing $a\leftrightarrow b$ and $q\leftrightarrow -q'$. In what follows we explicitly calculate only the s channel and make use of pion-crossing symmetry at the end to obtain the u-channel result. We perform the manipulations such that the result of a pseudoscalar coupling may also be read off. To that end, we need to apply the Dirac equation where, to the given accuracy, we make use of $m_N = m + \mathcal{O}(q^2)$. Using the Dirac equation, we rewrite

$$\slashed{q}\gamma_5 u(p) = (\slashed{p}' + \slashed{q}' - m + m - \slashed{p})\gamma_5 u(p) = (\slashed{p}' + \slashed{q}' - m)\gamma_5 u(p) + 2m\gamma_5 u(p)$$

and obtain

$$\mathcal{M}_s = i\frac{g_A^2}{4F^2}\bar{u}(p')(-\slashed{q}')\gamma_5\frac{1}{\slashed{p}' + \slashed{q}' - m}[(\slashed{p}' + \slashed{q}' - m) + 2m]\gamma_5\tau_b\tau_a u(p)$$

$$\overset{\gamma_5^2=1}{=} i\frac{g_A^2}{4F^2}\bar{u}(p')\left[(-\slashed{q}') + (-\slashed{q}')\gamma_5\frac{1}{\slashed{p}' + \slashed{q}' - m}2m\gamma_5\right]\tau_b\tau_a u(p).$$

We repeat the above procedure

$$\bar{u}(p')\slashed{q}'\gamma_5 = \bar{u}(p')[-2m\gamma_5 - \gamma_5(\slashed{p} + \slashed{q} - m)],$$

yielding

$$\mathcal{M}_s = i\frac{g_A^2}{4F^2}\bar{u}(p')[(-\slashed{q}') + \underbrace{4m^2\gamma_5\frac{1}{\slashed{p}' + \slashed{q}' - m}\gamma_5}_{\text{PS coupling}} + 2m]\tau_b\tau_a u(p), \tag{4.53}$$

where, for the identification of the PS-coupling result, one has to make use of the Goldberger-Treiman relation (see Sect. 4.3.1)

$$\frac{g_A}{F} = \frac{g_{\pi N}}{m},$$

where $g_{\pi N}$ denotes the pion-nucleon coupling constant in the chiral limit.

Since we only work at lowest-order tree level, we now replace $m \to m_N$, $g_A \to g_A$ etc. Using

$$s - m_N^2 = 2m_N(v - v_B),$$

we find

$$\bar{u}(p')\gamma_5\frac{1}{\slashed{p}' + \slashed{q}' - m_N}\gamma_5 u(p) = \bar{u}(p')\gamma_5\frac{\slashed{p}' + \slashed{q}' + m_N}{(p' + q')^2 - m_N^2}\gamma_5 u(p)$$

$$= \frac{1}{2m_N(v - v_B)}\left[-\frac{1}{2}\bar{u}(p')(\slashed{q} + \slashed{q}')u(p)\right],$$

where we again made use of the Dirac equation. We finally obtain for the s-channel pole contribution

$$\mathcal{M}_s = i\frac{g_A^2}{4F_\pi^2}\bar{u}(p')\left[2m_N + \frac{1}{2}(\slashed{q} + \slashed{q}')\left(-1 - \frac{2m_N}{v - v_B}\right)\right]\tau_b\tau_a u(p). \tag{4.54}$$

As noted above, the expression for the u channel results from the substitution $a \leftrightarrow b$ and $q \leftrightarrow -q'$,

Table 4.1 Tree-level contributions to the functions $A^\pm$ and $B^\pm$ of Eq. 4.42

Amplitude/origin	PS	ΔPV	Contact	Sum
A^+	0	$\frac{g_A^2 m_N}{F_\pi^2}$	0	$\frac{g_A^2 m_N}{F_\pi^2}$
A^-	0	0	0	0
B^+	$-\frac{g_A^2}{F_\pi^2}\frac{m_N v}{v^2-v_B^2}$	0	0	$-\frac{g_A^2}{F_\pi^2}\frac{m_N v}{v^2-v_B^2}$
B^-	$-\frac{g_A^2}{F_\pi^2}\frac{m_N v_B}{v^2-v_B^2}$	$-\frac{g_A^2}{2F_\pi^2}$	$\frac{1}{2F_\pi^2}$	$\frac{1-g_A^2}{2F_\pi^2}-\frac{g_A^2}{F_\pi^2}\frac{m_N v_B}{v^2-v_B^2}$

The second column (PS) denotes the result using the pseudoscalar pion-nucleon coupling (using the Goldberger-Treiman relation). The sum of the second and third column (PS + ΔPV) represents the result of the pseudovector pion-nucleon coupling. The contact term is specific to the chiral approach. The last column, the sum of the second, third, and fourth columns, is the lowest-order ChPT result

$$\mathcal{M}_u = i\frac{g_A^2}{4F_\pi^2}\bar{u}(p')\left[2m_N + \frac{1}{2}(\slashed{q} + \slashed{q}')\left(1 - \frac{2m_N}{v + v_B}\right)\right]\tau_a\tau_b u(p). \qquad (4.55)$$

We combine the s- and u-channel pole contributions using

$$\tau_b\tau_a = \frac{1}{2}\{\tau_b, \tau_a\} + \frac{1}{2}[\tau_b, \tau_a], \quad \tau_a\tau_b = \frac{1}{2}\{\tau_b, \tau_a\} - \frac{1}{2}[\tau_b, \tau_a],$$

and

$$\frac{1}{v - v_B} \pm \frac{1}{v + v_B} = \frac{\left\{\begin{array}{c} 2v \\ 2v_B \end{array}\right\}}{v^2 - v_B^2},$$

and summarize the contributions to the functions $A^\pm$ and $B^\pm$ of Eq. 4.42 in Table 4.1 (see also Eq. A.26 of Ref. [26]).

In order to extract the scattering lengths, let us consider threshold kinematics,

$$p^\mu = p'^\mu = (m_N, \vec{0}), \quad q^\mu = q'^\mu = (M_\pi, \vec{0}), \quad v|_{\text{thr}} = M_\pi, \quad v_B|_{\text{thr}} = -\frac{M_\pi^2}{2m_N}. \qquad (4.56)$$

Together with[10]

$$u(p) \to \sqrt{2m_N}\begin{pmatrix} \chi \\ 0 \end{pmatrix}, \quad \bar{u}(p') \to \sqrt{2m_N}(\chi'^\dagger\, 0)$$

we find for the threshold matrix element

$$T|_{\text{thr}} = 2m_N\chi'^\dagger[\delta_{ab}(A^+ + M_\pi B^+) - i\varepsilon_{abc}\tau_c(A^- + M_\pi B^-)]_{\text{thr}}\chi. \qquad (4.57)$$

Using

---

[10] Recall that we use the normalization $\bar{u}u = 2m_N$.

$$\left[v^2 - v_B^2\right]_{\text{thr}} = M_\pi^2\left(1 - \frac{\mu^2}{4}\right), \quad \mu = \frac{M_\pi}{m_N} \approx \frac{1}{7},$$

we obtain

$$T\big|_{\text{thr}} = 2m_N\chi'^\dagger\Bigg[\delta_{ab}\underbrace{\left(\frac{g_A^2 m_N}{F_\pi^2} + M_\pi\underbrace{\left(-\frac{g_A^2}{F_\pi^2}\right)\frac{m_N}{M_\pi}\frac{1}{1-\frac{\mu^2}{4}}}_{\text{PS}}\right)}_{\text{ChPT}\,=\,\text{PV}}$$

$$- i\varepsilon_{abc}\tau_c M_\pi\underbrace{\left(\frac{1}{2F_\pi^2} - \frac{g_A^2}{2F_\pi^2}\underbrace{\frac{g_A^2}{F_\pi^2}\left(-\frac{1}{2}\right)\frac{1}{1-\frac{\mu^2}{4}}}_{\text{PS}}\right)}_{\underbrace{}_{\text{PV}}}\Bigg]\chi, \tag{4.58}$$

$$\underbrace{}_{\text{ChPT}}$$

where we have indicated the results for the various coupling schemes.

Let us discuss the *s*-wave scattering lengths resulting from Eq. 4.58. Using the above normalization for the Dirac spinors, the differential cross section in the center-of-mass frame is given by [18]

$$\frac{d\sigma}{d\Omega} = \frac{|\vec{q}'|}{|\vec{q}|}\left(\frac{1}{8\pi\sqrt{s}}\right)^2 |T|^2, \tag{4.59}$$

which, at threshold, reduces to

$$\frac{d\sigma}{d\Omega}\bigg|_{\text{thr}} = \left(\frac{1}{8\pi(m_N + M_\pi)}\right)^2 |T|_{\text{thr}}^2 = |a|^2. \tag{4.60}$$

The *s*-wave scattering lengths are defined as[11]

$$a_{0+}^\pm \equiv \frac{1}{8\pi(m_N + M_\pi)}T^\pm\big|_{\text{thr}} = \frac{1}{4\pi(1+\mu)}\left[A^\pm + M_\pi B^\pm\right]_{\text{thr}}. \tag{4.61}$$

---

[11] The threshold parameters are defined in terms of a multipole expansion of the πN scattering amplitude [14]. The sign convention for the *s*-wave scattering parameters $a_{0+}^{(\pm)}$ is opposite to the convention of the effective-range expansion.

The subscript 0+ refers to the fact that the πN system is in an orbital s wave ($l = 0$) with total angular momentum $1/2 = 0 + 1/2$. Inserting the results of Table 4.1, we obtain at $\mathcal{O}(q)$,[12]

$$a_{0+}^- = \frac{M_\pi}{8\pi(1 + \mu)F_\pi^2}\left(1 + \frac{g_A^2\mu^2}{4}\frac{1}{1 - \frac{\mu^2}{4}}\right) = \frac{M_\pi}{8\pi(1 + \mu)F_\pi^2}[1 + \mathcal{O}(q^2)], \quad (4.62)$$

$$a_{0+}^+ = -\frac{g_A^2 M_\pi}{16\pi(1 + \mu)F_\pi^2}\frac{\mu}{1 - \frac{\mu^2}{4}} = \mathcal{O}(q^2), \quad (4.63)$$

where we have also indicated the chiral order. Taking the linear combinations $a^{\frac{1}{2}} = a_{0+}^+ + 2a_{0+}^-$ and $a^{\frac{3}{2}} = a_{0+}^+ - a_{0+}^-$ (see Eq. 4.46), we see that the results of Eqs. 4.62 and 4.63 at $\mathcal{O}(q)$ indeed satisfy the Weinberg-Tomozawa relation [60, 62]:[13]

$$a^I = -\frac{M_\pi}{8\pi(1 + \mu)F_\pi^2}\left[I(I + 1) - \frac{3}{4} - 2\right], \quad (4.64)$$

where I denotes the total isospin of the pion-nucleon system. As in $\pi\pi$ scattering, the scattering lengths vanish in the chiral limit reflecting the fact that the interaction of Goldstone bosons vanishes in the zero-energy limit. The pseudoscalar pion-nucleon interaction produces a scattering length a_{0+}^+ proportional to m_N instead of μM_π and is clearly in conflict with the requirements of chiral symmetry. Moreover, the scattering length a_{0+}^- of the pseudoscalar coupling is too large by a factor g_A^2 in comparison with the two-pion contact term of Eq. 4.51 (sometimes also referred to as the Weinberg-Tomozawa term) induced by the nonlinear realization of chiral symmetry. On the other hand, the pseudovector pion-nucleon interaction gives a completely wrong result for a_{0+}^-, because it misses the two-pion contact term of Eq. 4.51.

Using the values

$$g_A = 1.267, \quad F_\pi = 92.4\,\text{MeV},$$
$$m_N = m_p = 938.3\,\text{MeV}, \quad M_\pi = M_{\pi^+} = 139.6\,\text{MeV}, \quad (4.65)$$

the numerical results for the scattering lengths are given in Table 4.2. We have included the full results of Eqs. 4.62 and 4.63 and the consistent corresponding prediction at $\mathcal{O}(q)$. The empirical results quoted have been taken from low-energy partial-wave analyses [39, 43] and precision X-ray experiments on pionic hydrogen and deuterium [56]. For a discussion of πN scattering beyond tree level, see, e.g., Refs. [4, 21, 45].

---

[12] We do not expand the fraction $1/(1 + \mu)$, because the μ dependence is not of dynamical origin.

[13] The result, in principle, holds for a general target of isospin T (except for the pion) after replacing 3/4 by $T(T + 1)$ and μ by M_π/M_T.

Table 4.2 s-wave scattering lengths $a_{0+}^{\pm}$

Scattering length	a_{0+}^{+} [MeV^{-1}]	a_{0+}^{-} [MeV^{-1}]
Tree-level result	-6.80×10^{-5}	$+5.71 \times 10^{-4}$
ChPT $\mathcal{O}(q)$	0	$+5.66 \times 10^{-4}$
PS	-1.23×10^{-2}	$+9.14 \times 10^{-4}$
PV	-6.80×10^{-5}	$+5.06 \times 10^{-6}$
Empirical values [39]	$(-7 \pm 1) \times 10^{-5}$	$(6.6 \pm 0.1) \times 10^{-4}$
Empirical values [43]	$(2.04 \pm 1.17) \times 10^{-5}$	$(5.71 \pm 0.12) \times 10^{-4}$
		$(5.92 \pm 0.11) \times 10^{-4}$
Experiment [56]	$(-2.7 \pm 3.6) \times 10^{-5}$	$(+6.59 \pm 0.30) \times 10^{-4}$

4.4 The Next-to-Leading-Order Lagrangian

The next-to-leading-order pion-nucleon Lagrangian contains seven low-energy constants c_i [22, 26],

$$\mathcal{L}_{\pi N}^{(2)} = c_1 \text{Tr}(\chi_+)\bar{\Psi}\Psi - \frac{c_2}{4m^2}\text{Tr}(u_\mu u_\nu)(\bar{\Psi}D^\mu D^\nu \Psi + \text{H.c.})$$
$$+ \frac{c_3}{2}\text{Tr}(u^\mu u_\mu)\bar{\Psi}\Psi - \frac{c_4}{4}\bar{\Psi}\gamma^\mu\gamma^\nu[u_\mu, u_\nu]\Psi + c_5\bar{\Psi}\left[\chi_+ - \frac{1}{2}\text{Tr}(\chi_+)\right]\Psi$$
$$+ \bar{\Psi}\sigma^{\mu\nu}\left[\frac{c_6}{2}f_{\mu\nu}^+ + \frac{c_7}{2}v_{\mu\nu}^{(s)}\right]\Psi, \tag{4.66}$$

where H.c. refers to the Hermitian conjugate and

$$\chi_\pm = u^\dagger \chi u^\dagger \pm u\chi^\dagger u,$$
$$v_{\mu\nu}^{(s)} = \partial_\mu v_\nu^{(s)} - \partial_\nu v_\mu^{(s)},$$
$$f_{\mu\nu}^\pm = u f_{L\mu\nu} u^\dagger \pm u^\dagger f_{R\mu\nu} u,$$
$$f_{L\mu\nu} = \partial_\mu l_\nu - \partial_\nu l_\mu - i[l_\mu, l_\nu],$$
$$f_{R\mu\nu} = \partial_\mu r_\nu - \partial_\nu r_\mu - i[r_\mu, r_\nu].$$

In order to be able to make predictions for physical quantities, the values of the c_i have to be determined from comparisons with experiments. A first estimate of the low-energy constants $c_1, \ldots, c_4$ may be obtained from a tree-level fit [3] to the πN threshold parameters of Koch [39]:

$$c_1 = -0.9m_N^{-1}, \quad c_2 = 2.5m_N^{-1}, \quad c_3 = -4.2m_N^{-1}, \quad c_4 = 2.3m_N^{-1}. \tag{4.67}$$

In general, once calculations are performed beyond tree level and LECs are fitted to empirical data, it is necessary to specify the renormalization prescription to which the given values of the LECs refer. We will come back to this point in Sect. 4.5.3. When considering the chiral expansion of an observable, the coefficient at a given order does not depend on the renormalization scale μ (see Sect. 3.5.2 for an illustration in terms of the Goldstone-boson masses).

In particular, if a renormalization scheme is set up such that renormalized loop contributions start at $\mathcal{O}(q^3)$ (see Sect. 4.5.2) this means that the renormalized coefficients of $\mathscr{L}_{\pi N}^{(2)}$ do not depend on μ in this scheme.

Combining the analyses of several sources, the following estimates for the c_i have been compiled in Ref. [44] (values in GeV^{-1}):

$$c_1 = -0.9^{+0.5}_{-0.2}, \quad c_2 = 3.3 \pm 0.2, \quad c_3 = -4.7^{+1.2}_{-1.0}, \quad c_4 = 3.5^{+0.5}_{-0.2}. \quad (4.68)$$

The constant c_5 is related to the strong contribution to the neutron-proton mass difference and has been estimated to be $c_5 = -0.09 \pm 0.01$ [7].

Finally, the constants c_6 and c_7 are related to the isovector and isoscalar magnetic moments of the nucleon in the chiral limit. This is seen by considering the coupling to an external electromagnetic four-vector potential by setting

$$r_\mu = l_\mu = -e\mathscr{A}_\mu \frac{\tau_3}{2}, \quad v_\mu^{(s)} = -e\frac{1}{2}\mathscr{A}_\mu.$$

We then obtain

$$v_{\mu\nu}^{(s)} = -e\frac{1}{2}\mathscr{F}_{\mu\nu}, \quad \mathscr{F}_{\mu\nu} = \partial_\mu\mathscr{A}_\nu - \partial_\nu\mathscr{A}_\mu,$$

$$f_{L\mu\nu} = \partial_\mu l_\nu - \partial_\nu l_\mu - \underbrace{i\left[l_\mu, l_\nu\right]}_{=\,0} = -e\mathscr{F}_{\mu\nu}\frac{\tau_3}{2} = f_{R\mu\nu},$$

and thus

$$f_{\mu\nu}^+ = uf_{L\mu\nu}u^\dagger + u^\dagger f_{R\mu\nu}u = f_{L\mu\nu} + f_{R\mu\nu} + \cdots = -e\mathscr{F}_{\mu\nu}\tau_3 + \cdots.$$

Inserting these terms into the Lagrangian we see that the contributions without pion fields are given by

$$-\frac{e}{2}\bar{\Psi}\sigma^{\mu\nu}\left(c_6\tau_3 + \frac{1}{2}c_7\right)\Psi\mathscr{F}_{\mu\nu}.$$

Comparing with the interaction Lagrangian of a magnetic field with the *anomalous* magnetic moment of the nucleon,

$$-\frac{e}{4m_p}\bar{\Psi}\sigma^{\mu\nu}\frac{1}{2}\left(\kappa^{(s)} + \kappa^{(v)}\tau_3\right)\Psi\mathscr{F}_{\mu\nu},$$

we obtain

$$c_7 = \frac{\overset{\circ}{\kappa}^{(s)}}{2m}, \quad c_6 = \frac{\overset{\circ}{\kappa}^{(v)}}{4m},$$

where $\circ$ denotes the chiral limit. The physical values read

$$\kappa_p = \frac{1}{2}\left(\kappa^{(s)} + \kappa^{(v)}\right) = 1.793, \quad \kappa_n = \frac{1}{2}\left(\kappa^{(s)} - \kappa^{(v)}\right) = -1.913,$$

and thus $\kappa^{(s)} = -0.120$ and $\kappa^{(v)} = 3.706$. The results for $\kappa^{(s)}$ and $\kappa^{(v)}$ up to and including $\mathcal{O}(q^3)$ [6, 20],[14]

$$\kappa^{(s)} = \overset{\circ}{\kappa}^{(s)} + \mathcal{O}(q^4),$$

$$\kappa^{(v)} = \overset{\circ}{\kappa}^{(v)} - \frac{M_\pi m_N g_A^2}{4\pi F_\pi^2} + \mathcal{O}(q^4),$$

are used to express the parameters c_6 and c_7 in terms of physical quantities. Note that the numerical correction of -1.96 (parameters of Eq. 4.65) to the isovector anomalous magnetic moment is substantial.

4.5 Loop Diagrams: Renormalization and Power Counting

For the applications in Sects. 4.3.1 and 4.3.2 we only considered contributions at leading order, which meant that only tree-level diagrams had to be taken into account. To go beyond leading order, we will have to calculate loop diagrams as well. In Sect. 3.4.9 we saw that in the purely mesonic sector contributions of n-loop diagrams are at least of $\mathcal{O}(q^{2n+2})$, i.e., they are suppressed by q^{2n} in comparison with tree-level diagrams. An important ingredient in deriving this result was the fact that we treated the squared pion mass as a small quantity of order q^2. Such an approach is motivated by the observation that the masses of the Goldstone bosons must vanish in the chiral limit. In the framework of ordinary chiral perturbation theory $M_\pi^2 \sim m_q$ (see Eqs. 3.59 and 3.161), which translates into a momentum expansion of observables at fixed ratio m_q/q^2. On the other hand, there is no reason to believe that the masses of hadrons other than the Goldstone bosons should vanish or become small in the chiral limit. In other words, the nucleon mass entering the pion-nucleon Lagrangian of Eq. 4.17 should—as already anticipated in the discussion following Eq. 4.17—not be treated as a small quantity of, say, $\mathcal{O}(q)$.

Before discussing how this affects the calculation of loop diagrams and the construction of a consistent power counting, we recall the principles of renormalization. As we will see, the choice of a renormalization condition is intimately connected with the power counting.

---

[14] The calculations were performed in the heavy-baryon approach (see Sect. 4.6.1) in which the c_i are renormalization-scale independent.

4.5.1 Counter Terms of the Baryonic ChPT Lagrangian

In Sect. 3.4.8 we discussed how to renormalize a toy model by expressing all bare fields and parameters in terms of renormalized quantities. The procedure results in counter terms which are used to absorb divergences appearing in loop diagrams. We also pointed out that one is free to specify a renormalization condition, determining which finite parts of the loop integrals are absorbed in the counter terms. As this freedom to choose a renormalization condition plays a crucial role in baryon chiral perturbation theory, we briefly want to discuss the counter terms appearing in $\mathscr{L}_{\pi N}^{(1)}$, for simplicity only considering the free part in combination with the πN interaction term with the smallest number of pion fields. In terms of bare fields this part of the Lagrangian reads

$$\mathscr{L}_{\pi N}^{(1)} = \bar{\Psi}_B \left(i\gamma^\mu \partial_\mu - m_B - \frac{1}{2}\frac{g_{AB}}{F_B}\gamma^\mu \gamma_5 \partial_\mu \phi_{iB} \tau_i \right) \Psi_B + \cdots. \qquad (4.69)$$

After introducing renormalized fields as

$$\Psi = \frac{\Psi_B}{\sqrt{Z_\Psi}}, \quad \phi_i = \frac{\phi_{iB}}{\sqrt{Z_\phi}}, \qquad (4.70)$$

we express the field redefinition constants for the nucleon and pion fields as well as the bare quantities in terms of renormalized parameters:

$$\begin{aligned}
Z_\Psi &= 1 + \delta Z_\Psi(m, g_A, g_i, v), \\
Z_\phi &= 1 + \delta Z_\phi(m, g_A, g_i, v), \\
m_B &= m(v) + \delta m(m, g_A, g_i, v), \\
g_{AB} &= g_A(v) + \delta g_A(m, g_A, g_i, v),
\end{aligned} \qquad (4.71)$$

where g_i, $i = 1, \ldots, \infty$, collectively denote all the renormalized parameters which correspond to bare parameters g_{iB} of the full effective Lagrangian. As before, the dependence on the renormalization prescription is contained in the parameter v. Here we choose the renormalized mass parameter to be the nucleon mass in the chiral limit, $m(v) = m$. We emphasize that our choice is only one among an infinite number of possibilities. The basic and counter-term Lagrangians are given by

$$\mathscr{L}_{\text{basic}} = \bar{\Psi} \left(i\gamma^\mu \partial_\mu - m - \frac{1}{2}\frac{g_A}{F}\gamma^\mu \gamma_5 \partial_\mu \phi_i \tau_i \right) \Psi, \qquad (4.72)$$

$$\mathscr{L}_{\text{ct}} = \delta Z_\Psi \bar{\Psi} i\gamma^\mu \partial_\mu \Psi - \delta\{m\}\bar{\Psi}\Psi - \frac{1}{2}\delta\left\{\frac{g_A}{F}\right\}\bar{\Psi}\gamma^\mu \gamma_5 \partial_\mu \phi_i \tau_i \Psi, \qquad (4.73)$$

where we introduced the abbreviations

$$\delta\{m\} \equiv \delta Z_\Psi m + Z_\Psi \delta m,$$

$$\delta\left\{\frac{g_A}{F}\right\} \equiv \delta Z_\Psi \frac{g_A}{F}\sqrt{Z_\phi} + Z_\Psi \left(\frac{g_{AB}}{F_B} - \frac{g_A}{F}\right)\sqrt{Z_\phi} + \frac{g_A}{F}(\sqrt{Z_\phi} - 1).$$

As previously explained (see Sects. 3.4.4, 3.4.5, 4.2, and 4.3.1), m, g_A, and F denote the chiral limit values of the physical nucleon mass, the axial-vector coupling constant, and the pion-decay constant, respectively. Divergences appearing in loop diagrams can be absorbed by expanding the counter-term Lagrangian of Eq. 4.73 in powers of the renormalized coupling constants and by suitably adjusting the expansion coefficients.

4.5.2 Power Counting

Our goal is to obtain a power counting for tree-level and loop diagrams of the (relativistic) EFT for baryons which is analogous to that for mesons given in Sect. 3.4.9. It is important to note that the power counting applies to *renormalized* diagrams, i.e., the sum of diagrams generated with the basic Lagrangian and the corresponding contributions from the counter-term Lagrangian. The counter-term contribution not only absorbs any potential divergence, we can also implement different renormalization conditions by adjusting the finite pieces of the counter terms. Choosing a suitable renormalization condition will allow us to apply the following power counting: a loop integration in n dimensions counts as q^n, pion and fermion propagators count as q^{-2} and q^{-1}, respectively, vertices derived from $\mathscr{L}_{2k}$ and $\mathscr{L}_{\pi N}^{(k)}$ count as q^{2k} and q^k, respectively. Here, q generically denotes a small expansion parameter such as, e.g., the pion mass. In total this yields for the power D of a diagram the standard formula [17, 64]

$$D = nN_L - 2I_\pi - I_N + \sum_{k=1}^{\infty} 2kN_{2k}^\pi + \sum_{k=1}^{\infty} kN_k^N, \qquad (4.74)$$

where N_L, I_π, I_N, N_{2k}^π, and N_k^N denote the number of independent loop momenta, internal pion lines, internal nucleon lines, vertices originating from $\mathscr{L}_{2k}$, and vertices originating from $\mathscr{L}_{\pi N}^{(k)}$, respectively. We make use of the relation[15]

$$N_L = I_\pi + I_N - N_\pi - N_N + 1$$

---

[15] This relation can be understood as follows: For each internal line we have a propagator in combination with an integration with measure $d^4k/(2\pi)^4$. Therefore, there are $I_\pi + I_N$ integrations. However, at each vertex we have a four-momentum conserving delta function, reducing the number of integrations by $N_\pi + N_N - 1$, where the -1 is related to the overall four-momentum conserving delta function $\delta^4(P_f - P_i)$.

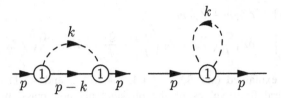

Fig. 4.4 One-loop contributions to the nucleon self energy

with N_π and N_N the total number of pionic and baryonic vertices, respectively, to eliminate I_π:

$$D = (n-2)N_L + I_N + 2 + \sum_{k=1}^{\infty} 2(k-1)N_{2k}^\pi + \sum_{k=1}^{\infty}(k-2)N_k^N.$$

Finally, for processes containing exactly one nucleon in the initial and final states we have[16] $N_N = I_N + 1$ and we thus obtain

$$D = 1 + (n-2)N_L + \sum_{k=1}^{\infty} 2(k-1)N_{2k}^\pi + \sum_{k=1}^{\infty}(k-1)N_k^N \qquad (4.75)$$

$$\geq 1 \text{ in four dimensions.}$$

According to Eq. 4.75, one-loop calculations in the single-nucleon sector should start contributing at $\mathcal{O}(q^{n-1})$. For example, let us consider the one-loop contribution of the first diagram of Fig. 4.4 to the nucleon self energy. According to Eq. 4.74, the renormalized result should be of order

$$D = n \cdot 1 - 2 \cdot 1 - 1 \cdot 1 + 1 \cdot 2 = n - 1. \qquad (4.76)$$

We will see below that a renormalization scheme that respects this power counting is more complex than in the mesonic sector.

4.5.3 One-Loop Correction to the Nucleon Mass

In the mesonic sector, the combination of dimensional regularization and the modified minimal subtraction scheme (see Eq. 3.124) leads to a straightforward correspondence between the chiral and loop expansions. By studying the one-loop contributions of Fig. 4.4 to the nucleon self energy, we will see that this correspondence, at first sight, seems to be lost in the baryonic sector.

---

[16] In the low-energy effective field theory there are no closed fermion loops. In other words, in the single-nucleon sector exactly one fermion line runs through the diagram connecting the initial and final states.

Exercise 4.10 In the following we will calculate the mass m_N of the nucleon up to and including $\mathcal{O}(q^3)$. As in the case of pions, the physical mass is defined through the pole of the full propagator (at $\not{p} = m_N$ for the nucleon). The full unrenormalized propagator of the nucleon is defined as the Fourier transform

$$S_B(p) = \int d^4x e^{ip \cdot x} S_B(x) \tag{4.77}$$

of the two-point function

$$S_B(x) = -i \langle 0|T[\Psi_B(x)\bar{\Psi}_B(0)]|0\rangle, \tag{4.78}$$

where Ψ_B denotes the bare nucleon field. We parameterize

$$S_B(p) = \frac{1}{\not{p} - m_B - \Sigma_B(\not{p})} \equiv \frac{1}{\not{p} - m - \Sigma(\not{p})}, \tag{4.79}$$

where m_B refers to the bare mass of Eq. 4.69 and Σ_B is the self energy calculated in terms of bare parameters. On the other hand, m is the nucleon mass in the chiral limit and Eq. 4.79 *defines* the quantity Σ. Here, $\Sigma_B(\not{p})$ and $\Sigma(\not{p})$ are matrix functions [36] which, using $\not{p}\not{p} = p^2$, can be parameterized as

$$\Sigma_B(x) = -xf_B(x^2) + m_B g_B(x^2)$$

with an analogous expression for Σ. To determine the mass, the equation

$$m_N - m_B - \Sigma_B(m_N) = m_N - m - \Sigma(m_N) = 0 \tag{4.80}$$

has to be solved, so the task is to calculate the nucleon self energy $\Sigma(\not{p})$.

According to the power counting specified above, we need to calculate the two types of one-loop contributions shown in Fig. 4.4 together with the corresponding counter-term contribution and a tree-level contribution. After renormalization, we would like the first loop diagram to be of chiral order $D = n \cdot 1 - 2 \cdot 1 - 1 \cdot 1 + 2 \cdot 1 = n - 1$, and the second loop diagram of order $D = n \cdot 1 - 2 \cdot 1 + 1 \cdot 1 = n - 1$.

(a) The πN Lagrangian at $\mathcal{O}(q^2)$ is given in Eq. 4.66. Which of these terms contain only the nucleon fields and therefore give a tree-level contribution to the self energy? Determine $-i\Sigma_2^{\text{tree}}(\not{p})$ from $i\langle\bar{\Psi}|\mathcal{L}_{\pi N}^{(2)}|\Psi\rangle$.

Remark There are no tree-level contributions from the Lagrangian $\mathcal{L}_{\pi N}^{(3)}$.

(b) By using the expansion of $\mathcal{L}_{\pi N}^{(1)}$ up to two pion fields (see Exercise 4.4), verify the following Feynman rules:[17]

---

[17] Note that we work with the basic Lagrangian.

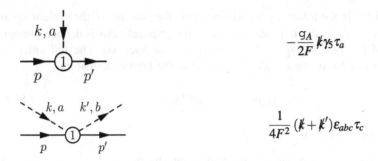

$$-\frac{g_A}{2F}\,\slashed{k}\gamma_5\tau_a$$

$$\frac{1}{4F^2}\,(\slashed{k}+\slashed{k}')\varepsilon_{abc}\tau_c$$

(c) Use the Feynman rules to show that the second diagram in Fig. 4.4 does not contribute to the self energy.

(d) Use the Feynman rules and the expressions for the propagators,

$$i\Delta_F(p) = \frac{i}{p^2 - M^2 + i0^+},$$

$$iS_F(p) = i\frac{\slashed{p} + m}{p^2 - m^2 + i0^+},$$

to verify that in dimensional regularization the first diagram in Fig. 4.4 gives the contribution

$$-i\Sigma^{\text{loop}}(\slashed{p}) = -\frac{3g_A^2}{4F^2}i\mu^{4-n}\int\frac{d^n k}{(2\pi)^n}\frac{\slashed{k}(\slashed{p}-m-\slashed{k})\slashed{k}}{[(p-k)^2 - m^2 + i0^+](k^2 - M^2 + i0^+)}.$$

$$(4.81)$$

(e) Show that the numerator can be rewritten as

$$-(\slashed{p}+m)(k^2 - M^2) - (\slashed{p}+m)M^2 + (p^2 - m^2)\slashed{k} - \left[(p-k)^2 - m^2\right]\slashed{k}, \quad (4.82)$$

which, when inserted in Eq. 4.81, gives

$$\Sigma^{\text{loop}}(\slashed{p}) = \frac{3g_A^2}{4F^2}\left\{ -(\slashed{p}+m)\mu^{4-n}i\int\frac{d^n k}{(2\pi)^n}\frac{1}{(p-k)^2 - m^2 + i0^+} \right.$$

$$- (\slashed{p}+m)M^2\mu^{4-n}i\int\frac{d^n k}{(2\pi)^n}\frac{1}{[(p-k)^2 - m^2 + i0^+](k^2 - M^2 + i0^+)}$$

$$+ (p^2 - m^2)\mu^{4-n}i\int\frac{d^n k}{(2\pi)^n}\frac{\slashed{k}}{[(p-k)^2 - m^2 + i0^+](k^2 - M^2 + i0^+)}$$

$$\left. -\mu^{4-n}i\int\frac{d^n k}{(2\pi)^n}\frac{\slashed{k}}{k^2 - M^2 + i0^+} \right\}.$$

$$(4.83)$$

Hint: $\{\gamma_\mu, \gamma_\nu\} = 2g_{\mu\nu}\mathbb{1}$, $\{\gamma_\mu, \gamma_5\} = 0$, $\gamma_5\gamma_5 = \mathbb{1}$, $\slashed{k}^2 = k^2 - M^2 + M^2$.

(f) The last term in Eq. 4.83 vanishes since the integrand is odd in k. We use the following convention for scalar loop integrals:

$$I_{N...\pi...}(p_1,\ldots,q_1,\ldots)$$

$$= \mu^{4-n} i \int \frac{d^n k}{(2\pi)^n} \frac{1}{[(k+p_1)^2 - m^2 + i0^+]\ldots[(k+q_1)^2 - M^2 + i0^+]\ldots}.$$

To determine the vector integral use the ansatz

$$\mu^{4-n} i \int \frac{d^n k}{(2\pi)^n} \frac{k_\mu}{[(p-k)^2 - m^2 + i0^+][k^2 - M^2 + i0^+]} = p_\mu C. \tag{4.84}$$

Multiply Eq. 4.84 by p^μ to show that C is given by

$$C = \frac{1}{2p^2}\left[I_N - I_\pi + (p^2 - m^2 + M^2)I_{N\pi}(-p,0)\right]. \tag{4.85}$$

Using the above convention, the loop contribution to the nucleon self energy reads

$$\Sigma^{\text{loop}}(\not{p}) = -\frac{3g_A^2}{4F^2}\left\{(\not{p}+m)I_N + (\not{p}+m)M^2 I_{N\pi}(-p,0)\right.$$

$$\left. - (p^2 - m^2)\frac{\not{p}}{2p^2}\left[I_N - I_\pi + (p^2 - m^2 + M^2)I_{N\pi}(-p,0)\right]\right\}. \tag{4.86}$$

The explicit expressions for the integrals are given by

$$I_\pi = \frac{M^2}{16\pi^2}\left[R + \ln\left(\frac{M^2}{\mu^2}\right)\right],$$

$$I_N = \frac{m^2}{16\pi^2}\left[R + \ln\left(\frac{m^2}{\mu^2}\right)\right],$$

$$I_{N\pi}(p,0) = \frac{1}{16\pi^2}\left[R + \ln\left(\frac{m^2}{\mu^2}\right) - 1 + \frac{p^2 - m^2 - M^2}{p^2}\ln\left(\frac{M}{m}\right) + \frac{2mM}{p^2}F(\Omega)\right],$$

$$\tag{4.87}$$

where R is given in Eq. 3.111, Ω is defined as

$$\Omega = \frac{p^2 - m^2 - M^2}{2mM},$$

and

$$F(\Omega) = \begin{cases} \sqrt{\Omega^2 - 1}\ln\left(-\Omega - \sqrt{\Omega^2 - 1}\right), & \Omega \leq -1, \\ \sqrt{1 - \Omega^2}\arccos(-\Omega), & -1 \leq \Omega \leq 1, \\ \sqrt{\Omega^2 - 1}\ln\left(\Omega + \sqrt{\Omega^2 - 1}\right) - i\pi\sqrt{\Omega^2 - 1}, & 1 \leq \Omega. \end{cases}$$

(g) The result for the self energy contains divergences as $n \to 4$ (the terms proportional to R), so it has to be renormalized. The counter-term Lagrangian must produce structures which precisely cancel the divergences, as otherwise the

Fig. 4.5 Renormalized one-
loop self-energy diagram

result for the nucleon mass will not be finite. For convenience, choose the
renormalization parameter $\mu = m$. In the modified minimal subtraction scheme
$\widetilde{MS}$ all contributions proportional to R are canceled by corresponding contributions generated by the counter-term Lagrangian of Eq. 4.73, as well as by
counter-term Lagrangians resulting from higher-order terms of the most general
effective πN Lagrangian. Operationally this means that we simply drop all terms
proportional to R and indicate the $\widetilde{MS}$-renormalized coupling constants by a
subscript r. Again, this is possible because we include in the Lagrangian all of the
infinite number of interactions allowed by symmetries [65]. The renormalized
diagram is depicted in Fig. 4.5, where the cross generically denotes counter-term
contributions. The $\widetilde{MS}$-renormalized self-energy contribution then reads

$$\Sigma_r^{\text{loop}}(\not{p}) = -\frac{3g_{Ar}^2}{4F_r^2}\left\{(\not{p}+m)M^2 I_{N\pi}^r(-p,0)\right.$$

$$\left.-(p^2-m^2)\frac{\not{p}}{2p^2}[(p^2-m^2+M^2)I_{N\pi}^r(-p,0)-I_\pi^r]\right\}, \quad (4.88)$$

where the superscript r on the integrals means that the terms proportional to R
have been dropped. Writing $\not{p} + m = 2m + (\not{p} - m)$ and comparing the first
term of Eq. 4.86 with Eq. 4.88, we note that, among other terms, the $\widetilde{MS}$
renormalization involves (even in the chiral limit) an infinite renormalization
yielding the relation between the bare and the renormalized mass [26]

$$m_B = m + \frac{3g_{Ar}^2}{32\pi^2 F_r^2}m^3 R + \cdots.$$

Using the definition of the integrals, show that Eq. 4.88 contains a term of
$\mathcal{O}(q^2)$. What does the presence of this term indicate about the applicability of
the $\widetilde{MS}$ scheme in baryon ChPT?
Hint: What chiral order did the power counting assign to the diagram from
which we calculated Σ^{loop}?

(h) We can now solve Eq. 4.80 for the nucleon mass,

$$m_N = m + \Sigma_{2r}^{\text{tree}}(m_N) + \Sigma_r^{\text{loop}}(m_N) = m - 4c_{1r}M^2 + \Sigma_r^{\text{loop}}(m_N). \quad (4.89)$$

We find $m_N - m = \mathcal{O}(q^2)$. Since our calculation is only valid up to $\mathcal{O}(q^3)$,[18]
determine $\Sigma_r^{\text{loop}}(m_N)$ to that order. Check that you only need an expansion of
$I_{N\pi}^r$ which, using

---

[18] For brevity, we use the expression "up to $\mathcal{O}(q^n)$" to mean "up to and including $\mathcal{O}(q^n)$" in the
following.

$$\arccos\left(-\Omega\right) = \frac{\pi}{2} + \cdots,$$

verify to be

$$I_{N\pi}^r = \frac{1}{16\pi^2}\left(-1 + \frac{\pi M}{m} + \cdots\right). \tag{4.90}$$

Show that this yields

$$m_N = m - 4c_{1r}M^2 + \frac{3g_{Ar}^2 M^2}{32\pi^2 F_r^2}m - \frac{3g_{Ar}^2 M^3}{32\pi F_r^2}. \tag{4.91}$$

(i) The solution to the power-counting problem is the observation that the term violating the power counting (the third on the right of Eq. 4.91) is *analytic* in small quantities and can thus be absorbed in counter terms. In addition to the $\widetilde{\text{MS}}$ scheme we have to perform an additional *finite* renormalization. Rewrite

$$c_{1r} = c_1 + \delta c_1 \tag{4.92}$$

in Eq. 4.91 and determine δc_1 so that the term violating the power counting is absorbed, which then gives the final result for the nucleon mass at $\mathcal{O}(q^3)$,

$$m_N = m - 4c_1 M^2 - \frac{3g_A^2 M^3}{32\pi F^2}. \tag{4.93}$$

Note that the renormalized LECs of Eq. 4.93 are, in general, different from those in the $\widetilde{\text{MS}}$ scheme, as they correspond to a renormalization scheme that does not violate the power counting.

We have seen in the exercise above that, for the case of the nucleon self energy, the expression for loop diagrams renormalized by applying dimensional regularization in combination with the $\widetilde{\text{MS}}$ scheme as in the mesonic sector contained terms not consistent with the power counting. The appearance of terms violating the power counting when using the $\widetilde{\text{MS}}$ scheme is a general feature of loop calculations in baryonic chiral perturbation theory [26].

4.6 Renormalization Schemes

In the calculation of the nucleon mass, a subtraction of suitable finite terms in addition to the $\widetilde{\text{MS}}$ scheme leads to expressions for loop diagrams consistent with the power counting of Sect. 4.5.2. The question arises whether this procedure can be generalized. We will discuss three approaches that solve the power-counting problem: the so-called heavy-baryon approach [5, 37], the infrared regularization of Becher and Leutwyler [3], and the extended on-mass-shell scheme of Ref. [24, 27]. The heavy-baryon formalism consists of an additional expansion of the Lagrangian in powers of the inverse nucleon mass in the chiral limit, $1/m$, similar to a Foldy-Wouthuysen expansion [23]. Both the infrared regularization and the extended on-mass-shell scheme are manifestly Lorentz-invariant approaches that result in a consistent power counting.

4.6.1 Heavy-Baryon Approach

The nucleon mass is not expected to vanish in the chiral limit, and it is roughly of the same size as the chiral-symmetry-breaking scale $4\pi F$ which appears in the calculations of pion loop contributions. This complicates the expansion of observables in powers of momenta, as the energy component of the nucleon four-momentum p is not small compared to $4\pi F$, i.e.,

$$\frac{p^0}{4\pi F} \sim 1, \tag{4.94}$$

and thus the ratio of Eq. 4.94 is not a good expansion parameter. The basic idea of heavy-baryon chiral perturbation theory (HBChPT) consists of separating the nucleon four-momentum p into a large piece close to on-shell kinematics and a soft residual contribution k_p,

$$p^\mu = mv^\mu + k_p^\mu. \tag{4.95}$$

The four-vector v^μ has the properties

$$v^2 = 1, \quad v^0 \geq 1, \tag{4.96}$$

and is often taken as $v^\mu = (1, 0, 0, 0)$ for convenience. Note that

$$v \cdot k_p = -\frac{k_p^2}{2m} \stackrel{v^\mu=(1,0,0,0)}{=} k_p^0 = E - m \ll m. \tag{4.97}$$

In addition, the relativistic nucleon field is decomposed into two velocity-dependent fields,

$$\Psi(x) = e^{-imv\cdot x}[\mathcal{N}_v(x) + \mathcal{H}_v(x)], \tag{4.98}$$

where

$$\mathcal{N}_v = e^{+imv\cdot x}P_{v+}\Psi, \quad \mathcal{H}_v = e^{+imv\cdot x}P_{v-}\Psi, \tag{4.99}$$

and we have defined the projection operators[19]

$$P_{v\pm} \equiv \frac{1}{2}(1 \pm \rlap{/}{v}).$$

(4.100)

Exercise 4.11 Using Eq. 4.96, show that

$$P_{v+} + P_{v-} = 1, \quad P_{v\pm}^2 = P_{v\pm}, \quad P_{v\pm}P_{v\mp} = 0.$$

Use these results to verify that

$$\rlap{/}{v}\mathcal{N}_v = \mathcal{N}_v, \quad \rlap{/}{v}\mathcal{H}_v = -\mathcal{H}_v.$$

(4.101)

Starting from the relativistic leading-order Lagrangian of Eq. 4.17,

$$\mathcal{L}_{\pi N}^{(1)} = \bar{\Psi}\left(i\rlap{/}{D} - m + \frac{g_A}{2}\gamma^\mu\gamma_5 u_\mu\right)\Psi,$$

(4.102)

we now derive the leading-order Lagrangian of HBChPT. While it is possible to employ the path-integral formalism [5, 42], we follow a different method using the equations of motion that does not require detailed knowledge of functional integrals.

Exercise 4.12 Our aim is to derive the leading-order Lagrangian in the heavy-baryon formalism.

(a) Use the Euler-Lagrange equation for the nucleon field,

$$-\partial_\mu \frac{\partial\mathcal{L}_{\pi N}^{(1)}}{\partial\partial_\mu\bar{\Psi}} + \frac{\partial\mathcal{L}_{\pi N}^{(1)}}{\partial\bar{\Psi}} = 0,$$

to derive the equation of motion (EOM)

$$\left(i\rlap{/}{D} - m + \frac{g_A}{2}\rlap{/}{u}\gamma_5\right)\Psi = 0.$$

(4.103)

(b) Insert the decomposition of the nucleon field of Eq. 4.98 into the EOM, use the properties of Eq. 4.101, and multiply the result with $e^{imv\cdot x}$ to obtain

$$\left(i\rlap{/}{D} + \frac{g_A}{2}\rlap{/}{u}\gamma_5\right)\mathcal{N}_v + \left(i\rlap{/}{D} - 2m + \frac{g_A}{2}\rlap{/}{u}\gamma_5\right)\mathcal{H}_v = 0.$$

(4.104)

(c) In order to separate the P_{v+} and P_{v-} components, use the algebra of gamma matrices, $\gamma^\mu\gamma^\nu + \gamma^\nu\gamma^\mu = 2g^{\mu\nu}1$, to verify the relations:

---

[19] Note that $P_{v\pm}$ do not define orthogonal projectors in the mathematical sense, because they do not satisfy $P_{v\pm}^\dagger = P_{v\pm}$, with the exception of the special case $v^\mu = (1,0,0,0)$.

$$P_{v+}\slashed{A}P_{v+} = v \cdot AP_{v+},$$

$$P_{v+}\slashed{A}P_{v-} = \slashed{A}_\perp P_{v-} = P_{v+}\slashed{A}_\perp,$$

$$P_{v-}\slashed{A}P_{v-} = -v \cdot AP_{v-},$$

$$P_{v-}\slashed{A}P_{v+} = \slashed{A}_\perp P_{v+} = P_{v-}\slashed{A}_\perp,$$

$$P_{v+}\slashed{B}\gamma_5 P_{v+} = \slashed{B}_\perp \gamma_5 P_{v+},$$

$$P_{v+}\slashed{B}\gamma_5 P_{v-} = v \cdot B\gamma_5 P_{v-} = v \cdot BP_{v+}\gamma_5,$$

$$P_{v-}\slashed{B}\gamma_5 P_{v-} = \slashed{B}_\perp \gamma_5 P_{v-},$$

$$P_{v-}\slashed{B}\gamma_5 P_{v+} = -v \cdot B\gamma_5 P_{v+} = -v \cdot BP_{v-}\gamma_5, \tag{4.105}$$

where A and B are arbitrary four-vectors and

$$A_\perp^\mu = A^\mu - v \cdot Av^\mu, \quad v \cdot A_\perp = 0, \quad \slashed{A}_\perp = A_\perp^\mu \gamma_\mu.$$

(d) We can now separate the projections onto the P_{v+} and P_{v-} parts of the EOM. Show that, using Eq. 4.105, the relations

$$\left(iv \cdot D + \frac{g_A}{2}\slashed{u}_\perp \gamma_5\right)\mathcal{N}_v + \left(i\slashed{D}_\perp + \frac{g_A}{2}v \cdot u\gamma_5\right)\mathcal{H}_v = 0, \tag{4.106}$$

$$\left(i\slashed{D}_\perp - \frac{g_A}{2}v \cdot u\gamma_5\right)\mathcal{N}_v + \left(-iv \cdot D - 2m + \frac{g_A}{2}\slashed{u}_\perp \gamma_5\right)\mathcal{H}_v = 0, \tag{4.107}$$

hold.

(e) In the next step, we isolate the EOM for the so-called light component $\mathcal{N}_v$. To this end, formally solve Eq. 4.107 for $\mathcal{H}_v$ and insert the result into Eq. 4.106. Show that the resulting EOM for $\mathcal{N}_v$ reads

$$\left(iv \cdot D + \frac{g_A}{2}\slashed{u}_\perp \gamma_5\right)\mathcal{N}_v + \left(i\slashed{D}_\perp + \frac{g_A}{2}v \cdot u\gamma_5\right)$$

$$\times \left(2m + iv \cdot D - \frac{g_A}{2}\slashed{u}_\perp \gamma_5\right)^{-1}\left(i\slashed{D}_\perp - \frac{g_A}{2}v \cdot u\gamma_5\right)\mathcal{N}_v = 0. \tag{4.108}$$

Equation 4.108 represents the EOM for the field $\mathcal{N}_v$. The same EOM can be derived by applying the variational principle to the Lagrangian

$$\mathcal{L}_{\text{eff}} = \bar{\mathcal{N}}_v\left(iv \cdot D + \frac{g_A}{2}\slashed{u}_\perp \gamma_5\right)\mathcal{N}_v + \bar{\mathcal{N}}_v\left(i\slashed{D}_\perp + \frac{g_A}{2}v \cdot u\gamma_5\right)$$

$$\times \left(2m + iv \cdot D - \frac{g_A}{2}\slashed{u}_\perp \gamma_5\right)^{-1}\left(i\slashed{D}_\perp - \frac{g_A}{2}v \cdot u\gamma_5\right)\mathcal{N}_v. \tag{4.109}$$

Note that the nucleon mass only appears in the denominator of the second term. This second term is therefore suppressed relative to the first term.[20]

---

[20] Because of Eq. 4.99, a partial derivative acting on $\mathcal{N}_v$ produces a small four-momentum.

While the above Lagrangian achieves the objective of isolating the field $\mathcal{N}_v$, it is not in the form commonly used in HBChPT. It is convenient to introduce the spin matrix S_v^μ, defined as

$$S_v^\mu \equiv \frac{i}{2}\gamma_5\sigma^{\mu\nu}v_\nu = -\frac{1}{2}\gamma_5(\gamma^\mu\not{v} - v^\mu), \quad S_v^{\mu\dagger} = \gamma_0 S_v^\mu \gamma_0. \tag{4.110}$$

Exercise 4.13 Show that, in four dimensions, S_v^μ obeys the relations

$$v \cdot S_v = 0, \quad \{S_v^\mu, S_v^\nu\} = \frac{1}{2}(v^\mu v^\nu - g^{\mu\nu}), \quad [S_v^\mu, S_v^\nu] = i\varepsilon^{\mu\nu\rho}{}_\sigma v_\rho S_v^\sigma. \tag{4.111}$$

Hint:

$$\gamma_5\sigma^{\sigma\tau} = -\frac{i}{2}\varepsilon^{\sigma\tau\alpha\beta}\sigma_{\alpha\beta}, \quad \varepsilon_{0123} = 1, \quad \varepsilon^{\mu\nu\rho}{}_\sigma\varepsilon^{\sigma\tau\alpha\beta} = \det\begin{pmatrix} g^{\mu\tau} & g^{\mu\alpha} & g^{\mu\beta} \\ g^{\nu\tau} & g^{\nu\alpha} & g^{\nu\beta} \\ g^{\rho\tau} & g^{\rho\alpha} & g^{\rho\beta} \end{pmatrix}.$$

The spin matrix S_v^μ allows us to rewrite the 16 combinations $\bar{\mathcal{N}}_v\Gamma\mathcal{N}_v$, where $\Gamma \in \{\mathbb{1}, \gamma^\mu, \gamma^5, \gamma^\mu\gamma^5, \sigma^{\mu\nu}\}$, as[21]

$$\begin{aligned} (\bar{\mathcal{N}}_v\mathbb{1}\mathcal{N}_v &= \bar{\mathcal{N}}_v\mathbb{1}\mathcal{N}_v,) \\ \bar{\mathcal{N}}_v\gamma_5\mathcal{N}_v &= 0, \\ \bar{\mathcal{N}}_v\gamma^\mu\mathcal{N}_v &= v^\mu\bar{\mathcal{N}}_v\mathcal{N}_v, \\ \bar{\mathcal{N}}_v\gamma^\mu\gamma_5\mathcal{N}_v &= 2\bar{\mathcal{N}}_v S_v^\mu\mathcal{N}_v, \\ \bar{\mathcal{N}}_v\sigma^{\mu\nu}\mathcal{N}_v &= 2\varepsilon^{\mu\nu\rho}{}_\sigma v_\rho\bar{\mathcal{N}}_v S_v^\sigma\mathcal{N}_v, \\ \bar{\mathcal{N}}_v\sigma^{\mu\nu}\gamma_5\mathcal{N}_v &= 2i(v^\mu\bar{\mathcal{N}}_v S_v^\nu\mathcal{N}_v - v^\nu\bar{\mathcal{N}}_v S_v^\mu\mathcal{N}_v). \end{aligned} \tag{4.112}$$

Exercise 4.14 Show that the relations of Eq. 4.112 hold.

Hint: Use Eq. 4.101. For example,

$$\bar{\mathcal{N}}_v\gamma_5\mathcal{N}_v = \bar{\mathcal{N}}_v\gamma_5 \not{v}\mathcal{N}_v = \cdots.$$

Equations 4.112 result in a nice simplification of the Dirac structures in the heavy-baryon approach, because only two groups of 4×4 matrices, the unit matrix and S_v^μ, instead of the original six groups on the left-hand side of Eq. 4.112 appear.

To obtain the final form of the leading-order Lagrangian in HBChPT, we formally expand Eq. 4.109 in inverse powers of the nucleon mass,

---

[21] We include the combination $\sigma^{\mu\nu}\gamma_5$ for convenience.

Fig. 4.6 s-channel pole
diagram of πN scattering

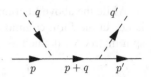

$$\mathscr{L}_{\text{eff}} = \bar{\mathscr{N}}_v\left(iv\cdot D + \frac{g_A}{2}\slashed{u}_\perp\gamma_5\right)\mathscr{N}_v + \sum_{n=1}^{\infty}\frac{1}{(2m)^n}\mathscr{L}_{\text{eff},n}, \tag{4.113}$$

and apply Eq. 4.112. The result for the leading-order term reads

$$\widehat{\mathscr{L}}_{\pi N}^{(1)} = \bar{\mathscr{N}}_v(iv\cdot D + g_A S_v\cdot u)\mathscr{N}_v, \tag{4.114}$$

where the symbol $\widehat{}$ indicates the heavy-baryon formalism. The nucleon mass has disappeared from the leading-order Lagrangian, in contrast to the relativistic Lagrangian of Eq. 4.17. It only appears in the terms of higher orders as powers of $1/m$. In the power-counting scheme $\widehat{\mathscr{L}}_{\pi N}^{(1)}$ counts as $\mathcal{O}(q)$, because the covariant derivative D_μ and the chiral vielbein u_μ both count as $\mathcal{O}(q)$. The heavy-baryon Feynman propagator derived from the free part of the Lagrangian of Eq. 4.114 is given by

$$G_{vF}(k) = \frac{P_{v+}}{v\cdot k + i0^+}. \tag{4.115}$$

The expansion of the Lagrangian of Eq. 4.109 generates terms that are suppressed by powers of $1/m$. In addition to these $1/m$ corrections, the Lagrangian at higher orders also contains terms that stem from additional chiral structures that do *not* contain inverse powers of the nucleon mass. The complete Lagrangian up to and including order q^4 is given in Ref. [22]. A wide variety of processes has been calculated in the heavy-baryon scheme. We refer the reader to Ref. [6] for a comprehensive overview.

While the heavy-baryon approach results in a consistent power counting and, similar to the mesonic sector, allows for the application of dimensional regularization in combination with the $\widetilde{\text{MS}}$ scheme, the complete $1/m$ expansion can create difficulties with analyticity under specific kinematics. To illustrate this point, consider the example of pion-nucleon scattering [2]. The invariant amplitudes have poles at $s = m_N^2$ and $u = m_N^2$ which, at tree level, can be understood in terms of the relativistic propagator (see Fig. 4.6),

$$\frac{1}{(p+q)^2 - m_N^2} = \frac{1}{2p\cdot q + M_\pi^2}. \tag{4.116}$$

The propagator of Eq. 4.116 has a pole at $2p\cdot q = -M_\pi^2$, which is equivalent to a pole at $s = m_N^2$. An analogous pole appears in the u-channel diagram at $u = m_N^2$.

Despite not being in the physical region of pion-nucleon scattering, analyticity of the invariant amplitudes requires both these poles to be present (see Sect. 4.3.2).

In the heavy-baryon approach, the relativistic propagator is expanded in inverse powers of m_N. Choosing $p = m_N v$ for simplicity,

$$\frac{1}{2p \cdot q + M_\pi^2} = \frac{1}{2m_N} \frac{1}{v \cdot q + \frac{M_\pi^2}{2m_N}} = \frac{1}{2m_N} \frac{1}{v \cdot q} \left(1 - \frac{M_\pi^2}{2m_N v \cdot q} + \cdots \right). \quad (4.117)$$

To any finite order in the heavy-baryon expansion, poles appear at $v \cdot q = 0$ instead of $v \cdot q = -M_\pi^2/(2m_N)$ being a single pole. Without summing an infinite number of diagrams, the heavy-baryon approach therefore does not generate the correct nucleon pole structure of the invariant amplitudes.

We saw in Exercise 4.10 that applying dimensional regularization in combination with the $\widetilde{MS}$ scheme does not produce a consistent power counting. The solution was to subtract the finite terms that violated the power-counting rules, which was possible since the relevant terms were analytic in small quantities. We will now present two methods that remove the terms that violate the power counting while keeping the analytic structure of amplitudes in the low-energy region intact, the infrared regularization of Ref. [3] and the extended on-mass-shell scheme of Ref. [24].

4.6.2 Infrared Regularization

The method of infrared regularization relies on the analytic structure of loop integrals in n dimensions, in particular on *infrared* singularities which arise for small momenta. To illustrate the existence of these infrared singularities consider the integral

$$
\begin{aligned}
H(p^2, m^2, M^2; n) &\equiv -i \int \frac{d^n k}{(2\pi)^n} \frac{1}{[(k-p)^2 - m^2 + i0^+](k^2 - M^2 + i0^+)} \\
&= -i \int \frac{d^n k}{(2\pi)^n} \frac{1}{[k^2 - 2p \cdot k + (p^2 - m^2) + i0^+](k^2 - M^2 + i0^+)},
\end{aligned}
$$

$$(4.118)$$

where, compared to the definition of $I_{N\pi}$ of Exercise 4.10, we are using the sign convention of Ref. [3] and have dropped the factor μ^{4-n}. We consider nucleon momenta close to the mass shell, $p^2 \approx m^2$, which means that $p^2 - m^2$ is counted as a small quantity of $\mathcal{O}(q)$. By counting powers of the loop momentum, we see that the integral converges for $n < 4$. In the chiral limit $M \to 0$, however, the integral becomes infrared singular when going to smaller n. For $n = 3$, the integral is infrared regular as long as $p^2 \neq m^2$, but exhibits an infrared singularity for $p^2 = m^2$, since the integrand behaves as k^2/k^3. For any smaller value of n, the integral is infrared singular even for $p^2 \neq m^2$. The infrared singularity stems from small loop momenta $k = \mathcal{O}(q)$. In this momentum region, the first factor of the integrand is of order q^{-1} for both $p^2 = m^2$ and $p^2 \neq m^2$, since we count $p^2 - m^2 = \mathcal{O}(q)$, while the second factor is $\mathcal{O}(q^{-2})$. For $n < 3$, the chiral expansion of the integral $H(p^2, m^2, M^2; n)$ therefore contains an infrared-singular term of $\mathcal{O}(q^{n-3})$ in the chiral limit.

Exercise 4.15 Verify by explicit calculation at threshold, i.e., $p^2 = p_{\text{thr}}^2 = (m + M)^2$, that in the limit $M \to 0$ the integral $H(p^2, m^2, M^2; n)$ of Eq. 4.118 develops an infrared singularity for $n \leq 3$.

(a) Use the Feynman parameterization

$$\frac{1}{ab} = \int\limits_0^1 dz \frac{1}{[az + b(1 - z)]^2},$$
(4.119)

with $a = (k - p)^2 - m^2 + i0^+$ and $b = k^2 - M^2 + i0^+$, to show that the integral can be written as

$$H(p^2, m^2, M^2; n) = -i \int\limits_0^1 dz \int \frac{d^n k}{(2\pi)^n} \frac{1}{[k^2 - A(z) + i0^+]^2},$$
(4.120)

with

$$A(z) = z^2 p^2 - z(p^2 - m^2 + M^2) + M^2.$$

Hint: Use the shift of variables $k \to k + zp$.

(b) The integral can be further simplified by use of the formula (see Exercise 3.22)

$$\int \frac{d^n k}{(2\pi)^n} \frac{(k^2)^p}{(k^2 - A + i0^+)^q} = \frac{i(-)^{p-q}}{(4\pi)^{\frac{n}{2}}} \frac{\Gamma\left(p + \frac{n}{2}\right)\Gamma\left(q - p - \frac{n}{2}\right)}{\Gamma\left(\frac{n}{2}\right)\Gamma(q)} (A - i0^+)^{p + \frac{n}{2} - q}.$$

Show that

$$H(p^2, m^2, M^2; n) = \frac{1}{(4\pi)^{\frac{n}{2}}} \Gamma\left(2 - \frac{n}{2}\right) \int\limits_0^1 dz [A(z) - i0^+]^{\frac{n}{2} - 2}.$$
(4.121)

(c) Show that the squared threshold momentum $p_{\text{thr}}^2 = (m + M)^2$ corresponds to $A(z) = [z(m + M) - M]^2$. Split the integration interval into two parts,

$$\int\limits_{0}^{1} = \int\limits_{0}^{z_0} + \int\limits_{z_0}^{1},$$

with $z_0 = M/(m+M)$, to show that for $n > 3$ the integral over $A(z)$ is given by

$$\int\limits_{0}^{1} dz [A(z)]^{\frac{n}{2}-2} = \frac{1}{(n-3)(m+M)}(M^{n-3} + m^{n-3}).$$

Why can the small imaginary part, $-i0^+$, be neglected? What is the sign of $A(z)$ in the considered z interval $[0,1]$?

(d) The result for arbitrary n can be obtained through analytic continuation and is given by

$$H\big((m+M)^2, m^2, M^2; n\big) = \frac{\Gamma\big(2-\frac{n}{2}\big)}{(4\pi)^{\frac{n}{2}}(n-3)}\left(\frac{M^{n-3}}{m+M} + \frac{m^{n-3}}{m+M}\right). \qquad (4.122)$$

The first term in Eq. 4.122, proportional to M^{n-3}, is called the infrared-singular part of H, while the second term, proportional to m^{n-3}, is the infrared-regular part. Show that for noninteger values of n the expansion of the infrared-singular part in small quantities contains only noninteger powers of M, while the expansion of the infrared-regular part only contains nonnegative integer powers of M.

The example above introduces the concept of infrared-singular and infrared-regular parts. We now turn to the formal definition of these terms for arbitrary momenta p close to the mass shell [3]. Let us introduce the dimensionless variables

$$\Omega = \frac{p^2 - m^2 - M^2}{2mM} = \mathcal{O}(q^0), \qquad \alpha = \frac{M}{m} = \mathcal{O}(q). \qquad (4.123)$$

In terms of these variables the integrand $A(z)$ in Eq. 4.120 is given by

$$A(z) = m^2 [z^2 - 2\alpha\Omega z(1-z) + \alpha^2(1-z)^2] \equiv m^2 C(z),$$

and the integral H can be written as (see Eq. 4.121)

$$H(p^2, m^2, M^2; n) = \kappa(m; n) \int\limits_{0}^{1} dz [C(z) - i0^+]^{\frac{n}{2}-2}, \qquad (4.124)$$

with

$$\kappa(m; n) = \frac{\Gamma\big(2-\frac{n}{2}\big)}{(4\pi)^{\frac{n}{2}}} m^{n-4}. \qquad (4.125)$$

If we consider $M \to 0$, which in the new dimensionless variables corresponds to $\alpha \to 0$, infrared singularities arise for small values of z, as the integrand $C(z)$ vanishes. Since we require both z and α to be small we perform the substitution $z = \alpha x$. The upper integration limit $z = 1$ then becomes $x = 1/\alpha \to \infty$ as $\alpha \to 0$. We can define a new integral I that has the same infrared singularities as H. It is identical to H with the exception that the upper integration limit is ∞ even for finite values of α:

$$I \equiv \kappa(m; n) \int_0^\infty dz [C(z) - i0^+]^{\frac{n}{2}-2} = \kappa(m; n)\alpha^{n-3} \int_0^\infty dx [D(x) - i0^+]^{\frac{n}{2}-2}, \quad (4.126)$$

where

$$D(x) = 1 - 2\Omega x + x^2 + 2\alpha x(\Omega x - 1) + \alpha^2 x^2.$$

The difference between H and I is the infrared-regular part R,

$$R \equiv -\kappa(m; n) \int_1^\infty dz [C(z) - i0^+]^{\frac{n}{2}-2}, \quad (4.127)$$

so that

$$H = I + R. \quad (4.128)$$

Exercise 4.16 We now show that these more general definitions of the infrared-singular and infrared-regular parts reproduce the behavior of the threshold integral of Eq. 4.122.

(a) Show that $\Omega_{thr} = 1$, and thus

$$I_{thr} = \kappa(m; n)\alpha^{n-3} \int_0^\infty dx \left\{ [(1+\alpha)x - 1]^2 - i0^+ \right\}^{\frac{n}{2}-2}.$$

For which values of n does the integral converge?

(b) Verify that the integrand can be rewritten as

$$\left\{ [(1+\alpha)x - 1]^2 - i0^+ \right\}^{\frac{n}{2}-2} = \frac{(1+\alpha)x - 1}{(1+\alpha)(n-4)} \frac{d}{dx} \left\{ [(1+\alpha)x - 1]^2 - i0^+ \right\}^{\frac{n}{2}-2}.$$

Using integration by parts, show that

$$\int_0^\infty dx \left\{ [(1+\alpha)x - 1]^2 - i0^+ \right\}^{\frac{n}{2}-2}$$

$$= \left[\frac{(1+\alpha)x - 1}{(1+\alpha)(n-4)} \left\{ [(1+\alpha)x - 1]^2 - i0^+ \right\}^{\frac{n}{2}-2} \right]_0^\infty$$

$$- \frac{1}{n-4} \int_0^\infty dx \left\{ [(1+\alpha)x - 1]^2 - i0^+ \right\}^{\frac{n}{2}-2}.$$

(c) Starting from $n < 3$, show that the integral can be analytically continued to give

$$\int_0^\infty dx \left\{ [(1+\alpha)x - 1]^2 - i0^+ \right\}^{\frac{n}{2}-2} = \frac{1}{(n-3)(1+\alpha)}.$$

Verify that the obtained expression for I_{thr} agrees with the infrared-singular part of Eq. 4.122.

(d) Show that $C_{\text{thr}}(z) = [z(1+\alpha) - \alpha]^2$. Use this result to evaluate R_{thr} for $n < 3$ and analytically continue to verify that the obtained expression agrees with the infrared-regular part of Eq. 4.122.

As seen in Eq. 4.126, the infrared-singular part contains an overall factor of α^{n-3}, so that for noninteger n the chiral expansion of I only consists of noninteger powers of the small expansion parameter,

$$I = \mathcal{O}(q^{n-3}) + \mathcal{O}(q^{n-2}) + \mathcal{O}(q^{n-1}) + \cdots, \tag{4.129}$$

while the infrared-regular term only contains nonnegative integer powers,

$$R = \mathcal{O}(q^0) + \mathcal{O}(q^1) + \mathcal{O}(q^2) + \cdots. \tag{4.130}$$

The method can be extended to general one-loop integrals [3]. It suffices to consider scalar integrals, as tensor integrals can be reduced to combinations of integrals of the type

$$H_{mn} = -i \int \frac{d^n k}{(2\pi)^n} \frac{1}{a_1 \ldots a_m} \frac{1}{b_1 \ldots b_n},$$

where $a_i = (q_i + k)^2 - M^2 + i0^+$ and $b_j = (p_j - k)^2 - m^2 + i0^+$ are inverse meson and nucleon propagators, respectively. The four-momenta q_i are $\mathcal{O}(q)$, while the four-momenta p_j are close to the nucleon mass shell, i.e., $p_j^2 - m^2 = \mathcal{O}(q)$. One first combines all meson propagators and nucleon propagators separately. In the case of the meson propagators, this can be done with use of the formula

$$\frac{1}{a_1 \ldots a_m} = \left(-\frac{\partial}{\partial M^2}\right)^{m-1} \int_0^1 dx_1 \ldots \int_0^1 dx_{m-1} \frac{X}{A}, \qquad (4.131)$$

where $X = 1$ for $m = 2$, while for $m > 2$ the numerator is given by

$$X = x_2(x_3)^2 \ldots (x_m)^{m-1}.$$

The denominator is quadratic in the loop momentum k,

$$A = \bar{A} - (k - \bar{q})^2 - i0^+,$$

where $\bar{A}$ is constant and $\bar{q}$ is a linear combination of the momenta q_i. A can be obtained with the recursion relation

$$A = A_m, \quad A_1 = a_1, \quad A_{p+1} = x_p A_p + (1 - x_p)a_{p+1}, \quad (p = 1, \ldots, m-1).$$

An analogous calculation for the nucleon propagators gives

$$\frac{1}{b_1 \ldots b_n} = \left(-\frac{\partial}{\partial m^2}\right)^{n-1} \int_0^1 dy_1 \ldots \int_0^1 dy_{n-1} \frac{Y}{B}, \qquad (4.132)$$

with

$$Y = y_2(y_3)^2 \ldots (y_n)^{n-1},$$

and

$$B = \bar{B} - (k - \bar{p})^2 - i0^+,$$

where again $\bar{B}$ is constant and $\bar{p}$ is a linear combination of the external nucleon momenta. The integrals H_{mn} can then be written as

$$H_{mn} = \left(-\frac{\partial}{\partial M^2}\right)^{m-1} \left(-\frac{\partial}{\partial m^2}\right)^{n-1} \int_0^1 dx\,dy\,(-i) \int \frac{d^n k}{(2\pi)^n} \frac{1}{AB}, \qquad (4.133)$$

where

$$\int_0^1 dx\,dy \equiv \int_0^1 dx_1 \ldots \int_0^1 dx_{m-1} \int_0^1 dy_1 \ldots \int_0^1 dy_{n-1}.$$

Since the denominators A and B have the same general form as single meson and nucleon propagators, respectively, the integral over the loop momentum k can be performed in complete analogy to the integral H considered above. One again combines the two terms in the denominator using the Feynman parameterization of

Eq. 4.119 and splits the original integral H_{mn} into an infrared-singular part I_{mn} and an infrared-regular part R_{mn},

$$H_{mn} = I_{mn} + R_{mn}. \tag{4.134}$$

Just as in the case of the nucleon self-energy integral H above, the infrared-singular part I_{mn} is identical to the original integral H_{mn} with the exception of the upper limit in the integration that combines the denominators A and B, with the integration running to 1 for H_{mn} and to ∞ for I_{mn}.

The chiral expansion of the infrared-singular part I_{mn} only contains noninteger powers for noninteger n, while the infrared-regular term R_{mn} can be expanded in an ordinary Taylor series. The infrared-regular term can therefore be absorbed in the counter terms of the most general Lagrangian. This is equivalent to replacing H_{mn} with its infrared-singular part I_{mn}, which is the infrared renormalization condition,

$$H_{mn}^r = I_{mn}.$$

All terms violating the power counting are contained in R_{mn} [3], therefore the renormalized expressions containing I_{mn} automatically satisfy the power counting.

Depending on the dimension n and the number of nucleon and meson propagators, the integral H_{mn} might contain an ultraviolet (UV) divergence, e.g., the integrand of H_{11} for $n = 4$ scales as k^3/k^4 in the UV limit, which results in a logarithmic divergence upon integration. We can thus write

$$H_{mn} = \frac{H_{mn}^{UV}}{\varepsilon} + \tilde{H}_{mn},$$

where as before $\varepsilon = 4 - n$. When separating the integral H_{mn} into its infrared-singular and infrared-regular parts, these terms might contain additional divergences that are not present in H_{mn} [3],

$$I_{mn} = \frac{I_{mn}^{add}}{\varepsilon} + \tilde{I}_{mn}, \quad R_{mn} = \frac{R_{mn}^{UV}}{\varepsilon} + \frac{R_{mn}^{add}}{\varepsilon} + \tilde{R}_{mn},$$

where R_{mn}^{UV} corresponds to the original UV divergence. Since the additional divergences are not present in the original integral H_{mn}, they have to cancel in the sum of I and R, i.e.,

$$I_{mn}^{add} = -R_{mn}^{add}.$$

The renormalized expression for H_{mn} is then given by

$$H_{mn}^r = \tilde{I}_{mn}.$$

While the infrared renormalization solves the power-counting problem, it would not be useful if the resulting expressions violated chiral symmetry. This would be manifest in a violation of the Ward identities of the theory. It can be shown, however, that chiral symmetry is preserved in infrared regularization [3]. Expressions containing the original integrals $H_{mn} = I_{mn} + R_{mn}$ satisfy the Ward identities since they are derived from a Lagrangian that is explicitly symmetric, and since dimensional regularization preserves the symmetries of the Lagrangian. As seen above, infrared-singular and infrared-regular parts differ in the analytic structure of their chiral expansions. Since the expansion of I_{mn} (R_{mn}) only contains noninteger (nonnegative integer) powers of the small quantity q for noninteger n, infrared-singular and infrared-regular terms have to satisfy the Ward identities separately. This means that replacing the original integral H_{mn} by its infrared-singular part I_{mn} preserves the Ward identities and thus chiral symmetry is preserved.

4.6.3 Extended On-Mass-Shell Scheme

While infrared regularization offers one solution to the power-counting problem, it is not the only one. We now turn to a different solution, the extended on-mass-shell (EOMS) scheme. This approach was first motivated in Ref. [27] and has been worked out in detail in Ref. [24]. In infrared regularization, the terms that violate the power counting are contained in the infrared-regular part of an integral. However, the chiral expansion of this infrared-regular part can also contain an infinite number of terms that do *not* violate the power counting. While the general principles of renormalization allow us to subtract these terms by absorbing them in counter terms in the Lagrangian, it is not necessary to do so. The idea behind the EOMS scheme is to absorb only those terms that violate the power counting by performing finite subtractions in addition to the $\widetilde{\text{MS}}$ scheme such that the resulting expressions for renormalized diagrams satisfy the power-counting rules. As was the case for infrared regularization, this procedure can be made systematic. It offers the additional advantage of allowing for the application to multi-loop diagrams and diagrams containing additional degrees of freedom such as vector mesons.

To illustrate the EOMS approach we consider the integral H in the chiral limit,

$$H(p^2, m^2, 0; n) = -i \int \frac{d^n k}{(2\pi)^n} \frac{1}{[(k-p)^2 - m^2 + i0^+](k^2 + i0^+)}.$$

Going to the chiral limit simplifies the calculations while keeping the main features of the method intact. According to the power-counting rules in Sect. 4.5.2, the renormalized integral is supposed to be of order $D = n - 1 - 2 = n - 3$. Since the nucleon momentum p is close to the mass shell, we define the small dimensionless quantity

$$\Delta = \frac{p^2 - m^2}{m^2} = \mathcal{O}(q).$$

Introducing $C(z, \Delta) = z^2 - \Delta z(1 - z) - i0^+$, the integral $H(p^2, m^2, 0; n)$ can be written as

$$H(p^2, m^2, 0; n) = \kappa(m; n) \int_0^1 dz [C(z, \Delta)]^{\frac{n}{2} - 2}, \qquad (4.135)$$

where $\kappa(m; n)$ is given in Eq. 4.125. To evaluate Eq. 4.135 we write

$$\int_0^1 dz [C(z, \Delta)]^{\frac{n}{2} - 2} = (-\Delta)^{\frac{n}{2} - 2} \int_0^1 dz\, z^{\frac{n}{2} - 2} \left(1 - \frac{1 + \Delta}{\Delta} z\right)^{\frac{n}{2} - 2}.$$

The integral on the right-hand side can be expressed in terms of the integral representation of the hypergeometric function [1],

$$F(a, b; c; z) = \frac{\Gamma(c)}{\Gamma(b)\Gamma(c - b)} \int_0^1 dt\, t^{b-1}(1 - t)^{c-b-1}(1 - tz)^{-a}, \operatorname{Re}(c) > \operatorname{Re}(b) > 0.$$

We substitute $a = 2 - \frac{n}{2}$, $b = \frac{n}{2} - 1$, $c = \frac{n}{2}$, and $z = (1 + \Delta)/\Delta$, make use of $\Gamma(1) = 1$, and obtain

$$H(p^2, m^2, 0; n) = \kappa(m; n) \frac{\Gamma\left(\frac{n}{2} - 1\right)}{\Gamma\left(\frac{n}{2}\right)} (-\Delta)^{\frac{n}{2} - 2} F\left(2 - \frac{n}{2}, \frac{n}{2} - 1; \frac{n}{2}; \frac{1 + \Delta}{\Delta}\right).$$

We apply the transformation formula [1]

$$F(a, b; c; z) = (1 - z)^{-a} F\left(a, c - b; c; \frac{z}{z - 1}\right)$$

and the symmetry property $F(a, b; c; z) = F(b, a; c; z)$ to obtain

$$H(p^2, m^2, 0; n) = \kappa(m; n) \frac{\Gamma\left(\frac{n}{2} - 1\right)}{\Gamma\left(\frac{n}{2}\right)} F\left(1, 2 - \frac{n}{2}; \frac{n}{2}; \frac{p^2}{m^2}\right). \qquad (4.136)$$

For nucleon momenta close to the mass shell the last argument in the hypergeometric function is close to unity, $p^2/m^2 \approx 1$, and therefore not a good expansion parameter. Fortunately, the properties of hypergeometric functions (see Eq. 15.3.6 of Ref. [1]) allow us to rewrite a hypergeometric function of argument z as a combination of other hypergeometric functions of argument $1 - z$. In our case, this corresponds to

$$1 - \frac{p^2}{m^2} = \frac{m^2 - p^2}{m^2} = -\Delta.$$

Equation 4.136 then reads

$$H(p^2, m^2, 0; n) = \frac{m^{n-4}}{(4\pi)^{\frac{n}{2}}} \left[\frac{\Gamma\left(2 - \frac{n}{2}\right)}{n-3} F\left(1, 2 - \frac{n}{2}; 4 - n; -\Delta\right) \right.$$

$$\left. + (-\Delta)^{n-3} \Gamma\left(\frac{n}{2} - 1\right) \Gamma(3-n) F\left(\frac{n}{2} - 1, n-2; n-2; -\Delta\right) \right].$$

$$(4.137)$$

Since Δ counts as a small quantity of $\mathcal{O}(q)$, we can now use the expansion of $F(a, b; c; z)$ for $|z| < 1$,

$$F(a, b; c; z) = 1 + \frac{ab}{c} z + \frac{a(a+1)b(b+1)}{c(c+1)} \frac{z^2}{2} + \cdots. \qquad (4.138)$$

Since in our case $z = -\Delta$, the expansion of the hypergeometric functions results in terms with only nonnegative integer powers of Δ. While the second term of Eq. 4.137 thus only contains terms of $\mathcal{O}(q^{n-3})$ and higher as dictated by the power counting, we see that the first term contains a contribution which does not satisfy the power counting, i.e., which is not proportional to $\mathcal{O}(q)$ as $n \to 4$. For $n \to 4$ we obtain

$$H = \frac{m^{n-4}}{(4\pi)^{\frac{n}{2}}} \left[\frac{\Gamma\left(2 - \frac{n}{2}\right)}{n-3} + \left(1 - \frac{p^2}{m^2}\right) \ln\left(1 - \frac{p^2}{m^2}\right) \right.$$

$$\left. + \left(1 - \frac{p^2}{m^2}\right)^2 \ln\left(1 - \frac{p^2}{m^2}\right) + \cdots \right], \qquad (4.139)$$

where ... refers to terms which are at least of $\mathcal{O}(q^3)$ or $O(n-4)$. Terms of the type $-\Delta \ln(-\Delta)$ are counted as $\mathcal{O}(q)$, i.e., as a small quantity just as $-\Delta$.

While Eq. 4.139 contains terms with logarithmic dependence on the nucleon momentum, these terms satisfy the power counting. The term that violates the power counting is local in the external momentum, which means that it is a polynomial in p^2 (here of zeroth order), and can be absorbed in a finite number of counter terms in the Lagrangian. We can thus subtract

$$\frac{m^{n-4}}{(4\pi)^{\frac{n}{2}}} \frac{\Gamma\left(2 - \frac{n}{2}\right)}{n-3} \qquad (4.140)$$

from Eq. 4.139 to obtain the renormalized integral

$$H_R(p^2, m^2, 0; n) = \frac{m^{n-4}}{(4\pi)^{n/2}} \left[\left(1 - \frac{p^2}{m^2}\right) \ln\left(1 - \frac{p^2}{m^2}\right) \right.$$
$$\left. + \left(1 - \frac{p^2}{m^2}\right)^2 \ln\left(1 - \frac{p^2}{m^2}\right) + \cdots \right]. \tag{4.141}$$

Exercise 4.17 Show that the only finite terms appearing in H up to $\mathcal{O}(q^3)$ and $O(n-4)$ as $n \to 4$ are proportional to $\ln(-\Delta)$.

(a) Use Eq. 4.138 to show that the expansion of the first hypergeometric function in Eq. 4.137 is given by

$$F\left(1, 2 - \frac{n}{2}; 4 - n; -\Delta\right) = 1 - \frac{\Delta}{2} + \frac{6-n}{5-n}\frac{\Delta^2}{4} + \mathcal{O}(q^3). \tag{4.142}$$

Using the parameter $\varepsilon \equiv 4 - n$ introduced in Sect. 3.4.7, this expression is rewritten as

$$1 - \frac{\Delta}{2} + \frac{2+\varepsilon}{1+\varepsilon}\frac{\Delta^2}{4} + \mathcal{O}(q^3). \tag{4.143}$$

(b) Show that the expansion of Eq. 4.143 to order ε is given by

$$1 - \frac{\Delta}{2} + (2 - \varepsilon)\frac{\Delta^2}{4} + \cdots. \tag{4.144}$$

(c) We see that the term independent of Δ reproduces the first term in Eq. 4.139. For the terms proportional to Δ and Δ^2 we need to expand the coefficient of the hypergeometric function about $n = 4$. Using $\Gamma(x+1) = x\Gamma(x)$, show that

$$\frac{\Gamma\left(2 - \frac{n}{2}\right)}{n-3} = \frac{2}{\varepsilon} + 2 + \Gamma'(1) + \cdots. \tag{4.145}$$

(d) Using these results, show that the terms proportional to Δ and Δ^2 of the first term in Eq. 4.137 are given by

$$-\left(\frac{1}{\varepsilon} + 1 + \frac{\Gamma'(1)}{2}\right)\Delta + \left(\frac{1}{\varepsilon} + \frac{1}{2} - \frac{\Gamma'(1)}{2}\right)\Delta^2. \tag{4.146}$$

(e) Show that

$$(-\Delta)^{n-3} = -\Delta + \varepsilon\Delta \ln(-\Delta) + \cdots. \tag{4.147}$$

Recall: $a^b = e^{b \ln a}$.

(f) Performing an analysis for the second term in Eq. 4.137 analogous to the calculation above, verify that all finite terms appearing in the expression for H for $n \to 4$ are proportional to $\ln(-\Delta)$.

In order to identify and subtract the terms that violate the power counting, we have explicitly calculated the integral H. While this did not pose too great of a problem for the case of H in the chiral limit, our aim is to find a method to determine the subtraction terms even in those cases where the explicit calculation of the integrals is more difficult. As seen in Eq. 4.137, the result for $H(p^2, m^2, 0; n)$ is of the form

$$H \sim F(n, \Delta) + \Delta^{n-3} G(n, \Delta), \qquad (4.148)$$

where F and G are proportional to hypergeometric functions that are analytic in Δ for arbitrary n. The term of interest, i.e., the subtraction term, is contained in F. Central to the development of a systematic scheme is the observation that the expansion of F can be obtained by first expanding the integrand of H and then exchanging integration and summation, i.e., integrating each term in the expansion separately [28] (see Sect. 4.6.4 for an illustration of the general method). After applying a conventional $\widetilde{\text{MS}}$ renormalization scheme we can identify and subtract the terms that violate the power counting without having to calculate the complete integral. In essence we work with a modified integrand which is obtained from the original integrand by subtracting a suitable number of counter terms. To explain what we mean by suitable consider the series

$$\sum_{l=0}^{\infty} \frac{(p^2 - m^2)^l}{l!} \left[\left(\frac{1}{2p^2} p_\mu \frac{\partial}{\partial p_\mu} \right)^l \frac{1}{[k^2 - 2k \cdot p + (p^2 - m^2) + i0^+](k^2 + i0^+)} \right]_{p^2 = m^2}$$

$$= \frac{1}{(k^2 - 2k \cdot p + i0^+)(k^2 + i0^+)} \bigg|_{p^2 = m^2}$$

$$+ (p^2 - m^2) \left[\frac{1}{2m^2} \frac{1}{(k^2 - 2k \cdot p + i0^+)^2} - \frac{1}{2m^2} \frac{1}{(k^2 - 2k \cdot p + i0^+)(k^2 + i0^+)} \right.$$

$$\left. - \frac{1}{(k^2 - 2k \cdot p + i0^+)^2 (k^2 + i0^+)} \right]_{p^2 = m^2} + \cdots, \qquad (4.149)$$

where $[\ldots]_{p^2 = m^2}$ means that the *coefficients* of $(p^2 - m^2)^l$ are considered only for four-momenta p satisfying the on-mass-shell condition. While the coefficients in this series still depend on the direction of p_μ, performing the integration over loop momenta k and evaluating the corresponding coefficients for $p^2 = m^2$ results in a series that is a function of only p^2. In fact, it was shown in Ref. [28] that the integrated series exactly reproduces the first term in Eq. 4.137.

The EOMS scheme is then defined as follows: We subtract from the integrand of $H(p^2, m^2, 0; n)$ those terms of the series of Eq. 4.149 that violate the power counting. These terms are analytic in the small expansion parameter and do not contain infrared singularities. In our example only the first term in Eq. 4.137 has to be subtracted. All higher-order terms contain infrared singularities, such as generated by the last term in the second coefficient: for small k the integrand scales as

k^3/k^4 for $n = 4$. The integral of the first term of Eq. 4.149 is given by Eq. 4.140, and our result for the renormalized integral is

$$H_R = H - H_{\text{subtr}} = \mathcal{O}(q^{n-3}).$$

Since the subtraction point is $p^2 = m^2$, the renormalization condition is denoted "extended on-mass-shell" (EOMS) scheme in analogy with the on-mass-shell renormalization scheme in renormalizable theories.

So far we have considered the special case of an integral in the chiral limit, but the method can be generalized to the case of a finite pion mass. Instead of Eq. 4.149 one now considers a simultaneous expansion in $p^2 - m^2$ and M^2,

$$\frac{1}{(k^2 - 2k \cdot p + i0^+)(k^2 + i0^+)}\Big|_{p^2 = m^2}$$

$$+ (p^2 - m^2)\left[\frac{1}{2m^2}\frac{1}{(k^2 - 2k \cdot p + i0^+)^2} + \cdots\right]_{p^2 = m^2}$$

$$+ M^2 \frac{1}{(k^2 - 2k \cdot p + i0^+)(k^2 + i0^+)^2}\Big|_{p^2 = m^2} + \cdots.$$

Since all terms of order q satisfy the power counting, the contribution resulting from the first term is still the only one that is subtracted.

While the original formulation of the infrared regularization was specific to one-loop integrals with pion and nucleon propagators, the EOMS scheme can be straightforwardly extended to include other degrees of freedom, such as vector mesons [25] or the $\Delta(1232)$ resonance [33], and it can be applied to multi-loop diagrams [55].

Moreover, the infrared regularization can be reformulated in a form analogous to the EOMS scheme and can thus be applied to multi-loop diagrams with an arbitrary number of particles with arbitrary masses [54]. After combining the meson and baryon propagators as explained in Sect. 4.6.2 and performing the integration over the loop momentum, an arbitrary integral can be written as[22]

$$H = \int_0^1 dz f(z),$$

where $f(z)$ is a function depending on the external momenta, masses, and the space-time dimension n. The chiral expansion of the infrared-regular part R, with

$$R = -\int_1^\infty dz f(z),$$

---

[22] For notational convenience we suppress the subscripts m and n of Eq. 4.134.

can be performed before the z-integration, resulting in integrals of the type

$$R^{(i)} = -\int_1^\infty dz\, z^{n+i},\tag{4.150}$$

where i is a nonnegative integer. The integrals can be calculated by analytic continuation from the domain of n in which they converge,

$$R^{(i)} = -\frac{z^{n+i+1}}{n+i+1}\bigg|_1^\infty = \frac{1}{n+i+1}.\tag{4.151}$$

One can reproduce the result of Eq. 4.151 without splitting the original integral H into two parts. Instead, we perform the chiral expansion of the integrand in H and interchange summation and integration. The result thus only contains terms that are analytic in small parameters. However, since in most cases the original integral H also contains nonanalytic terms, this procedure does *not* reproduce the chiral expansion of H. Summation and integration only commute as long as H converges absolutely. The series resulting from expanding the integrand of H and integrating each term separately contains the same coefficients as the chiral expansion of R, but the integrals $R^{(i)}$ are replaced by integrals of the type

$$J^{(i)} = \int_0^1 dz\, z^{n+i}.\tag{4.152}$$

Again performing an analytic continuation, the integrals are given by

$$J^{(i)} = \frac{z^{n+i+1}}{n+i+1}\bigg|_0^1 = \frac{1}{n+i+1}.\tag{4.153}$$

Comparing Eqs. 4.151 and 4.153, we see that the chiral expansion of the infrared-regular part R can be obtained by reducing H to an integral over Feynman parameters, expanding the resulting expression in small quantities, and interchanging summation and integration.

Other approaches to the extension of infrared regularization are given in Refs. [10, 11, 41].

4.6.4 Dimensional Counting Analysis

While we have been able to find closed-form expressions for the loop integrals we have considered so far, analytic solutions to more complex integrals, such as two- or multi-loop integrals containing two or more masses, are increasingly difficult to obtain. Since we are often interested in the chiral expansion of observables, we can avoid having to analytically solve integrals and use the method of dimensional counting analysis instead [28]. A closely related method of calculating loop integrals is the so-called "method of regions" [57]. While particularly useful for two- and multi-loop integrals, the advantage of dimensional counting analysis for one-loop integrals lies in its applicability to dimensionally regulated integrals containing several different masses, such as the pion mass and the nucleon mass in the chiral limit. We provide an illustration of dimensional counting analysis in terms of the one-loop integral of Eq. 4.118,

$$H(p^2, m^2, M^2; n) = -i \int \frac{d^n k}{(2\pi)^n} \frac{1}{k^2 - 2p \cdot k + p^2 - m^2 + i0^+} \frac{1}{k^2 - M^2 + i0^+}.$$

$$(4.154)$$

One would like to know how the integral behaves for small values of M and/or $p^2 - m^2$ as a function of n. If we consider, for fixed $p^2 \neq m^2$, the limit $M \to 0$, the integral H can be represented as

$$H(p^2, m^2, M^2; n) = \sum_i M^{\beta_i} F_i(p^2, m^2, M^2; n),$$

$$(4.155)$$

where the functions F_i are analytic in M^2 and are obtained as follows. First, one rewrites the integration variable as $k = M^{\alpha_i} \tilde{k}$, where α_i is an arbitrary nonnegative real number. Next, one isolates the overall factor of M^{β_i} so that the remaining integrand can be expanded in positive powers of M^2 and interchanges integration and summation. The resulting series represents the expansion of $F_i(p^2, m^2, M^2; n)$ in powers of M^2. The sum of all possible rescalings with subsequent expansions with nontrivial coefficients then reproduces the expansion of the result of the original integral.

To be specific, let us apply this program to H:

$$H(p^2, m^2, M^2; n)$$
$$= -i \int \frac{M^{n\alpha_i} d^n \tilde{k}}{(2\pi)^n} \frac{1}{\tilde{k}^2 M^{2\alpha_i} - 2p \cdot \tilde{k} M^{\alpha_i} + p^2 - m^2 + i0^+} \frac{1}{\tilde{k}^2 M^{2\alpha_i} - M^2 + i0^+}.$$

$$(4.156)$$

From Eq. 4.156 we see that the first fraction does not contribute to the overall factor M^{β_i} for any α_i. It will be expanded in (positive) powers of $(\tilde{k}^2 M^{2\alpha_i} - 2p \cdot \tilde{k} M^{\alpha_i})$ except for $\alpha_i = 0$. For $0 < \alpha_i < 1$, we rewrite the second fraction as

$$\frac{1}{M^{2\alpha_i}} \frac{1}{(\tilde{k}^2 - M^{2-2\alpha_i} + i0^+)} = \frac{1}{M^{2\alpha_i}} \frac{1}{\tilde{k}^2 + i0^+} \left(1 + \frac{M^{2-2\alpha_i}}{\tilde{k}^2 + i0^+} + \cdots\right). \quad (4.157)$$

On the other hand, if $1 < \alpha_i$ we rewrite the second fraction as

$$\frac{1}{M^2}\frac{1}{(\tilde{k}^2 M^{2\alpha_i-2} - 1 + i0^+)} = -\frac{1}{M^2}(1 + \tilde{k}^2 M^{2\alpha_i-2} + \cdots). \tag{4.158}$$

In both cases one obtains integrals of the type $\int d^n\tilde{k}\,\tilde{k}^{\mu_1}\ldots\tilde{k}^{\mu_m}$ as the coefficients of the expansion. However, such integrals vanish in dimensional regularization. Therefore, the only nontrivial terms in the sum of Eq. 4.155 correspond to either $\alpha_i = 0$ or $\alpha_i = 1$. Thus we obtain

$$H(p^2, m^2, M^2; n) = H^{(0)}(p^2, m^2, M^2; n) + H^{(1)}(p^2, m^2, M^2; n), \tag{4.159}$$

where

$$
\begin{aligned}
&H^{(0)}(p^2, m^2, M^2; n) \\
&= -i\sum_{j=0}^{\infty}(M^2)^j \int \frac{d^n k}{(2\pi)^n}\frac{1}{k^2 - 2p\cdot k + p^2 - m^2 + i0^+}\frac{1}{(k^2 + i0^+)^{j+1}},
\end{aligned} \tag{4.160}
$$

and

$$
\begin{aligned}
&H^{(1)}(p^2, m^2, M^2; n) \\
&= -i\frac{M^{n-2}}{p^2 - m^2 + i0^+}\sum_{j=0}^{\infty}\frac{(-1)^j M^j}{(p^2 - m^2 + i0^+)^j}\int \frac{d^n\tilde{k}}{(2\pi)^n}\frac{(\tilde{k}^2 M - 2p\cdot\tilde{k})^j}{\tilde{k}^2 - 1 + i0^+}.
\end{aligned} \tag{4.161}
$$

A comparison with the direct calculation of H shows that the dimensional-counting method indeed leads to the correct expressions [28]. While the loop integrals of Eq. 4.161 have a simple analytic structure in $p^2 - m^2$, the same technique can be repeated for the loop integrals of Eq. 4.160 when $p^2 - m^2 \to 0$, now using the change of variable $k = (p^2 - m^2)^{\gamma_i}\tilde{k}$ with arbitrary nonnegative real numbers γ_i.

4.7 The Delta Resonance

So far we have discussed the lowest-lying states in baryon ChPT with particular emphasis on the nucleon in the two-flavor sector. However, it is a well-known fact that the $\Delta(1232)$ resonance $[I(J^P) = \frac{3}{2}(\frac{3}{2}^+)]$ plays an important role in the phenomenological description of low- and medium-energy processes such as pion-nucleon scattering, electromagnetic pion production, Compton scattering, etc. This is due to the rather small mass gap between the $\Delta(1232)$ and the nucleon, the strong coupling of the $\Delta(1232)$ to the πN channel, and its relatively large photon decay amplitudes.

In ordinary baryon ChPT the effects of resonances are implicitly taken into account through the values of the LECs. A close-by resonance such as the $\Delta(1232)$ may then result in a rather slow convergence for observables sensitive to the quantum numbers of the given resonance. Therefore, it seems natural to ask whether the chiral effective field theory can be extended to also include resonances as dynamical degrees of freedom. One thereby not only hopes to improve the convergence by essentially reordering an infinite number of higher-order terms which contribute to higher-order LECs in the standard formulation, but also to extend the kinematic range of the EFT. If one succeeds in defining a consistent expansion scheme one may even be able to perform calculations of processes which involve center-of-mass energies covering the resonance region and thus study properties of the particular resonance.

As in the baryonic sector, a consistent expansion scheme was first developed in the heavy-baryon approach (see, e.g., Refs. [12, 34, 38]). More recently, the discussion has focussed on a manifestly Lorentz-invariant approach (see, e.g., Refs. [8, 33, 51]). In a Lorentz-invariant formulation of a field theory involving particles of higher spin ($s \geq 1$), one necessarily introduces unphysical degrees of freedom [46, 53]. Therefore, one has to impose constraints which specify the physical degrees of freedom. A detailed treatment of systems with constraints is beyond the scope of these lecture notes (see, e.g., Refs. [16, 31, 35]) and we restrict ourselves to a basic introduction.

4.7.1 The Free Lagrangian of a Spin-3/2 System

The Rarita-Schwinger formalism [53] allows for a covariant field-theoretical description of systems with spin $\frac{3}{2}$. The field is represented by a so-called vector spinor denoted by Ψ^μ ($\mu = 0, 1, 2, 3$), where each Ψ^μ is a Dirac field. Under a proper orthochronous Lorentz transformation[23] $x'^\mu = \Lambda^\mu{}_\nu x^\nu$, the Rarita-Schwinger field has the mixed transformation properties of a four-vector field and a four-component Dirac field,

$$\Psi'^\mu(x') = \Lambda^\mu{}_\nu S(\Lambda)\Psi^\nu(x),$$

where $S(\Lambda)$ is the usual matrix representation acting on Dirac spinors. For a relativistic description of spin $\frac{3}{2}$ we need $2 \cdot 4 = 8$ independent complex fields, where the factor of two accounts for the description of particles and antiparticles, and the factor of four results from four spin projections in the rest frame. In other words, we need to generate 8 complex conditions among the $4 \cdot 4 = 16$ complex fields of the vector spinor in order to eliminate the additional degrees of freedom.

The most general free Lagrangian serving that purpose reads [46]

$$\mathscr{L}_{\frac{3}{2}} = \bar{\Psi}_\mu \Lambda^{\mu\nu}(A)\Psi_\nu, \tag{4.162}$$

where[24]

---

[23] $\det(\Lambda) = 1$ and $\Lambda^0{}_0 \geq 1$.

[24] It is common practice to denote both Lorentz transformations and the tensor describing the Δ with the same symbol Λ.

$$\Lambda^{\mu\nu}(A) = - \left[(i\slashed{\partial} - m_\Delta)g^{\mu\nu} + iA(\gamma^\mu\partial^\nu + \gamma^\nu\partial^\mu) \right.$$
$$\left. + \frac{i}{2}(3A^2 + 2A + 1)\gamma^\mu \slashed{\partial}\gamma^\nu + m_\Delta(3A^2 + 3A + 1)\gamma^\mu\gamma^\nu \right], \qquad (4.163)$$

with $A \neq -\frac{1}{2}$ an arbitrary real parameter and m_Δ the mass of the Δ.[25] The Lagrangian introduced by Rarita and Schwinger [53] corresponds to $A = -\frac{1}{3}$. From the Euler-Lagrange equation,

$$\frac{\partial \mathscr{L}_{\frac{3}{2}}}{\partial \bar{\Psi}_\mu} - \partial_\rho \underbrace{\frac{\partial \mathscr{L}_{\frac{3}{2}}}{\partial \partial_\rho \bar{\Psi}_\mu}}_{= 0} = 0,$$

we obtain the equation of motion (EOM)

$$\Lambda^{\mu\nu}(A)\Psi_\nu = 0. \qquad (4.164)$$

In addition to the EOM, the fields Ψ^μ satisfy the equations

$$(i\slashed{\partial} - m_\Delta)\Psi^\mu = 0, \qquad (4.165)$$

$$\gamma_\mu\Psi^\mu = 0, \qquad (4.166)$$

$$\partial_\mu\Psi^\mu = 0. \qquad (4.167)$$

Each of the Eqs. 4.166 and 4.167 generate four complex (subsidiary) conditions. Therefore we end up with the correct number of $16 - 4 - 4 = 8$ independent components. Note that Eq. 4.165 does not reduce the number of independent fields: given that the subsidiary conditions hold, it may rather be interpreted as the equation of motion.

Exercise 4.18 Consider the Lagrangian of Eq. 4.162 for $A = -1$.

(a) Derive the EOM.
(b) Contract the EOM with γ_μ and verify

$$2i\partial_\mu\Psi^\mu - 2i\slashed{\partial}\gamma_\mu\Psi^\mu - 3m_\Delta\gamma_\mu\Psi^\mu = 0 \qquad (4.168)$$

for solutions of the EOM.
(c) Contract the EOM with ∂_μ and verify

$$m_\Delta\partial_\mu\Psi^\mu - m_\Delta\slashed{\partial}\gamma_\mu\Psi^\mu = 0 \qquad (4.169)$$

for solutions of the EOM.

---

[25] Note that m_Δ denotes the leading-order contribution to the mass of the Δ in an expansion in small quantities.

(d) Substitute Eq. 4.169 into Eq. 4.168 and verify Eq. 4.166,

$$\gamma_\mu \Psi^\mu = 0, \tag{4.170}$$

for solutions of the EOM.

(e) Substitute Eq. 4.170 into Eq. 4.169 and verify Eq. 4.167,

$$\partial_\mu \Psi^\mu = 0, \tag{4.171}$$

for solutions of the EOM.

(f) Substitute Eqs. 4.170 and 4.171 into the EOM and verify

$$(i\slashed{\partial} - m_\Delta)\Psi^\mu = 0.$$

Hint:

$$\gamma^\mu \gamma^\nu + \gamma^\nu \gamma^\mu = 2g^{\mu\nu}\mathbb{1}.$$

While, using the same techniques, the results may also be verified for general A, the actual calculation is more elaborate.

For the application of dimensional regularization with n space-time dimensions the generalization of the Lagrangian is (see, e.g., Ref. [52])

$$\mathscr{L}_{\frac{3}{2}} = \bar{\Psi}_\mu \Lambda^{\mu\nu}(A, n)\Psi_\nu, \tag{4.172}$$

where

$$\Lambda^{\mu\nu}(A, n) = -\Big\{ (i\slashed{\partial} - m_\Delta)g^{\mu\nu} + iA(\gamma^\mu \partial^\nu + \gamma^\nu \partial^\mu)$$

$$+ \frac{i}{n-2}\big[(n-1)A^2 + 2A + 1\big]\gamma^\mu \slashed{\partial}\gamma^\nu$$

$$+ \frac{m_\Delta}{(n-2)^2}\big[n(n-1)A^2 + 4(n-1)A + n\big]\gamma^\mu \gamma^\nu \Big\}, \quad n \neq 2. \tag{4.173}$$

In the special case of $A = -1$, Eq. 4.172 does not explicitly depend on n.

The free Lagrangian of Eq. 4.172 is invariant under the set of transformations

$$\Psi_\mu \mapsto \Psi_\mu + \frac{4a}{n}\gamma_\mu \gamma_\nu \Psi^\nu,$$

$$A \mapsto \frac{An - 8a}{n(1 + 4a)}, \quad a \neq -\frac{1}{4}, \tag{4.174}$$

which are often referred to as a point transformation [49]. The invariance under the point transformation guarantees that physical quantities do not depend on the so-called "off-shell parameter" A [49, 58], provided that the interaction terms are also invariant under the point transformation.

4.7.2 Isospin

So far, we have only discussed the transformation properties under the Lorentz group. In order to address the transformation properties under $SU(2)_L \times SU(2)_R \times U(1)_V$, we first need a convenient representation of the isospin group $SU(2)_V$. Once we have found such a representation, we will generate a realization of $SU(2)_L \times SU(2)_R \times U(1)_V$ by applying the procedure discussed in Sect. 4.1 [13, 15].

The $\Delta(1232)$ resonance has isospin $I = \frac{3}{2}$ and comes in four charged states: Δ^{++}, Δ^+, Δ^0, and Δ^-. In the following we make use of the isovector-isospinor formalism, i.e., we consider the Δ states as the $I = \frac{3}{2}$ components of the tensor product of $I = 1$ and $I = \frac{1}{2}$ states. Let X and Y denote Hilbert spaces carrying isospin representations with $I = 1$ and $I = \frac{1}{2}$, respectively. Elements of X and Y are written as

$$|x\rangle = \sum_{i=1}^{3} x_i |i\rangle = \sum_{m=-1}^{1} (-1)^m x_{-m} |1, m\rangle, \quad |y\rangle = \sum_{r=-\frac{1}{2}}^{\frac{1}{2}} y_r \left|\frac{1}{2}, r\right\rangle.$$

For $|x\rangle$ we have displayed both the Cartesian and spherical decompositions.[26] Later on, the complex components will be replaced by the vector-spinor fields of the previous section. Under an $SU(2)$ transformation V the vectors $|x\rangle$ and $|y\rangle$ transform according to the adjoint and fundamental representations, respectively. For the components x_i and y_r this means

$$x_i' = \sum_{j=1}^{3} D_{ij}(V) x_j, \quad D_{ij}(V) = \frac{1}{2} \mathrm{Tr}\left(\tau_i V \tau_j V^\dagger\right),$$

$$y_r' = \sum_{s=-\frac{1}{2}}^{\frac{1}{2}} V_{rs} y_s.$$

Exercise 4.19 Consider an infinitesimal $SU(2)$ transformation

$$V = 1 - i\varepsilon_a \frac{\tau_a}{2}.$$

In the adjoint representation the infinitesimal transformation reads

$$D = 1 - i\varepsilon_a T_a^{\mathrm{ad}},$$

---

[26] The Cartesian notation is convenient for displaying final results in a compact form while the spherical notation is used to apply angular momentum coupling methods. Recall $x_{-1} = (x_1 - ix_2)/\sqrt{2}$, $x_0 = x_3$, and $x_{+1} = -(x_1 + ix_2)/\sqrt{2}$.

where the 3×3 matrices T_a^{ad} are given in Eq. 1.68. Verify that $D_{ij}(V)$ defines the adjoint representation.

Hint: $\text{Tr}(\tau_a \tau_b) = 2\delta_{ab}$, $[\tau_a, \tau_b] = 2i\varepsilon_{abc}\tau_c$.

Now consider an element of the tensor product $Z = X \otimes Y$,

$$|z\rangle = \sum_{i=1}^{3} \sum_{r=-\frac{1}{2}}^{\frac{1}{2}} z_{i,r}|i\rangle \otimes \left|\frac{1}{2}, r\right\rangle = \sum_{m=-1}^{1} \sum_{r=-\frac{1}{2}}^{\frac{1}{2}} (-1)^m z_{-m,r}|1, m\rangle \otimes \left|\frac{1}{2}, r\right\rangle. \quad (4.175)$$

Using the Clebsch-Gordan decomposition, the tensor product may be decomposed into a direct sum, $Z = Z_{\frac{3}{2}} \oplus Z_{\frac{1}{2}}$. The isospin-$\frac{3}{2}$ states live in the first space and we therefore need projection operators $P_{\frac{3}{2}}$ and $P_{\frac{1}{2}}$ projecting onto the corresponding subspaces. The basis states of $Z_{\frac{3}{2}}$ and $Z_{\frac{1}{2}}$ are given in terms of the uncoupled basis by[27]

$$\left|\left(1\frac{1}{2}\right)\frac{3}{2}, M\right\rangle = \sum_{m=-1}^{1} \sum_{r=-\frac{1}{2}}^{\frac{1}{2}} \left(1, m; \frac{1}{2}, r \middle| \frac{3}{2}, M\right) |1, m\rangle \left|\frac{1}{2}, r\right\rangle, \quad (4.176)$$

$$\left|\left(1\frac{1}{2}\right)\frac{1}{2}, M\right\rangle = \sum_{m=-1}^{1} \sum_{r=-\frac{1}{2}}^{\frac{1}{2}} \left(1, m; \frac{1}{2}, r \middle| \frac{1}{2}, M\right) |1, m\rangle \left|\frac{1}{2}, r\right\rangle, \quad (4.177)$$

where $(j_1, m_1; j_2, m_2 | J, M)$ are Clebsch-Gordan coefficients. The corresponding projection operators for $Z_{\frac{3}{2}}$ and $Z_{\frac{1}{2}}$ read

$$P_{\frac{3}{2}} = \sum_{M=-\frac{3}{2}}^{\frac{3}{2}} \left|\left(1\frac{1}{2}\right)\frac{3}{2}, M\right\rangle\left\langle\left(1\frac{1}{2}\right)\frac{3}{2}, M\right|, \quad (4.178)$$

$$P_{\frac{1}{2}} = \sum_{M=-\frac{1}{2}}^{\frac{1}{2}} \left|\left(1\frac{1}{2}\right)\frac{1}{2}, M\right\rangle\left\langle\left(1\frac{1}{2}\right)\frac{1}{2}, M\right|. \quad (4.179)$$

Given the representations

$$|1, 1\rangle = \begin{pmatrix} 1 \\ 0 \\ 0 \end{pmatrix}, \quad |1, 0\rangle = \begin{pmatrix} 0 \\ 1 \\ 0 \end{pmatrix}, \quad |1, -1\rangle = \begin{pmatrix} 0 \\ 0 \\ 1 \end{pmatrix},$$

$$\left|\frac{1}{2}, \frac{1}{2}\right\rangle = \chi_{\frac{1}{2}} = \begin{pmatrix} 1 \\ 0 \end{pmatrix}, \quad \left|\frac{1}{2}, -\frac{1}{2}\right\rangle = \chi_{-\frac{1}{2}} = \begin{pmatrix} 0 \\ 1 \end{pmatrix}$$

---

[27] We now follow common practice in physics and omit the $\otimes$ symbol.

for the basis states, the matrix representation $\xi^{\frac{1}{2}}_{sph}$ of $P_{\frac{1}{2}}$ with respect to the spherical basis reads

$$\xi^{\frac{1}{2}}_{sph} = \frac{1}{3}\begin{pmatrix} \mathbb{1} - \tau_3 & -\frac{1}{\sqrt{2}}(\tau_1 - i\tau_2) & 0 \\ -\frac{1}{\sqrt{2}}(\tau_1 + i\tau_2) & \mathbb{1} & -\frac{1}{\sqrt{2}}(\tau_1 - i\tau_2) \\ 0 & -\frac{1}{\sqrt{2}}(\tau_1 + i\tau_2) & \mathbb{1} + \tau_3 \end{pmatrix}. \tag{4.180}$$

Exercise 4.20 Verify Eq. 4.180.

(a) Insert Eq. 4.177 into Eq. 4.179. Make use of the Clebsch-Gordan coefficients

$$\left(1,0;\frac{1}{2},-\frac{1}{2}\Big|\frac{1}{2},-\frac{1}{2}\right) = -\left(1,0;\frac{1}{2},\frac{1}{2}\Big|\frac{1}{2},\frac{1}{2}\right) = \frac{1}{\sqrt{3}},$$

$$\left(1,1;\frac{1}{2},-\frac{1}{2}\Big|\frac{1}{2},\frac{1}{2}\right) = -\left(1,-1;\frac{1}{2},\frac{1}{2}\Big|\frac{1}{2},-\frac{1}{2}\right) = \sqrt{\frac{2}{3}}.$$

The remaining Clebsch-Gordan coefficients vanish because of the selection rule for the projections.

(b) Express terms of the type $|1,m\rangle\langle 1,m'|$ in terms of 3×3 matrices. For example,

$$|1,0\rangle\langle 1,0| = \begin{pmatrix} 0 \\ 1 \\ 0 \end{pmatrix}(0 \quad 1 \quad 0) = \begin{pmatrix} 0 & 0 & 0 \\ 0 & 1 & 0 \\ 0 & 0 & 0 \end{pmatrix}.$$

(c) Finally, express terms of the type $|\frac{1}{2},r\rangle\langle\frac{1}{2},r'|$ in terms of Pauli matrices and the unit matrix $\mathbb{1}$. For example,

$$\left|\frac{1}{2},\frac{1}{2}\right\rangle\left\langle\frac{1}{2},\frac{1}{2}\right| = \begin{pmatrix} 1 \\ 0 \end{pmatrix}(1 \quad 0) = \begin{pmatrix} 1 & 0 \\ 0 & 0 \end{pmatrix} = \frac{1}{2}(\mathbb{1} + \tau_3).$$

With the transformation matrix

$$T = \begin{pmatrix} -\frac{1}{\sqrt{2}} & \frac{i}{\sqrt{2}} & 0 \\ 0 & 0 & 1 \\ \frac{1}{\sqrt{2}} & \frac{i}{\sqrt{2}} & 0 \end{pmatrix}$$

the transition to Cartesian coordinates yields the matrix representation $\xi^{\frac{1}{2}}$ of $P_{\frac{1}{2}}$,

$$\xi^{\frac{1}{2}} = T^{\dagger}\xi^{\frac{1}{2}}_{sph}T = \frac{1}{3}\begin{pmatrix} \mathbb{1} & i\tau_3 & -i\tau_2 \\ -i\tau_3 & \mathbb{1} & i\tau_1 \\ i\tau_2 & -i\tau_1 & \mathbb{1} \end{pmatrix}, \tag{4.181}$$

or in compact notation

$$\xi_{ij}^{\frac{1}{2}} = \frac{1}{3}\tau_i\tau_j. \tag{4.182}$$

Note that the entries of $\xi^{\frac{1}{2}}$ are 2×2 matrices acting on the isospinors χ_r. Sometimes it may be helpful to also specify the isospinor indices rs of $\xi^{\frac{1}{2}}$. For example $\xi_{11,\frac{1}{2}\frac{1}{2}}^{\frac{1}{2}} = \frac{1}{3}$. Either by explicit calculation or using the property of projection operators, $\xi^{\frac{1}{2}} + \xi^{\frac{3}{2}} = 1$, one obtains the matrix representation of the second projection operator,

$$\xi_{ij}^{\frac{3}{2}} = \delta_{ij} - \frac{1}{3}\tau_i\tau_j. \tag{4.183}$$

For a vector $|\Delta\rangle$ of the subspace $Z_{\frac{3}{2}}$,

$$|\Delta\rangle = \sum_{M=-\frac{3}{2}}^{\frac{3}{2}} \Delta_M \left|\frac{3}{2}, M\right\rangle = \Delta^- \left|\frac{3}{2}, -\frac{3}{2}\right\rangle + \cdots,$$

the scalar product $(\langle 1, m|\langle \frac{1}{2}, r|)|\Delta\rangle$ generates the component $(-)^m z_{-m,r}^{(1)}$ of the state $|\Delta\rangle$ in terms of the Clebsch-Gordan coefficient $(1, m; \frac{1}{2}, r|\frac{3}{2}, M)$ and the components Δ_M. Reexpressing the spherical components in terms of Cartesian components, we then obtain, in terms of the projection operator $\xi^{\frac{3}{2}}$ of Eq. 4.183,

$$\begin{aligned}
\xi_{1j}^{\frac{3}{2}} z_j &= \frac{1}{\sqrt{2}} \begin{pmatrix} \frac{1}{\sqrt{3}}\Delta^0 - \Delta^{++} \\ \Delta^- - \frac{1}{\sqrt{3}}\Delta^+ \end{pmatrix}, \\
\xi_{2j}^{\frac{3}{2}} z_j &= -\frac{i}{\sqrt{2}} \begin{pmatrix} \frac{1}{\sqrt{3}}\Delta^0 + \Delta^{++} \\ \Delta^- + \frac{1}{\sqrt{3}}\Delta^+ \end{pmatrix}, \\
\xi_{3j}^{\frac{3}{2}} z_j &= \sqrt{\frac{2}{3}} \begin{pmatrix} \Delta^+ \\ \Delta^0 \end{pmatrix}.
\end{aligned} \tag{4.184}$$

This phase convention agrees with Ref. [58] but is opposite to Ref. [34].

4.7.3 Leading-Order Lagrangian of the $\Delta(1232)$ Resonance

In the above discussion, the components of the vectors were complex numbers which we now interpret as fields by adjusting the notation accordingly, i.e., $z_{i,r} \rightarrow \Psi_{i,r}$ etc. We suppress indices referring to the Lorentz-transformation properties until the very end. Under $SU(2)_V$ these fields transform as[28]

---

[28] We return to the repeated-index summation convention, because the ranges of summation should now be clear.

$$\Psi_{i,r} \mapsto \Psi'_{i,r} = D_{ij}(V)V_{rs}\Psi_{j,s}.$$

A realization of $SU(2)_L \times SU(2)_R \times U(1)_V$ is then obtained as in Sect. 4.1: we first replace V by $K(L, R, U)$ of Eq. 4.8 and then promote global transformations to local transformations (see Sect. 4.2). Moreover, we take into account that the Δ has baryon number $+1$. The field components therefore transform as [58]

$$\Psi_{i,r}(x) \mapsto \Psi'_{i,r}(x) = \exp[-i\Theta(x)]\mathscr{K}_{ij,rs}[V_L(x), V_R(x), U(x)]\Psi_{j,s}(x), \quad (4.185)$$

where

$$\mathscr{K}_{ij,rs} = \frac{1}{2}\text{Tr}(\tau_i K \tau_j K^\dagger)K_{rs}, \quad (4.186)$$

with K defined in Eq. 4.8. The corresponding covariant derivative is given by

$$(D_\mu \Psi)_{i,r} \equiv \mathscr{D}_{\mu,ij,rs}\Psi_{j,s},$$
$$\mathscr{D}_{\mu,ij,rs} = \partial_\mu \delta_{ij}\delta_{rs} - 2i\varepsilon_{ijk}\Gamma_{\mu,k}\delta_{rs} + \delta_{ij}\Gamma_{\mu,rs} - iv_\mu^{(s)}\delta_{ij}\delta_{rs},$$

where we parameterized the chiral connection Γ_μ of Eq. 4.13 as $\Gamma_\mu = \Gamma_{\mu,k}\tau_k$. The leading-order Lagrangian is given by [34][29]

$$\mathscr{L}_{\pi\Delta}^{(1)} = \bar{\Psi}_\mu \xi^{\frac{3}{2}}\Lambda_{\pi\Delta}^{(1)\mu\nu}\xi^{\frac{3}{2}}\Psi_\nu, \quad (4.187)$$

where

$$\begin{aligned}
\Lambda_{\pi\Delta}^{(1)\mu\nu} = -\Big[& (i\not{D} - m_\Delta)g^{\mu\nu} + iA(\gamma^\mu D^\nu + \gamma^\nu D^\mu) \\
& + \frac{i}{2}(3A^2 + 2A + 1)\gamma^\mu \not{D}\gamma^\nu + m_\Delta(3A^2 + 3A + 1)\gamma^\mu\gamma^\nu \\
& + \frac{g_1}{2}\not{u}\gamma_5 g^{\mu\nu} + \frac{g_2}{2}(\gamma^\mu u^\nu + u^\mu\gamma^\nu)\gamma_5 + \frac{g_3}{2}\gamma^\mu \not{u}\gamma_5\gamma^\nu \Big]. \quad (4.188)
\end{aligned}$$

Similar to the case of the QCD Lagrangian, Eq. 4.187 represents an extremely compact notation. The vector-spinor isovector-isospinor field Ψ contains $4 \cdot 4 \cdot 3 \cdot 2 = 96$ fields $\Psi_{\mu,\alpha,i,r}$, where μ denotes the Lorentz-vector index, α the Dirac-spinor index, i the isovector index, and r the isospinor index. The projection operator $\xi^{\frac{3}{2}}$ is responsible for the fact that only the isospin-$\frac{3}{2}$ component of the isovector-isospinor field enters the Lagrangian. In comparison with the free Lagrangian of Eqs. 4.162 and 4.163, we notice that the ordinary partial derivative has been replaced by the covariant derivative. In addition, terms involving the chiral vielbein of Eq. 4.16 have been constructed. Note that at first sight there seem to exist three independent structures of this type. Application of Dirac's constraint analysis [16] shows that the Lagrangian of Eq. 4.187 only leads to a

---

[29] We have explicitly included the projection operator in the definition of the Lagrangian.

consistent theory provided certain relations hold among the coupling constants g_1, g_2, and g_3 [66].

4.7.4 Consistent Interactions

We have seen in Eqs. 4.166 and 4.167 that the free Lagrangian describes a system with constraints. The same is true for the Lagrangian of Eq. 4.187, which now also contains interactions with pions and external fields. The interesting question arises under which conditions the interacting system still has the correct number of dynamical degrees of freedom.

Applying a method described in Chap. 1 of Ref. [16], one may analyze the theory including interactions within the Hamiltonian formalism. For a finite number of degrees of freedom an outline of the method is as follows (for a more detailed description see, e.g., Refs. [16, 31, 35]). Let us consider a classical system with N degrees of freedom q_i and velocities $\dot{q}_i = dq_i/dt$ described by the Lagrange function $L(q, \dot{q})$. Here, we assume that L contains the $\dot{q}$'s at the most quadratically. The Hamilton function is obtained using the Legendre transform

$$H(q, p) = \sum_{i=1}^{N} p_i \dot{q}_i - L(q, \dot{q}), \qquad (4.189)$$

where the p_i are the canonical momenta defined by

$$p_i \equiv \frac{\partial L(q, \dot{q})}{\partial \dot{q}_i}, \quad i = 1, \dots, N. \qquad (4.190)$$

Since H is a function of q and p, the velocities $\dot{q}_i$ have to be replaced using Eq. 4.190. If, according to Eq. 4.190, this is not possible because

$$\det A = 0, \text{ with } A_{ij} = \frac{\partial p_i}{\partial \dot{q}_j}, \qquad (4.191)$$

we are dealing with a singular system [35]. With a suitable change of coordinates, the Lagrange function can be written as a linear function of the unsolvable new velocities $\dot{q}_i'$. In the following the new coordinates are again denoted by q_i. Let the unsolvable $\dot{q}_i$ be the first n velocities $\dot{q}_1, \dots, \dot{q}_n$. The so-called primary constraints occur as follows. The Lagrange function L can be written as

$$L(q, \dot{q}) = \sum_{i=1}^{n} F_i(q) \dot{q}_i + G(q, \dot{q}_{n+1}, \dots, \dot{q}_N), \qquad (4.192)$$

from which we obtain as the canonical momenta

$$
p_i = \begin{cases} F_i(q) & \text{for } i = 1, \ldots, n, \\ \frac{\partial G(q, \dot{q}_{n+1}, \ldots, \dot{q}_N)}{\partial \dot{q}_i} & \text{for } i = n+1, \ldots, N. \end{cases} \tag{4.193}
$$

The first part of Eq. 4.193 can be reexpressed in terms of the relations

$$
\phi_i(q, p) = p_i - F_i(q) \approx 0, \quad i = 1, \ldots, n, \tag{4.194}
$$

which are referred to as the primary constraints. Here, $\phi_i \approx 0$ denotes a weak equation in Dirac's sense, namely that one must not use one of these constraints before working out a Poisson bracket [16]. Using Eq. 4.189, we consider the so-called total Hamilton function [16]

$$
H_T(q, p) = \sum_{j=n+1}^{N} p_j \dot{q}_j(p, q) - G(q, \dot{q}_{n+1}(p, q), \ldots, \dot{q}_N(p, q)) + \sum_{i=1}^{n} \lambda_i \phi_i(q, p)
$$

$$
= H(q, p) + \sum_{i=1}^{n} \lambda_i \phi_i(q, p), \tag{4.195}
$$

where the λ's are Lagrange multipliers taking care of the primary constraints and the $\dot{q}_i(p, q)$ are the solutions to Eq. 4.193 for $i = n+1, \ldots, N$. The constraints ϕ_i have to be zero throughout all time. For consistency, also $\dot{\phi}_i$ must be zero. The time evolution of the primary constraints ϕ_i is given by the Poisson bracket with the Hamilton function, leading to the consistency conditions

$$
\{\phi_i, H_T\} = \{\phi_i, H\} + \sum_{j=1}^{n} \lambda_j \{\phi_i, \phi_j\} \approx 0. \tag{4.196}
$$

Either all the λ's can be determined from these equations, or new constraints arise. The number of these secondary constraints corresponds to the number of λ's (or linear combinations thereof) which could not be determined. Again one demands the conservation in time of these (new) constraints and tries to solve the remaining λ's from these equations, etc. The number of physical degrees of freedom is given by the initial number of degrees of freedom (coordinates plus momenta) minus the number of constraints. In order for a theory to be consistent, the chain of new constraints has to terminate such that at the end of the procedure the correct number of degrees of freedom has been generated.

The application of this program to the Lagrangian of Eq. 4.187 leads, after a lengthy calculation, to the following relations among the coupling constants [66]:

$$
g_2 = A g_1, \quad g_3 = -\frac{1 + 2A + 3A^2}{2} g_1. \tag{4.197}
$$

In other words, what seem to be independent interaction terms from the point of view of constructing the most general Lagrangian [34], turn out to be related once the self-consistency conditions are imposed.

The Lagrangian of Eq. 4.187 with the couplings of Eq. 4.197 is invariant under the set of transformations of Eq. 4.174 for $n = 4$. However, this invariance is an outcome of, rather than an input to the constraint analysis. Demanding the invariance under the point transformation alone is not sufficient to obtain the relations of Eq. 4.197.

The effective Lagrangian of Eq. 4.187 is also invariant under the following local transformations

$$\Psi_{\mu,i}(x) \mapsto \Psi_{\mu,i}(x) + \tau_i \alpha_\mu(x), \tag{4.198}$$

where α_μ is an arbitrary vector-spinor isospinor function. This is due to the fact that we use six isospin degrees of freedom $\Psi_{\mu,\alpha,i}(x)$ instead of four physical isospin degrees of freedom. The quantization of the effective Lagrangian of Eq. 4.187 with the gauge fixing condition $\tau_i \Psi_{\mu,i} = 0$ leads to the following free-Δ Feynman propagator[30]

$$S^{\mu\nu}_{F,ij,\alpha\beta}(p) = \xi^{\frac{3}{2}}_{ij,\alpha\beta} S^{\mu\nu}_F(p), \tag{4.199}$$

where

$$S^{\mu\nu}_F(p) = -\frac{\slashed{p} + m_\Delta}{p^2 - m^2_\Delta + i0^+}\left[g^{\mu\nu} - \frac{1}{3}\gamma^\mu\gamma^\nu + \frac{1}{3m_\Delta}(p^\mu\gamma^\nu - \gamma^\mu p^\nu) - \frac{2}{3m^2_\Delta}p^\mu p^\nu\right]$$
$$+ \frac{1}{3m^2_\Delta}\frac{1+A}{1+2A}\left\{\left[\frac{A}{1+2A}m_\Delta - \frac{1+A}{2(1+2A)}\slashed{p}\right]\gamma^\mu\gamma^\nu - \gamma^\mu p^\nu - \frac{A}{1+2A}p^\mu\gamma^\nu\right\}.$$

In particular, choosing $A = -1$ results in the most convenient expression for the free-Δ Feynman propagator.

The leading-order $\pi N\Delta$ interaction Lagrangian can be written as

$$\mathscr{L}^{(1)}_{\pi N\Delta} = g\bar{\Psi}_{\mu,i}\xi^{\frac{3}{2}}_{ij}(g^{\mu\nu} + \tilde{z}\gamma^\mu\gamma^\nu)u_{\nu,j}\Psi + \text{H.c.}, \tag{4.200}$$

where we parameterized $u_\mu = u_{\mu,k}\tau_k$, and g and $\tilde{z}$ are coupling constants. The analysis of the structure of constraints yields

$$\tilde{z} = \frac{3A+1}{2}. \tag{4.201}$$

Again, the interaction term of Eq. 4.200 with the coupling constants g and $\tilde{z}$ constrained by Eq. 4.201 is invariant under the point transformation of Eq. 4.174.

---

[30] With this choice we associate a factor $iS^{\mu\nu}_F(p)$ with an internal Δ line of momentum p.

A simple estimate of the couplings g_1 and g of Eqs. 4.187 and 4.200, respectively, is obtained as follows. Consider the z component of the third axial-vector current in the nonrelativistic quark model [59],

$$A_{z,3} = \sum_{i=1}^{3} \sigma_z(i) \frac{\tau_3(i)}{2}.$$

The evaluation of $A_{z,3}$ between quark-model spin-flavor states of the nucleon and the Δ yields:

$$\left\langle p, \frac{1}{2} \middle| A_{z,3} \middle| p, \frac{1}{2} \right\rangle = \frac{5}{6} \equiv \frac{g_A}{2},$$

$$\left\langle \Delta^{++}, \frac{3}{2} \middle| A_{z,3} \middle| \Delta^{++}, \frac{3}{2} \right\rangle = \frac{3}{2} = \frac{9}{5} \frac{g_A}{2},$$

$$\left\langle \Delta^{+}, \frac{1}{2} \middle| A_{z,3} \middle| p, \frac{1}{2} \right\rangle = \frac{2}{3}\sqrt{2} = \frac{4}{5}\sqrt{2} \frac{g_A}{2}.$$

By comparing the ratios with the corresponding matrix elements originating from Eqs. 4.17, 4.187, and 4.200, one finds [34]

$$g_1 = \frac{9}{5} g_A, \quad g = \frac{3}{5}\sqrt{2} g_A. \tag{4.202}$$

Note that Eq. 4.202 is only a model-dependent estimate for the size of these couplings. In the spirit of EFT they have to be treated as independent LECs [34].

In summary, the lowest-order Lagrangian for the description of the pion-nucleon-Delta system is given by

$$\mathscr{L}_{\text{eff}} = \mathscr{L}_2 + \mathscr{L}_{\pi N}^{(1)} + \mathscr{L}_{\pi \Delta}^{(1)} + \mathscr{L}_{\pi N \Delta}^{(1)}, \tag{4.203}$$

where the individual Lagrangians are given in Eqs. 3.77, 4.17, 4.187, and 4.200, respectively. This Lagrangian contains in total seven LECs: F and B from the mesonic sector, g_A and m from $\mathscr{L}_{\pi N}^{(1)}$, g_1 and m_Δ from $\mathscr{L}_{\pi \Delta}^{(1)}$, and g from the $\pi N \Delta$ interaction Lagrangian.

Perturbative calculations including the $\Delta(1232)$ resonance may be organized by applying the "standard" power counting of Refs. [17, 64] to the renormalized diagrams, i.e., an interaction vertex obtained from an $\mathcal{O}(q^n)$ Lagrangian counts as q^n, a pion propagator as q^{-2}, a nucleon propagator as q^{-1}, and the integration of a loop as q^4. Here, q generically denotes a small expansion parameter such as, e.g., the pion mass. Note that this does not apply to the δ expansion, which is discussed below. The rules for the Δ propagator are more complicated. If the Δ propagator is part of a loop integral it counts as q. The same is true for tree diagrams of channels where no real resonance can be generated such as, e.g., the u-channel Δ-pole diagram in pion-nucleon scattering. On the other hand, in a resonance-generating channel, such as the s-channel Δ-pole diagram in pion-nucleon scattering, we dress the Δ propagator by resumming the self-energy insertions. We count the dressed propagator as q^{-3}, because the self energy starts at $\mathcal{O}(q^3)$. In the so-called small-scale expansion the mass difference $\delta \equiv m_\Delta - m$ is also counted as $\mathcal{O}(q)$ [34]. In a different counting [51]—the so-called δ expansion—one introduces a single small parameter, $\delta = (m_\Delta - m)/\Lambda$, where $\Lambda \sim 1$ GeV stands for the "high-energy scale" (nucleon mass or chiral-symmetry-breaking scale Λ_χ), and regards the ratio M_π/Λ as $\mathcal{O}(\delta^2)$.

References

1. Abramowitz, M., Stegun, I.A. (eds.): Handbook of Mathematical Functions. Dover, New York (1972)
2. Becher, T. In: Bernstein, A.M., Goity, J.L., Meißner, U.-G. (eds.) Chiral Dynamics: Theory and Experiment III, World Scientific, Singapore (2002)
3. Becher, T., Leutwyler, H.: Eur. Phys. J. C **9**, 643 (1999)
4. Becher, T., Leutwyler, H.: JHEP **0106**, 017 (2001)
5. Bernard, V., Kaiser, N., Kambor, J., Meißner, U.-G.: Nucl. Phys. B **388**, 315 (1992)
6. Bernard, V., Kaiser, N., Meißner, U.-G.: Int. J. Mod. Phys. E **4**, 193 (1995)
7. Bernard, V., Kaiser, N., Meißner, U.-G.: Nucl. Phys. A **615**, 483 (1997)
8. Bernard, V., Hemmert, T.R., Meißner, U.-G.: Phys. Lett. B **565**, 137 (2003)
9. Borasoy, B.: Phys. Rev. D **59**, 054021 (1999)
10. Bruns, P.C., Meißner, U.-G.: Eur. Phys. J. C **40**, 97 (2005)
11. Bruns, P.C., Meißner, U.-G.: Eur. Phys. J. C **58**, 407 (2008)
12. Butler, M.N., Savage, M.J., Springer, R.P.: Nucl. Phys. B **399**, 69 (1993)
13. Callan, C.G., Coleman, S.R., Wess, J., Zumino, B.: Phys. Rev. **177**, 2247 (1969)
14. Chew, G.F., Goldberger, M.L., Low, F.E., Nambu, Y.: Phys. Rev. **106**, 1337 (1957)
15. Coleman, S.R., Wess, J., Zumino, B.: Phys. Rev. **177**, 2239 (1969)
16. Dirac, P.A.M.: Lectures on Quantum Mechanics. Dover, Mineola (2001)
17. Ecker, G.: Prog. Part. Nucl. Phys. **35**, 1 (1995)
18. Ericson, T.E., Weise, W.: Pions and Nuclei. Clarendon, Oxford (1988). Apps 3 and 8
19. Fearing, H.W., Poulis, G.I., Scherer, S.: Nucl. Phys. A **570**, 657 (1994)
20. Fearing, H.W., Lewis, R., Mobed, N., Scherer, S.: Phys. Rev. D **56**, 1783 (1997)
21. Fettes, N., Meißner, U.-G., Steininger, S.: Nucl. Phys. A **640**, 199 (1998)
22. Fettes, N., Meißner, U.-G., Mojžiš, M., Steininger, S.: Ann. Phys. **283**, 273 (2001) [Erratum, ibid **288**, 249 (2001)]
23. Foldy, L.L., Wouthuysen, S.A.: Phys. Rev. **78**, 29 (1950)
24. Fuchs, T., Gegelia, J., Japaridze, G., Scherer, S.: Phys. Rev. D **68**, 056005 (2003)
25. Fuchs, T., Schindler, M.R., Gegelia, J., Scherer, S.: Phys. Lett. B **575**, 11 (2003)
26. Gasser, J., Sainio, M.E., Švarc, A.: Nucl. Phys. B **307**, 779 (1988)
27. Gegelia, J., Japaridze, G.: Phys. Rev. D **60**, 114038 (1999)
28. Gegelia, J., Japaridze, G.S., Turashvili, K.S.: Theor. Math. Phys. **101**, 1313 (1994) [Teor. Mat. Fiz. **101**, 225 (1994)]
29. Gell-Mann, M., Low, F.: Phys. Rev. **84**, 350 (1951)
30. Georgi, H.: Weak Interactions and Modern Particle Theory. Benjamin/Cummings, Menlo Park (1984)
31. Gitman, D.M., Tyutin, I.V.: Quantization of Fields with Constraints. Springer, Berlin (1990)
32. Goldberger, M.L., Treiman, S.B.: Phys. Rev. **110**, 1178 (1958)
33. Hacker, C., Wies, N., Gegelia, J., Scherer, S.: Phys. Rev. C **72**, 055203 (2005)
34. Hemmert, T.R., Holstein, B.R., Kambor, J.: J. Phys. G **24**, 1831 (1998)

35. Henneaux, M., Teitelboim, C.: Quantization of Gauge Systems. Princeton University Press, Princeton (1992)
36. Itzykson, C., Zuber, J.B.: Quantum Field Theory. McGraw-Hill, New York (1980)
37. Jenkins, E., Manohar, A.V.: Phys. Lett. B **255**, 558 (1991)
38. Jenkins, E., Manohar, A.V.: Phys. Lett. B **259**, 353 (1991)
39. Koch, R.: Nucl. Phys. A **448**, 707 (1986)
40. Krause, A.: Helv. Phys. Acta **63**, 3 (1990)
41. Lehmann, D., Prezeau, G.: Phys. Rev. D **65**, 016001 (2002)
42. Mannel, T., Roberts, W., Ryzak, Z.: Nucl. Phys. B **368**, 204 (1992)
43. Matsinos, E.: Phys. Rev. C **56**, 3014 (1997)
44. Meißner, U.-G.: PoS (LAT2005), 009 (2005)
45. Mojžiš, M.: Eur. Phys. J. C **2**, 181 (1998)
46. Moldauer, P.A., Case, K.M.: Phys. Rev. **102**, 279 (1956)
47. Nakamura, K. et al.: [Particle Data Group], J. Phys. G **37**, 075021 (2010)
48. Nambu, Y.: Phys. Rev. Lett. **4**, 380 (1960)
49. Nath, L.M., Etemadi, B., Kimel, J.D.: Phys. Rev. D **3**, 2153 (1971)
50. Pagels, H.: Phys. Rev. **179**, 1337 (1969)
51. Pascalutsa, V., Phillips, D.R.: Phys. Rev. C **67**, 055202 (2003)
52. Pilling, T.: Int. J. Mod. Phys. A **20**, 2715 (2005)
53. Rarita, W., Schwinger, J.S.: Phys. Rev. **60**, 61 (1941)
54. Schindler, M.R., Gegelia, J., Scherer, S.: Phys. Lett. B **586**, 258 (2004)
55. Schindler, M.R., Gegelia, J., Scherer, S.: Nucl. Phys. B **682**, 367 (2004)
56. Schröder, H.C. et al.: Eur. Phys. J. C **21**, 473 (2001)
57. Smirnov, V.A.:Applied Asymptotic Expansions in Momenta and Masses, Springer Tracts Mod. Phys. **177**, 1 (2001)
58. Tang, H.B., Ellis, P.J.: Phys. Lett. B **387**, 9 (1996)
59. Thomas, A.W.: Adv. Nucl. Phys. **13**, 1 (1984)
60. Tomozawa, Y.: Nuovo Cimento **46A**, 707 (1966)
61. Weinberg, S.: Phys. Rev. **112**, 1375 (1958)
62. Weinberg, S.: Phys. Rev. Lett. **17**, 616 (1966)
63. Weinberg, S.: Phys. Rev. **166**, 1568 (1968)
64. Weinberg, S.: Nucl. Phys. B **363**, 3 (1991)
65. Weinberg, S.: The Quantum Theory of Fields. Foundations, vol. 1. Cambridge University Press, Cambridge (1995). Chap. 12
66. Wies, N., Gegelia, J., Scherer, S.: Phys. Rev. D **73**, 094012 (2006)

Chapter 5
Applications and Outlook

5.1 Nucleon Mass and Sigma Term

In this section we will address the quark-mass expansion of the nucleon mass. Starting with a calculation to $\mathcal{O}(q^3)$ in HBChPT, we will recover the result of Eq. 4.93 for the nucleon mass. We will then extend the discussion to $\mathcal{O}(q^4)$ in the EOMS scheme. Next, we will consider the nucleon mass within a framework containing the Δ resonance as an explicit dynamical degree of freedom. Finally, we will discuss some aspects of a two-loop calculation up to and including $\mathcal{O}(q^6)$.

5.1.1 Nucleon Mass to $\mathcal{O}(q^3)$ in the Heavy-Baryon Formalism

As an application of the heavy-baryon formulation, we calculate the nucleon mass to $\mathcal{O}(q^3)$, the lowest order at which loop diagrams contribute. The calculation proceeds analogously to the one in Sect. 4.5.3. We will see how the power counting is automatically satisfied in the heavy-baryon formalism when using dimensional regularization in combination with the modified minimal subtraction scheme $(\widetilde{\text{MS}})$ of ChPT.[1] The physical mass is given by the pole of the full heavy-baryon propagator

$$G_v(k_p) = \frac{P_{v+}}{v \cdot k_p - \Sigma(p) + i0^+} = \frac{P_{v+}}{v \cdot p - m - \Sigma(p) + i0^+}, \tag{5.1}$$

where we have used the decomposition of the nucleon four-momentum $p = mv + k_p$ (see Eq. 4.95).

---

[1] The existence of a consistent power counting in HBChPT relies on specifying the renormalization scheme. See Sect. V of Ref. [53] for a discussion of this point.

S. Scherer and M. R. Schindler, *A Primer for Chiral Perturbation Theory*,
Lecture Notes in Physics 830, DOI: 10.1007/978-3-642-19254-8_5,
© Springer-Verlag Berlin Heidelberg 2012

Exercise 5.1

(a) Determine the tree contributions $\Sigma^{\text{tree}}(p)$ to $\mathcal{O}(q^2)$. The relevant terms of the second-order Lagrangian are given by[2]

$$\widehat{\mathscr{L}}_{\pi N}^{(2)} = \bar{\mathcal{N}}_v \left[-\frac{D^2}{2m} + c_1 \,\text{Tr}(\chi_+) + \cdots \right] \mathcal{N}_v. \tag{5.2}$$

The first term originates from the $1/m$ correction of Eq. 4.113. The term proportional to c_1 is the analogue of the c_1 term in the Lagrangian of Eq. 4.66. In terms of Sect. 4.5.1, we treat Eq. 5.2 as part of a basic Lagrangian. There are no contributions to $\Sigma^{\text{tree}}(p)$ at $\mathcal{O}(q^3)$.

(b) Using the leading-order Lagrangian of Eq. 4.114, show that the Feynman rule for an incoming pion with four-momentum q and Cartesian isospin index a is given by

$$-\frac{g_A}{F} S_v \cdot q \tau_a; \tag{5.3}$$

and that for an incoming pion with q, a and outgoing pion with q', b by

$$\frac{v \cdot (q + q')}{4F^2} \varepsilon_{abc} \tau_c. \tag{5.4}$$

As in the case of Exercise 4.10, the second Feynman rule implies that the loop diagram of Fig. 5.1b vanishes.

(c) Calculate the loop diagram of Fig. 5.1a. Note that, for convenience, we have chosen a slightly different momentum assignment. Show that the self-energy contribution is given by[3]

$$-i\Sigma^{\text{loop}}(p) = -i\frac{3g_A^2}{F^2} S_v^\mu S_v^\nu i \mu^{4-n} \int \frac{d^n q}{(2\pi)^n} \frac{q_\mu q_\nu}{(q^2 - M^2 + i0^+)[v \cdot (k_p + q) + i0^+]}. \tag{5.5}$$

---

[2] The corrections of first order in $1/m$ in Eq. 4.113 contain a piece of the type

$$\frac{1}{2m} \bar{\mathcal{N}}_v \left[(v \cdot D)^2 - D^2 \right] \mathcal{N}_v.$$

Using the field redefinition [65]

$$\mathcal{N}_v \rightarrow \left[1 + \frac{iv \cdot D}{4m} - \frac{g_A S_v \cdot u}{4m} \right] \mathcal{N}_v,$$

the term containing $v \cdot D$ can be eliminated. As in the case of the two-flavor mesonic Lagrangian at $\mathcal{O}(q^4)$ (see Exercise 3.25), one finds *equivalent parameterizations of* $\widehat{\mathscr{L}}_{\pi N}^{(2)}$ (and also of the higher-order Lagrangians) in the baryonic sector.

[3] In the remaining part of this section, we adopt the common practice of leaving out the projector P_{v+} in the propagator and (possibly) in vertices with the understanding that all operators act only in the projected subspace.

Fig. 5.1 One-loop
contributions to the nucleon
self energy

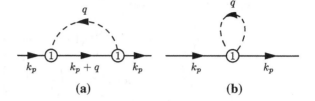

(a)   (b)

The nucleon propagator at leading order is given in Eq. 4.115.

(d) The tensor integral can be parameterized as

$$
i\mu^{4-n} \int \frac{d^n q}{(2\pi)^n} \frac{q_\mu q_\nu}{(q^2 - M^2 + i0^+)(v \cdot q + \omega + i0^+)}
= v_\mu v_\nu C_{20}(\omega, M^2) + g_{\mu\nu} C_{21}(\omega, M^2),
$$

where $\omega = v \cdot k_p$. Contracting with $v^\mu v^\nu$ and $g^{\mu\nu}$, show that the integrals
$C_{20}(\omega, M^2)$ and $C_{21}(\omega, M^2)$ can be determined from the equations

$$
C_{20}(\omega, M^2) + C_{21}(\omega, M^2) = -\omega I_\pi(0) + \omega^2 J_{\pi N}(0; \omega),
$$
$$
C_{20}(\omega, M^2) + n C_{21}(\omega, M^2) = M^2 J_{\pi N}(0; \omega),
$$

where

$$
I_\pi(0) = i\mu^{4-n} \int \frac{d^n q}{(2\pi)^n} \frac{1}{q^2 - M^2 + i0^+},
$$

$$
J_{\pi N}(0; \omega) = i\mu^{4-n} \int \frac{d^n q}{(2\pi)^n} \frac{1}{(q^2 - M^2 + i0^+)(v \cdot q + \omega + i0^+)}.
$$

Hint: $g^{\mu\nu} g_{\mu\nu} = n$ and

$$
i\mu^{4-n} \int \frac{d^n q}{(2\pi)^n} \frac{q_\mu}{q^2 - M^2 + i0^+} = 0, \quad
i\mu^{4-n} \int \frac{d^n q}{(2\pi)^n} \frac{1}{v \cdot q + \omega + i0^+} = 0.
$$

(e) Using the results above, as well as $S_v \cdot v = 0$ and $S_v^2 = (1 - n)/4$, verify that
the loop contribution to the self energy is given by

$$
\Sigma^{\text{loop}}(p) = -\frac{3g_A^2}{4F^2}\left[(M^2 - \omega^2) J_{\pi N}(0; \omega) + \omega I_\pi(0)\right]. \tag{5.6}
$$

(f) The explicit expression for $I_\pi(0)$ is given in Eq. 4.87, and

$$
J_{\pi N}(0; \omega) = \frac{\omega}{8\pi^2}\left[R + \ln\left(\frac{M^2}{\mu^2}\right) - 1\right] + \frac{1}{4\pi^2}\sqrt{M^2 - \omega^2}\arccos\left(-\frac{\omega}{M}\right) + O(n-4)
$$

for $\omega^2 < M^2$, where R is given in Eq. 3.111. Verify that the expression for the
self energy is given by

$$\Sigma(p) = \Sigma^{\text{tree}}(p) + \Sigma^{\text{loop}}(p)$$

$$= -\frac{k_p^2}{2m} - 4c_1 M^2 - \frac{3g_A^2}{(4\pi F)^2}\left((M^2 - \omega^2)^{\frac{3}{2}}\arccos\left(-\frac{\omega}{M}\right)\right.$$

$$+ \left.\frac{\omega}{4}\left\{(3M^2 - 2\omega^2)\left[R + \ln\left(\frac{M^2}{\mu^2}\right)\right] - \frac{1}{2}(M^2 - \omega^2)\right\}\right). \qquad (5.7)$$

(g) To determine the nucleon mass, we need to evaluate the self energy for p on the mass shell, which corresponds to

$$p = m_N v.$$

Show that this condition corresponds to $\omega = v \cdot k_p = m_N - m$, so that

$$m_N = m + \Sigma(m_N v)$$

$$= m - \frac{(m_N - m)^2}{2m} - 4c_1 M^2 + \Sigma^{\text{loop}}(m_N v). \qquad (5.8)$$

Given that $\Sigma^{\text{loop}}(m_N v)$ is at least $\mathcal{O}(M^2)$, this implies that $m_N - m = \mathcal{O}(M^2)$. Since our calculation is only valid to $\mathcal{O}(q^3)$, we can neglect the second term on the right-hand side of Eq. 5.8 and can set $\omega = 0$ in the loop contribution. Verify the final result for the nucleon mass to $\mathcal{O}(q^3)$:

$$m_N = m - 4c_1 M^2 - \frac{3\pi g_A^2 M^3}{2(4\pi F)^2}.$$

The loop contribution is of $\mathcal{O}(q^3)$ as predicted by the power counting. It is therefore not necessary to perform any additional finite subtractions. It is exactly this feature which distinguishes the heavy-baryon formulation from the original, manifestly Lorentz-invariant approach of Ref. [83] discussed in Sect. 4.5.3. Both calculations make use of dimensional regularization with the modified minimal subtraction scheme of ChPT, but only in the heavy-baryon case does this renormalization condition lead to a consistent power counting. The result for the nucleon mass agrees with the expression of Eq. 4.93, obtained in the manifestly Lorentz-invariant calculation with the additional subtraction.

5.1.2 Nucleon Mass and Sigma Term at $\mathcal{O}(q^4)$

We now turn to a full one-loop calculation of the nucleon mass at $\mathcal{O}(q^4)$ in the EOMS approach [76]. In addition to the loop diagrams of Fig. 4.4 and the tree-level contribution originating from $\mathscr{L}_{\pi N}^{(2)}$, we need to consider the diagrams shown in Fig. 5.2. Note that $\mathscr{L}_{\pi N}^{(2)}$ does not generate a contribution to the πNN vertex.

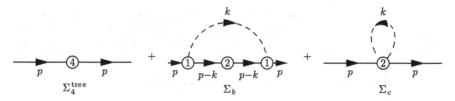

Fig. 5.2 Contributions to the nucleon self energy at $\mathcal{O}(q^4)$. The number n in the interaction blobs refers to $\mathscr{L}_{\pi N}^{(n)}$. The Lagrangian $\mathscr{L}_{\pi N}^{(2)}$ does not produce a contribution to the πNN vertex

Therefore, a diagram with the topology of the first diagram of Fig. 4.4, where one of the two vertices is replaced by an $\mathcal{O}(q^2)$ vertex, does not exist. The tree-level contribution at $\mathcal{O}(q^4)$ reads

$$\Sigma_4^{\text{tree}} = -\hat{e}_1 M^4, \tag{5.9}$$

where $\hat{e}_1 = 16e_{38} + 2e_{115} + 2e_{116}$ is a linear combination of $\mathcal{O}(q^4)$ LECs [72], and the subscript 4 denotes chiral order four. In order to facilitate comparison with Refs. [10, 76], let us denote the loop contribution of the first diagram of Fig. 4.4 by Σ_a,

$$\Sigma_a = \frac{3g_A^2}{4F^2} i\mu^{4-n} \int \frac{d^n k}{(2\pi)^n} \not{k}\gamma_5 \frac{1}{\not{p} - \not{k} - m + i0^+} \not{k}\gamma_5 \frac{1}{k^2 - M^2 + i0^+}. \tag{5.10}$$

Applying Feynman rules, we obtain for the two one-loop contributions of Fig. 5.2

$$\Sigma_4^{\text{loop}} = \Sigma_b + \Sigma_c, \tag{5.11}$$

where

$$\begin{aligned}
\Sigma_b &= -4M^2 c_1 \frac{3g_A^2}{4F^2} i\mu^{4-n} \int \frac{d^n k}{(2\pi)^n} \not{k}\gamma_5 \left(\frac{1}{\not{p} - \not{k} - m + i0^+}\right)^2 \not{k}\gamma_5 \frac{1}{k^2 - M^2 + i0^+} \\
&= -4M^2 c_1 \frac{\partial \Sigma_a}{\partial m}, \tag{5.12}
\end{aligned}$$

$$\Sigma_c = 3\frac{M^2}{F^2}\left(2c_1 - c_3 - \frac{p^2}{m^2}\frac{c_2}{n}\right) i\mu^{4-n} \int \frac{d^n k}{(2\pi)^n} \frac{1}{k^2 - M^2 + i0^+}. \tag{5.13}$$

In general, the chiral orders assigned by the power counting will not hold until the corresponding subtractions have been performed.

The renormalization of the loop diagrams is performed in two steps. First, we render the diagrams finite by applying the modified minimal subtraction scheme of ChPT ($\widetilde{\text{MS}}$). We choose $\mu = m$ for the 't Hooft parameter. In a second step, we then perform additional *finite* subtractions for integrals which contain nucleon propagators with the purpose of imposing the power-counting scheme. In fact, in order to apply the $\widetilde{\text{MS}}$ subtraction in practical calculations, we do not actually need to explicitly write down the corresponding counter terms. We simply subtract all loop diagrams and tag the coupling constants with a subscript r indicating the $\widetilde{\text{MS}}$ scheme.

The nucleon mass is determined by solving Eq. 4.80,

$$m_N - m - \Sigma(m_N) = 0.$$

Using the $\widetilde{\text{MS}}$-renormalized expressions for the integrals of Eq. 4.87, we obtain for the mass in the $\widetilde{\text{MS}}$ scheme,

$$m_N = m - 4c_{1r}M^2 + \frac{3g_{Ar}^2}{32\pi^2 F_r^2}m(1 + 8c_{1r}m)M^2 - \frac{3g_{Ar}^2}{32\pi F_r^2}M^3$$

$$+ \frac{3}{32\pi^2 F_r^2}\left(8c_{1r} - c_{2r} - 4c_{3r} - \frac{g_{Ar}^2}{m}\right)M^4 \ln\left(\frac{M}{m}\right)$$

$$+ \frac{3g_{Ar}^2}{32\pi^2 F_r^2 m}(1 + 4c_{1r}m)M^4 + \left(\frac{3}{128\pi^2 F_r^2}c_{2r} - \hat{e}_{1r}\right)M^4 + \mathcal{O}(M^5), \quad (5.14)$$

where "r" refers to $\widetilde{\text{MS}}$-renormalized quantities. When solving Eq. 4.80, we expanded the results of the loop integrals and consistently omitted terms which count as $O(\hbar^2)$ in the loop expansion, i.e., terms proportional to $(g_A/F)^4$, as well as terms proportional to $(c_{1r})^2$. The third term on the right-hand side of Eq. 5.14 violates the power counting because it is of $\mathcal{O}(M^2)$. It receives contributions from both Σ_a and Σ_b.

In order to perform the second step, namely another *finite* renormalization, a given $\widetilde{\text{MS}}$-renormalized diagram is written as the sum of a subtracted diagram which, through the application of the subtraction scheme described in Sect. 4.6.3, satisfies the power counting, and a remainder which violates the power counting and thus still needs to be subtracted. For the case at hand, we determine the terms to be subtracted from Σ_a and Σ_b by first expanding the integrands and coefficients in Eqs. 5.10 and 5.12 in powers of M^2, $\not{p} - m$, and $p^2 - m^2$. In this expansion we keep all the terms having a chiral order which is smaller than what is suggested by the power counting for the given diagram. We then obtain

$$\Sigma_{r,a+b}^{\text{subtr}} = \frac{3g_{Ar}^2}{32\pi^2 F_r^2}\left[mM^2 - \frac{(p^2 - m^2)^2}{4m}\right] + \frac{3c_{1r}g_{Ar}^2 M^2}{8\pi^2 F_r^2}\left[m(\not{p} + m) - \frac{3}{2}(p^2 - m^2)\right].$$

$$(5.15)$$

Equation 5.15 specifies which parts of the self-energy diagrams at $\mathcal{O}(q^3)$ and $\mathcal{O}(q^4)$ need to be subtracted. We fix the corresponding counter terms so that they exactly cancel the expression given by Eq. 5.15. Since the most general Lagrangian contains all the structures consistent with the symmetries of the theory, it also provides the required counter terms. Finally, the renormalized self-energy expression is obtained by subtracting Eq. 5.15 from the $\widetilde{\text{MS}}$-subtracted version of Eqs. 5.10 (see Eq. 4.88) and 5.12 and replacing the $\widetilde{\text{MS}}$-renormalized couplings with the ones of the EOMS scheme. We note that the $\widetilde{\text{MS}}$-subtracted version for Σ_c needs no further subtraction because it already is of $\mathcal{O}(q^4)$.

The correction to the nucleon mass resulting from the counter terms is calculated by substituting $\not{p} = m_N$ in the negative of Eq. 5.15. Recall that Eq. 5.15 has to be subtracted. We thus obtain the following expression for the contribution to the mass,

$$\Delta m_{\text{c.t.}} = -\frac{3g_{Ar}^2}{32\pi^2 F_r^2} m(1 + 8c_{1r}m)M^2, \tag{5.16}$$

where the subscript c.t. refers to counter term. Comparing with Eq. 5.14, we see that the subtraction term of Eq. 5.15 indeed cancels the power-counting-violating contributions in Eq. 5.14. Finally, we express the physical mass of the nucleon up to and including order q^4 as [128, 162][4]

$$m_N = m + k_1 M^2 + k_2 M^3 + k_3 M^4 \ln\left(\frac{M}{m}\right) + k_4 M^4 + \mathcal{O}(M^5), \tag{5.17}$$

where the coefficients k_i in the EOMS scheme read [76]

$$k_1 = -4c_1, \quad k_2 = -\frac{3g_A^2}{32\pi F^2}, \quad k_3 = -\frac{3}{32\pi^2 F^2 m}(g_A^2 - 8c_1 m + c_2 m + 4c_3 m),$$

$$k_4 = \frac{3g_A^2}{32\pi^2 F^2 m}(1 + 4c_1 m) + \frac{3}{128\pi^2 F^2}c_2 - \hat{e}_1. \tag{5.18}$$

A comparison with the results using the infrared regularization [10] shows that the lowest-order correction (k_1 term) and those terms which are nonanalytic in the quark mass $\hat{m}$ (k_2 and k_3 terms) are identical. On the other hand, the analytic k_4 term ($\sim M^4$) is different. This is not surprising; although both renormalization schemes satisfy the power counting specified in Sect. 4.5.2, the use of different renormalization conditions is compensated by different values of the renormalized parameters.

For an estimate of the various contributions of Eq. 5.17 to the nucleon mass, we make use of the numerical values of Eq. 4.65 for g_A, etc., and the parameter set of Eq. 4.67 for the c_i. Note that using the physical values for g_A and F_π instead of their chiral limit values is consistent up to the order considered here, as $g_A = \overset{\circ}{g}_A[1 + \mathcal{O}(M^2)]$ and $F_\pi = F[1 + \mathcal{O}(M^2)]$. As has been discussed, e.g., in Ref. [10], a fully consistent description would also require to determine the low-energy coupling constant c_1 from a complete $\mathcal{O}(q^4)$ calculation of, say, πN

---

[4] In our convention, k_3 is larger by a factor of two than in Refs. [128, 162], because we use $\ln(M/m)$ instead of $\ln(M^2/m^2)$.

scattering. One obtains for the mass of nucleon in the chiral limit (at fixed $m_s \neq 0$):

$$m = m_N - \Delta m$$
$$= (938.3 - 74.8 + 15.3 + 4.7 + 1.6 - 2.3 \pm 4)\,\text{MeV}$$
$$= (882.8 \pm 4)\,\text{MeV}, \tag{5.19}$$

with $\Delta m = (55.5 \pm 4)\,\text{MeV}$. Here, we have made use of an estimate for $\hat{e}_1 M^4 = (2.3 \pm 4)\,\text{MeV}$ obtained from the σ term (see below). Note that errors due to higher-order corrections are not taken into account.

Sigma terms provide a sensitive measure of explicit chiral symmetry breaking in QCD because "they are corrections to a null result in the chiral limit rather than small corrections to a non-trivial result" [150] (see, e.g., Refs. [93, 159] for a review). In the three-flavor sector, the so-called sigma commutator is defined as

$$\sigma_{ab}(x) \equiv [Q_{Aa}(x_0), [Q_{Ab}(x_0), \mathscr{H}_{sb}(x)]], \tag{5.20}$$

where $Q_{Ac} = Q_{Rc} - Q_{Lc}$ denotes one of the eight axial-charge operators of Eq. 3.9 and

$$\mathscr{H}_{sb} = \bar{q}\mathscr{M}q = \hat{m}(\bar{u}u + \bar{d}d) + m_s\bar{s}s$$

is the chiral-symmetry-breaking mass term of the QCD Hamiltonian in the isospin-symmetrical limit. Using equal-time anticommutation relations (see Eqs. 1.103 and 3.20), Eq. 5.20 can be written as [132]

$$\sigma_{ab}(x) = \bar{q}(x)\left\{\frac{\lambda_a}{2}, \left\{\frac{\lambda_b}{2}, \mathscr{M}\right\}\right\}q(x), \tag{5.21}$$

yielding for the flavor-diagonal pieces,

$$\sigma_{11} = \sigma_{22} = \sigma_{33} = \hat{m}(\bar{u}u + \bar{d}d),$$
$$\sigma_{44} = \sigma_{55} = \frac{\hat{m} + m_s}{2}(\bar{u}u + \bar{s}s),$$
$$\sigma_{66} = \sigma_{77} = \frac{\hat{m} + m_s}{2}(\bar{d}d + \bar{s}s), \tag{5.22}$$

$$\sigma_{88} = \frac{1}{3}[\hat{m}(\bar{u}u + \bar{d}d) + 4m_s\bar{s}s],$$
$$\sigma_{38} = \sigma_{83} = \frac{\hat{m}}{\sqrt{3}}(\bar{u}u - \bar{d}d).$$

Exercise 5.2 Verify Eqs. 5.21 and 5.22.

In the following, we restrict ourselves to the two-flavor case. In terms of the $SU(2)_L \times SU(2)_R$-chiral-symmetry-breaking mass term of the QCD Hamiltonian,

$$\mathscr{H}_{sb} = \hat{m}(\bar{u}u + \bar{d}d), \tag{5.23}$$

the pion-nucleon sigma term is defined as the proton matrix element

$$\sigma = \frac{1}{2m_p} \langle p(p,s)|\mathscr{H}_{sb}(0)|p(p,s)\rangle \tag{5.24}$$

at zero momentum transfer.[5] The sigma term may either be obtained by explicit calculation or through the application of the Hellmann-Feynman theorem.

Exercise 5.3 Consider a Hermitian operator $H(\lambda)$ depending smoothly on a real parameter λ. Let $|\alpha(\lambda)\rangle$ denote a normalized eigenstate with eigenvalue $E(\lambda)$,

$$H(\lambda)|\alpha(\lambda)\rangle = E(\lambda)|\alpha(\lambda)\rangle,$$
$$\langle \alpha(\lambda)|\alpha(\lambda)\rangle = 1.$$

Verify the Hellmann-Feynman theorem,

$$\frac{\partial E(\lambda)}{\partial \lambda} = \left\langle \alpha(\lambda)\left|\frac{\partial H(\lambda)}{\partial \lambda}\right|\alpha(\lambda)\right\rangle. \tag{5.25}$$

In the present context, we multiply Eq. 5.25 by λ and perform the substitutions

$$\lambda \to \hat{m},$$
$$|\alpha(\lambda)\rangle \to |N(\hat{m})\rangle,$$
$$E(\lambda) \to m_N(\hat{m}),$$
$$\frac{\partial H}{\partial \lambda} \to \frac{\partial \mathscr{H}_{QCD}}{\partial \hat{m}} = \bar{u}u + \bar{d}d.$$

Note that $M^2 = 2B\hat{m}$ and thus [83]

$$\sigma = M^2 \frac{\partial m_N}{\partial M^2}. \tag{5.26}$$

The quark-mass expansion of the σ term reads

$$\sigma = \sigma_1 M^2 + \sigma_2 M^3 + \sigma_3 M^4 \ln\left(\frac{M}{m}\right) + \sigma_4 M^4 + \mathcal{O}(M^5), \tag{5.27}$$

with

$$\sigma_1 = -4c_1, \quad \sigma_2 = -\frac{9g_A^2}{64\pi F^2}, \quad \sigma_3 = -\frac{3}{16\pi^2 F^2 m}(g_A^2 - 8c_1 m + c_2 m + 4c_3 m),$$

$$\sigma_4 = \frac{3}{8\pi^2 F^2 m}\left[\frac{3g_A^2}{8} + c_1 m(1 + 2g_A^2) - \frac{c_3 m}{2}\right] - 2\hat{e}_1. \tag{5.28}$$

---

[5] In the linear sigma model with explicit symmetry breaking (see Sect. 2.4), the double commutator $\sigma_{11}(x)$ is proportional to the sigma field. This is the origin of the name sigma term.

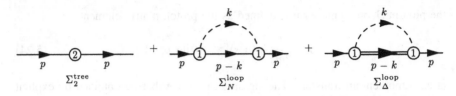

Fig. 5.3 Contributions to the nucleon self energy to $\mathcal{O}(q^3)$. The number n in the interaction blobs refers to $\mathscr{L}_{\pi N}^{(n)}$ and $\mathscr{L}_{\pi N\Delta}^{(n)}$. The Δ is represented by a double line

Exercise 5.4 Using the coefficients k_i of Eq. 5.18, verify Eq. 5.28 by applying the Hellmann-Feynman theorem of Eq. 5.26.

We obtain (with $\hat{e}_1 = 0$ in Eq. 5.28)

$$\sigma = (74.8 - 22.9 - 9.4 - 2.0)\,\text{MeV} = 40.5\,\text{MeV}. \tag{5.29}$$

The result of Eq. 5.29 has to be compared with, e.g., the dispersive analysis $\sigma = (45 \pm 8)\,\text{MeV}$ of Ref. [84] which would imply, neglecting higher-order terms, $-2\hat{e}_1 M^4 \approx (4.5 \pm 8)\,\text{MeV}$. Note that c_1 has been estimated in terms of an $\mathcal{O}(q^2)$ tree-level calculation of πN scattering, whereas a fully consistent description would require determining c_1 from a complete $\mathcal{O}(q^4)$ calculation.

5.1.3 Nucleon Mass Including the Delta Resonance

In this section we discuss the result for the nucleon mass to order q^3 within the EFT of Sect. 4.7 including the $\Delta(1232)$ resonance as an explicit degree of freedom. We will make use of the small-scale expansion, treating both the pion mass M and the mass difference $\delta = m_\Delta - m$ as $\mathcal{O}(q)$. We will fix the renormalization condition such that δ denotes the mass difference in the chiral limit between the pole mass of the Δ and the nucleon mass. However, as will be seen in Eq. 5.34, we do not identify the parameter m with the nucleon mass in the chiral limit.

The relevant Feynman diagrams for the self energy are shown in Fig. 5.3. At $\mathcal{O}(q^2)$, we obtain a constant tree-level contribution $-4\tilde{c}_1 M^2$ to the self energy, where $\tilde{c}_1$ refers to the coupling constant in the theory explicitly including Δ degrees of freedom. The EOMS-renormalized one-loop contribution resulting from Σ_N^{loop} of Fig. 5.3 is given by the same expression as in Sect. 4.5.3,

$$\Sigma_N^{\text{loop}}(\not{p} = m_N) = -\frac{3g_A^2 M^3}{32\pi F^2}. \tag{5.30}$$

Finally, the EOMS-renormalized one-loop contribution of the Δ resonance resulting from $\Sigma_\Delta^{\text{loop}}$ of Fig. 5.3 reads [98]

$$\Sigma_\Delta^{\text{loop}}(\not{p} = m_N)$$

$$= \frac{g^2}{288\pi^2 F^2}\left[35\delta^3 + 6\delta M^2(1 + 25\tilde{c}_1 m) + 96(\delta^2 - M^2)^{3/2}\ln\left(\frac{\delta - \sqrt{\delta^2 - M^2}}{M}\right)\right]$$

$$- \frac{g^2}{6\pi^2 F^2}(2\delta^3 - 3M^2\delta)\ln\left(\frac{M}{m}\right). \tag{5.31}$$

Combining the tree-level result at $\mathcal{O}(q^2)$ with the $\mathcal{O}(q^3)$ one-loop contributions results in the following expression for the nucleon mass:

$$m_N = m - 4\tilde{c}_1 M^2 - \frac{3g_A^2 M^3}{32\pi F^2} + \frac{g^2}{288\pi^2 F^2}\left[35\delta^3 + 6\delta M^2(1 + 25\tilde{c}_1 m)\right.$$

$$\left. + 96(\delta^2 - M^2)^{3/2}\ln\left(\frac{\delta - \sqrt{\delta^2 - M^2}}{M}\right)\right]$$

$$- \frac{g^2}{6\pi^2 F^2}(2\delta^3 - 3M^2\delta)\ln\left(\frac{M}{m}\right) + \mathcal{O}(q^4). \tag{5.32}$$

The nonanalytic part of Eq. 5.32 agrees with a covariant calculation in the framework of infrared regularization [21]. The analytic terms differ because of different renormalization conditions and a different choice for the interaction terms.

By explicitly including the spin-3/2 degrees of freedom, terms of higher order in the *chiral* expansion have been resummed. In order to obtain a numerical value for these terms, let us expand Eq. 5.32 in powers of M.

Exercise 5.5 Consider $M \ll \delta$ and introduce $x = M/\delta$. Verify

$$(\delta^2 - M^2)^{\frac{3}{2}}\ln\left(\frac{\delta - \sqrt{\delta^2 - M^2}}{M}\right) = \delta^3\left[\ln\left(\frac{x}{2}\right) - \frac{3}{2}x^2\ln\left(\frac{x}{2}\right) + \frac{x^2}{4} + \mathcal{O}(x^4)\right].$$

$$\tag{5.33}$$

Using the result of Eq. 5.33, we match the terms of orders M^0 and M^2 in Eq. 5.32 to the corresponding quantities of the EFT without explicit spin-3/2 degrees of freedom (see Eq. 4.93). Taking into account that there are no tree-level Δ contributions to c_1 [10], we obtain

$$\overset{\circ}{m} = m + \frac{g^2\delta^3}{3\pi^2 F^2}\ln\left(\frac{m}{2\delta}\right) + \frac{35g^2\delta^3}{288\pi^2 F^2}, \tag{5.34}$$

$$c_1 = \tilde{c}_1 - (1 + 5\tilde{c}_1 m)\frac{5g^2\delta}{192\pi^2 F^2} + \frac{g^2\delta}{8\pi^2 F^2}\ln\left(\frac{m}{2\delta}\right), \tag{5.35}$$

where, for the purpose of this section, $\overset{\circ}{m}$ denotes the nucleon mass in the chiral limit and c_1 refers to the theory without spin-3/2 degrees of freedom. Using Eqs. 5.34 and 5.35, the nucleon mass of Eq. 5.32 can be rewritten as

$$m_N = \overset{\circ}{m} - 4c_1 M^2 - \frac{3g_A^2 M^3}{32\pi F^2} + \tilde{m}_N, \tag{5.36}$$

where $\tilde{m}_N$ is of order M^4 and contains an infinite number of terms if expanded in powers of M/δ.

In order to obtain an estimate for $\tilde{m}_N$, we make use of $g = 1.127$ as obtained from a fit to the $\Delta \to \pi N$ decay width [98],[6] and take the numerical values

$$
\begin{aligned}
g_A &= 1.267, \quad F_\pi = 92.4\,\text{MeV}, \quad m_N = m_p = 938.3\,\text{MeV}, \\
M_\pi &= M_{\pi^+} = 139.6\,\text{MeV}, \quad m_\Delta = 1210\,\text{MeV}, \quad \delta = m_\Delta - m_N.
\end{aligned}
\tag{5.37}
$$

Substituting the above values in the expression for $\tilde{m}_N$ results in

$$\tilde{m}_N = -5.7\,\text{MeV}. \tag{5.38}$$

We recall that the analysis of the nucleon mass up to and including order M^4 of Eq. 5.19 yields $(882.8 + 74.8 - 15.3)\,\text{MeV} = 942.3\,\text{MeV}$ for the first three terms of Eq. 5.36. In other words, the explicit inclusion of the spin-3/2 degrees of freedom does not have a significant impact on the nucleon mass at the physical pion mass.

Applying the Hellmann-Feynman theorem in the form of Eq. 5.26 to the nucleon mass, we obtain for the pion-nucleon sigma term to order q^3,

$$
\begin{aligned}
\sigma = &-4\tilde{c}_1 M^2 - \frac{9g_A^2 M^3}{64\pi F^2} + \frac{5g^2(1 + 5\tilde{c}_1 m)\delta M^2}{48\pi^2 F^2} \\
&- \frac{g^2(\delta^2 - M^2)^{\frac{1}{2}} M^2}{2\pi^2 F^2} \ln\left(\frac{\delta - \sqrt{\delta^2 - M^2}}{M}\right) + \frac{g^2 \delta M^2}{2\pi^2 F^2} \ln\left(\frac{M}{m}\right).
\end{aligned}
\tag{5.39}
$$

Again, expanding Eq. 5.39 in powers of M and using Eq. 5.35, we rewrite σ as

$$\sigma = -4c_1 M^2 - \frac{9g_A^2 M^3}{64\pi F^2} + \tilde{\sigma}, \tag{5.40}$$

where $\tilde{\sigma}$ is of order M^4 and contains an infinite number of terms if expanded in powers of M/δ. With the numerical values of Eq. 5.37 we obtain from Eq. 5.39

$$\tilde{\sigma} = -10.2\,\text{MeV}, \tag{5.41}$$

while the first two terms of Eq. 5.40 yield $(74.8 - 22.9)\,\text{MeV} = 51.9\,\text{MeV}$. These numbers have to be compared with the empirical values of the sigma term

---

[6] The quark-model estimate of Eq. 4.202 yields $g = 1.075$.

extracted from data on pion-nucleon scattering: 40 MeV [38], (45 ± 5) MeV [84], and (64 ± 7) MeV [151]. Equation 5.41 indicates that the explicit inclusion of the spin-3/2 degrees of freedom plays a more important role for the sigma term than for the nucleon mass. However, one has to keep in mind that the sigma term only starts at order M^2 and thus, on a relative scale, is automatically more sensitive to higher-order corrections.

5.1.4 Nucleon Mass to $\mathcal{O}(q^6)$

In the previous sections we discussed the nucleon mass up to and including $\mathcal{O}(q^4)$. Using estimates for various low-energy couplings, we found good convergence at the physical pion mass. However, the convergence of the chiral expansion of a physical quantity is also of interest when unphysical values of the parameter M are considered. Lattice QCD presents a numerical approach in which correlation functions are calculated from the QCD Lagrangian by discretizing space-time [47, 87, 99, 160, 184]. One of the factors that determine the amount of resources required to perform these calculations is the size of the quark masses, with small quark masses corresponding to higher calculational costs. While lattice QCD has made tremendous progress towards calculations performed at the physical quark masses, in general calculations have been performed at a series of unphysical values. Observables at the physical quark masses are then extrapolated. Since chiral perturbation theory corresponds to an expansion in the quark, or equivalently the pion, mass, it is a crucial tool in performing these extrapolations. The range of pion masses that can be used for reliable extrapolations is determined by the convergence properties of ChPT [174].

So far only a few calculations in the baryon sector have been performed beyond $\mathcal{O}(q^4)$. These include the calculation of the nucleon mass in the heavy-baryon formalism to fifth order [136], and a determination of the leading nonanalytic contributions to the axial-vector coupling constant g_A at the two-loop level using so-called renormalization group techniques [22]. In the following we will discuss some aspects of the chiral expansion of the nucleon mass up to and including $\mathcal{O}(q^6)$ in the reformulated infrared regularization scheme, based on the work of Refs. [170, 171]. According to the power counting, only tree-level and one-loop diagrams have to be considered in a calculation up to $\mathcal{O}(q^4)$, while starting at $\mathcal{O}(q^5)$, one also has to take into account two-loop diagrams. As the calculation of all diagrams is too involved to be presented here in detail, we will focus on a few relevant aspects and explore some implications.

Before discussing details specific to infrared regularization and our power counting, we give a brief description of the renormalization of two-loop diagrams in general. The discussion follows Ref. [43]. In order to keep track of the number of loop integrations, we make explicit the dependence on $\hbar$, where each power of $\hbar$ corresponds to one loop integration. At the two-loop level, one has to distinguish

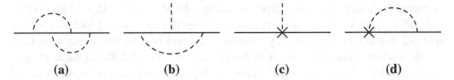

Fig. 5.4 Two-loop diagram with corresponding subdiagram and counter-term diagram

between overall divergences, which occur when both loop momenta become large, and so-called subdivergences, in which only one loop momentum is large while the other remains finite. As an example, consider the diagram in Fig. 5.4a. If one loop momentum is kept fixed, the integration of the other momentum corresponds to a one-loop subdiagram as shown in Fig. 5.4b. This subdiagram can contain a divergence, which has to be subtracted with a counter term of order $\hbar$ (Fig. 5.4c). Since we are working at the two-loop level, i.e. $O(\hbar^2)$, there are so-called counter-term diagrams, in which one vertex corresponds to a counter term, see Fig. 5.4d. Taking into account the sum of a two-loop diagram and all its corresponding counter-term diagrams ensures that any remaining divergence is local and can thus be absorbed into counter terms in the Lagrangian.

In the calculation of the nucleon mass in baryon ChPT we have to ensure that, in addition to subtracting all divergences, the resulting expressions satisfy the power counting and the relevant Ward identities. To demonstrate the subtleties involved, consider the example of a two-loop integral H that can be written as the product of two one-loop integrals H_1 and H_2,[7]

$$H = H_1 H_2. \qquad (5.42)$$

The chiral order of the two-loop integral is simply the sum of the chiral orders of the two one-loop integrals. Each of the one-loop integrals can be separated into an infrared-singular and an infrared-regular part, and we obtain

$$H = I_1 I_2 + I_1 R_2 + R_1 I_2 + R_1 R_2. \qquad (5.43)$$

As discussed, we need to add the contribution of counter-term diagrams. The subtraction terms for the one-loop integral H_1 are given by its infrared-regular part R_1. The unrenormalized counter-term integral is thus

$$-R_1 H_2, \qquad (5.44)$$

with an analogous expression for the other counter-term integral. Previously, when going to the limit $n \to 4$, it was only necessary to discuss terms of order $1/\varepsilon$ and ε^0, where $\varepsilon = 4 - n$. However, if the integral H_2 contains divergences, these can be multiplied by terms $\sim \varepsilon$ in the subtraction term R_1, resulting in finite contributions. It turns out to be crucial for the preservation of chiral symmetry to

---

[7] Note that not all two-loop integrals can be decomposed in this way. However, this special case is sufficient for our considerations.

include in the subtraction terms not only divergent and finite terms, but also all terms of positive power in ε. A detailed discussion can be found in Ref. [171]. The properly renormalized expression for the two-loop integral H^{ir} has the particularly simple form

$$H^{ir} = \tilde{I}_1 \tilde{I}_2, \tag{5.45}$$

where the $\tilde{}$ indicates that all additional divergences in the infrared-singular parts of the one-loop integrals have been dropped (see Sect. 4.6.2).

The chiral expansion of the nucleon mass up to $\mathcal{O}(q^6)$ is given by

$$m_N = m + k_1 M^2 + k_2 M^3 + k_3 M^4 \ln\left(\frac{M}{\mu}\right) + k_4 M^4 + k_5 M^5 \ln\left(\frac{M}{\mu}\right) + k_6 M^5$$
$$+ k_7 M^6 \ln^2\left(\frac{M}{\mu}\right) + k_8 M^6 \ln\left(\frac{M}{\mu}\right) + k_9 M^6. \tag{5.46}$$

The lengthy expressions for all of the coefficients k_i are given in Ref. [171]. While we refrain from displaying them here, we want to discuss a few aspects and implications. The coefficients k_5 and k_6 are given by

$$k_5 = \frac{3g_A^2}{1024\pi^3 F^4}(16g_A^2 - 3),$$
$$k_6 = \frac{3g_A^2}{256\pi^3 F^4}\left[g_A^2 + \frac{\pi^2 F^2}{m^2} - 8\pi^2(3l_3^r - 2l_4^r) - \frac{32\pi^2 F^2}{g_A}(2d_{16} - d_{18})\right].$$

Note that while k_5 receives contributions from a number of diagrams with various low-energy couplings, it only depends on the parameters of the lowest-order Lagrangian, i.e. g_A and F. The term $k_5 M^5 \ln\left(\frac{M}{\mu}\right)$ is the leading chiral logarithm at two-loop order, and its value is constrained by renormalization group equations and thus only depends on the lowest-order constants [39]. The coupling k_6 on the other hand depends on the mesonic LECs l_3, l_4 of the Lagrangian at $\mathcal{O}(q^4)$, and the baryonic LECs d_{16}, d_{18} of the Lagrangian at $\mathcal{O}(q^3)$. Also note that the coefficient k_5 has to be the same in all renormalization schemes.

How do these higher-order contributions affect the convergence of the chiral expansion? Unfortunately, most of the coefficients cannot be evaluated numerically as the values of various LECs are not known. However, as seen above, k_5 only depends on the axial-vector coupling constant g_A and the pion-decay constant F. While their values should be taken in the chiral limit, in order to get an estimate for higher-order contributions, we choose to evaluate them at their physical values $g_A = 1.2694$ and $F_\pi = 92.42\,\text{MeV}$. Setting $\mu = m_N$, where $m_N = (m_p + m_n)/2 = 938.92\,\text{MeV}$, we obtain $k_5 M^5 \ln(M/m_N) = -4.8\,\text{MeV}$ at the physical pion mass $M = M_{\pi^+} = 139.57\,\text{MeV}$. This corresponds to approximately 31% of the leading nonanalytic contribution at one-loop order, $k_2 M^3$. As mentioned above, the convergence at unphysical values of the pion mass is also of great interest. Figure 5.5

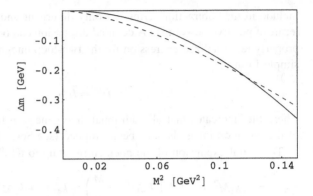

Fig. 5.5 Pion-mass dependence of the term $k_5 M^5 \ln\left(\frac{M}{\mu}\right)$ (*solid line*) for $M < 400$ MeV. The *dashed line* shows the term $k_2 M^3$ for comparison

shows the pion-mass dependence of the term $k_5 M^5 \ln(M/m_N)$ (solid line) in comparison with the term $k_2 M^3$ (dashed line) for $M < 400$ MeV. Chiral extrapolations are considered to be applicable in the shown pion mass range [54, 141]. We see that already at $M \approx 360$ MeV the fifth-order term $k_5 M^5 \ln(M/m_N)$ becomes as large as $k_2 M^3$. This comparison does not present a strict study of the convergence properties of the chiral expansion, as not all contributions at a specific chiral order are considered. For example, we have not taken into account the contributions from $k_6 M^5$, which might cancel parts of the k_5 term. However, these results indicate the importance of higher-order terms at larger pion masses, and they are in agreement with the convergence estimates determined with other methods [54, 141].

5.2 Nucleon Electromagnetic Form Factors to $\mathcal{O}(q^4)$

As mentioned in Sect. 1.4.2, the matrix element of the electromagnetic current operator,

$$J^\mu(x) = \frac{2}{3}\bar{u}(x)\gamma^\mu u(x) - \frac{1}{3}\bar{d}(x)\gamma^\mu d(x),$$

evaluated between single-nucleon states is related to the nucleon electromagnetic form factors. Imposing the relevant symmetries such as translational invariance, Lorentz covariance, the discrete symmetries, and current conservation, the nucleon matrix element of the electromagnetic current operator can be parameterized in terms of two form factors,

$$\langle N(p',s')|J^\mu(0)|N(p,s)\rangle = \bar{u}(p',s')\left[F_1^N(Q^2)\gamma^\mu + i\frac{\sigma^{\mu\nu}q_\nu}{2m_p}F_2^N(Q^2)\right]u(p,s), \quad (5.47)$$

where $q = p' - p$, $Q^2 = -q^2$, and $N = p, n$.[8] In principle, a third form factor $F_3^N(Q^2)$ proportional to q^μ exists which, however, vanishes for on-shell nucleons due to current conservation as well as time-reversal invariance. The Dirac and Pauli form factors F_1^N and F_2^N are normalized such that at $Q^2 = 0$ they reduce to the charge and anomalous magnetic moment [in units of e and the nuclear magneton $e/(2m_p)$], respectively,

$$F_1^p(0) = 1, \quad F_1^n(0) = 0, \quad F_2^p(0) = 1.793, \quad F_2^n(0) = -1.913.$$

In the actual calculation, it is more convenient to work in the isospin basis (s for isoscalar and v for isovector)

$$F_i^{(s)} = F_i^p + F_i^n, \quad F_i^{(v)} = F_i^p - F_i^n, \quad i = 1, 2, \tag{5.48}$$

so that the electromagnetic form factors may be combined in a 2×2 matrix as follows,

$$\mathbb{F}_i = \frac{1}{2} F_i^{(s)} \mathbb{1} + \frac{1}{2} F_i^{(v)} \tau_3, \quad i = 1, 2.$$

Experimental results are commonly presented in terms of the electric and magnetic Sachs form factors $G_E(Q^2)$ and $G_M(Q^2)$, which are related to the Dirac and Pauli form factors via

$$G_E^N(Q^2) = F_1^N(Q^2) - \frac{Q^2}{4m_p^2} F_2^N(Q^2), \quad G_M^N(Q^2) = F_1^N(Q^2) + F_2^N(Q^2), \quad N = p, n.$$

In the nonrelativistic limit, the Fourier transforms of the Sachs form factors are often interpreted as the distribution of charge and magnetization inside the nucleon. For a covariant interpretation in terms of the transverse charge density see Refs. [43, 143].

As they are experimentally well-studied, the description of the electromagnetic form factors provides a stringent test for any theory or model of the strong interactions. As baryon ChPT is a low-energy approximation of QCD in the one-nucleon sector, one would expect ChPT calculations to show good agreement with data. These have been performed in the early relativistic approach [83], the heavy-baryon approach [15, 71], the small-scale expansion [18], the infrared regularization [119], and the EOMS scheme [79]. These calculations have in common that they describe the form factors for momentum transfers up to around $Q^2 = 0.1\,\text{GeV}^2$ (see Fig. 5.6), which corresponds to a small expansion parameter of $q \sim 350\,\text{MeV}$, in agreement with the breakdown of the chiral expansion of the nucleon mass. While the complete calculation of the form factors even up to only $\mathcal{O}(q^3)$, the first order at which loop

---

[8] Since we discuss the form factors in the space-like region, here we adopt the convention of taking $Q^2 = -q^2$ as the argument of the form factors as is common practice in the context of electron scattering.

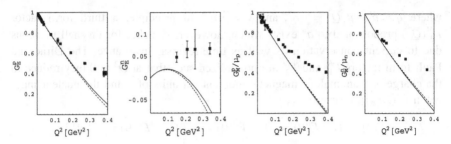

Fig. 5.6 The Sachs form factors of the nucleon in manifestly Lorentz-invariant chiral perturbation theory at $\mathcal{O}(q^4)$. *Full lines*: results in the extended on-mass-shell scheme; *dashed lines*: results in infrared regularization. The experimental data are taken from Ref. [75]

diagrams enter, is somewhat involved, we will discuss a few features of the calculation in a manifestly covariant renormalization scheme in the following exercise.

Exercise 5.6 Diagrams contributing to the electromagnetic form factors to $\mathcal{O}(q^4)$ are shown in Fig. 5.7. There are, in fact, additional diagrams with an insertion of the vertex proportional to c_1 from $\mathscr{L}_{\pi N}^{(2)}$ in the nucleon propagator. These can be included in the calculation of the shown diagrams by using $m_2 = m - 4c_1 M^2$ as the mass in the nucleon propagator instead of the lowest-order mass m (see Sect. 10 of Ref. [10]). Evaluating the diagrams, we obtain the invariant amplitude $\mathcal{M}$, which is related to the matrix element of Eq. 5.47 by

$$\mathcal{M} = -ie\varepsilon_\mu \langle N(p', s')|J^\mu(0)|N(p, s)\rangle, \tag{5.49}$$

where ε is the polarization four-vector of the virtual photon. Note that a calculation of diagrams to $\mathcal{O}(q^D)$ determines the Dirac form factor to $\mathcal{O}(q^{D-1})$ and the Pauli form factor to $\mathcal{O}(q^{D-2})$, as both the polarization four-vector ε and the four-momentum transfer q count as small quantities, and the Γ matrices γ^μ and $\sigma^{\mu\nu}$ as $\mathcal{O}(q^0)$ (see Eq. 4.19).

To consider a coupling to an external electromagnetic four-vector potential (see Eq. 1.165) we set

$$v_\mu = r_\mu = l_\mu = -e\mathscr{A}_\mu \frac{\tau_3}{2}, \quad v_\mu^{(s)} = -e\frac{1}{2}\mathscr{A}_\mu,$$

in the Lagrangians.

(a) Using the Lagrangians of Eqs. 4.17 and 4.66, as well as the relevant terms of the third-order Lagrangian,[9]

$$\mathscr{L}_{\pi N}^{(3)} = \cdots + \frac{i}{2m}d_6\bar{\Psi}\left[D^\mu, f_{\mu\nu}^+\right]D^\nu\Psi + \text{H.c.} + \frac{2i}{m}d_7\bar{\Psi}\left(\partial^\mu v_{\mu\nu}^{(s)}\right)D^\nu\Psi + \text{H.c.} + \cdots,$$

---

[9] The Lagrangian corresponds to the one of Ref. [72] with the replacements $\tilde{F}_{\mu\nu}^+ \rightarrow f_{\mu\nu}^+$ and $\text{Tr}(F_{\mu\nu}^+) \rightarrow 4v_{\mu\nu}^{(s)}$.

Fig. 5.7 Diagrams contributing to the nucleon electromagnetic form factors to $\mathcal{O}(q^4)$. Nucleons are denoted by *solid lines*, pions by *dashed lines*, and the *wavy line* stands for a coupling to an external electromagnetic four-vector potential

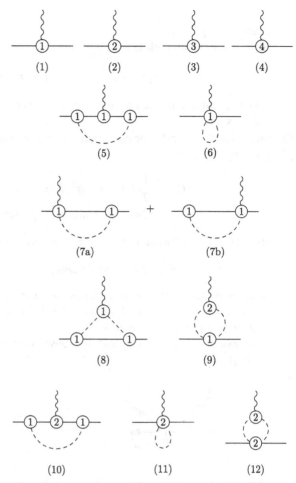

show the following Feynman rules:

$$-ie\varepsilon_\mu\gamma^\mu\frac{1}{2}(\mathbb{1}+\tau_3),$$

$$e\varepsilon_\mu\sigma^{\mu\nu}q_\nu\frac{1}{2}(2c_6\tau_3+c_7\mathbb{1}),$$

$$ie\varepsilon_\mu\left(q^2P^\mu-q^\mu q\cdot P\right)\left(\frac{1}{2m}d_6\tau_3+\frac{1}{m}d_7\mathbb{1}\right),$$

where $P^\mu = p'^\mu + p^\mu$.

Hint: The field-strength tensors are given by

$$v^{(s)}_{\mu\nu} = \partial_\mu v^{(s)}_\nu - \partial_\nu v^{(s)}_\mu, \quad f^+_{\mu\nu} = u f_{L\mu\nu} u^+ + u^+ f_{R\mu\nu} u,$$

with $f_{R\mu\nu}$ and $f_{L\mu\nu}$ defined as the two-flavor versions of Eqs. 3.66 and 3.67, respectively.

(b) Show that the form-factor contributions from tree-level diagrams to $\mathcal{O}(q^3)$ are given by[10]

$$\mathbb{F}_1(Q^2) = \frac{1}{2}(\mathbb{1} + \tau_3) - q^2(d_6\tau_3 + 2d_7\mathbb{1}),$$
$$\mathbb{F}_2(Q^2) = 2m_N c_6\tau_3 + m_N c_7\mathbb{1} + q^2(d_6\tau_3 + 2d_7\mathbb{1}).$$
(5.50)

Hint: Use $p^\mu = g^{\mu\nu}p_\nu$ and $\gamma^\mu\gamma^\nu = g^{\mu\nu} - i\sigma^{\mu\nu}$ to show that[11]

$$\bar{u}(p')P^\mu u(p) = \bar{u}(p')(2m_N\gamma^\mu - i\sigma^{\mu\nu}q_\nu)u(p).$$

Also verify that $q \cdot P = 0$ for nucleons on the mass shell. In Eq. 5.50, we have replaced a factor m_N/m by 1 because the difference is of higher order in the contribution to the form factors.

(c) As an example of a loop diagram, we consider diagram (7a) of Fig. 5.7. Verify the Feynman rule

$$ie\varepsilon_\mu \frac{g_A}{2F} \gamma^\mu\gamma_5\varepsilon_{3ab}\tau_b.$$

(d) Show that the invariant amplitude $\mathcal{M}$ of diagram (7a) is given by

$$\mathcal{M} = -ie\varepsilon_\mu\bar{u}(p')\frac{g_A^2}{2F^2}\tau_3\gamma_5 i\mu^{4-n}\int\frac{d^n k}{(2\pi)^n}\,\slashed{k}S_F(p'-k)\Delta_F(k)\gamma^\mu\gamma_5 u(p),$$

where the nucleon and pion Feynman propagators are given by

$$S_F(p) = \frac{1}{\slashed{p} - m + i0^+} = \frac{\slashed{p} + m}{p^2 - m^2 + i0^+},$$
$$\Delta_F(k) = \frac{1}{k^2 - M^2 + i0^+}.$$

(e) In order to avoid tensor integrals of higher rank, verify that

---

[10] Since we work in the isospin-symmetric limit we set $m_p = m_n = m_N$.

[11] In the following, spin and isospin quantum numbers as well as isospinors are suppressed.

$$\not{k} = -S_F^{-1}(p'-k) + (\not{p}' - m),$$

and use this relation to show that the invariant amplitude can be written as

$$
\mathcal{M} = -ie\varepsilon_\mu \bar{u}(p') \frac{g_A^2}{2F^2} \tau_3 i \mu^{4-n} \int \frac{d^n k}{(2\pi)^n} \left\{ \frac{1}{k^2 - M^2 + i0^+} \right.
$$
$$
- \frac{m_N^2 - m^2}{[(p'-k)^2 - m^2 + i0^+](k^2 - M^2 + i0^+)}
$$
$$
\left. + \frac{(m_N + m)\not{k}}{[(p'-k)^2 - m^2 + i0^+](k^2 - M^2 + i0^+)} \right\} \gamma^\mu u(p).
$$

Hints: Make use of the Dirac equation, $\bar{u}(p')\not{p}' = m_N\bar{u}(p')$. $\{\gamma^\mu, \gamma_5\} = 0$, $\gamma_5^2 = 1$.

(f) Using the integrals of Exercise 4.10, show that

$$
\mathcal{M} = -ie\varepsilon_\mu \bar{u}(p') \frac{g_A^2}{2F^2} \tau_3 \left\{ I_\pi - (m_N^2 - m^2) I_{N\pi} \right.
$$
$$
\left. + \frac{m_N + m}{2m_N} \left[I_N - I_\pi + (m_N^2 - m^2 + M^2) I_{N\pi} \right] \right\} \gamma^\mu u(p),
$$

where

$$I_{N\pi} = I_{N\pi}(-p', 0)\big|_{p'^2 = m_N^2}.$$

The integrals I_π, I_N, and $I_{N\pi}$ are given in Eq. 4.87. The unrenormalized contribution of diagram (7a) to the isovector form factor $F_1^{(v)}$ is then given by

$$
F_1^{(v)} = \frac{g_A^2}{F^2} \left\{ I_\pi - (m_N^2 - m^2) I_{N\pi} + \frac{m_N + m}{2m_N} \left[I_N - I_\pi + (m_N^2 - m^2 + M^2) I_{N\pi} \right] \right\}.
$$

In order to obtain the results in infrared regularization, one has to replace all integrals by their infrared-singular parts. Replacing the physical nucleon mass with its chiral expansion, we see that in a calculation of the form factors to $\mathcal{O}(q^3)$, the term proportional to $m_N^2 - m^2$ is of higher order and can be neglected, while it has to be taken into account for calculations of $\mathcal{O}(q^4)$ and higher. Setting the 't Hooft parameter $\mu = m$, in the EOMS scheme no additional subtraction beyond $\widetilde{\text{MS}}$ is necessary, because the $\widetilde{\text{MS}}$-renormalized expression in combination with the polarization vector is of $\mathcal{O}(q^3)$. This corresponds to the order assigned to the renormalized diagram by the power counting.

The results for the remaining diagrams are given in Refs. [79, 119].[12] Once all diagrams are evaluated and the wavefunction renormalization constant is taken

---

[12] Note that different notations for the loop integrals are used in the literature.

into account, both isospin components of the Dirac form factor F_1 produce the correct values at $Q^2 = 0$, namely, the isoscalar and isovector charges of $+1$.

Figure 5.6 shows that a calculation at the one-loop level using nucleon and pion degrees of freedom only is not sufficient to describe the form factors for $Q^2 \geq 0.1\,\mathrm{GeV}^2$ and that higher-order contributions must play an important role. Moreover, up to and including $\mathcal{O}(q^4)$, the most general effective Lagrangian provides sufficiently many independent parameters such that the empirical values of the anomalous magnetic moments and the charge and magnetic radii are fitted rather than predicted. We will now discuss how introducing additional dynamical degrees of freedom may improve the description of the electromagnetic form factors. We will focus on the inclusion of the vector mesons ρ, ω, and ϕ, because the importance of vector mesons for the interactions between photons and hadrons was established already a long time ago. In the original vector-meson-dominance picture (see Ref. [163]) the coupling of a virtual photon to the matrix element of the isovector current operator between hadronic states is dominated by a $\gamma\rho^0$ transition, propagation of the ρ^0, and a subsequent (strong) transition induced by the interaction with the ρ^0.

In ChPT, the contributions from vector mesons, as well as other heavy particles, are included implicitly in the values of the LECs. Symbolically, this can be understood from the expansion of a vector-meson propagator,

$$\frac{1}{q^2 - M_V^2} = -\frac{1}{M_V^2}\left[1 + \frac{q^2}{M_V^2} + \left(\frac{q^2}{M_V^2}\right)^2 + \mathcal{O}(q^6)\right],$$

where M_V is the vector-meson mass, in combination with the relevant vector-meson vertices. The contributions from the expanded propagator are included order by order in the ChPT couplings. It was shown in Ref. [119] that the inclusion of vector mesons as explicit degrees of freedom in an EFT results in the resummation of a subset of higher-order contributions that turned out to be important for the description of the nucleon electromagnetic form factors. However, no diagrams with *internal* vector-meson lines were considered because a generalization of ChPT which fully includes the effects of vector mesons as intermediate states in loops was not yet available. The EOMS scheme [76], the reformulated version of the infrared regularization of Ref. [168], and the extension of infrared regularization of Refs. [36, 37] all provide a framework to systematically include virtual vector mesons in the domain of applicability of baryon chiral perturbation theory [78]. This means that there is a power counting that predicts the relative size of diagrams, even for those including internal vector-meson lines.

In Ref. [169] the electromagnetic form factors were calculated with ρ, ω, and ϕ mesons as explicit degrees of freedom. While originally vector mesons were described in terms of antisymmetric tensor fields [62, 81], Ref. [169] employs the vector-field representation, which was shown to be equivalent in Ref. [63] provided certain conditions hold. In this formalism, the ρ meson is represented by

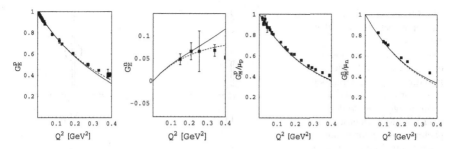

Fig. 5.8 The Sachs form factors of the nucleon in manifestly Lorentz-invariant chiral perturbation theory at $\mathcal{O}(q^4)$ including ρ, ω, and ϕ mesons as explicit degrees of freedom. *Full lines*: results in the extended on-mass-shell scheme; *dashed lines*: results in infrared regularization. The experimental data are taken from Ref. [75]

$\rho^\mu = \rho_i^\mu \tau_i$, and the ω and ϕ mesons by ω^μ and ϕ^μ, respectively. The coupling of vector mesons to pions and external fields is at least of $\mathcal{O}(q^3)$ [63],

$$\mathscr{L}_{\pi V}^{(3)} = -f_\rho \, \mathrm{Tr}(\rho^{\mu\nu} f_{\mu\nu}^+) - f_\omega \omega^{\mu\nu} f_{\mu\nu}^{(s)} - f_\phi \phi^{\mu\nu} f_{\mu\nu}^{(s)} + \cdots, \tag{5.51}$$

where the vector-meson field-strength tensors are given by

$$\rho^{\mu\nu} = \nabla^\mu \rho^\nu - \nabla^\nu \rho^\mu, \quad \nabla^\mu \rho^\nu = \partial^\mu \rho^\nu + [\Gamma^\mu, \rho^\nu],$$

with Γ^μ the chiral connection of Eq. 4.13, and

$$\omega^{\mu\nu} = \partial^\mu \omega^\nu - \partial^\nu \omega^\mu, \quad \phi^{\mu\nu} = \partial^\mu \phi^\nu - \partial^\nu \phi^\mu.$$

The lowest-order Lagrangian for the coupling to the nucleon is given by

$$\mathscr{L}_{VN}^{(0)} = \frac{1}{2} \sum_{V=\rho,\omega,\phi} g_V \bar{\Psi} \gamma^\mu V_\mu \Psi, \tag{5.52}$$

and the $\mathcal{O}(q)$ Lagrangian reads

$$\mathscr{L}_{VN}^{(1)} = \frac{1}{4} \sum_{V=\rho,\omega,\phi} G_V \bar{\Psi} \sigma^{\mu\nu} V_{\mu\nu} \Psi. \tag{5.53}$$

The coupling constants f_V, g_V, and G_V, with $V = \{\rho, \omega, \phi\}$, are not constrained by chiral symmetry and have to be determined by comparison with data.

The inclusion of additional degrees of freedom also requires additional power-counting rules, which for the vector mesons state that vertices from $\mathscr{L}_{\pi V}^{(3)}$ count as $\mathcal{O}(q^3)$ and vertices from $\mathscr{L}_{VN}^{(i)}$ as $\mathcal{O}(q^i)$, respectively, while the vector-meson propagators count as $\mathcal{O}(q^0)$.

Calculations of the form factors in both the EOMS scheme and in infrared regularization [169] result in an improved description of the data even for higher values of Q^2, as expected on phenomenological grounds (see Fig. 5.8). The

Fig. 5.9 Feynman diagrams including vector mesons that contribute to the electromagnetic form factors of the nucleon up to and including $\mathcal{O}(q^4)$. External-leg corrections are not shown. *Solid*, *wiggly*, and *double lines* refer to nucleons, photons, and vector mesons, respectively. The *numbers* in the interaction blobs denote the order of the Lagrangian from which they are obtained. The direct coupling of the photon to the nucleon is obtained from $\mathscr{L}_{\pi N}^{(1)}$ and $\mathscr{L}_{\pi N}^{(2)}$

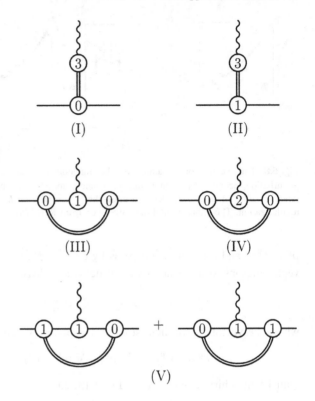

parameters of the vector-meson Lagrangian of Eq. 5.51 for the coupling to external fields have been taken from Ref. [63], and those of Eqs. 5.52 and 5.53 for the coupling of vector mesons to the nucleon from the dispersion relations of Refs. [101, 142]. The small difference between the two renormalization schemes is due to the different treatment of regular higher-order terms of loop integrals. Numerically, the results are similar to those of Ref. [119], which indicates that contributions from diagrams with internal vector-meson lines are small. In fact, in infrared renormalization diagrams that do not contain internal pion lines vanish, which is the case for all vector-meson loop diagrams to $\mathcal{O}(q^4)$, shown in Fig. 5.9. One could therefore interpret these results as providing a firmer theoretical basis to the vector-meson-dominance model, in which only tree-level couplings are considered. It should be noted that, in a strict chiral expansion in terms of small external momenta q and quark masses m_q at a fixed ratio m_q/q^2, up to and including $\mathcal{O}(q^4)$ the results with and without explicit vector mesons are completely equivalent. Contributions from vector mesons as explicit degrees of freedom are compensated by different values of the LECs common to the theories with and without vector mesons. On the other hand, the inclusion of vector-meson degrees of freedom in the present framework results in a reordering of terms which, in an ordinary chiral expansion, would contribute at higher orders beyond $\mathcal{O}(q^4)$. It is these terms which change the form factor results favorably for larger values of Q^2.

Note that this re-organization proceeds according to well-defined rules so that a controlled, order-by-order calculation of corrections is made possible.

5.3 Advanced Applications and Outlook

5.3.1 Chiral Extrapolations

As mentioned in Sect. 5.1.4, chiral perturbation theory is of interest to lattice QCD calculations since it predicts the quark-mass dependence of physical observables, while lattice QCD calculations are routinely performed at unphysical quark masses and results have to be extrapolated to the physical point. In return, lattice QCD in principle provides a way to determine the low-energy constants of ChPT from the underlying theory. However, ChPT as described in the previous sections is the effective field theory of continuum QCD in an infinite volume, while lattice calculations discretize space-time with a finite lattice spacing a and are restricted to some finite volume V. In addition, the symmetries of the discretized version of QCD are different from those in the continuum. This is most easily seen for the case of rotational symmetry, which translates into a hypercubic symmetry on the lattice. In addition, the implementation of chiral symmetry in lattice formulations of QCD is a complex and well-studied problem. Therefore, ChPT should in principle only be used for extrapolations in the quark masses after the lattice QCD results have been extrapolated to the continuum and infinite volume limits. A different approach is to formulate effective field theories that amount to modifications of ChPT to systematically take into account the effects of symmetry breakings, finite lattice spacings, and finite volumes. These have been studied for a variety of lattice actions, including the partially-quenched and so-called mixed-action approaches, in which different masses and, in addition, different discretizations, respectively, are employed for valence and sea quarks. Introductions to applications of ChPT to lattice QCD can be found, e.g., in Refs. [5, 94, 174] and references therein.

5.3.2 Pion Photo- and Electroproduction

Besides pion-nucleon scattering discussed in Sect. 4.3.2, electromagnetic production of pions on the nucleon is one of the most prominent examples of the application of baryon ChPT. A particular advantage of this type of reactions is the fact that very precise experimental data are available close to production threshold (see Ref. [61] and references therein). The special interest in neutral pion photoproduction at threshold arose from the fact that experimental data [12, 135] pointed to a large deviation from predictions for the s-wave electric dipole amplitude E_{0+} based on current algebra and PCAC [48]. In Ref. [13] an explanation for this discrepancy was given: pion loops, which are beyond the current-algebra framework, generate infrared singularities in the scattering amplitude which then modify the predicted low-energy expansion of E_{0+} at next-to-leading order $[\mathcal{O}(q^3)]$. For an overview of numerous subsequent activities, see Ref. [24]. The so-called Adler-Gilman relation [3] provides a chiral Ward identity establishing a connection between charged pion electroproduction at threshold and the isovector axial-vector current evaluated between single-nucleon states (see, e.g., Refs. [77, 164] for more details). Via this relation, the axial form factor has been investigated in terms of pion electroproduction experiments [129]. A systematic difference between the values for the axial mass M_A extracted from such experiments and neutrino scattering experiments was explained in heavy-baryon chiral perturbation theory [16]. It was shown that at $\mathcal{O}(q^3)$ pion loop contributions modify the momentum dependence of the electric dipole amplitude from which the axial mass is extracted. These contributions result in a change of the axial mass of $\Delta M_A = 0.056\,\text{GeV}$, bringing the neutrino scattering and pion electroproduction results for the axial mass into agreement (see Ref. [20] for further details).

5.3.3 Compton Scattering and Polarizabilities

Based on the requirement of gauge invariance, Lorentz invariance, crossing symmetry, and the discrete symmetries, the famous low-energy theorem of Low [130] and Gell-Mann and Goldberger [91] uniquely specifies the low-energy Compton scattering amplitude up to and including terms linear in the photon momentum. The coefficients of this expansion are expressed in terms of global properties of the nucleon: its mass, charge, and magnetic moment. It is only terms of second order which contain new information on the structure of the nucleon specific to Compton scattering. For a general target, these effects can be parameterized in terms of two constants, the electric and magnetic polarizabilities α and β, respectively. The predictions of HBChPT at $\mathcal{O}(q^3)$ [15], generating the leading $1/M_\pi$ singularity, are surprisingly close to the empirical values (see, e.g., Refs. [61, 108, 173] for an overview of the experimental status). These predictions contain no unknown LECs, i.e., they are given in terms of the pion mass, the axial-vector coupling constant, and the pion-decay constant. Higher-order calculations have been performed in the heavy-baryon framework [6], the ε expansion including the Δ resonance [103], and a covariant calculation including the Δ resonance [125]. For a discussion of how to extract the neutron polarizabilities see, e.g., Ref. [154] and references therein. Generalizations of the static polarizabilities α and β in terms of dynamical polarizabilities are discussed in Ref. [95].

Including the spin of the nucleon introduces, at third order in the photon momentum, four so-called spin polarizabilities $\gamma_1, \gamma_2, \gamma_3$, and γ_4 into the Compton scattering amplitude [158]. In a heavy-baryon calculation at $\mathcal{O}(q^3)$ [17], the nucleon spin polarizabilities are isoscalar, i.e., the same for proton and neutron, and behave as $1/M_\pi^2$.[13] As for the spin-independent polarizabilities α and β, at $\mathcal{O}(q^3)$ the nucleon spin polarizabilities are entirely given in terms of pion-nucleon loop diagrams and are thus expressed in terms of M_π, g_A, and F_π. Full one-loop calculations to $\mathcal{O}(q^4)$ have been performed in Refs. [90, 181]. No new LECs, except for the anomalous magnetic moments of the nucleon, enter at this order, but the degeneracy between proton and neutron polarizabilities is lifted. Unfortunately, the next-to-leading-order contributions turn out to be very large, calling the convergence of the expansion into question [181]. Predictions for the nucleon spin polarizabilities including the $\Delta(1232)$ excitation have been discussed in Refs. [103, 105, 149]. For a comparison with experimental results we refer the reader to Refs. [61, 108, 173]. The status of dispersion-theoretic analyses can be found in Ref. [60].

5.3.4 Virtual Compton Scattering

In virtual Compton scattering (VCS) one or even both photons are allowed to be virtual. The corresponding amplitude for the proton may be tested in reactions such as $e^-p \to e^-p\gamma, \gamma p \to pe^+e^-$ or $e^-p \to e^-pe^+e^-$.[14] The possibilities to investigate the structure of the target increase substantially if virtual photons are used since (a) photon energy and momentum can be varied independently and (b) longitudinal components of the transition current are accessible. For the nucleon, the model-independent properties of the low-energy VCS amplitude have been identified in Refs. [96, 165]. In Ref. [96] the model-dependent part beyond the low-energy theorem was analyzed in terms of a multipole expansion. Keeping only terms linear in the energy of the final photon, and imposing the constraints due to charge-conjugation invariance [58, 59], the corresponding amplitude may be parameterized in terms of six generalized polarizabilities (GPs), which are functions of the three-momentum transfer of the virtual photon in the VCS process (for an overview, see Ref. [97]). Predictions for the GPs of the nucleon have been obtained in HBChPT at $\mathcal{O}(q^3)$ [102, 104] and $\mathcal{O}(q^4)$ [111, 112], as well as the small-scale expansion at $\mathcal{O}(q^3)$ [106]. While the electromagnetic polarizabilities α and β of real Compton scattering characterize the global response of hadrons to soft external electric and magnetic fields, the use of a virtual photon in the initial state and a real low-energy photon in the final state allows for a local resolution of the induced electric polarization and magnetization. In Ref. [131] it was shown that three generalized dipole polarizabilities are required in order to fully

---

[13] The π^0-exchange graph driven by the WZW term of Sect. 3.5.3 results in an isovector contribution which is usually subtracted.

[14] In principle, the VCS amplitude $\gamma^*\pi \to \gamma\pi$ can be investigated in the reaction $\pi e \to \pi e\gamma$ [179].

reconstruct local polarizations induced by soft external fields in a hadron. These spatial distributions were determined at large distances $r \sim 1/M_\pi$ for pions, kaons, and octet baryons by use of ChPT. For an overview of the experimental status of generalized polarizabilities, see Refs. [52, 61, 74, 108].

5.3.5 Isospin-Symmetry Breaking

In these lecture notes we always assumed isospin symmetry, i.e. $m_u = m_d$. Moreover, the electromagnetic interaction, breaking isospin symmetry, was always treated in terms of external fields, i.e., without loop corrections involving virtual photons. Besides the mass differences within isospin multiplets of a given strangeness (see, e.g., Figs. 3.3 and 3.4), there are various dynamical manifestations of isospin-symmetry breaking. For example, the decay of an η into three pions can only proceed via isospin-symmetry-breaking effects [82]. Cusp effects such as in neutral pion photoproduction on the proton close to threshold [12, 19, 70, 172] or $K \rightarrow 3\pi$ decays [6, 33, 41, 42] are generated by the nucleon and pion mass differences. The inclusion of virtual photons in mesonic and baryonic chiral perturbation theory was discussed in Refs. [180, 145] and [144], respectively. An additional inclusion of virtual leptons allows for a full treatment of isospin-symmetry-breaking effects in semileptonic decays of pions and kaons [117]. In the baryonic sector, the general two-flavor pion-nucleon Lagrangian including both virtual photons and leptons was constructed in Ref. [177]. There have been numerous investigations concerning isospin-symmetry breaking in both mesonic and baryonic sectors and we refer the interested reader to Ref. [161] for a recent overview. Finally, one is often interested in separating electromagnetic and strong contributions to a physical quantity. However, as discussed in Ref. [85], the splitting of the Hamiltonian of QCD + γ into a strong and an electromagnetic piece is ambiguous due to the ultraviolet divergences generated by photon loops. A systematic method for the "purification of physical matrix elements from electromagnetic effects" has been proposed in Ref. [85].

5.3.6 Three-Flavor Calculations

In the mesonic three-flavor sector many calculations have been performed at the two-loop level and fitted to experimental results (see Ref. [29] for a comprehensive overview and Ref. [30] for an update). In comparison to the two-flavor sector including pions only, the number of physical observables is considerably larger in the three-flavor case and, due to the presence of different masses in the loops, calculational effort and difficulty increase. Since $m_s \gg m_u, m_d$, the convergence is expected to be slower and higher-order terms are expected to be more important. Three-flavor ChPT seems to work fairly well in most cases but there also appear to be exceptions such as the α parameter of the Dalitz plot for $\eta \to 3\pi^0$ [30]. In the baryonic three-flavor sector, the convergence properties are more controversial (see Refs. [17, 24] for a review). For example, the results for the individual contributions to the masses of the baryon octet differ strongly depending on which renormalization condition is applied and which approximation is chosen for keeping or neglecting higher-order terms in a given framework [34, 57, 67, 122]. The slow convergence is a combination of various circumstances: in the baryonic sector the chiral order increases in steps of $\mathcal{O}(q)$ as opposed to $\mathcal{O}(q^2)$ in the mesonic case; the ratio of the kaon mass to the chiral-symmetry-breaking scale, $M_K/(4\pi F_0) \approx 0.42$, is rather large raising some doubt on the validity of a perturbative treatment at low orders; in some channels resonances such as the $\Lambda(1405)$ and $\Sigma(1385)$ lie below the $N\bar{K}$ threshold. For calculations of other observables such as magnetic moments or electromagnetic form factors see, e.g., Refs. [92, 100, 120, 138]. Alternative methods of discussing properties of hyperons include two-flavor chiral perturbation theory [178] and chiral unitary approaches (see below).

5.3.7 Chiral Unitary Approaches

The extension of chiral perturbation theory to higher energies as described in Sects. 4.7 and 5.2 consists of the explicit inclusion of particular additional degrees of freedom in the Lagrangian. This method relies on a perturbative expansion of physical observables, and the domain of applicability is governed by the existence of an underlying scale such as the mass difference to the lightest state not included in the Lagrangian. A different approach to study the impact of chiral symmetry on phenomena of the strong interaction at higher energies is based on constraints provided by the unitarity of the S-matrix, see, e.g., Refs. [109, 110, 139, 148]. At low energies, chiral perturbation theory is used to describe meson-meson and meson-baryon interactions. These results are then non-perturbatively extended to higher energies while implementing exact unitarity and possibly further constraints by causality and electromagnetic gauge invariance. These methods have been applied to meson-meson and meson-baryon scattering in the SU(2) and SU(3) cases as well as meson photo- and electroproduction (see, e.g., Refs. [35, 80, 118, 133, 147] and references therein).

5.3.8 Complex-Mass Scheme

In Sect. 5.2 we saw how the inclusion of virtual vector mesons generates an improved description of the electromagnetic form factors, for which ordinary chiral perturbation theory does not produce sufficient curvature. So far the inclusion of virtual vector mesons has been restricted to low-energy processes in which the vector mesons cannot be generated explicitly. One would also like to investigate the properties of hadronic resonances such as their masses and widths [107, 184] as well as their electromagnetic properties. Since the main decay of the ρ meson involves two pions with vanishing masses in the chiral limit, loop diagrams develop large imaginary parts for energies of the order of the ρ-meson mass. These power-counting-violating contributions, being imaginary, cannot be absorbed in the redefinition of the parameters of the Lagrangian as long as the usual renormalization procedure is used.

An extension of chiral effective field theory to the momentum region near the complex pole corresponding to the vector mesons was proposed in Ref. [55], in which the power-counting problem was addressed by applying the complex-mass scheme (CMS) [1, 2, 49, 50, 176] to the effective field theory. The CMS originates from the Standard Model where it was developed to derive properties of W, Z_0, and Higgs bosons obtained from resonant processes. In the CMS, complex gauge-boson masses are used in tree-level and loop calculations, necessitating the introduction of complex counter terms in the Lagrangian. In the framework of EFT, the method has been applied to the quark-mass expansion of the pole mass and the width of the ρ meson, which are of particular interest in the context of lattice extrapolations [123, 124], as well as the chiral structure of the Roper resonance [56].

5.3.9 Chiral Effective Theory for Two- and Few-Nucleon Systems

The extension of the methods described in the previous chapters to systems of two and more nucleons was first suggested by Weinberg in Refs. [182, 183]. Interactions between two nucleons arise from the Lagrangians of the pion and one-nucleon sectors via one- and multiple-pion exchanges, supplemented by NN contact interactions. The existence of nuclear bound states such as the deuteron implies that loop contributions are not necessarily suppressed in the two- and few-nucleon sectors, as one cannot obtain bound states by considering only a finite number of scattering diagrams. Weinberg therefore suggested to apply the power counting to an effective potential, which is defined as the sum of all diagrams that do not contain purely nucleonic intermediate states. The potential is then iterated with n-nucleon intermediate states to generate an infinite number of diagrams. In the two-nucleon case, the effective potential consists of all two-nucleon-irreducible diagrams, and observables can be calculated using the Lippmann-Schwinger or Schrödinger equations.

It has been argued that while this approach might produce phenomenologically satisfactory results, there are issues whether and how the theory can be properly renormalized [113]. An alternative was proposed in which pions are treated perturbatively [114, 115]. However, it was shown that the resulting expansion has problematic convergence properties [73, 88],[15] and the correct implementation of a

---

[15] See Ref. [9] for a different approach to include pions perturbatively.

chiral EFT program for two and more nucleons is still being debated (see, e.g., Refs. [8, 31, 69, 89, 146, 152] and references therein).

In addition to the two-nucleon sector, three- and four-nucleon interactions have been studied in the chiral effective-field-theory approach, and a number of few-nucleon observables have been calculated based on these interactions. For a recent review and an extensive list of the relevant literature see Ref. [68]. A pedagogical introduction is given in Ref. [153].

If one only considers energies well below M_π^2/m, it is possible to construct a different EFT in which pions are integrated out and the only dynamical degrees of freedom are nucleons interacting via contact terms. This pionless EFT reproduces the results of the effective range expansion, while also allowing for the consistent coupling to electromagnetic and weak external currents (see, e.g., Refs. [7, 14, 157] and references therein). Calculations in light nuclei up to $A = 6$ have been performed within the framework of pionless EFT [116, 175].

Chiral perturbation theory has been a very active field in the last 25 years. Readers who wish to supplement this monograph with additional literature or who are interested in the present status of applications are referred to lecture notes and review articles [17, 23, 24, 28, 29, 32, 40, 44, 51, 64, 66, 86, 121, 126, 127, 134, 137, 154–156, 166, 167] as well as conference proceedings [4, 25–27, 45, 140].

References

1. Actis, S., Passarino, G.: Nucl. Phys. B **777**, 100 (2007)
2. Actis, S., Passarino, G., Sturm, C., Uccirati, S.: Phys. Lett. B **669**, 62 (2008)
3. Adler, S.L., Gilman, F.J.: Phys. Rev. **152**, 1460 (1966)
4. Ahmed, M.W., Gao, H., Weller, H.R., Holstein, B.R. (eds.) Chiral Dynamics 2006. Proceedings, Workshop, Durham/Chapel Hill, USA, 18–22 September 2006. World Scientific, Singapore (2007)
5. Bär, O.: Nucl. Phys. Proc. Suppl. **140**, 106 (2005)
6. Batley, J.R., et al. NA48/2 Collaboration: Phys. Lett. B **633**, 173 (2006)
7. Beane, S.R., Bedaque, P.F., Haxton, W.C., Phillips, D.R., Savage, M.J.: In: Shifman, M., Ioffe, B. (eds.) At the Frontier of Particle Physics. Handbook of QCD, vol. 1, pp. 133–269. World Scientific, Singapore (2001). arXiv:nucl-th/0008064
8. Beane, S.R., Bedaque, P.F., Savage, M.J., van Kolck, U.: Nucl. Phys. A **700**, 377 (2002)
9. Beane, S.R., Kaplan, D.B., Vuorinen, A.: Phys. Rev. C **80**, 011001 (2009)
10. Becher, T., Leutwyler, H.: Eur. Phys. J. C **9**, 643 (1999)
11. Becher, T., Leutwyler, H.: JHEP **0106**, 017 (2001)
12. Beck, R., et al.: Phys. Rev. Lett. **65**, 1841 (1990)
13. Bernard, V., Kaiser, N., Gasser, J., Meißner, U.-G.: Phys. Lett. B **268**, 291 (1991)
14. Bedaque, P.F., van Kolck, U.: Annu. Rev. Nucl. Part. Sci. **52**, 339 (2002)
15. Bernard, V., Kaiser, N., Kambor, J., Meißner, U.-G.: Nucl. Phys. B **388**, 315 (1992)
16. Bernard, V., Kaiser, N., Meißner, U.-G.: Phys. Rev. Lett. **69**, 1877 (1992)
17. Bernard, V., Kaiser, N., Meißner, U.-G.: Int. J. Mod. Phys. E **4**, 193 (1995)
18. Bernard, V., Fearing, H.W., Hemmert, T.R., Meißner, U.-G.: Nucl. Phys. A **635**, 121 (1998) [Erratum, ibid. A **642**, 563]
19. Bernard, V., Kaiser, N., Meißner, U.-G.: Eur. Phys. J. A **11**, 209 (2001)
20. Bernard, V., Elouadrhiri, L., Meißner, U.-G.: J. Phys. G **28**, R1 (2002)

21. Bernard, V., Hemmert, T.R., Meißner, U.-G.: Phys. Lett. B **565**, 137 (2003)
22. Bernard, V., Meißner, U.-G.: Phys. Lett. B **639**, 278 (2006)
23. Bernard, V., Meißner, U.-G.: Annu. Rev. Nucl. Part. Sci. **57**, 33 (2007)
24. Bernard, V.: Prog. Part. Nucl. Phys. **60**, 82 (2008)
25. Bernstein, A.M., Holstein, B.R. (eds.) Chiral Dynamics: Theory and Experiment. Proceedings, Workshop, Cambridge, USA, 25–29 July 1994. Lect. Notes Phys. **452**. Springer, Berlin (1995)
26. Bernstein, A.M., Drechsel, D., Walcher, Th. (eds.) Chiral Dynamics: Theory and Experiment. Proceedings, Workshop, Mainz, Germany, 1–5 September 1997. Lect. Notes Phys. **513**. Springer, Berlin (1998)
27. Bernstein, A.M., Goity, J.L., Meißner, U.-G. (eds.) Chiral Dynamics: Theory and Experiment III. Proceedings, Workshop, Jefferson Laboratory, USA, 17–20 July 2000. World Scientific, Singapore (2002)
28. Bijnens, J.: Int. J. Mod. Phys. A **8**, 3045 (1993)
29. Bijnens, J.: Prog. Part. Nucl. Phys. **58**, 521 (2007)
30. Bijnens, J.: In: PoS (CD09), 031 (2009)
31. Birse, M.C.: Phys. Rev. C **74**, 014003 (2006)
32. Birse, M.C., McGovern, J.: In: Close, F., Donnachie, S., Shaw, G. (eds.) Electromagnetic Interactions and Hadronic Structure, pp. 229–270. Cambridge University Press, Cambridge (2007)
33. Bissegger, M., Fuhrer, A., Gasser, J., Kubis, B., Rusetsky, A.: Phys. Lett. B **659**, 576 (2008)
34. Borasoy, B., Meißner, U.-G.: Ann. Phys. **254**, 192 (1997)
35. Borasoy, B., Bruns, P.C., Meißner, U.-G., Nissler, R.: Eur. Phys. J. A **34**, 161 (2007)
36. Bruns, P.C., Meißner, U.-G.: Eur. Phys. J. C **40**, 97 (2005)
37. Bruns, P.C., Meißner, U.-G.: Eur. Phys. J. C **58**, 407 (2008)
38. Buettiker, P., Meißner, U.-G.: Nucl. Phys. A **668**, 97 (2000)
39. Büchler, M., Colangelo, G.: Eur. Phys. J. C **32**, 427 (2003)
40. Burgess, C.P.: Phys. Rep. **330**, 193 (2000)
41. Cabibbo, N.: Phys. Rev. Lett. **93**, 121801 (2004)
42. Cabibbo, N., Isidori, G.: JHEP **0503**, 021 (2005)
43. Carlson, C.E., Vanderhaeghen, M.: Phys. Rev. Lett. **100**, 032004 (2008)
44. Colangelo, G., Isidori G.: arXiv:hep-ph/0101264
45. Colangelo, G., et al. (eds.) Chiral Dynamics. Proceedings, Workshop, Bern, Switzerland, 6–10 July 2009. Proceedings of Science. PoS CD09 (2009)
46. Collins, J.C.: Renormalization. Cambridge University Press, Cambridge (1984)
47. Creutz, M.: Quarks, Gluons and Lattices. Cambridge University Press, Cambridge (1983)
48. De Baenst, P.: Nucl. Phys. B **24**, 633 (1970)
49. Denner, A., Dittmaier, S., Roth, M., Wackeroth, D.: Nucl. Phys. B **560**, 33 (1999)
50. Denner, A., Dittmaier, S., Roth, M., Wieders, L.H.: Nucl. Phys. B **724**, 247 (2005)
51. de Rafael, E.: CP violation and the limits of the standard model. In: Donoghue, J.F. (ed.) Proceedings of the 1994 Advanced Theoretical Study Institute in Elementary Particle Physics, Boulder, Colorado, 29 May–24 June 1994. World Scientific, Singapore (1995)
52. d'Hose, N., et al.: Prog. Part. Nucl. Phys. **44**, 371 (2000)
53. Djukanovic, D., Schindler, M.R., Gegelia, J., Scherer, S.: Phys. Rev. D **72**, 045002 (2005)
54. Djukanovic, D., Gegelia, J., Scherer, S.: Eur. Phys. J. A **29**, 337 (2006)
55. Djukanovic, D., Gegelia, J., Keller, A., Scherer, S.: Phys. Lett. B **680**, 235 (2009)
56. Djukanovic, D., Gegelia, J., Scherer, S.: Phys. Lett. B **690**, 123 (2010)
57. Donoghue, F., Holstein, B.R., Borasoy, B.: Phys. Rev. D **59**, 036002 (1999)
58. Drechsel, D., Knöchlein, G., Metz, A., Scherer, S.: Phys. Rev. C **55**, 424 (1997)
59. Drechsel, D., Knöchlein, G., Korchin, A.Y., Metz, A., Scherer, S.: Phys. Rev. C **57**, 941 (1998)
60. Drechsel, D., Pasquini, B., Vanderhaeghen, M.: Phys. Rep. **378**, 99 (2003)
61. Drechsel, D., Walcher, Th.: Rev. Mod. Phys. **80**, 731 (2008)
62. Ecker, G., Gasser, J., Pich, A., de Rafael, E.: Nucl. Phys. B **321**, 311 (1989)

63. Ecker, G., Gasser, J., Leutwyler, H., Pich, A., de Rafael, E.: Phys. Lett. B **223**, 425 (1989)
64. Ecker, G.: Prog. Part. Nucl. Phys. **35**, 1 (1995)
65. Ecker, G., Mojžiš, M.: Phys. Lett. B **365**, 312 (1996)
66. Ecker, G.: arXiv:hep-ph/9805500
67. Ellis, P.J., Torikoshi, K.: Phys. Rev. C **61**, 015205 (2000)
68. Epelbaum, E., Hammer, H.W., Meißner, U.-G.: Rev. Mod. Phys. **81**, 1773 (2009)
69. Epelbaum, E., Gegelia, J.: Eur. Phys. J. A **41**, 341 (2009)
70. Fäldt, G.: Nucl. Phys. A **333**, 357 (1980)
71. Fearing, H.W., Lewis, R., Mobed, N., Scherer, S.: Phys. Rev. D **56**, 1783 (2007)
72. Fettes, N., Meißner, U.-G., Mojžiš, M., Steininger, S.: Ann. Phys. **283**, 273 (2001) [Erratum, ibid. **288**, 249 (2001)]
73. Fleming, S., Mehen, T., Stewart, I.W.: Nucl. Phys. A **677**, 313 (2000)
74. Fonvieille, H.: Prog. Part. Nucl. Phys. **55**, 198 (2005)
75. Friedrich, J., Walcher, Th.: Eur. Phys. J. A **17**, 607 (2003)
76. Fuchs, T., Gegelia, J., Japaridze, G., Scherer, S.: Phys. Rev. D **68**, 056005 (2003)
77. Fuchs, T., Scherer, S.: Phys. Rev. C **68**, 055501 (2003)
78. Fuchs, T., Schindler, M.R., Gegelia, J., Scherer, S.: Phys. Lett. B **575**, 11 (2003)
79. Fuchs, T., Gegelia, J., Scherer, S.: J. Phys. G **30**, 1407 (2004)
80. Gasparyan, A., Lutz, M.F.M.: Nucl. Phys. A **848**, 126 (2010)
81. Gasser, J., Leutwyler, H.: Ann. Phys. **158**, 142 (1984)
82. Gasser, J., Leutwyler, H.: Nucl. Phys. B **250**, 539 (1985)
83. Gasser, J., Sainio, M.E., Švarc, A.: Nucl. Phys. B **307**, 779 (1988)
84. Gasser, J., Leutwyler, H., Sainio, M.E.: Phys. Lett. B **253**, 252 (1991)
85. Gasser, J., Rusetsky, A., Scimemi, I.: Eur. Phys. J. C **32**, 97 (2003)
86. Gasser, J.: Lect. Notes Phys. **629**, 1 (2004)
87. Gattringer, C., Lang, C.B.: Quantum Chromodynamics on the Lattice: An Introductory Presentation. Lect. Notes Phys. **788**. Springer, Berlin (2010)
88. Gegelia, J.: arXiv:nucl-th/9806028
89. Gegelia, J., Scherer, S.: Int. J. Mod. Phys. A **21**, 1079 (2006)
90. Gellas, G.C., Hemmert, T.R., Meißner, U.-G.: Phys. Rev. Lett. **85**, 14 (2000)
91. Gell-Mann, M., Goldberger, M.L.: Phys. Rev. **96**, 1433 (1954)
92. Geng, L.S., Martin Camalich, J., Alvarez-Ruso, L., Vacas, M.J.V.: Phys. Rev. Lett. **101**, 222002 (2008)
93. Gensini, P.M.: πN Newslett. **6**, 21 (1998)
94. Golterman, M.: arXiv:0912.4042 [hep-lat]
95. Grießhammer, H.W., Hemmert, T.R.: Phys. Rev. C **65**, 045207 (2002)
96. Guichon, P.A.M., Liu, G.Q., Thomas, A.W.: Nucl. Phys. A **591**, 606 (1995)
97. Guichon, P.A.M., Vanderhaeghen, M.: Prog. Part. Nucl. Phys. **41**, 125 (1998)
98. Hacker, C., Wies, N., Gegelia, J., Scherer, S.: Phys. Rev. C **72**, 055203 (2005)
99. Hägler, Ph.: Phys. Rep. **490**, 99 (2010)
100. Hammer, H.W., Puglia, S.J., Ramsey-Musolf, M.J., Zhu, S.L.: Phys. Lett. B **562**, 208 (2003)
101. Hammer, H.W., Meißner, U.-G.: Eur. Phys. J. A **20**, 469 (2004)
102. Hemmert, T.R., Holstein, B.R., Knöchlein, G., Scherer, S.: Phys. Rev. D **55**, 2630 (1997)
103. Hemmert, T.R., Holstein, B.R., Kambor, J.: Phys. Rev. D **55**, 5598 (1997)
104. Hemmert, T.R., Holstein, B.R., Knöchlein, G., Scherer, S.: Phys. Rev. Lett. **79**, 22 (1997)
105. Hemmert, T.R., Holstein, B.R., Kambor, J., Knöchlein, G.: Phys. Rev. D **57**, 5746 (1998)
106. Hemmert, T.R., Holstein, B.R., Knöchlein, G., Drechsel, D.: Phys. Rev. D **62**, 014013 (2000)
107. Höhler, G.: Against Breit-Wigner parameters—a pole-emic. In: Caso, C., et al. (eds.) Particle Data Group, Eur. Phys. J. C **3**, 624 (1998)
108. Hyde-Wright, C.E., de Jager, K.: Annu. Rev. Nucl. Part. Sci. **54**, 217 (2004)
109. Kaiser, N., Siegel, P.B., Weise, W.: Nucl. Phys. A **594**, 325 (1995)
110. Kaiser, N., Siegel, P.B., Weise, W.: Phys. Lett. B **362**, 23 (1995)

111. Kao, C.W., Vanderhaeghen, M.: Phys. Rev. Lett. **89**, 272002 (2002)
112. Kao, C.W., Pasquini, B., Vanderhaeghen, M.: Phys. Rev. D **70**, 114004 (2004)
113. Kaplan, D.B., Savage, M.J., Wise, M.B.: Nucl. Phys. B **478**, 629 (1996)
114. Kaplan, D.B., Savage, M.J., Wise, M.B.: Phys. Lett. B **424**, 390 (1998)
115. Kaplan, D.B., Savage, M.J., Wise, M.B.: Nucl. Phys. B **534**, 329 (1998)
116. Kirscher, J., Grießhammer, H.W., Shukla, D., Hofmann, H.M.: Eur. Phys. J. A **44**, 239 (2010)
117. Knecht, M., Neufeld, H., Rupertsberger, H., Talavera, P.: Eur. Phys. J. C **12**, 469 (2000)
118. Kolomeitsev, E.E., Lutz, M.F.M.: Phys. Lett. B **585**, 243 (2004)
119. Kubis, B., Meißner, U.-G.: Nucl. Phys. A **679**, 698 (2001)
120. Kubis, B., Meißner, U.-G.: Eur. Phys. J. C **18**, 747 (2001)
121. Kubis, B.: arXiv:hep-ph/0703274
122. Lehnhart, B.C., Gegelia, J., Scherer, S.: J. Phys. G **31**, 89 (2005)
123. Leinweber, D.B., Cohen, T.D.: Phys. Rev. D **49**, 3512 (1994)
124. Leinweber, D.B., Thomas, A.W., Tsushima, K., Wright, S.V.: Phys. Rev. D **64**, 094502 (2001)
125. Lensky, V., Pascalutsa, V.: Eur. Phys. J. C **65**, 195 (2010)
126. Leutwyler, H.: In: Ellis, R.K., Hill, C.T., Lykken, J.D. (eds.) Perspectives in the Standard Model. Proceedings of the 1991 Theoretical Advanced Study Institute in Elementary Particle Physics, Boulder, Colorado, 2–28 June 1991. World Scientific, Singapore (1992)
127. Leutwyler, H.: In: Herscovitz, V.E. (ed.) Hadron Physics 94: Topics on the Structure and Interaction of Hadronic Systems, Proceedings, Workshop, Gramado, Brasil. World Scientific, Singapore (1995)
128. Leutwyler, H.: πN Newslett. **15**, 1 (1999)
129. Liesenfeld, A., et al., A1 Collaboration: Phys. Lett. B **468**, 20 (1999)
130. Low, F.E.: Phys. Rev. **96**, 1428 (1954)
131. L'vov, A.I., Scherer, S., Pasquini, B., Unkmeir, C., Drechsel, D.: Phys. Rev. C **64**, 015203 (2001)
132. Lyubovitskij, V.E., Gutsche, T., Faessler, A., Drukarev, E.G.: Phys. Rev. D **63**, 054026 (2001)
133. Lutz, M.F.M., Kolomeitsev, E.E.: Nucl. Phys. A **730**, 392 (2004)
134. Manohar, A.V.: Lectures given at 35th Int. Universitätswochen für Kern- und Teilchenphysik: Perturbative and Nonperturbative Aspects of Quantum Field Theory. Schladming, Austria 2–9 March 1996. arXiv:hep-ph/9606222 (1996)
135. Mazzucato, E., et al.: Phys. Rev. Lett. **57**, 3144 (1986)
136. McGovern, J.A., Birse, M.C.: Phys. Lett. B **446**, 300 (1999)
137. Meißner, U.-G.: Rep. Prog. Phys. **56**, 903 (1993)
138. Meißner, U.-G., Steininger, S.: Nucl. Phys. B **499**, 349 (1997)
139. Meißner, U.-G., Oller, J.A.: Nucl. Phys. A **673**, 311 (2000)
140. Meißner, U.-G., Hammer, H.W., Wirzba, A. (eds.) Chiral Dynamics: Theory and Experiment (CD2003). Mini-Proceedings, Workshop, Bonn, Germany, 8–13 September 2003, arXiv:hep-ph/0311212 (2003)
141. Meißner, U.-G.: In: PoS (LAT2005), 009 (2005)
142. Mergell, P., Meißner, U.-G., Drechsel, D.: Nucl. Phys. A **596**, 367 (1996)
143. Miller, G.A.: Phys. Rev. Lett. **99**, 112001 (2007)
144. Müller, G., Meißner, U.-G.: Nucl. Phys. B **556**, 265 (1999)
145. Neufeld, H., Rupertsberger, H.: Z. Phys. C **71**, 131 (1996)
146. Nogga, A., Timmermans, R.G.E., van Kolck, U.: Phys. Rev. C **72**, 054006 (2005)
147. Oller, J.A., Oset, E., Ramos, A.: Prog. Part. Nucl. Phys. **45**, 157 (2000)
148. Oset, E., Ramos, A.: Nucl. Phys. A **635**, 99 (1998)
149. Pascalutsa, V., Phillips, D.R.: Phys. Rev. C **68**, 055205 (2003)
150. Pagels, H.: Phys. Rep. **16**, 219 (1975)
151. Pavan, M.M., Strakovsky, I.I., Workman, R.L., Arndt, R.A.: πN Newslett. **16**, 110 (2002)
152. Pavon Valderrama, M., Ruiz Arriola, E.: Phys. Rev. C **70**, 044006 (2004)

153. Phillips, D.R.: Czech. J. Phys. **52**, B49 (2002)
154. Phillips, D.R.: J. Phys. G **36**, 104004 (2009)
155. Pich, A.: Rept. Prog. Phys. **58**, 563 (1995)
156. Pich, A.: In: Gupta, R., Morel, A., de Rafael, E., David, F. (eds.) Probing the Standard Model of Particle Interactions. Proceedings of the Les Houches Summer School in Theoretical Physics, Session 68, Les Houches, France, 28 July–5 September 1997. Elsevier, Amsterdam (1999)
157. Platter, L.: Few Body Syst. **46**, 139 (2009)
158. Ragusa, S.: Phys. Rev. D **47**, 3757 (1993)
159. Reya, E.: Rev. Mod. Phys. **46**, 545 (1974)
160. Rothe, H.J.: Lattice Gauge Theories: An Introduction, 3rd edn. World Sci. Lect. Notes Phys., vol. 74. World Scientific, Singapore (2005)
161. Rusetsky, A.: In: PoS (CD09) 071 (2009)
162. Sainio, M.E.: πN Newslett. **16**, 138 (2002)
163. Sakurai, J.J.: Currents and Mesons. University of Chicago Press, Chicago (1969)
164. Scherer, S., Koch, J.H.: Nucl. Phys. A **534**, 461 (1991)
165. Scherer, S., Korchin, A.Y., Koch, J.H.: Phys. Rev. C **54**, 904 (1996)
166. Scherer, S.: Adv. Nucl. Phys. **27**, 277 (2003)
167. Scherer, S.: Prog. Part. Nucl. Phys. **64**, 1 (2010)
168. Schindler, M.R., Gegelia, J., Scherer, S.: Phys. Lett. B **586**, 258 (2004)
169. Schindler, M.R., Gegelia, J., Scherer, S.: Eur. Phys. J. A **26**, 1 (2005)
170. Schindler, M.R., Djukanovic, D., Gegelia, J., Scherer, S.: Phys. Lett. B **649**, 390 (2007)
171. Schindler, M.R., Djukanovic, D., Gegelia, J., Scherer, S.: Nucl. Phys. A **803**, 68 (2008)
172. Schmidt, A., et al.: Phys. Rev. Lett. **87**, 232501 (2001)
173. Schumacher, M.: Prog. Part. Nucl. Phys. **55**, 567 (2005)
174. Sharpe, S.R.: arXiv:hep-lat/0607016
175. Stetcu, I., Rotureau, J., Barrett, B.R., van Kolck, U.: J. Phys. G **37**, 064033 (2010)
176. Stuart, R.G.: In: Tran Thanh Van, J. (ed.) Z^0 Physics, p. 41, Editions Frontieres, Gif-sur-Yvette (1990)
177. Supanam, N., Fearing, H.W., Yan, Y.: JHEP **1011**, 124 (2010)
178. Tiburzi, B.C., Walker-Loud, A.: Phys. Lett. B **669**, 246 (2008)
179. Unkmeir, C., Ocherashvili, A., Fuchs, T., Moinester, M.A., Scherer, S.: Phys. Rev. C **65**, 015206 (2002)
180. Urech, R.: Nucl. Phys. B **433**, 234 (1995)
181. Vijaya Kumar, K.B., McGovern, J.A., Birse, M.C.: Phys. Lett. B **479**, 167 (2000)
182. Weinberg, S.: Phys. Lett. B **251**, 288 (1990)
183. Weinberg, S.: Nucl. Phys. B **363**, 3 (1991)
184. Wittig, H.: In: Landolt-Börnstein - Group I Elementary Particles, Nuclei and Atoms, vol. 21 A: Elementary Particles. Springer, Berlin (2008)
185. Workman, R.: Phys. Rev. C **59**, 3441 (1999)

Appendix A
Pauli and Dirac Matrices

A.1 Pauli Matrices

The Hermitian, traceless Pauli matrices τ_i $(i = 1, 2, 3)$[1] are the generators of the group SU(2). They are given by

$$\tau_1 = \begin{pmatrix} 0 & 1 \\ 1 & 0 \end{pmatrix}, \quad \tau_2 = \begin{pmatrix} 0 & -i \\ i & 0 \end{pmatrix}, \quad \tau_3 = \begin{pmatrix} 1 & 0 \\ 0 & -1 \end{pmatrix}, \tag{A.1}$$

and satisfy the commutation relations

$$[\tau_i, \tau_j] = 2i\varepsilon_{ijk}\tau_k, \tag{A.2}$$

where ε_{ijk} is the completely antisymmetric tensor. Furthermore,

$$\tau_i^2 = 1. \tag{A.3}$$

The anticommutator of two Pauli matrices is given by

$$\{\tau_i, \tau_j\} = 2\delta_{ij}1, \tag{A.4}$$

and therefore

$$\tau_i\tau_j = i\varepsilon_{ijk}\tau_k + \delta_{ij}1. \tag{A.5}$$

Two useful relations are given by

$$\frac{1}{3}\tau_i\tau_i = 1, \quad \tau_i\tau_j\tau_i = -\tau_j, \tag{A.6}$$

where we have summed over repeated indices. From Eq. A.5 we obtain for the trace of the product of two Pauli matrices

$$\text{Tr}(\tau_i\tau_j) = 2\delta_{ij}. \tag{A.7}$$

---

[1] We adopt the convention to use the notation τ_i for Pauli matrices in isospin space, while σ_i denotes Pauli matrices in spin space.

S. Scherer and M. R. Schindler, *A Primer for Chiral Perturbation Theory*,
Lecture Notes in Physics 830, DOI: 10.1007/978-3-642-19254-8,
© Springer-Verlag Berlin Heidelberg 2012

A.2 Dirac Matrices

The Dirac matrices satisfy the relation

$$\gamma^\mu\gamma^\nu + \gamma^\nu\gamma^\mu = 2g^{\mu\nu}\mathbb{1}. \tag{A.8}$$

There are several different representations of the Dirac matrices, see, e.g., Ref. [1]. Independent of the chosen representation, further important properties of the Dirac matrices are given by

$$(\gamma^0)^2 = \mathbb{1}, \quad (\gamma^i)^2 = -\mathbb{1}, \quad (\gamma^0)^\dagger = \gamma^0, \quad (\gamma^i)^\dagger = -\gamma^i. \tag{A.9}$$

The chirality matrix γ^5 is defined as

$$\gamma_5 = \gamma^5 \equiv i\gamma^0\gamma^1\gamma^2\gamma^3, \tag{A.10}$$

and

$$\gamma^5\gamma^\mu + \gamma^\mu\gamma^5 = 0, \quad (\gamma^5)^2 = \mathbb{1}, \quad (\gamma^5)^\dagger = \gamma^5. \tag{A.11}$$

It is common to define a quantity $\sigma^{\mu\nu}$ as

$$\sigma^{\mu\nu} \equiv \frac{i}{2}(\gamma^\mu\gamma^\nu - \gamma^\nu\gamma^\mu). \tag{A.12}$$

The generalization of the Dirac matrices to n dimensions as needed in dimensional regularization results in

$$\begin{aligned}
\gamma^\mu\gamma^\alpha\gamma_\mu &= (2-n)\gamma^\alpha, \\
\gamma^\mu\gamma^\alpha\gamma^\beta\gamma_\mu &= 4g^{\alpha\beta}\mathbb{1} + (n-4)\gamma^\alpha\gamma^\beta, \\
\gamma^\mu\gamma^\alpha\gamma^\beta\gamma^\delta\gamma_\mu &= -2\gamma^\delta\gamma^\beta\gamma^\alpha + (4-n)\gamma^\alpha\gamma^\beta\gamma^\delta.
\end{aligned} \tag{A.13}$$

A number of further useful relations can be found, e.g., in Ref. [1].

Reference

1. Borodulin, V.I., Rogalev, R.N., Slabospitsky, S.R.: CORE: COmpendium of RElations. arXiv:hep-ph/9507456

Appendix B
Functionals and Local Functional Derivatives

Here we collect a few properties of functionals and local functional derivatives which are used in the main text (for a thorough discussion, see Ref. [1]). Local functional derivatives are natural generalizations of classical partial derivatives to infinite dimensions. For the purpose of illustration, let $\mathscr{F}$ denote the set of all functions $j : \mathbb{R}^n \to \mathbb{K}$ ($\mathbb{K} = \mathbb{R}$ or $\mathbb{C}$). If necessary, we may require additional restrictions such as continuous functions j, smooth functions, integrable functions, and so on. A real (complex) functional is a map $j \mapsto Z[j]$ from $\mathscr{F}$ to $\mathbb{R}$ ($\mathbb{C}$), which assigns a real (complex) number $Z[j]$ to each function j. A typical example is given by an integral of the type

$$F[j] = \int d^n x \, g(j(x)),$$

with g an integrable function. We choose the convention of writing the arguments of functionals inside square brackets. Moreover, let j be a function of two sets of variables, collectively denoted by x and y. Then $F[j(y)]$ denotes a functional which depends on the values of j for all x at fixed y. Finally, a functional may depend on several, independent functions j_i.

In the following we consider a definition of partial functional derivatives based on the Dirac delta function,

$$\delta_y : \begin{cases} \mathbb{R}^n \to \mathbb{R}, \\ x \mapsto \delta_y(x) = \delta^n(x - y). \end{cases}$$

In terms of the Dirac delta function the partial functional derivative is defined as

$$\frac{\delta F[f]}{\delta f(y)} \equiv \lim_{\varepsilon \to 0} \frac{F[f + \varepsilon \delta_y] - F[f]}{\varepsilon}. \tag{B.1}$$

S. Scherer and M. R. Schindler, *A Primer for Chiral Perturbation Theory*,
Lecture Notes in Physics 830, DOI: 10.1007/978-3-642-19254-8,
© Springer-Verlag Berlin Heidelberg 2012

Note the analogy to the partial derivative of an ordinary function,

$$\frac{\partial f(x)}{\partial x_i} \equiv \lim_{\varepsilon \to 0} \frac{f(x + \varepsilon e_i) - f(x)}{\varepsilon}.$$

As discussed in Ref. [1], experience shows that the definition of Eq. B.1 leads to the same results as a rigorous mathematical approach.

Partial functional derivatives share basic properties with ordinary partial derivatives, namely,

$$\frac{\delta}{\delta f(x)} (\alpha_1 F_1[f] + \alpha_2 F_2[f]) = \alpha_1 \frac{\delta F_1[f]}{\delta f(x)} + \alpha_2 \frac{\delta F_2[f]}{\delta f(x)} \quad \text{(linearity)},$$

$$\frac{\delta}{\delta f(x)} (F_1[f] F_2[f]) = \frac{\delta F_1[f]}{\delta f(x)} F_2[f] + F_1[f] \frac{\delta F_2[f]}{\delta f(x)} \quad \text{(product rule)},$$

$$\frac{\delta}{\delta f(x)} F[g(f)] = g'(f(x)) \frac{\delta F}{\delta h(x)} [h = g(f)] \quad \text{(chain rule)}.$$

An important rule for the local functional derivative of a function is

$$\frac{\delta f(y)}{\delta f(x)} = \delta^n(y - x). \tag{B.2}$$

Exercise B.1. Verify Eq. B.2.
Hint: Define $f(y)$ as the functional

$$f(y) = F_y[f] = \int d^n z \, \delta^n(y - z) f(z)$$

and apply the definition of the local functional derivative.

Analogously we have

$$\frac{\delta g(f(y))}{\delta f(x)} = \delta^n(y - x) g'(f(y))$$

and

$$\frac{\delta^k g(f(y))}{\delta f(x_k) \dots \delta f(x_1)} = \delta^n(y - x_k) \dots \delta^n(y - x_1) g^{(k)}(f(y)).$$

One of the prime applications of functionals and partial functional derivatives is the generating functional of Green functions. As a simple, pedagogical illustration let us consider the Green functions

$$G_n(x_1, \dots, x_n) = \langle 0 | T[\phi(x_1) \dots \phi(x_n)] | 0 \rangle$$

of a real scalar field operator ϕ whose dynamics is determined by a Lagrangian $\mathcal{L}$. In very much the same way as the element a_n of the series $(a_0, a_1, a_2, \dots)$ may be obtained from the generating function

$$f(x) = a_0 + a_1 x + \frac{1}{2} a_2 x^2 + \frac{1}{3!} a_3 x^3 + \cdots$$

by calculating the derivative

$$\frac{d^n f}{dx^n}(x = 0) = a_n,$$

the generating functional for the Green functions G_n is given by

$$\exp(iZ[j]) = \langle 0|T \exp\left[i \int d^4 x \mathscr{L}_{\text{ext}}(x)\right]|0\rangle$$

$$= 1 + i \int d^4 x j(x) \langle 0|\phi(x)|0\rangle$$

$$+ \sum_{k=2}^{\infty} \frac{i^k}{k!} \int d^4 x_1 \ldots d^4 x_k j(x_1) \ldots j(x_k) \langle 0|T[\phi(x_1) \ldots \phi(x_k)]|0\rangle,$$

where

$$\mathscr{L}_{\text{ext}} = j(x)\phi(x).$$

Remarks

1. Many textbooks use the nomenclature $Z[j]$ for our $\exp(iZ[j])$ and $W[j]$ for our $Z[j]$. We follow the convention and nomenclature of Gasser and Leutwyler.

2. Note that j represents a function and can thus be taken out of the matrix element, e.g.,

$$\langle 0|T[j(x_1)\phi(x_1)j(x_2)\phi(x_2)]|0\rangle$$

$$= \langle 0|[j(x_1)\phi(x_1)j(x_2)\phi(x_2)\Theta(x_1^0 - x_2^0) + j(x_2)\phi(x_2)j(x_1)\phi(x_1)\Theta(x_2^0 - x_1^0)]|0\rangle$$

$$= j(x_1)j(x_2)\Theta(x_1^0 - x_2^0)\langle 0|\phi(x_1)\phi(x_2)|0\rangle$$

$$+ j(x_1)j(x_2)\Theta(x_2^0 - x_1^0)\langle 0|\phi(x_2)\phi(x_1)|0\rangle$$

$$= j(x_1)j(x_2)\langle 0|T[\phi(x_1)\phi(x_2)]|0\rangle.$$

3. The underlying dynamics is hidden in the fact that both the ground state and Green functions depend on the dynamics in terms of the equation of motion.

As an example, let us discuss how the Green function $G_2(x_1, x_2)$ results from evaluating the second partial functional derivative,

$$G_2(x_1, x_2) = \langle 0|T[\phi(x_1)\phi(x_2)]|0\rangle = (-i)^2 \frac{\delta^2 \exp(iZ[j])}{\delta j(x_1)\delta j(x_2)}\bigg|_{j=0}.$$

In order to obtain a nonzero result, the number of and, in the case of several different (combinations of) fields, type of partial functional derivatives must match the number and type of fields in the Green function of interest. In the present context this means that

$$1, \quad i \int d^4x j(x) \langle 0|\phi(x)|0 \rangle$$

contain too few terms and

$$\frac{i^k}{k!} \int d^4x_1 \ldots d^4x_k j(x_1) \ldots j(x_k) \langle 0|T[\phi(x_1)\ldots\phi(x_k)]|0 \rangle, \quad k \geq 3,$$

too many terms, because j is set equal to 0 at the end. Therefore, the Green function $G_2(x_1, x_2)$ is obtained by the second partial functional derivative of the generating functional.

Reference

1. Zeidler, E.: Quantum Field Theory I: Basics in Mathematics and Physics. Springer, Berlin (2006)

Solutions to Exercises

In the following we provide solutions to all exercises. We strongly encourage the readers to solve the problems on their own and to only use these solutions to check their own work. There are often several ways to solve an exercise, and our calculations simply represent one possible solution. While we sometimes omit intermediate steps, our hope is that readers who have worked through the exercises can easily follow the solutions outlined here. In certain cases we have deliberately not chosen the shortest available explanation to allow readers with a wide range of backgrounds to follow the solutions given here.

Problems of Chapter 1

1.1

$$\text{Tr}([\lambda_a, \lambda_b]\lambda_c) = 2if_{abd}\text{Tr}(\lambda_d\lambda_c) = 4if_{abd}\delta_{dc} = 4if_{abc} \Rightarrow f_{abc} = \frac{1}{4i}\text{Tr}([\lambda_a, \lambda_b]\lambda_c).$$

1.2 $f_{abc} = -f_{bac}$ is obvious, because $[\lambda_a, \lambda_b] = -[\lambda_b, \lambda_a]$. Using the cyclic property of the trace, the remaining relations follow from

$$\begin{aligned}
\text{Tr}([\lambda_a, \lambda_b]\lambda_c) &= \text{Tr}(\lambda_a\lambda_b\lambda_c - \lambda_b\lambda_a\lambda_c) \\
&= \text{Tr}(\lambda_b\lambda_c\lambda_a - \lambda_c\lambda_b\lambda_a) = \text{Tr}([\lambda_b, \lambda_c]\lambda_a) \\
&= \text{Tr}(\lambda_c\lambda_a\lambda_b - \lambda_a\lambda_c\lambda_b) = \text{Tr}([\lambda_c, \lambda_a]\lambda_b).
\end{aligned}$$

1.3

$$\text{Tr}(\{\lambda_a, \lambda_b\}\lambda_c) = \frac{4}{3}\delta_{ab}\underbrace{\text{Tr}(\mathbb{1}\lambda_c)}_{=0} + 2d_{abd}\text{Tr}(\lambda_d\lambda_c) = 4d_{abc}$$

$$\Rightarrow d_{abc} = \frac{1}{4}\text{Tr}(\{\lambda_a, \lambda_b\}\lambda_c).$$

S. Scherer and M. R. Schindler, *A Primer for Chiral Perturbation Theory*,
Lecture Notes in Physics 830, DOI: 10.1007/978-3-642-19254-8,
© Springer-Verlag Berlin Heidelberg 2012

The d symbols are totally symmetric because

$$
\begin{aligned}
\mathrm{Tr}(\{A,B\}C) &= \mathrm{Tr}(\{B,A\}C) \\
&= \mathrm{Tr}(ABC + BAC) = \mathrm{Tr}(BCA + CBA) \\
&= \mathrm{Tr}(\{B,C\}A) \\
&= \mathrm{Tr}(CAB + ACB) = \mathrm{Tr}(\{C,A\}B).
\end{aligned}
$$

1.4

$$
\mathrm{Tr}(\lambda_a\lambda_b\lambda_c) = \mathrm{Tr}\left[\left(\frac{2}{3}\delta_{ab}\mathbb{1} + h_{abd}\lambda_d\right)\lambda_c\right] = \frac{2}{3}\delta_{ab}\underbrace{\mathrm{Tr}(\lambda_c)}_{=\,0} + h_{abd}\underbrace{\mathrm{Tr}(\lambda_d\lambda_c)}_{=\,2\delta_{dc}}
$$

$$
= 2h_{abc},
$$

$$
\mathrm{Tr}(\lambda_a\lambda_b\lambda_c\lambda_d) = \mathrm{Tr}\left[\lambda_a\lambda_b\left(\frac{2}{3}\delta_{cd}\mathbb{1} + h_{cde}\lambda_e\right)\right]
$$

$$
= \frac{2}{3}\delta_{cd}\underbrace{\mathrm{Tr}(\lambda_a\lambda_b)}_{=\,2\delta_{ab}} + \underbrace{h_{cde}}_{=\,h_{ecd}}\underbrace{\mathrm{Tr}(\lambda_a\lambda_b\lambda_e)}_{=\,2h_{abe}}
$$

$$
= \frac{4}{3}\delta_{ab}\delta_{cd} + 2h_{abe}h_{ecd},
$$

$$
\mathrm{Tr}(\lambda_a\lambda_b\lambda_c\lambda_d\lambda_e) = \mathrm{Tr}\left[\lambda_a\lambda_b\lambda_c\left(\frac{2}{3}\delta_{de}\mathbb{1} + h_{def}\lambda_f\right)\right]
$$

$$
= \frac{4}{3}h_{abc}\delta_{de} + h_{def}\left(\frac{4}{3}\delta_{ab}\delta_{cf} + 2h_{abg}h_{gcf}\right)
$$

$$
= \frac{4}{3}h_{abc}\delta_{de} + \frac{4}{3}\delta_{ab}h_{cde} + 2h_{abf}\,h_{fcg}h_{gde}.
$$

1.5

$$
D_\mu q_f \mapsto D'_\mu q'_f = \left[\partial_\mu + ig_3\left(U\mathscr{A}_\mu U^\dagger + \frac{i}{g_3}\partial_\mu UU^\dagger\right)\right]Uq_f
$$

$$
= \partial_\mu Uq_f + U\partial_\mu q_f + ig_3 U\mathscr{A}_\mu \underbrace{U^\dagger U}_{=\,\mathbb{1}}q_f + ig_3\frac{i}{g_3}\underbrace{\partial_\mu U\,U^\dagger U}_{=\,\mathbb{1}}q_f
$$

$$
\underbrace{\phantom{ig_3\frac{i}{g_3}\partial_\mu U\,U^\dagger U q_f}}_{=\,-\partial_\mu Uq_f}
$$

$$
= U\left(\partial_\mu + ig_3\mathscr{A}_\mu\right)q_f = UD_\mu q_f.
$$

1.6

$$\mathscr{G}_{\mu\nu} = \partial_\mu \mathscr{A}_\nu - \partial_\nu \mathscr{A}_\mu + ig_3[\mathscr{A}_\mu, \mathscr{A}_\nu]$$

$$\mapsto \partial_\mu \left(U\mathscr{A}_\nu U^\dagger + \frac{i}{g_3} \partial_\nu U U^\dagger \right) - \partial_\nu \left(U\mathscr{A}_\mu U^\dagger + \frac{i}{g_3} \partial_\mu U U^\dagger \right)$$

$$+ ig_3 \left[U\mathscr{A}_\mu U^\dagger + \frac{i}{g_3} \partial_\mu U U^\dagger, U\mathscr{A}_\nu U^\dagger + \frac{i}{g_3} \partial_\nu U U^\dagger \right]$$

$$= \underbrace{\partial_\mu U\mathscr{A}_\nu U^\dagger}_{(1)} + \underbrace{U\partial_\mu \mathscr{A}_\nu U^\dagger}_{(2)} + \underbrace{U\mathscr{A}_\nu \partial_\mu U^\dagger}_{(3)} + \underbrace{\frac{i}{g_3} \partial_\mu \partial_\nu U U^\dagger}_{(4)} + \underbrace{\frac{i}{g_3} \partial_\nu U \partial_\mu U^\dagger}_{(5)}$$

$$- \underbrace{\partial_\nu U\mathscr{A}_\mu U^\dagger}_{(6)} - \underbrace{U\partial_\nu \mathscr{A}_\mu U^\dagger}_{(7)} - \underbrace{U\mathscr{A}_\mu \partial_\nu U^\dagger}_{(8)} - \underbrace{\frac{i}{g_3} \partial_\nu \partial_\mu U U^\dagger}_{(9)} - \underbrace{\frac{i}{g_3} \partial_\mu U \partial_\nu U^\dagger}_{(10)}$$

$$+ ig_3 \underbrace{[U\mathscr{A}_\mu U^\dagger, U\mathscr{A}_\nu U^\dagger]}_{(11)} - \underbrace{[U\mathscr{A}_\mu U^\dagger, \partial_\nu U U^\dagger]}_{(12)}$$

$$- \underbrace{[\partial_\mu U U^\dagger, U\mathscr{A}_\nu U^\dagger]}_{(13)} - \underbrace{\frac{i}{g_3} [\partial_\mu U U^\dagger, \partial_\nu U U^\dagger]}_{(14)}.$$

Make use of:

$$(4) - (9) = 0,$$

$$-(12) = -U\mathscr{A}_\mu \underbrace{U^\dagger \partial_\nu U U^\dagger}_{= -\partial_\nu U^\dagger} + \partial_\nu U U^\dagger U\mathscr{A}_\mu U^\dagger$$

$$= (8) + (6),$$

$$-(13) = [U\mathscr{A}_\nu U^\dagger, \partial_\mu U U^\dagger] = -U\mathscr{A}_\nu \partial_\mu U^\dagger - \partial_\mu U\mathscr{A}_\nu U^\dagger$$

$$= -(3) - (1),$$

$$-(14) = -\frac{i}{g_3} [\partial_\mu U U^\dagger, \partial_\nu U U^\dagger]$$

$$= -\frac{i}{g_3} \partial_\mu U U^\dagger \partial_\nu U U^\dagger + \frac{i}{g_3} \partial_\nu U U^\dagger \partial_\mu U U^\dagger$$

$$= \frac{i}{g_3} \partial_\mu U \partial_\nu U^\dagger - \frac{i}{g_3} \partial_\nu U \partial_\mu U^\dagger$$

$$= (10) - (5).$$

We thus have

$$\cdots = (2) - (7) + (11) = U\left(\partial_\mu \mathscr{A}_\nu - \partial_\nu \mathscr{A}_\mu + ig_3[\mathscr{A}_\mu, \mathscr{A}_\nu]\right)U^\dagger = U\mathscr{G}_{\mu\nu}U^\dagger.$$

1.7

$$P_R = \frac{1}{2}(\mathbb{1} + \gamma_5) = \frac{1}{2}(\mathbb{1} + \gamma_5^\dagger) = P_R^\dagger,$$

$$P_L = \frac{1}{2}(\mathbb{1} - \gamma_5) = \frac{1}{2}(\mathbb{1} - \gamma_5^\dagger) = P_L^\dagger,$$

$$P_R + P_L = \frac{1}{2}(\mathbb{1} + \gamma_5) + \frac{1}{2}(\mathbb{1} - \gamma_5) = \mathbb{1},$$

$$P_R^2 = \frac{1}{4}(\mathbb{1} + \gamma_5)(\mathbb{1} + \gamma_5) = \frac{1}{4}(\mathbb{1} + 2\gamma_5 + \gamma_5^2) = \frac{1}{2}(\mathbb{1} + \gamma_5) = P_R,$$

$$P_L^2 = \frac{1}{4}(\mathbb{1} - \gamma_5)(\mathbb{1} - \gamma_5) = \frac{1}{4}(\mathbb{1} - 2\gamma_5 + \gamma_5^2) = \frac{1}{2}(\mathbb{1} - \gamma_5) = P_L,$$

$$P_R P_L = \frac{1}{4}(\mathbb{1} + \gamma_5)(\mathbb{1} - \gamma_5) = \underbrace{\frac{1}{4}(\mathbb{1} - \gamma_5^2)}_{= P_L P_R} = 0.$$

1.8

$$P_R u_+ = \frac{1}{2}\begin{pmatrix} \mathbb{1}_{2\times2} & \mathbb{1}_{2\times2} \\ \mathbb{1}_{2\times2} & \mathbb{1}_{2\times2} \end{pmatrix}\sqrt{E}\begin{pmatrix} \chi_+ \\ \chi_+ \end{pmatrix} = \sqrt{E}\begin{pmatrix} \chi_+ \\ \chi_+ \end{pmatrix} = u_+,$$

$$P_L u_+ = \frac{1}{2}\begin{pmatrix} \mathbb{1}_{2\times2} & -\mathbb{1}_{2\times2} \\ -\mathbb{1}_{2\times2} & \mathbb{1}_{2\times2} \end{pmatrix}\sqrt{E}\begin{pmatrix} \chi_+ \\ \chi_+ \end{pmatrix} = 0,$$

$$P_R u_- = \frac{1}{2}\begin{pmatrix} \mathbb{1}_{2\times2} & \mathbb{1}_{2\times2} \\ \mathbb{1}_{2\times2} & \mathbb{1}_{2\times2} \end{pmatrix}\sqrt{E}\begin{pmatrix} \chi_- \\ -\chi_- \end{pmatrix} = 0,$$

$$P_L u_- = \frac{1}{2}\begin{pmatrix} \mathbb{1}_{2\times2} & -\mathbb{1}_{2\times2} \\ -\mathbb{1}_{2\times2} & \mathbb{1}_{2\times2} \end{pmatrix}\sqrt{E}\begin{pmatrix} \chi_- \\ -\chi_- \end{pmatrix} = u_-.$$

1.9 We start from $\Gamma = (P_R + P_L)\Gamma(P_R + P_L)$ and make use of $\{\Gamma, \gamma_5\} = 0$ for $\Gamma \in \Gamma_1$ and $[\Gamma, \gamma_5] = 0$ for $\Gamma \in \Gamma_2$ to obtain

$$\Gamma \in \Gamma_1 : P_R\Gamma P_R = \Gamma P_L P_R = 0, \quad P_L\Gamma P_L = \Gamma P_R P_L = 0,$$
$$\Gamma \in \Gamma_2 : P_R\Gamma P_L = \Gamma P_R P_L = 0, \quad P_L\Gamma P_R = \Gamma P_L P_R = 0.$$

Sandwich between $\bar{q}$ and q using $q_R = P_R q$, $q_L = P_L q$, $\bar{q}_R = \bar{q}P_L$, and $\bar{q}_L = \bar{q}P_R$.

1.10

$$Q_a(t) = \int d^3x J_a^0(t, \vec{x}).$$

For a time-independent infinite volume

$$\frac{dQ_a(t)}{dt} = \int d^3x \frac{\partial J_a^0(t, \vec{x})}{\partial t}.$$

Assuming that the current density $\vec{J}_a(t, \vec{x})$ vanishes faster than $1/r^2$ for $r = |\vec{x}| \to \infty$, we make use of the divergence theorem as

$$\int d^3x \vec{\nabla} \cdot \vec{J}_a = \int d\vec{F} \cdot \vec{J}_a = \lim_{R \to \infty} R^2 \int d\Omega \, \hat{e}_r \cdot \vec{J}_a = 0.$$

$$\cdots = \int d^3x \left[\frac{\partial J_a^0(t, \vec{x})}{\partial t} + \vec{\nabla} \cdot \vec{J}_a(t, \vec{x}) \right] = \int d^3x \partial_\mu J_a^\mu(t, \vec{x}) = \int d^3x \frac{\partial \delta \mathscr{L}}{\partial \varepsilon_a}$$

$$= 0 \quad \text{for} \quad \delta \mathscr{L} = 0.$$

1.11 (a)

$$\delta \mathscr{L} = \frac{\partial \mathscr{L}}{\partial \Phi_i} \delta \Phi_i + \frac{\partial \mathscr{L}}{\partial \partial_\mu \Phi_i} \partial_\mu \delta \Phi_i$$

$$= \underbrace{-m^2 \Phi_1 [-\varepsilon(x)] \Phi_2 - m^2 \Phi_2 \varepsilon(x) \Phi_1}_{= 0}$$

$$\underbrace{- \lambda (\Phi_1^2 + \Phi_2^2) \{ \Phi_1 [-\varepsilon(x)] \Phi_2 + \Phi_2 \varepsilon(x) \Phi_1 \}}_{= 0}$$

$$+ \partial^\mu \Phi_1 \partial_\mu [-\varepsilon(x) \Phi_2] + \partial^\mu \Phi_2 \partial_\mu [\varepsilon(x) \Phi_1]$$

$$= \partial_\mu \varepsilon(x) (-\partial^\mu \Phi_1 \Phi_2 + \Phi_1 \partial^\mu \Phi_2).$$

(b)

$$J^\mu = \frac{\partial \delta \mathscr{L}}{\partial \partial_\mu \varepsilon} = \Phi_1 \partial^\mu \Phi_2 - \partial^\mu \Phi_1 \Phi_2,$$

$$\partial_\mu J^\mu = \frac{\partial \delta \mathscr{L}}{\partial \varepsilon} = 0.$$

1.12

$$[Q_a(t), \Phi_k(t, \vec{y})]$$

$$= -it_{a,ij} \int d^3x [\Pi_i(t, \vec{x}) \Phi_j(t, \vec{x}), \Phi_k(t, \vec{y})]$$

$$= -it_{a,ij} \int d^3x \left(\Pi_i(t, \vec{x}) \underbrace{[\Phi_j(t, \vec{x}), \Phi_k(t, \vec{y})]}_{= 0} + \underbrace{[\Pi_i(t, \vec{x}), \Phi_k(t, \vec{y})]}_{= -i\delta^3(\vec{x} - \vec{y})\delta_{ik}} \Phi_j(t, \vec{x}) \right)$$

$$= -t_{a,kj} \Phi_j(t, \vec{y}).$$

1.13

$$[Q_a(t), Q_b(t)] = -\int d^3x d^3y [\Pi_i(t,\vec{x})t_{a,ij}\Phi_j(t,\vec{x}), \Pi_k(t,\vec{y})t_{b,kl}\Phi_l(t,\vec{y})]$$

$$\overset{(1.52)}{=} -t_{a,ij}t_{b,kl}\int d^3x d^3y \Big(\Pi_i(t,\vec{x})[\Phi_j(t,\vec{x}), \Pi_k(t,\vec{y})]\Phi_l(t,\vec{y})$$

$$+ \Pi_k(t,\vec{y})[\Pi_i(t,\vec{x}), \Phi_l(t,\vec{y})]\Phi_j(t,\vec{x})\Big)$$

$$\overset{(1.52)}{=} -t_{a,ij}t_{b,kl}\int d^3x d^3y \Big(\Pi_i(t,\vec{x})i\delta^3(\vec{x}-\vec{y})\delta_{jk}\Phi_l(t,\vec{y})$$

$$+ \Pi_k(t,\vec{y})[-i\delta^3(\vec{x}-\vec{y})]\delta_{il}\Phi_j(t,\vec{x})\Big)$$

$$= -it_{a,ij}t_{b,kl}\int d^3x \Big(\Pi_i(t,\vec{x})\Phi_l(t,\vec{x})\delta_{jk} - \Pi_k(t,\vec{x})\Phi_j(t,\vec{x})\delta_{il}\Big)$$

$$= -i\int d^3x \Big(\Pi_i(t,\vec{x})t_{a,ij}t_{b,jl}\Phi_l(t,\vec{x}) - \Pi_k(t,\vec{x})t_{b,kl}t_{a,lj}\Phi_j(t,\vec{x})\Big)$$

$$= -i(t_{a,ij}t_{b,jk} - t_{b,ij}t_{a,jk})\int d^3x \Pi_i(t,\vec{x})\Phi_k(t,\vec{x}).$$

1.14 (a) In the following, the ellipses refer to terms of higher order in ε.

$$\bar{\Psi}(i\slashed{\partial} - m_N)\Psi$$

$$\mapsto \bar{\Psi}\left[\mathbb{1} + i\varepsilon_j(x)\frac{\tau_j}{2}\right](i\gamma^\mu\partial_\mu - m_N)\left\{\left[\mathbb{1} - i\varepsilon_i(x)\frac{\tau_i}{2}\right]\Psi\right\}$$

$$= \bar{\Psi}\left[\mathbb{1} + i\varepsilon_j(x)\frac{\tau_j}{2}\right]\left[\mathbb{1} - i\varepsilon_i(x)\frac{\tau_i}{2}\right](i\slashed{\partial} - m_N)\Psi + \bar{\Psi}\gamma^\mu\partial_\mu\varepsilon_i(x)\frac{\tau_i}{2}\Psi + \cdots$$

$$= \bar{\Psi}(i\slashed{\partial} - m_N)\Psi + \partial_\mu\varepsilon_i(x)\bar{\Psi}\gamma^\mu\frac{\tau_i}{2}\Psi + \cdots,$$

$$\frac{1}{2}\big(\partial_\mu\Phi_i\partial^\mu\Phi_i - M_\pi^2\Phi_i\Phi_i\big)$$

$$\mapsto \frac{1}{2}\big\{\partial_\mu[\Phi_i + \varepsilon_{iaj}\varepsilon_a(x)\Phi_j]\partial^\mu[\Phi_i + \varepsilon_{ibk}\varepsilon_b(x)\Phi_k]$$

$$- M_\pi^2[\Phi_i + \varepsilon_{iaj}\varepsilon_a(x)\Phi_j][\Phi_i + \varepsilon_{ibk}\varepsilon_b(x)\Phi_k]\big\}$$

$$= \frac{1}{2}\big(\partial_\mu\Phi_i\partial^\mu\Phi_i - M_\pi^2\Phi_i\Phi_i\big) + \varepsilon_{iaj}\partial_\mu\varepsilon_a(x)\Phi_j\partial^\mu\Phi_i + \cdots,$$

$$= \frac{1}{2}\big(\partial_\mu\Phi_i\partial^\mu\Phi_i - M_\pi^2\Phi_i\Phi_i\big) + \varepsilon_{ijk}\partial_\mu\varepsilon_i(x)\Phi_j\partial^\mu\Phi_k + \cdots,$$

$$-ig\bar{\Psi}\gamma_5\vec{\Phi}\cdot\vec{\tau}\Psi$$

$$\mapsto -ig\bar{\Psi}\left[\mathbb{1}+i\varepsilon_b(x)\frac{\tau_b}{2}\right]\gamma_5\left[\Phi_i+\varepsilon_{icj}\varepsilon_c(x)\Phi_j\right]\tau_i\left[\mathbb{1}-i\varepsilon_a(x)\frac{\tau_a}{2}\right]\Psi$$

$$=-ig\bar{\Psi}\gamma_5\vec{\Phi}\cdot\vec{\tau}\Psi$$

$$\quad -ig\bar{\Psi}\gamma_5\left[\Phi_i\tau_i(-i)\varepsilon_a(x)\frac{\tau_a}{2}+\varepsilon_{icj}\varepsilon_c(x)\Phi_j\tau_i+i\varepsilon_b(x)\frac{\tau_b}{2}\Phi_i\tau_i\right]\Psi+\cdots$$

$$=-ig\bar{\Psi}\gamma_5\vec{\Phi}\cdot\vec{\tau}\Psi-ig\bar{\Psi}\gamma_5\left[-\frac{i}{2}\Phi_i\varepsilon_a(x)(\tau_i\tau_a-\tau_a\tau_i)+\varepsilon_{iaj}\varepsilon_a(x)\Phi_j\tau_i\right]\Psi+\cdots$$

$$=-ig\bar{\Psi}\gamma_5\vec{\Phi}\cdot\vec{\tau}\Psi-ig\bar{\Psi}\gamma_5\left[\Phi_i\varepsilon_a(x)\varepsilon_{iaj}\tau_j+\varepsilon_{iaj}\varepsilon_a(x)\Phi_j\tau_i\right]\Psi+\cdots$$

$$=-ig\bar{\Psi}\gamma_5\vec{\Phi}\cdot\vec{\tau}\Psi+\cdots.$$

The variation of the Lagrangian thus reads

$$\delta\mathscr{L}=\partial_\mu\varepsilon_i(x)\left(\bar{\Psi}\gamma^\mu\frac{\tau_i}{2}\Psi+\varepsilon_{ijk}\Phi_j\partial^\mu\Phi_k\right).$$

(b)

$$[ab,cd]=abcd-cdab$$

$$=abcd+\underbrace{acbd-acbd}_{=0}\underbrace{-acdb+acdb}_{=0}+\underbrace{cadb-cadb}_{=0}-cdab$$

$$=\underbrace{abcd+acbd}_{=a\{b,c\}d}\underbrace{-acdb-acdb}_{=-ac\{b,d\}}+\underbrace{acdb+cadb}_{=\{a,c\}db}\underbrace{-cadb-cdab}_{=-c\{a,d\}b}$$

$$=a\{b,c\}d-ac\{b,d\}+\{a,c\}db-c\{a,d\}b.$$

$$\left[\Psi^\dagger_{\alpha,r}(t,\vec{x})\Psi_{\beta,s}(t,\vec{x}),\Psi^\dagger_{\gamma,t}(t,\vec{y})\Psi_{\delta,u}(t,\vec{y})\right]$$

$$\overset{(1.79,1.80,1.86)}{=}\quad \Psi^\dagger_{\alpha,r}(t,\vec{x})\left\{\Psi_{\beta,s}(t,\vec{x}),\Psi^\dagger_{\gamma,t}(t,\vec{y})\right\}\Psi_{\delta,u}(t,\vec{y})$$

$$\quad -\Psi^\dagger_{\gamma,t}(t,\vec{y})\left\{\Psi^\dagger_{\alpha,r}(t,\vec{x}),\Psi_{\delta,u}(t,\vec{y})\right\}\Psi_{\beta,s}(t,\vec{x})$$

$$\overset{(1.78)}{=}\quad \Psi^\dagger_{\alpha,r}(t,\vec{x})\Psi_{\delta,u}(t,\vec{y})\delta^3(\vec{x}-\vec{y})\delta_{\beta\gamma}\delta_{st}-\Psi^\dagger_{\gamma,t}(t,\vec{y})\Psi_{\beta,s}(t,\vec{x})\delta^3(\vec{x}-\vec{y})\delta_{\alpha\delta}\delta_{ru}.$$

(c)

$$A_{ij}=\int d^3x\mathbb{1}_{\alpha\beta}\left(\frac{\tau_i}{2}\right)_{rs}\mathbb{1}_{\gamma\delta}\left(\frac{\tau_j}{2}\right)_{tu}\left(\Psi^\dagger_{\alpha,r}(x)\delta_{\beta\gamma}\delta_{st}\Psi_{\delta,u}(x)-\Psi^\dagger_{\gamma,t}(x)\delta_{\alpha\delta}\delta_{ru}\Psi_{\beta,s}(x)\right)$$

$$=\int d^3x\Psi^\dagger(x)\left(\frac{\tau_i}{2}\frac{\tau_j}{2}-\frac{\tau_j}{2}\frac{\tau_i}{2}\right)\Psi(x)$$

$$=i\varepsilon_{ijk}\int d^3x\Psi^\dagger(x)\frac{\tau_k}{2}\Psi(x).$$

(d)

$$[ab, cd] = abcd - cdab,$$
$$= abcd \underbrace{- acbd + acbd}_{= 0} \underbrace{- acdb + acdb}_{= 0} \underbrace{- cadb + cadb}_{= 0} - cdab$$
$$= \underbrace{abcd - acbd}_{= a[b,c]d} + \underbrace{acbd - acdb}_{= ac[b,d]} + \underbrace{acdb - cadb}_{= [a,c]db} + \underbrace{cadb - cdab}_{= c[a,d]b}$$
$$= a\,[b,c]d + ac[b,d] + [a,c]db + c[a,d]b.$$

(e)

$$[\Phi_k(t,\vec{x})\Pi_l(t,\vec{x}), \Phi_m(t,\vec{y})\Pi_n(t,\vec{y})]$$

$$\overset{(1.82,1.83,1.88)}{=} \quad \Phi_k(t,\vec{x})[\Pi_l(t,\vec{x}), \Phi_m(t,\vec{y})]\Pi_n(t,\vec{y}) + \Phi_m(t,\vec{y})[\Phi_k(t,\vec{x}), \Pi_n(t,\vec{y})]\Pi_l(t,\vec{x})$$

$$\overset{(1.81)}{=} \quad -i\Phi_k(t,\vec{x})\Pi_n(t,\vec{y})\delta^3(\vec{x} - \vec{y})\delta_{lm} + i\Phi_m(t,\vec{y})\Pi_l(t,\vec{x})\delta^3(\vec{x} - \vec{y})\delta_{kn}.$$

(f)

$$B_{ij} = -i\varepsilon_{ikl}\varepsilon_{jmn} \int d^3x [\Phi_k(x)\Pi_n(x)\delta_{lm} - \Phi_m(x)\Pi_l(x)\delta_{kn}]$$
$$= -i \int d^3x [\Phi_k(x)\Pi_n(x)(\delta_{in}\delta_{kj} - \delta_{ij}\delta_{kn}) - \Phi_m(x)\Pi_l(x)(\delta_{lj}\delta_{im} - \delta_{lm}\delta_{ij})]$$
$$= -i \int d^3x [\Phi_j(x)\Pi_i(x) - \delta_{ij}\Phi_k(x)\Pi_k(x) - \Phi_i(x)\Pi_j(x) + \delta_{ij}\Phi_m(x)\Pi_m(x)]$$
$$= i\varepsilon_{ijk} \int d^3x \varepsilon_{klm}\Phi_l(x)\Pi_m(x).$$

We made use of

$$\varepsilon_{ijk}\varepsilon_{klm} = \delta_{il}\delta_{jm} - \delta_{im}\delta_{jl}.$$

1.15 The λ matrices are suppressed as they are not relevant for the argument.

$$V^\mu = \bar{q}_R\gamma^\mu q_R + \bar{q}_L\gamma^\mu q_L \overset{(1.38)}{=} \bar{q}\gamma^\mu q,$$
$$A^\mu = \bar{q}_R\gamma^\mu q_R - \bar{q}_L\gamma^\mu q_L = \bar{q}\frac{1}{2}(\mathbb{1} - \gamma_5)\gamma^\mu q_R - \bar{q}\frac{1}{2}(\mathbb{1} + \gamma_5)\gamma^\mu q_L$$
$$= \bar{q}\gamma^\mu \underbrace{\frac{1}{2}(\mathbb{1} + \gamma_5)q_R}_{= \frac{1}{2}(\mathbb{1} + \gamma_5)q} - \bar{q}\gamma^\mu \underbrace{\frac{1}{2}(\mathbb{1} - \gamma_5)q_L}_{= \frac{1}{2}(\mathbb{1} - \gamma_5)q} = \bar{q}\gamma^\mu\gamma_5 q.$$

1.16 Make use of $P_R^2 = P_R$ and $P_L P_R = P_R P_L = 0$.

$[Q_{Ra}, Q_{Rb}]$

$$= \int d^3x\, d^3y \left[q^\dagger(t,\vec{x}) P_R \frac{\lambda_a}{2} q(t,\vec{x}), q^\dagger(t,\vec{y}) P_R \frac{\lambda_b}{2} q(t,\vec{y}) \right]$$

$$= \int d^3x\, d^3y\, \delta^3(\vec{x}-\vec{y}) \left(q^\dagger(t,\vec{x}) P_R \frac{\lambda_a}{2} \frac{\lambda_b}{2} q(t,\vec{y}) - q^\dagger(t,\vec{y}) P_R \frac{\lambda_b}{2} \frac{\lambda_a}{2} q(t,\vec{x}) \right)$$

$$= \int d^3x\, q^\dagger(t,\vec{x}) P_R \left[\frac{\lambda_a}{2}, \frac{\lambda_b}{2} \right] q(t,\vec{x})$$

$$= i f_{abc} \int d^3x\, q^\dagger(t,\vec{x}) P_R \frac{\lambda_c}{2} q(t,\vec{x})$$

$$= i f_{abc} Q_{Rc},$$

$[Q_{La}, Q_{Rb}]$

$$= \int d^3x\, d^3y \left[q^\dagger(t,\vec{x}) P_L \frac{\lambda_a}{2} q(t,\vec{x}), q^\dagger(t,\vec{y}) P_R \frac{\lambda_b}{2} q(t,\vec{y}) \right]$$

$$= \int d^3x\, d^3y\, \delta^3(\vec{x}-\vec{y}) \left(q^\dagger(t,\vec{x}) \underbrace{P_L P_R}_{=0} \frac{\lambda_a}{2} \frac{\lambda_b}{2} q(t,\vec{y}) - q^\dagger(t,\vec{y}) \underbrace{P_R P_L}_{=0} \frac{\lambda_b}{2} \frac{\lambda_a}{2} q(t,\vec{x}) \right)$$

$$= 0,$$

$[Q_{L,Ra}, Q_V]$

$$= \int d^3x\, d^3y \left[q^\dagger(t,\vec{x}) P_{L,R} \frac{\lambda_a}{2} q(t,\vec{x}), q^\dagger(t,\vec{y}) q(t,\vec{y}) \right]$$

$$= \int d^3x\, d^3y\, \delta^3(\vec{x}-\vec{y}) \left(q^\dagger(t,\vec{x}) P_{L,R} \frac{\lambda_a}{2} q(t,\vec{y}) - q^\dagger(t,\vec{y}) P_{L,R} \frac{\lambda_a}{2} q(t,\vec{x}) \right)$$

$$= 0.$$

1.17 Since $\mathcal{M}$ is diagonal, we only need to consider λ_0, λ_3, and λ_8:

$$\mathcal{M} = m_0 \lambda_0 + m_3 \lambda_3 + m_8 \lambda_8,$$

$$m_0 = \frac{1}{2} \text{Tr}(\lambda_0 \mathcal{M}) = \frac{1}{2}\sqrt{\frac{2}{3}}(m_u + m_d + m_s) = \frac{m_u + m_d + m_s}{\sqrt{6}},$$

$$m_3 = \frac{1}{2} \text{Tr}(\lambda_3 \mathcal{M}) = \frac{m_u - m_d}{2},$$

$$m_8 = \frac{1}{2} \text{Tr}(\lambda_8 \mathcal{M}) = \frac{1}{2}\frac{1}{\sqrt{3}}(m_u + m_d - 2m_s) = \frac{\frac{m_u + m_d}{2} - m_s}{\sqrt{3}}.$$

1.18

$$T[\Phi(x)J^\mu(y)\Phi^\dagger(z)] =$$
$$\Phi(x)J^\mu(y)\Phi^\dagger(z)\Theta(x_0 - y_0)\Theta(y_0 - z_0) + \Phi(x)\Phi^\dagger(z)J^\mu(y)\Theta(x_0 - z_0)\Theta(z_0 - y_0)$$
$$+ J^\mu(y)\Phi(x)\Phi^\dagger(z)\Theta(y_0 - x_0)\Theta(x_0 - z_0) + \Phi^\dagger(z)\Phi(x)J^\mu(y)\Theta(z_0 - x_0)\Theta(x_0 - y_0)$$
$$+ J^\mu(y)\Phi^\dagger(z)\Phi(x)\Theta(y_0 - z_0)\Theta(z_0 - x_0) + \Phi^\dagger(z)J^\mu(y)\Phi(x)\Theta(z_0 - y_0)\Theta(y_0 - x_0).$$

$$\partial_\mu^y T[\Phi(x)J^\mu(y)\Phi^\dagger(z)] = T[\Phi(x)\underbrace{\partial_\mu^y J^\mu(y)}_{=0}\Phi^\dagger(z)]$$

$$- \Phi(x)J^0(y)\Phi^\dagger(z)\delta(x_0 - y_0)\Theta(y_0 - z_0) + \Phi(x)J^0(y)\Phi^\dagger(z)\Theta(x_0 - y_0)\delta(y_0 - z_0)$$
$$- \Phi(x)\Phi^\dagger(z)J^0(y)\Theta(x_0 - z_0)\delta(z_0 - y_0) + J^0(y)\Phi(x)\Phi^\dagger(z)\delta(y_0 - x_0)\Theta(x_0 - z_0)$$
$$- \Phi^\dagger(z)\Phi(x)J^0(y)\Theta(z_0 - x_0)\delta(x_0 - y_0) + J^0(y)\Phi^\dagger(z)\Phi(x)\delta(y_0 - z_0)\Theta(z_0 - x_0)$$
$$- \Phi^\dagger(z)J^0(y)\Phi(x)\delta(z_0 - y_0)\Theta(y_0 - x_0) + \Phi^\dagger(z)J^0(y)\Phi(x)\Theta(z_0 - y_0)\delta(y_0 - z_0)$$
$$= [-\Phi(x)J^0(y)\delta(x_0 - y_0)\Theta(y_0 - z_0) + J^0(y)\Phi(x)\delta(y_0 - x_0)\Theta(x_0 - z_0)]\Phi^\dagger(z)$$
$$+ \Phi(x)[J^0(y)\Phi^\dagger(z)\Theta(x_0 - y_0)\delta(y_0 - z_0) - \Phi^\dagger(z)J^0(y)\Theta(x_0 - z_0)\delta(z_0 - y_0)]$$
$$+ \Phi^\dagger(z)[-\Phi(x)J^0(y)\Theta(z_0 - y_0)\delta(x_0 - y_0) + J^0(y)\Phi(x)\Theta(z_0 - y_0)\delta(y_0 - z_0)]$$
$$+ [J^0(y)\Phi^\dagger(z)\delta(y_0 - z_0)\Theta(z_0 - x_0) - \Phi^\dagger(z)J^0(y)\delta(z_0 - y_0)\Theta(y_0 - x_0)]\Phi(x)$$
$$= \delta^4(y - x)\Phi(x)\Phi^\dagger(z)\Theta(x_0 - z_0) - \delta^4(y - z)\Phi(x)\Phi^\dagger(z)\Theta(x_0 - z_0)$$
$$+ \delta^4(y - x)\Phi^\dagger(z)\Phi(x)\Theta(z_0 - x_0) - \delta^4(y - z)\Phi^\dagger(z)\Phi(x)\Theta(z_0 - x_0)$$
$$= \delta^4(y - x)T[\Phi(x)\Phi^\dagger(z)] - \delta^4(y - z)T[\Phi(x)\Phi^\dagger(z)].$$

$$\Rightarrow \partial_\mu^y G^\mu(x, y, z) = [\delta^4(y - x) - \delta^4(y - z)]\langle 0|T[\Phi(x)\Phi^\dagger(z)]|0\rangle.$$

1.19 (a) Make use of Eq. 1.38:

$$\bar{q}\gamma^\mu q = \bar{q}_R\gamma^\mu q_R + \bar{q}_L\gamma^\mu q_L,$$
$$\bar{q}\gamma^\mu\gamma_5 q = \bar{q}_R\gamma^\mu \underbrace{\gamma_5 q_R}_{= q_R} + \bar{q}_L\gamma^\mu \underbrace{\gamma_5 q_L}_{= -q_L} = \bar{q}_R\gamma^\mu q_R - \bar{q}_L\gamma^\mu q_L.$$

Thus

$$\bar{q}\gamma^\mu\left(v_\mu + \frac{1}{3}v_\mu^{(s)} + \gamma_5 a_\mu\right)q = \bar{q}_R\gamma^\mu\left(v_\mu + a_\mu + \frac{1}{3}v_\mu^{(s)}\right)q_R$$
$$+ \bar{q}_L\gamma^\mu\left(v_\mu - a_\mu + \frac{1}{3}v_\mu^{(s)}\right)q_L$$
$$= \bar{q}_R\gamma^\mu\left(r_\mu + \frac{1}{3}v_\mu^{(s)}\right)q_R + \bar{q}_L\gamma^\mu\left(l_\mu + \frac{1}{3}v_\mu^{(s)}\right)q_L.$$

(b) Make use of Eq. 1.38:

$$\bar{q}q = \bar{q}_R q_L + \bar{q}_L q_R,$$
$$\bar{q}\gamma_5 q = \bar{q}_R \underbrace{\gamma_5 q_L}_{= -q_L} + \bar{q}_L \underbrace{\gamma_5 q_R}_{= q_R} = -\bar{q}_R q_L + \bar{q}_L q_R.$$

Thus

$$\bar{q}(s - i\gamma_5 p)q = \bar{q}_L(s - ip)q_R + \bar{q}_R(s + ip)q_L.$$

1.20 Using

$$q_L \mapsto \left(\mathbb{1} - i\frac{\varepsilon_{Li}\tau_i}{2}\right)q_L, \quad q_R \mapsto \left(\mathbb{1} - i\frac{\varepsilon_{Ri}\tau_i}{2}\right)q_R,$$

one obtains

$$\delta\mathcal{L}_{\text{ext}} = i\bar{q}_R\gamma^\mu\left[\frac{\varepsilon_{Ri}\tau_i}{2}, v_\mu\right]q_R + i\bar{q}_L\gamma^\mu\left[\frac{\varepsilon_{Li}\tau_i}{2}, v_\mu\right]q_L$$
$$+ i\bar{q}_R\gamma^\mu\gamma_5\left[\frac{\varepsilon_{Ri}\tau_i}{2}, a_\mu\right]q_R + i\bar{q}_L\gamma^\mu\gamma_5\left[\frac{\varepsilon_{Li}\tau_i}{2}, a_\mu\right]q_L$$
$$- i\bar{q}_R\frac{\varepsilon_{Ri}\tau_i}{2}sq_L + i\bar{q}_L s\frac{\varepsilon_{Ri}\tau_i}{2}q_R - i\bar{q}_L\frac{\varepsilon_{Li}\tau_i}{2}sq_R + i\bar{q}_R s\frac{\varepsilon_{Li}\tau_i}{2}q_L$$
$$- \bar{q}_R\gamma_5\frac{\varepsilon_{Ri}\tau_i}{2}pq_L + \bar{q}_L\gamma_5 p\frac{\varepsilon_{Ri}\tau_i}{2}q_R - \bar{q}_L\gamma_5\frac{\varepsilon_{Li}\tau_i}{2}pq_R + \bar{q}_R\gamma_5 p\frac{\varepsilon_{Li}\tau_i}{2}q_L,$$

$$\partial_\mu R_i^\mu = \frac{\partial\delta\mathcal{L}_{\text{ext}}}{\partial\varepsilon_{Ri}}$$
$$= i\bar{q}_R\gamma^\mu\left[\frac{\tau_i}{2}, v_\mu\right]q_R + i\bar{q}_R\gamma^\mu\gamma_5\left[\frac{\tau_i}{2}, a_\mu\right]q_R$$
$$- i\bar{q}_R\frac{\tau_i}{2}sq_L + i\bar{q}_L s\frac{\tau_i}{2}q_R - \bar{q}_R\gamma_5\frac{\tau_i}{2}pq_L + \bar{q}_L\gamma_5 p\frac{\tau_i}{2}q_R,$$

$$\partial_\mu L_i^\mu = \frac{\partial\delta\mathcal{L}_{\text{ext}}}{\partial\varepsilon_{Li}}$$
$$= i\bar{q}_L\gamma^\mu\left[\frac{\tau_i}{2}, v_\mu\right]q_L + i\bar{q}_L\gamma^\mu\gamma_5\left[\frac{\tau_i}{2}, a_\mu\right]q_L$$
$$- i\bar{q}_L\frac{\tau_i}{2}sq_R + i\bar{q}_R s\frac{\tau_i}{2}q_L - \bar{q}_L\gamma_5\frac{\tau_i}{2}pq_R + \bar{q}_R\gamma_5 p\frac{\tau_i}{2}q_L.$$

For the divergence of the vector current make use of Eq. 1.38:

$$\partial_\mu V_i^\mu = \partial_\mu R_i^\mu + \partial_\mu L_i^\mu$$
$$= i\bar{q}\gamma^\mu\left[\frac{\tau_i}{2}, v_\mu\right]q + i\bar{q}\gamma^\mu\gamma_5\left[\frac{\tau_i}{2}, a_\mu\right]q - i\bar{q}\left[\frac{\tau_i}{2}, s\right]q - \bar{q}\gamma_5\left[\frac{\tau_i}{2}, p\right]q.$$

For the divergence of the axial-vector current make use of $\gamma_5 q_R = q_R$ and $\gamma_5 q_L = -q_L$ and of Eq. 1.38 to rewrite

$$\bar{q}_R\gamma^\mu q_R - \bar{q}_L\gamma^\mu q_L = \bar{q}\gamma^\mu\gamma_5 q,$$
$$\bar{q}_R\gamma^\mu\gamma_5 q_R - \bar{q}_L\gamma^\mu\gamma_5 q_L = \bar{q}\gamma^\mu q,$$
$$\bar{q}_R q_L - \bar{q}_L q_R = -\bar{q}\gamma_5 q,$$
$$\bar{q}_R\gamma_5 q_L - \bar{q}_L\gamma_5 q_R = -\bar{q}q.$$

Thus

$$\partial_\mu A_i^\mu = \partial_\mu R_i^\mu - \partial_\mu L_i^\mu$$
$$= i\bar{q}\gamma^\mu\gamma_5\left[\frac{\tau_i}{2}, v_\mu\right]q + i\bar{q}\gamma^\mu\left[\frac{\tau_i}{2}, a_\mu\right]q + i\bar{q}\gamma_5\left\{\frac{\tau_i}{2}, s\right\}q + \bar{q}\left\{\frac{\tau_i}{2}, p\right\}q.$$

1.21 Insert

$$v_\mu = -e\mathscr{A}_\mu\frac{\tau_3}{2}, \quad a_\mu = 0, \quad s = \hat{m}\mathbb{1}, \quad p = 0,$$

into Eqs. 1.169 and 1.170, respectively, and make use of

$$\left[\frac{\tau_i}{2}, v_\mu\right] = -e\mathscr{A}_\mu\left[\frac{\tau_i}{2}, \frac{\tau_3}{2}\right] = ie\mathscr{A}_\mu\varepsilon_{3ij}\frac{\tau_j}{2},$$

to obtain

$$\partial_\mu V_i^\mu = -\varepsilon_{3ij}e\mathscr{A}_\mu\bar{q}\gamma^\mu\frac{\tau_j}{2}q = -\varepsilon_{3ij}e\mathscr{A}_\mu V_j^\mu,$$
$$\partial_\mu A_i^\mu = -e\mathscr{A}_\mu\varepsilon_{3ij}\bar{q}\gamma^\mu\gamma_5\frac{\tau_j}{2}q + 2\hat{m}i\bar{q}\gamma_5\frac{\tau_i}{2}q = -e\mathscr{A}_\mu\varepsilon_{3ij}A_j^\mu + \hat{m}P_i.$$

Problems of Chapter 2

2.1

$$\tilde{\mathscr{V}} = \frac{m^2}{2}(\pm\Phi_0 + \Phi')^2 + \frac{\lambda}{4}(\pm\Phi_0 + \Phi')^4$$
$$= \frac{m^2}{2}(\Phi_0^2 \pm 2\Phi_0\Phi' + \Phi'^2) + \frac{\lambda}{4}(\Phi_0^4 \pm 4\Phi_0^3\Phi' + 6\Phi_0^2\Phi'^2 \pm 4\Phi_0\Phi'^3 + \Phi'^4).$$

Make use of $\Phi_0^2 = -m^2/\lambda$ and consider powers of Φ':

$$\sim \text{const}: \frac{m^2}{2}\Phi_0^2 + \frac{\lambda}{4}\Phi_0^4 = -\frac{\lambda}{2}\Phi_0^4 + \frac{\lambda}{4}\Phi_0^4 = -\frac{\lambda}{4}\Phi_0^4,$$

$$\sim \Phi': \frac{m^2}{2}(\pm 2\Phi_0) + \frac{\lambda}{4}(\pm 4\Phi_0^3) = \pm m^2\Phi_0 \pm \lambda\Phi_0^3 = \mp\lambda\Phi_0^3 \pm \lambda\Phi_0^3 = 0,$$

$$\sim \Phi'^2: \frac{m^2}{2} + \frac{\lambda}{4}6\Phi_0^2 = \frac{m^2}{2} - \frac{3m^2}{2} = \frac{1}{2}(-2m^2).$$

$$\Rightarrow \tilde{\mathscr{V}} = -\frac{\lambda}{4}\Phi_0^4 + \frac{1}{2}(-2m^2)\Phi'^2 \pm \lambda\Phi_0\Phi'^3 + \frac{\lambda}{4}\Phi'^4.$$

2.2 Define $x = \Phi_i\Phi_i \geq 0$ and determine minimum of

$$V(x) = \frac{m^2}{2}x + \frac{\lambda}{4}x^2.$$

$$V'(x) = \frac{m^2}{2} + \frac{\lambda}{2}x = 0 \quad \Rightarrow \quad x = -\frac{m^2}{\lambda}.$$

The solution is indeed a minimum, because $V''(x) = \frac{\lambda}{2} > 0$. As a result, constant fields with $|\vec{\Phi}_{\min}| = \sqrt{\frac{-m^2}{\lambda}}$ minimize $\mathscr{V}(\Phi_1, \Phi_2, \Phi_3)$.

2.3 Insert $\Phi_3 = v + \eta$ into the potential,

$$\tilde{\mathscr{V}} = \frac{m^2}{2}\left[\Phi_1^2 + \Phi_2^2 + (\eta + v)^2\right] + \frac{\lambda}{4}\left[\Phi_1^2 + \Phi_2^2 + (\eta + v)^2\right]^2$$

$$= \frac{m^2}{2}(\Phi_1^2 + \Phi_2^2 + \eta^2 + 2\eta v + v^2) + \frac{\lambda}{4}(\Phi_1^2 + \Phi_2^2 + \eta^2 + 2\eta v + v^2)^2$$

$$= \frac{m^2}{2}(\Phi_1^2 + \Phi_2^2 + \eta^2 + 2\eta v + v^2) + \frac{\lambda}{4}(\Phi_1^2 + \Phi_2^2 + \eta^2)^2$$

$$+ \frac{\lambda}{4}\left[2(\Phi_1^2 + \Phi_2^2 + \eta^2)(2\eta v + v^2) + (2\eta v + v^2)^2\right].$$

Make use of $v^2 = -m^2/\lambda$ and consider powers of $\Phi_1^2 + \Phi_2^2$ and η:

$$\sim \Phi_1^2 + \Phi_2^2 : \frac{m^2}{2} + \frac{\lambda}{2}v^2 = 0,$$

$$\sim \eta^2 : \frac{m^2}{2} + \frac{\lambda}{2}v^2 + \lambda v^2 = -m^2,$$

$$\sim \eta : m^2 v + \lambda v^3 = 0,$$

$$\sim \text{const} : \frac{m^2}{2}v^2 + \frac{\lambda}{4}v^4 = -\frac{\lambda}{4}v^4.$$

$$\tilde{\mathscr{V}} = \frac{1}{2}(-2m^2)\eta^2 + \lambda v\eta(\Phi_1^2 + \Phi_2^2 + \eta^2) + \frac{\lambda}{4}(\Phi_1^2 + \Phi_2^2 + \eta^2)^2 - \frac{\lambda}{4}v^4.$$

2.4 For notational convenience rename $\Phi_3 \to \Phi$, $\Phi_3^{(0)} \to \Phi_0$, and $\Phi_3^{(1)} \to \Phi_1$. The criterion for an extremum reads

$$\mathscr{V}'(\Phi) = \lambda\Phi^3 + m^2\Phi + a = 0.$$

Ansatz $\langle\Phi\rangle = \Phi_0 + a\Phi_1 + O(a^2) \Rightarrow$

$$\lambda \underbrace{[\Phi_0 + a\Phi_1 + O(a^2)]^3}_{= \Phi_0^3 + 3a\Phi_0^2\Phi_1 + O(a^2)} + m^2[\Phi_0 + a\Phi_1 + O(a^2)] + a = 0.$$

$$O(a^0) : \lambda\Phi_0^3 + m^2\Phi_0 = 0.$$

$\Phi_0 = 0$ corresponds to maximum and $\Phi_0 = \pm\sqrt{-m^2/\lambda}$ correspond to the two minima without explicit symmetry breaking.

$$O(a) : 3\lambda a\Phi_0^2\Phi_1 + m^2 a\Phi_1 + a = 0,$$

or, using $\Phi_0^2 = -m^2/\lambda$,

$$-3m^2\Phi_1 + m^2\Phi_1 + 1 = 0 \quad\Rightarrow\quad \Phi_1 = \frac{1}{2m^2}.$$

For both values of Φ_0 corresponding to the two minima without explicit symmetry breaking we find

$$\langle\Phi\rangle = \pm\sqrt{\frac{-m^2}{\lambda}} + \frac{a}{2m^2} + O(a^2).$$

We determine the values of the potential by explicit calculation:

$$\begin{aligned}
\mathscr{V}(\langle\Phi\rangle) &= \mathscr{V}\left(\Phi_0 + a\Phi_1 + O(a^2)\right) \\
&= \mathscr{V}(\Phi_0) + a\Phi_1 \underbrace{\mathscr{V}'(\Phi_0)}_{= O(a)} + O(a^2) \\
&= \mathscr{V}(\Phi_0) + O(a^2) \\
&= \mathscr{V}_0(\Phi_0) + a\Phi_0 + O(a^2) \\
&= -\frac{1}{4}\frac{(-m^2)^2}{\lambda} \pm \sqrt{\frac{-m^2}{\lambda}}a + O(a^2).
\end{aligned}$$

Problems of Chapter 3

3.1 (a) The Feynman rule for a vertex is obtained by evaluating $i\mathscr{L}_{\text{int}}$ between states. Plane waves, normalization factors, and (possibly) spinors are dropped at the end. The rule may also be applied to momenta which do not satisfy mass-shell conditions.

$$\langle\varphi|i\mathscr{L}_{\text{int}}(x)|\varphi,\phi\rangle = \left\langle\varphi\left|-i\frac{\lambda}{2}\phi(x)\varphi^2(x)\right|\varphi,\phi\right\rangle \Rightarrow -i\lambda,$$

since two possibilities to contract φ fields.

(b)

$$\mathscr{M}_{(a)} = -i\lambda\frac{i}{(p_1+p_2)^2 - M^2 + i0^+}(-i\lambda) = -\lambda^2\frac{i}{s - M^2 + i0^+},$$

$$\mathscr{M}_{(b)} = -i\lambda\frac{i}{(p_1-p_3)^2 - M^2 + i0^+}(-i\lambda) = -\lambda^2\frac{i}{t - M^2 + i0^+},$$

$$\mathscr{M}_{(c)} = -i\lambda\frac{i}{(p_1-p_4)^2 - M^2 + i0^+}(-i\lambda) = -\lambda^2\frac{i}{u - M^2 + i0^+}.$$

(c) For $\{s,|t|,|u|\} \ll M^2$ we can expand

$$\frac{i}{\{s,t,u\} - M^2 + i0^+} = -\frac{i}{M^2}\frac{1}{1 - \frac{\{s,t,u\}}{M^2} - i0^+} = -\frac{i}{M^2}\left[1 + O\left(\frac{\{s,t,u\}}{M^2}\right)\right].$$

(d) Feynman rule:

$$\langle\varphi\varphi|i\tilde{\lambda}\varphi^4(x)|\varphi\varphi\rangle \Rightarrow i(4!)\tilde{\lambda} = i(4!)\frac{\lambda^2}{8M^2} = \frac{3i\lambda^2}{M^2},$$

from which one can read off $\mathscr{M}_{\text{eff}}$.

3.2

$$\begin{aligned}
[Q_{Va}, Q_{Vb}] &= [Q_{Ra} + Q_{La}, Q_{Rb} + Q_{Lb}]\\
&= [Q_{Ra}, Q_{Rb}] + [Q_{Ra}, Q_{Lb}] + [Q_{La}, Q_{Rb}] + [Q_{La}, Q_{Lb}]\\
&= if_{abc}Q_{Rc} + 0 + 0 + if_{abc}Q_{Lc}\\
&= if_{abc}(Q_{Rc} + Q_{Lc})\\
&= if_{abc}Q_{Vc}.
\end{aligned}$$

3.3

$$\begin{aligned}
[Q_{Va}, Q_{Ab}] &= [Q_{Ra} + Q_{La}, Q_{Rb} - Q_{Lb}]\\
&= [Q_{Ra}, Q_{Rb}] - [Q_{Ra}, Q_{Lb}] + [Q_{La}, Q_{Rb}] - [Q_{La}, Q_{Lb}]\\
&= if_{abc}Q_{Rc} + 0 + 0 - if_{abc}Q_{Lc}\\
&= if_{abc}(Q_{Rc} - Q_{Lc})\\
&= if_{abc}Q_{Ac},
\end{aligned}$$

$$[Q_{Aa}, Q_{Ab}] = [Q_{Ra} - Q_{La}, Q_{Rb} - Q_{Lb}]$$
$$= [Q_{Ra}, Q_{Rb}] - [Q_{Ra}, Q_{Lb}] - [Q_{La}, Q_{Rb}] + [Q_{La}, Q_{Lb}]$$
$$= if_{abc}Q_{Rc} + 0 + 0 + if_{abc}Q_{Lc}$$
$$= if_{abc}(Q_{Rc} + Q_{Lc})$$
$$= if_{abc}Q_{Vc}.$$

3.4 Note: $\gamma_5 P_R = P_R$ and $\gamma_5 P_L = -P_L$.

$$S_a = \bar{q}\lambda_a q = \bar{q}_R \lambda_a q_L + \bar{q}_L \lambda_a q_R$$
$$\mapsto \bar{q}_R U_R^\dagger \lambda_a U_L q_L + \bar{q}_L U_L^\dagger \lambda_a U_R q_R,$$
$$P_a = i\bar{q}\gamma_5 \lambda_a q = i\bar{q}_R \gamma_5 \lambda_a q_L + i\bar{q}_L \gamma_5 \lambda_a q_R$$
$$= -i\bar{q}_R \lambda_a q_L + i\bar{q}_L \lambda_a q_R$$
$$\mapsto -i\bar{q}_R U_R^\dagger \lambda_a U_L q_L + i\bar{q}_L U_L^\dagger \lambda_a U_R q_R.$$

3.5

$$\phi = \sum_{a=1}^{8} \phi_a \lambda_a = \begin{pmatrix} \phi_3 + \frac{1}{\sqrt{3}}\phi_8 & \phi_1 - i\phi_2 & \phi_4 - i\phi_5 \\ \phi_1 + i\phi_2 & -\phi_3 + \frac{1}{\sqrt{3}}\phi_8 & \phi_6 - i\phi_7 \\ \phi_4 + i\phi_5 & \phi_6 + i\phi_7 & -\frac{2}{\sqrt{3}}\phi_8 \end{pmatrix}$$
$$\equiv \begin{pmatrix} \pi^0 + \frac{1}{\sqrt{3}}\eta & \sqrt{2}\pi^+ & \sqrt{2}K^+ \\ \sqrt{2}\pi^- & -\pi^0 + \frac{1}{\sqrt{3}}\eta & \sqrt{2}K^0 \\ \sqrt{2}K^- & \sqrt{2}\bar{K}^0 & -\frac{2}{\sqrt{3}}\eta \end{pmatrix},$$

i.e.

$$\pi^0 = \phi_3, \qquad\qquad \eta = \phi_8,$$
$$\pi^+ = \frac{1}{\sqrt{2}}(\phi_1 - i\phi_2), \quad \pi^- = \frac{1}{\sqrt{2}}(\phi_1 + i\phi_2),$$
$$K^+ = \frac{1}{\sqrt{2}}(\phi_4 - i\phi_5), \quad K^- = \frac{1}{\sqrt{2}}(\phi_4 + i\phi_5),$$
$$K^0 = \frac{1}{\sqrt{2}}(\phi_6 - i\phi_7), \quad \bar{K}^0 = \frac{1}{\sqrt{2}}(\phi_6 + i\phi_7).$$

3.6

$$\mathrm{Tr}(\partial_\mu U U^\dagger)$$
$$= \mathrm{Tr}\left[\left(i\frac{\partial_\mu \phi}{F_0} + \frac{i\partial_\mu \phi i\phi + i\phi i\partial_\mu \phi}{2F_0^2} + \frac{i\partial_\mu \phi(i\phi)^2 + i\phi i\partial_\mu \phi i\phi + (i\phi)^2 i\partial_\mu \phi}{3!F_0^3} + \cdots\right)U^\dagger\right]$$
$$= \mathrm{Tr}\left[i\frac{\partial_\mu \phi}{F_0}U^\dagger + \frac{i\partial_\mu \phi i\phi + i\phi i\partial_\mu \phi}{2F_0^2}U^\dagger + \frac{i\partial_\mu \phi(i\phi)^2 + i\phi i\partial_\mu \phi i\phi + (i\phi)^2 i\partial_\mu \phi}{3!F_0^3}U^\dagger + \cdots\right]$$

$$\overset{*}{=} \mathrm{Tr}\left[i\frac{\partial_\mu\phi}{F_0}U^\dagger + i\frac{\partial_\mu\phi}{F_0}i\frac{\phi}{F_0}U^\dagger + i\frac{\partial_\mu\phi}{F_0}\frac{1}{2}\frac{(i\phi)^2}{F_0^2}U^\dagger + \cdots\right]$$

$$= \mathrm{Tr}\left[i\frac{\partial_\mu\phi}{F_0}\left(\mathbb{1} + i\frac{\phi}{F_0} + \frac{1}{2}\frac{(i\phi)^2}{F_0^2} + \cdots\right)U^\dagger\right]$$

$$= \mathrm{Tr}\left[i\frac{\partial_\mu\phi}{F_0}UU^\dagger\right]$$

$$= \mathrm{Tr}\left[i\frac{\partial_\mu\phi}{F_0}\right]$$

$$= \frac{i\partial_\mu\phi_a}{F_0}\mathrm{Tr}(\Lambda_a)$$

$$= 0.$$

$*$: $[\phi, U^\dagger] = 0$ and cyclic property of trace.

3.7 (a)

$$\mathscr{L}_{\mathrm{s.b.}} = \frac{F_0^2 B_0}{2}\mathrm{Tr}\left(\mathscr{M}U^\dagger + U\mathscr{M}^\dagger\right) = \frac{F_0^2 B_0}{2}\mathrm{Tr}\left[\mathscr{M}(U^\dagger + U)\right],$$

where

$$U^\dagger = \mathbb{1} - i\frac{\phi}{F_0} - \frac{\phi^2}{2F_0^2} + i\frac{\phi^3}{6F_0^3} + \frac{\phi^4}{24F_0^4} + \cdots,$$

$$U = \mathbb{1} + i\frac{\phi}{F_0} - \frac{\phi^2}{2F_0^2} - i\frac{\phi^3}{6F_0^3} + \frac{\phi^4}{24F_0^4} + \cdots,$$

$$U^\dagger + U = 2\left(\mathbb{1} - \frac{\phi^2}{2F_0^2} + \frac{\phi^4}{24F_0^4} + \cdots\right).$$

With $\phi = \phi_a\lambda_a$:

$$\frac{\partial}{\partial\phi_a}\phi^2 = \lambda_a\phi + \phi\lambda_a$$

and analogously for ϕ^4 term.

(b)

$$\frac{1}{F_0^2}(\lambda_a\phi + \phi\lambda_a) = \frac{1}{F_0^2}\left[\lambda_a\left(\phi_0 + \frac{1}{F_0^2}\phi_2 + \cdots\right) + \left(\phi_0 + \frac{1}{F_0^2}\phi_2 + \cdots\right)\lambda_a\right]$$

$$= \frac{1}{F_0^2}(\lambda_a\phi_0 + \phi_0\lambda_a) + \frac{1}{F_0^4}(\lambda_a\phi_2 + \phi_2\lambda_a) + O\left(\frac{1}{F_0^6}\right),$$

$$\frac{1}{F_0^4}(\lambda_a\phi^3 + \phi\lambda_a\phi^2 + \phi^2\lambda_a\phi + \phi^3\lambda_a) = \frac{1}{F_0^4}(\lambda_a\phi_0^3 + \phi_0\lambda_a\phi_0^2 + \phi_0^2\lambda_a\phi_0 + \phi_0^3\lambda_a)$$

$$+ O\left(\frac{1}{F_0^6}\right).$$

(c)

$$\mathrm{Tr}[\mathscr{M}(\lambda_a\phi_0 + \phi_0\lambda_a)] = \mathrm{Tr}[\mathscr{M}(\lambda_a\lambda_b + \lambda_b\lambda_a)]\phi_{0b}$$
$$= \mathrm{Tr}\left[\mathscr{M}\left(\frac{4}{3}\mathbb{1}\delta_{ab} + 2d_{abc}\lambda_c\right)\right]\phi_{0b}$$
$$= \frac{4}{3}(m_u + m_d + m_s)\phi_{0a} + 4(m_3 d_{ab3} + m_8 d_{ab8})\phi_{0b},$$

where we used $\mathrm{Tr}(\lambda_a) = 0, \mathrm{Tr}(\lambda_a\lambda_b) = 2\delta_{ab}$. Set $a = 1$ and use $d_{1b3} = 0, d_{1b8} = d_{118}\delta_{b1} = \frac{1}{\sqrt{3}}\delta_{b1}$:

$$\cdots = \left[\frac{4}{3}(m_u + m_d + m_s) + \frac{4}{\sqrt{3}}m_8\right]\phi_{01} = 2(m_u + m_d)\phi_{01}.$$

Since we want ϕ to minimize $\mathscr{H}_{\mathrm{eff}}$, we set $\mathrm{Tr}(\ldots) = 0 \Rightarrow 2(m_u + m_d)\phi_{01} = 0$. For nonvanishing quark masses therefore $\phi_{01} = 0$.

(d) With $\phi_0 = 0$

$$\mathrm{Tr}\left[\mathscr{M}\left(\lambda_a\phi_2 + \phi_2\lambda_a - \frac{\lambda_a\phi_0^3 + \phi_0\lambda_a\phi_0^2 + \phi_0^2\lambda_a\phi_0 + \phi_0^3\lambda_a}{12}\right)\right]$$
$$= \mathrm{Tr}[\mathscr{M}(\lambda_a\phi_2 + \phi_2\lambda_a)],$$

and calculation for ϕ_2 as above for ϕ_0.

3.8

$$\mathscr{L}_{\mathrm{s.b.}} = \frac{F_0^2 B_0}{2}\mathrm{Tr}(\mathscr{M}U^\dagger + U\mathscr{M}^\dagger)$$
$$= \frac{F_0^2 B_0}{2}\mathrm{Tr}[\mathscr{M}(U^\dagger + U)]$$
$$= \frac{F_0^2 B_0}{2}\mathrm{Tr}\left[\mathscr{M}\left(\mathbb{1} - \frac{\phi^2}{F_0^2} + \cdots\right)\right]$$
$$= \frac{F_0^2 B_0(m_u + m_d + m_s)}{2} - \frac{B_0}{2}\mathrm{Tr}(\mathscr{M}\phi^2) + \cdots.$$

We need

$$\mathrm{Tr}(\mathscr{M}\phi^2) = m_u(\phi^2)_{11} + m_d(\phi^2)_{22} + m_s(\phi^2)_{33}.$$

$$(\phi^2)_{11} = \left(\phi_3 + \frac{1}{\sqrt{3}}\phi_8\right)^2 + (\phi_1 - i\phi_2)(\phi_1 + i\phi_2) + (\phi_4 - i\phi_5)(\phi_4 + i\phi_5)$$
$$= \phi_3^2 + \frac{2}{\sqrt{3}}\phi_3\phi_8 + \frac{1}{3}\phi_8^2 + \phi_1^2 + \phi_2^2 + \phi_4^2 + \phi_5^2,$$

$$(\phi^2)_{22} = (\phi_1 + i\phi_2)(\phi_1 - i\phi_2) + \left(-\phi_3 + \frac{1}{\sqrt{3}}\phi_8\right)^2 + (\phi_6 - i\phi_7)(\phi_6 + i\phi_7)$$
$$= \phi_1^2 + \phi_2^2 + \phi_3^2 - \frac{2}{\sqrt{3}}\phi_3\phi_8 + \frac{1}{3}\phi_8^2 + \phi_6^2 + \phi_7^2,$$

$$(\phi^2)_{33} = (\phi_4 + i\phi_5)(\phi_4 - i\phi_5) + (\phi_6 + i\phi_7)(\phi_6 - i\phi_7) + \frac{4}{3}\phi_8^2$$
$$= \phi_4^2 + \phi_5^2 + \phi_6^2 + \phi_7^2 + \frac{4}{3}\phi_8^2.$$

Note: One could also start with ϕ^2 expressed in terms of physical fields.

$$-\frac{B_0}{2}\mathrm{Tr}\left(\mathscr{M}\phi^2\right)$$
$$= -\frac{B_0}{2}\left[(m_u + m_d)(\phi_1^2 + \phi_2^2 + \phi_3^2) + \frac{2}{\sqrt{3}}(m_u - m_d)\phi_3\phi_8\right.$$
$$\left. + (m_u + m_s)(\phi_4^2 + \phi_5^2) + (m_d + m_s)(\phi_6^2 + \phi_7^2) + \frac{1}{3}(m_u + m_d + 4m_s)\phi_8^2\right]$$
$$= -B_0(m_u + m_d)\pi^+\pi^- - \frac{1}{2}B_0(m_u + m_d)\pi^0\pi^0 - \frac{B_0}{\sqrt{3}}(m_u - m_d)\pi^0\eta$$
$$- B_0(m_u + m_s)K^+K^- - B_0(m_d + m_s)K^0\bar{K}^0 - \frac{1}{2}B_0\frac{m_u + m_d + 4m_s}{3}\eta^2.$$

3.9

$$D_\mu A \equiv \partial_\mu A - ir_\mu A + iAl_\mu$$
$$\mapsto \partial_\mu(V_R A V_L^\dagger) - i(V_R r_\mu V_R^\dagger + iV_R \partial_\mu V_R^\dagger)V_R A V_L^\dagger + iV_R A V_L^\dagger(V_L l_\mu V_L^\dagger + iV_L \partial_\mu V_L^\dagger)$$
$$= \partial_\mu V_R A V_L^\dagger + V_R \partial_\mu A V_L^\dagger + V_R A \partial_\mu V_L^\dagger$$
$$- iV_R r_\mu A V_L^\dagger + \underbrace{V_R \partial_\mu V_R^\dagger V_R A V_L^\dagger}_{= -\partial_\mu V_R A V_L^\dagger} + iV_R A l_\mu V_L^\dagger - V_R A \partial_\mu V_L^\dagger$$
$$= V_R(\partial_\mu A - ir_\mu A + iAl_\mu)V_L^\dagger$$
$$= V_R(D_\mu A)V_L^\dagger.$$

3.10
$$\mathrm{Tr}[(D_\mu D_\nu U)U^\dagger] = \mathrm{Tr}[(\partial_\mu(D_\nu U) - ir_\mu(D_\nu U) + i(D_\nu U)l_\mu)U^\dagger]$$
$$= \partial_\mu \underbrace{\mathrm{Tr}[(D_\nu U)U^\dagger]}_{= 0} - \mathrm{Tr}[(D_\nu U)\partial_\mu U^\dagger]$$
$$- i\mathrm{Tr}[r_\mu(D_\nu U)U^\dagger] + i\mathrm{Tr}[(D_\nu U)l_\mu U^\dagger]$$
$$= -\mathrm{Tr}[(D_\nu U)(\partial_\mu U^\dagger + iU^\dagger r_\mu - il_\mu U^\dagger)]$$
$$= -\mathrm{Tr}[(D_\nu U)(D_\mu U)^\dagger]$$
$$= -\mathrm{Tr}[(\partial_\nu U - ir_\nu U + iUl_\nu)(D_\mu U)^\dagger]$$
$$= -\partial_\nu \underbrace{\mathrm{Tr}[U(D_\mu U)^\dagger]}_{= 0} + \mathrm{Tr}(U\partial_\nu(D_\mu U)^\dagger + (ir_\nu U - iUl_\nu)(D_\mu U)^\dagger]$$
$$= \mathrm{Tr}[U(D_\nu D_\mu U)^\dagger].$$

3.11 (a)

$$\phi = \begin{pmatrix} \pi^0 + \frac{1}{\sqrt{3}}\eta & \sqrt{2}\pi^+ & \sqrt{2}K^+ \\ \sqrt{2}\pi^- & -\pi^0 + \frac{1}{\sqrt{3}}\eta & \sqrt{2}K^0 \\ \sqrt{2}K^- & \sqrt{2}\bar{K}^0 & -\frac{2}{\sqrt{3}}\eta \end{pmatrix}$$

$$\mapsto \begin{pmatrix} \pi^0 + \frac{1}{\sqrt{3}}\eta & \sqrt{2}\pi^- & \sqrt{2}K^- \\ \sqrt{2}\pi^+ & -\pi^0 + \frac{1}{\sqrt{3}}\eta & \sqrt{2}\bar{K}^0 \\ \sqrt{2}K^+ & \sqrt{2}K^0 & -\frac{2}{\sqrt{3}}\eta \end{pmatrix}$$

$$= \phi^T.$$

(b) Verify: $(A^T)^n = (A^n)^T$.

$$n = 1 : A^T = A^T, \checkmark$$
$$n \to n+1 : (A^T)^{n+1} = (A^T)^n A^T = (A^n)^T A^T = (AA^n)^T = (A^{n+1})^T.$$

$$U = \exp\left(i\frac{\phi}{F_0}\right) = \sum_{n=0}^{\infty} \frac{1}{n!}\left(\frac{i}{F_0}\right)^n \phi^n$$

$$\mapsto \sum_{n=0}^{\infty} \frac{1}{n!}\left(\frac{i}{F_0}\right)^n (\phi^T)^n = \sum_{n=0}^{\infty} \frac{1}{n!}\left(\frac{i}{F_0}\right)^n (\phi^n)^T = U^T.$$

(c)

$$r_\mu = v_\mu + a_\mu \mapsto -v_\mu^T + a_\mu^T = -l_\mu^T,$$
$$l_\mu = v_\mu - a_\mu \mapsto -v_\mu^T - a_\mu^T = -r_\mu^T,$$
$$\chi = 2B_0(s + ip) \mapsto \chi^T,$$
$$\chi^\dagger = 2B_0(s - ip) \mapsto \chi^{\dagger^T}.$$

(d)

$$D_\mu U = \partial_\mu U - ir_\mu U + iUl_\mu$$
$$\mapsto \partial_\mu U^T - i(-l_\mu^T)U^T + iU^T(-r_\mu^T)$$
$$= \partial_\mu U^T + i(Ul_\mu)^T - i(r_\mu U)^T$$
$$= (D_\mu U)^T.$$

(e)

$$\text{Tr}\Big[D_\mu U (D^\mu U)^\dagger\Big] \mapsto \text{Tr}\Big\{(D_\mu U)^T \big[(D^\mu U)^T\big]^\dagger\Big\}$$

$$= \text{Tr}\Big\{(D_\mu U)^T \big[(D^\mu U)^\dagger\big]^T\Big\}$$

$$= \text{Tr}\Big\{\big[(D^\mu U)^\dagger D_\mu U\big]^T\Big\}$$

$$= \text{Tr}\Big[(D^\mu U)^\dagger D_\mu U\Big]$$

$$= \text{Tr}\Big[D_\mu U (D^\mu U)^\dagger\Big],$$

$$\text{Tr}\big(\chi U^\dagger + U\chi^\dagger\big) \mapsto \text{Tr}\Big(\chi^T U^{\dagger T} + U^T \chi^{\dagger T}\Big)$$

$$= \text{Tr}\Big(\chi^T U^{\dagger T} + U^T \chi^{\dagger T}\Big)$$

$$= \text{Tr}\Big[(U^\dagger \chi)^T + (\chi^\dagger U)^T\Big]$$

$$= \text{Tr}(U^\dagger \chi + \chi^\dagger U)$$

$$= \text{Tr}(\chi U^\dagger + U\chi^\dagger).$$

(f)

$$\text{Tr}\Big[D_\mu U (D^\mu U)^\dagger D_\nu U (D^\nu U)^\dagger\Big]$$

$$\mapsto \text{Tr}\Big[(D_\mu U)^T (D^\mu U)^{T\dagger} (D_\nu U)^T (D^\nu U)^{T\dagger}\Big]$$

$$= \text{Tr}\Big[(D_\mu U)^T (D^\mu U)^{\dagger T} (D_\nu U)^T (D^\nu U)^{\dagger T}\Big]$$

$$= \text{Tr}\Big[\big((D^\nu U)^\dagger D_\nu U (D^\mu U)^\dagger D_\mu U\big)^T\Big]$$

$$= \text{Tr}\Big[(D^\nu U)^\dagger D_\nu U (D^\mu U)^\dagger D_\mu U\Big]$$

$$= \text{Tr}\Big[(D_\mu U)^\dagger D^\mu U (D_\nu U)^\dagger D^\nu U\Big]$$

$$= \text{Tr}\Big[-U^\dagger (D_\mu U) U^\dagger D^\mu U (-U^\dagger)(D_\nu U) U^\dagger D^\nu U\Big]$$

$$= \text{Tr}\Big[(D_\mu U) U^\dagger (D^\mu U) U^\dagger (D_\nu U) U^\dagger (D^\nu U) U^\dagger\Big]$$

$$= \text{Tr}\Big[D_\mu U (D^\mu U)^\dagger D_\nu U (D^\nu U)^\dagger\Big].$$

3.12

$$\frac{F_0^2}{4}\text{Tr}\Big[D_\mu U (D^\mu U)^\dagger\Big] = \frac{F_0^2}{4}\text{Tr}\big[(\partial_\mu U + iUl_\mu)(\partial^\mu U^\dagger - il^\mu U^\dagger)\big]$$

$$= \cdots + i\frac{F_0^2}{4}\text{Tr}(Ul_\mu \partial^\mu U^\dagger - l^\mu \underbrace{U^\dagger \partial_\mu U}_{= -\partial_\mu U^\dagger U}) + \cdots$$

$$= i\frac{F_0^2}{2}\text{Tr}(l_\mu \partial^\mu U^\dagger U) + \cdots.$$

3.13 (a)

$$\bar{u}_{\nu_\mu}(p_\nu)\,\not{p}_\nu = 0, \quad (*)$$

$$\{\gamma_\rho, \gamma_5\} = 0, \quad (**)$$

$$\not{p}_\mu v_{\mu^+}(p_\mu, s_\mu) = -m_\mu v_{\mu^+}(p_\mu, s_\mu). \quad (***)$$

$$\bar{u}_{\nu_\mu}(p_\nu)(p_\nu + p_\mu)_\rho \gamma^\rho (1 - \gamma_5) v_{\mu^+}(p_\mu, s_\mu)$$

$$\stackrel{(*)}{=} \bar{u}_{\nu_\mu}(p_\nu) p_{\mu\rho} \gamma^\rho (1 - \gamma_5) v_{\mu^+}(p_\mu, s_\mu)$$

$$\stackrel{(**)}{=} \bar{u}_{\nu_\mu}(p_\nu)(1 + \gamma_5) p_{\mu\rho} \gamma^\rho v_{\mu^+}(p_\mu, s_\mu)$$

$$\stackrel{(***)}{=} -m_\mu \bar{u}_{\nu_\mu}(p_\nu)(1 + \gamma_5) v_{\mu^+}(p_\mu, s_\mu).$$

(b)

$$[\bar{u}_{\nu_\mu}(p_\nu)(p_\nu + p_\mu)_\rho \gamma^\rho (1 - \gamma_5) v_{\mu^+}(p_\mu, s_\mu)]$$

$$\times [\bar{u}_{\nu_\mu}(p_\nu)(p_\nu + p_\mu)_\sigma \gamma^\sigma (1 - \gamma_5) v_{\mu^+}(p_\mu, s_\mu)]^*$$

$$= m_\mu^2 \bar{u}_{\nu_\mu}(p_\nu)(1 + \gamma_5) v_{\mu^+}(p_\mu, s_\mu) \bar{v}_{\mu^+}(p_\mu, s_\mu)(1 - \gamma_5) u_{\nu_\mu}(p_\nu)$$

$$= m_\mu^2 \mathrm{Tr}[u_{\nu_\mu}(p_\nu) \bar{u}_{\nu_\mu}(p_\nu)(1 + \gamma_5) v_{\mu^+}(p_\mu, s_\mu) \bar{v}_{\mu^+}(p_\mu, s_\mu)(1 - \gamma_5)]$$

$$= m_\mu^2 \mathrm{Tr}\left[(1 - \gamma_5)\not{p}_\nu(1 + \gamma_5)(\not{p}_\mu - m_\mu)\frac{1 + \gamma_5 \not{s}_\mu}{2}\right]$$

$$= m_\mu^2 \mathrm{Tr}\left[2\not{p}_\nu(1 + \gamma_5)(\not{p}_\mu - m_\mu)\frac{1 + \gamma_5 \not{s}_\mu}{2}\right]$$

$$= m_\mu^2 \mathrm{Tr}\left[\not{p}_\nu(1 + \gamma_5)(\not{p}_\mu - m_\mu)(1 + \gamma_5 \not{s}_\mu)\right]$$

$$\stackrel{*}{=} m_\mu^2 \mathrm{Tr}\left[\not{p}_\nu(1 + \gamma_5)(\not{p}_\mu - m_\mu \gamma_5 \not{s}_\mu)\right]$$

$$= m_\mu^2 \mathrm{Tr}\left[\not{p}_\nu \not{p}_\mu - m_\mu \not{p}_\nu \gamma_5 \not{s}_\mu + \not{p}_\nu \gamma_5 \not{p}_\mu - m_\mu \not{p}_\nu \gamma_5 \gamma_5 \not{s}_\mu\right]$$

$$= m_\mu^2 \mathrm{Tr}\left[\not{p}_\nu \not{p}_\mu - m_\mu \not{p}_\nu \not{s}_\mu\right]$$

$$= 4m_\mu^2(p_\nu \cdot p_\mu - m_\mu p_\nu \cdot s_\mu).$$

$*$: Only even number of gamma matrices.
Make use of four-momentum conservation:

$$p = p_\mu + p_\nu \Rightarrow p^2 = M_\pi^2 = p_\mu^2 + 2p_\mu \cdot p_\nu + p_\nu^2 = m_\mu^2 + 2p_\mu \cdot p_\nu$$

$$\Rightarrow p_\mu \cdot p_\nu = \frac{M_\pi^2 - m_\mu^2}{2}.$$

$$\cdots = 4m_\mu^2 M_\pi^2 \left[\frac{1}{2}\left(1 - \frac{m_\mu^2}{M_\pi^2}\right) - \frac{m_\mu p_\nu \cdot s_\mu}{M_\pi^2}\right].$$

(c) Sum over muon spins:

$$\sum_{\pm s_\mu} \left[\frac{1}{2} \left(1 - \frac{m_\mu^2}{M_\pi^2} \right) - \frac{m_\mu p_\nu \cdot s_\mu}{M_\pi^2} \right] = 1 - \frac{m_\mu^2}{M_\pi^2}.$$

The first term does not depend on the spin projections and simply yields a factor of 2. The second term adds up to zero. This is most easily seen in the rest frame of the muon ($p_\nu \cdot s_\mu$ is a Lorentz scalar) where the spin four-vector is given by $s_{\mu R} = (0, \vec{s}_{\mu R})$:

$$\sum_{\pm s_\mu} p_\nu \cdot s_\mu = - \sum_{\pm \vec{s}_{\mu R}} \vec{p}_{\nu R} \cdot \vec{s}_{\mu R} = 0.$$

Integration with respect to the unobserved neutrino yields

$$d\Gamma = \frac{1}{8\pi^2} G_F^2 V_{ud}^2 F_0^2 m_\mu^2 M_\pi \left(1 - \frac{m_\mu^2}{M_\pi^2} \right) \int \frac{d^3 p_\mu}{E_\mu E_\nu} \delta(M_\pi - E_\mu - E_\nu).$$

Now consider

$$\int \frac{d^3 p_\mu}{E_\mu E_\nu} \delta(M_\pi - E_\mu - E_\nu) = \int \frac{p_\mu^2 dp_\mu d\Omega_\mu}{E_\mu E_\nu} \delta\left(M_\pi - \sqrt{m_\mu^2 + p_\mu^2} - p_\mu \right).$$

Make use of

$$\int dx f(x) \delta(g(x)) = \sum_i \frac{f(x_i)}{|g'(x_i)|},$$

where $g(x)$ is supposed to have only simple zeroes for $x = x_i$. Here,

$$g(x) = M_\pi - \sqrt{m_\mu^2 + x^2} - x.$$

The zero is given by

$$x_0 = \frac{M_\pi^2 - m_\mu^2}{2M_\pi},$$

with

$$g'(x_0) = -\frac{x_0}{E_\mu(x_0)} - 1.$$

$$\int \frac{p_\mu^2 dp_\mu d\Omega_\mu}{E_\mu E_\nu} \delta\left(M_\pi - \sqrt{m_\mu^2 + p_\mu^2} - p_\mu\right)$$

$$= 4\pi \frac{x_0^2}{E_\mu(x_0) x_0} \frac{E_\mu(x_0)}{E_\mu(x_0) + x_0} = \frac{4\pi x_0}{E_\mu(x_0) + x_0} = 4\pi \frac{M_\pi^2 - m_\mu^2}{2M_\pi} \frac{1}{M_\pi} = 2\pi\left(1 - \frac{m_\mu^2}{M_\pi^2}\right).$$

The final result for the decay rate is given in Eq. 3.89.

3.14

(a) We need to investigate the behavior under $\phi \mapsto -\phi$ or $U \leftrightarrow U^\dagger$:

$$\mathcal{L}_2 = \frac{F^2}{4}\mathrm{Tr}\left(\partial_\mu U \partial^\mu U^\dagger\right) + \frac{F^2 B\hat{m}}{2}\mathrm{Tr}\left(U^\dagger + U\right)$$

$$\overset{U \leftrightarrow U^\dagger}{\mapsto} \frac{F^2}{4}\mathrm{Tr}\left(\partial_\mu U^\dagger \partial^\mu U\right) + \frac{F^2 B\hat{m}}{2}\mathrm{Tr}\left(U + U^\dagger\right) = \mathcal{L}_2.$$

(b)

$$U = \mathbb{1} + i\frac{\phi}{F} - \frac{1}{2}\frac{\phi^2}{F^2} - \frac{i}{6}\frac{\phi^3}{F^3} + \frac{1}{24}\frac{\phi^4}{F^4} + \cdots,$$

$$U^\dagger = \mathbb{1} - i\frac{\phi}{F} - \frac{1}{2}\frac{\phi^2}{F^2} + \frac{i}{6}\frac{\phi^3}{F^3} + \frac{1}{24}\frac{\phi^4}{F^4} + \cdots.$$

We need to collect the terms containing 4 fields. Contribution from second term:

$$\frac{F^2 B}{2}\mathrm{Tr}\left[\mathcal{M}\left(\frac{1}{24}\frac{\phi^4}{F^4} + \frac{1}{24}\frac{\phi^4}{F^4}\right)\right] = \frac{B}{24F^2}\mathrm{Tr}(\mathcal{M}\phi^4).$$

Since both $\partial_\mu U$ and $\partial^\mu U^\dagger$ are $O(\phi)$, we only need their expansion to third order:

$$\partial_\mu U = i\frac{\partial_\mu \phi}{F} - \frac{1}{2}\frac{\partial_\mu \phi \phi + \phi \partial_\mu \phi}{F^2} - \frac{i}{6}\frac{\partial_\mu \phi \phi^2 + \phi \partial_\mu \phi \phi + \phi^2 \partial_\mu \phi}{F^3} + \cdots,$$

$$\partial^\mu U^\dagger = -i\frac{\partial^\mu \phi}{F} - \frac{1}{2}\frac{\partial^\mu \phi \phi + \phi \partial^\mu \phi}{F^2} + \frac{i}{6}\frac{\partial^\mu \phi \phi^2 + \phi \partial^\mu \phi \phi + \phi^2 \partial^\mu \phi}{F^3} + \cdots.$$

We obtain

$$\frac{F^2}{4}\mathrm{Tr}\left(\partial_\mu U \partial^\mu U^\dagger\right)$$

$$= \frac{F^2}{4}\mathrm{Tr}\left[\left(i\frac{\partial_\mu \phi}{F} - \frac{1}{2}\frac{\partial_\mu \phi \phi + \phi \partial_\mu \phi}{F^2} - \frac{i}{6}\frac{\partial_\mu \phi \phi^2 + \phi \partial_\mu \phi \phi + \phi^2 \partial_\mu \phi}{F^3} + \cdots\right)\right.$$

$$\left. \times \left(-i\frac{\partial^\mu \phi}{F} - \frac{1}{2}\frac{\partial^\mu \phi \phi + \phi \partial^\mu \phi}{F^2} + \frac{i}{6}\frac{\partial^\mu \phi \phi^2 + \phi \partial^\mu \phi \phi + \phi^2 \partial^\mu \phi}{F^3} + \cdots\right)\right]$$

$$= \cdots + \frac{1}{4F^2} \text{Tr} \left[-\frac{1}{6} (\partial_\mu \phi \partial^\mu \phi \phi^2 + \partial_\mu \phi \phi \partial^\mu \phi \phi + \partial_\mu \phi \phi^2 \partial^\mu \phi) \right.$$

$$+ \frac{1}{4} (\partial_\mu \phi \phi + \phi \partial_\mu \phi)(\partial^\mu \phi \phi + \phi \partial^\mu \phi)$$

$$\left. - \frac{1}{6} (\partial_\mu \phi \phi^2 \partial^\mu \phi + \phi \partial_\mu \phi \phi \partial^\mu \phi + \phi^2 \partial_\mu \phi \partial^\mu \phi) \right] + \cdots .$$

Under the trace two distinct orderings:

$$-\frac{1}{6} - \frac{1}{6} + \frac{1}{4} + \frac{1}{4} - \frac{1}{6} - \frac{1}{6} = -\frac{1}{6},$$

$$-\frac{1}{6} + \frac{1}{4} + \frac{1}{4} - \frac{1}{6} = \frac{1}{6}.$$

$$= \cdots + \frac{1}{24F^2} \text{Tr}([\phi, \partial_\mu \phi] \phi \partial^\mu \phi) + \cdots = \frac{1}{48F^2} \text{Tr}([\phi, \partial_\mu \phi][\phi, \partial^\mu \phi]) + \cdots .$$

(c) Insert $\phi = \phi_i \tau_i$. Making use of

$$[\phi, \partial_\mu \phi] = 2i\varepsilon_{ijk} \phi_i \partial_\mu \phi_j \tau_k,$$

$$\text{Tr}(\tau_k \tau_n) = 2\delta_{kn},$$

$$\varepsilon_{ijk} \varepsilon_{lmk} = \delta_{il} \delta_{jm} - \delta_{im} \delta_{jl},$$

$$\phi^2 = \phi_i \phi_i,$$

we obtain

$$\mathcal{L}_2^{4\phi} = -\frac{1}{6F^2} \varepsilon_{ijk} \phi_i \partial_\mu \phi_j \varepsilon_{lmk} \phi_l \partial^\mu \phi_m + \frac{M^2}{24F^2} \phi_i \phi_i \phi_j \phi_j$$

$$= \frac{1}{6F^2} (\phi_i \partial^\mu \phi_i \partial_\mu \phi_j \phi_j - \phi_i \phi_i \partial_\mu \phi_j \partial^\mu \phi_j) + \frac{M^2}{24F^2} \phi_i \phi_i \phi_j \phi_j,$$

where $M^2 = 2B\hat{m}$.

(d) The Feynman rule for Cartesian isospin indices a, b, c, and d is obtained from "$i\mathcal{L}$". For example,

$$\langle p_c, c; p_d, d | \phi_i \partial^\mu \phi_i \partial_\mu \phi_j \phi_j | p_a, a; p_b, b \rangle$$

results in 24 combinations of combining 4 fields with 4 quanta, e.g.,

$$\langle p_c, c; p_d, d | \phi_i \partial^\mu \phi_i \partial_\mu \phi_j \phi_j | p_a, a; p_b, b \rangle$$

$$\Rightarrow \delta_{ic} i p_d^\mu \delta_{id} (-i p_{a\mu}) \delta_{ja} \delta_{jb} = p_a \cdot p_d \, \delta_{ab} \delta_{cd}.$$

The complete result reads

$$
\begin{aligned}
\mathcal{M} = i \Bigg[& \frac{1}{6F^2} \Big(2[\delta_{ab}\delta_{cd}(-ip_a - ip_b) \cdot (ip_c + ip_d) \\
& + \delta_{ac}\delta_{bd}(-ip_a + ip_c) \cdot (-ip_b + ip_d) \\
& + \delta_{ad}\delta_{bc}(-ip_a + ip_d) \cdot (-ip_b + ip_c)] \\
& - 4\{\delta_{ab}\delta_{cd}[(-ip_a) \cdot (-ip_b) + (ip_c) \cdot (ip_d)] \\
& + \delta_{ac}\delta_{bd}[(-ip_a) \cdot (ip_c) + (-ip_b) \cdot (ip_d)] \\
& + \delta_{ad}\delta_{bc}[(-ip_a) \cdot (ip_d) + (-ip_b) \cdot (ip_c)]\} \Big) \\
& + \frac{M^2}{24F^2} 8 (\delta_{ab}\delta_{cd} + \delta_{ac}\delta_{bd} + \delta_{ad}\delta_{bc}) \Bigg] \\
= \frac{i}{3F^2} \Big\{ & \delta_{ab}\delta_{cd}[(p_a + p_b)^2 + 2p_a \cdot p_b + 2p_c \cdot p_d + M^2] \\
& + \delta_{ac}\delta_{bd}[(p_a - p_c)^2 - 2p_a \cdot p_c - 2p_b \cdot p_d + M^2] \\
& + \delta_{ad}\delta_{bc}[(p_a - p_d)^2 - 2p_a \cdot p_d - 2p_b \cdot p_c + M^2] \Big\} \\
= \frac{i}{3F^2} \Big[& \delta_{ab}\delta_{cd}(3s - p_a^2 - p_b^2 - p_c^2 - p_d^2 + M^2) \\
& + \delta_{ac}\delta_{bd}(3t - p_a^2 - p_c^2 - p_b^2 - p_d^2 + M^2) \\
& + \delta_{ad}\delta_{bc}(3u - p_a^2 - p_d^2 - p_b^2 - p_c^2 + M^2) \Big] \\
= i \Big[& \delta_{ab}\delta_{cd}\frac{s - M^2}{F^2} + \delta_{ac}\delta_{bd}\frac{t - M^2}{F^2} + \delta_{ad}\delta_{bc}\frac{u - M^2}{F^2} \Big] \\
& - \frac{i}{3F^2}(\delta_{ab}\delta_{cd} + \delta_{ac}\delta_{bd} + \delta_{ad}\delta_{bc})(\Lambda_a + \Lambda_b + \Lambda_c + \Lambda_d),
\end{aligned}
$$

where $\Lambda_k = p_k^2 - M^2$.

(e)

$$
\begin{aligned}
s + t + u &= (p_a + p_b)^2 + (p_a - p_c)^2 + (p_a - p_d)^2 \\
&= p_a^2 + 2p_a \cdot p_b + p_b^2 + p_a^2 - 2p_a \cdot p_c + p_c^2 + p_a^2 - 2p_a \cdot p_d + p_d^2 \\
&= 3p_a^2 + p_b^2 + p_c^2 + p_d^2 + 2p_a \cdot (p_b \underbrace{-p_c - p_d}_{= -p_a - p_b}) \\
&= 3p_a^2 + p_b^2 + p_c^2 + p_d^2 - 2p_a^2 = p_a^2 + p_b^2 + p_c^2 + p_d^2.
\end{aligned}
$$

3.15 At threshold

$$s_{\mathrm{thr}} = (2M_\pi)^2,$$

and thus

$$A(s_{\text{thr}}, t_{\text{thr}}, u_{\text{thr}}) = \frac{3M_\pi^2}{F_\pi^2}.$$

- $I = 0$:

$$
\begin{aligned}
32\pi a_0^0 &= T^{I=0}|_{\text{thr}} \\
&= [3A(s,t,u) + A(t,u,s) + A(u,s,t)]_{\text{thr}} \\
&= [2A(s,t,u) + A(s,t,u) + A(t,u,s) + A(u,s,t)]_{\text{thr}} \\
&= \frac{6M_\pi^2}{F_\pi^2} + \frac{[s+t+u-3M_\pi^2]_{\text{thr}}}{F_\pi^2} \\
&= \frac{7M_\pi^2}{F_\pi^2}.
\end{aligned}
$$

- $I = 2$:

$$
\begin{aligned}
32\pi a_0^2 &= T^{I=2}|_{\text{thr}} \\
&= [A(t,u,s) + A(u,s,t)]_{\text{thr}} \\
&= [A(t,u,s) + A(u,s,t) + A(s,t,u) - A(s,t,u)]_{\text{thr}} \\
&= \frac{M_\pi^2}{F_\pi^2} - \frac{3M_\pi^2}{F_\pi^2} \\
&= -\frac{2M_\pi^2}{F_\pi^2}.
\end{aligned}
$$

3.16 Consider

$$U(x) = \frac{1}{F}[\sigma(x)\mathbb{1} + i\vec{\pi}(x)\cdot\vec{\tau}], \quad \sigma(x) = \sqrt{F^2 - \vec{\pi}^2(x)}.$$

The substitution $\vec{\pi} \mapsto -\vec{\pi}$ corresponds to $U \leftrightarrow U^\dagger$. As in the solution to Exercise 3.14 (a), $\mathcal{L}_2$ is invariant.

In terms of the pion fields, the Lagrangian reads

$$
\begin{aligned}
\mathcal{L}_2 &= \frac{F^2}{4}\text{Tr}\left[\frac{(\partial_\mu\sigma\mathbb{1} + i\partial_\mu\vec{\pi}\cdot\vec{\tau})}{F}\frac{(\partial^\mu\sigma\mathbb{1} - i\partial^\mu\vec{\pi}\cdot\vec{\tau})}{F}\right] \\
&\quad + \frac{F^2 B\hat{m}}{2}\text{Tr}\left(\frac{\sigma\mathbb{1} + i\vec{\pi}\cdot\vec{\tau}}{F} + \frac{\sigma\mathbb{1} - i\vec{\pi}\cdot\vec{\tau}}{F}\right) \\
&= \frac{1}{2}(\partial_\mu\sigma\partial^\mu\sigma + \partial_\mu\vec{\pi}\cdot\partial^\mu\vec{\pi}) + 2B\hat{m}F\sigma.
\end{aligned}
$$

Making use of

$$\partial_\mu \sigma = -\frac{\vec{\pi} \cdot \partial_\mu \vec{\pi}}{\sigma},$$

yields

$$\mathcal{L}_2 = \frac{1}{2}\partial_\mu \vec{\pi} \cdot \partial^\mu \vec{\pi} + \frac{1}{2}\frac{\vec{\pi} \cdot \partial_\mu \vec{\pi}\vec{\pi} \cdot \partial^\mu \vec{\pi}}{F^2 - \vec{\pi}^2} + 2B\hat{m}F^2\sqrt{1 - \frac{\vec{\pi}^2}{F^2}}$$

$$= 2B\hat{m}F^2 + \frac{1}{2}\partial_\mu \vec{\pi} \cdot \partial^\mu \vec{\pi} - B\hat{m}\vec{\pi}^2 + \frac{\vec{\pi} \cdot \partial_\mu \vec{\pi}\vec{\pi} \cdot \partial^\mu \vec{\pi}}{2F^2} - \frac{B\hat{m}(\vec{\pi}^2)^2}{4F^2} + \cdots.$$

Note that the dependence of $\mathcal{L}_2$ on the fields π_i differs from that on the ϕ_i in Exercise 3.14. Nonetheless, both versions generate identical observables. The Feynman rule obtained from

$$\mathcal{L}_2^{4\pi} = \frac{\vec{\pi} \cdot \partial_\mu \vec{\pi}\vec{\pi} \cdot \partial^\mu \vec{\pi}}{2F^2} - \frac{M^2(\vec{\pi}^2)^2}{8F^2}$$

reads

$$\begin{aligned}
\mathcal{M} = i\Big(&\frac{1}{2F^2}\{ip_c \cdot [\delta_{cd}\delta_{ab}(-ip_a - ip_b) + \delta_{ca}\delta_{db}(ip_d - ip_b) + \delta_{cb}\delta_{ad}(ip_d - ip_a)] \\
&+ ip_d \cdot [\delta_{cd}\delta_{ab}(-ip_a - ip_b) + \delta_{da}\delta_{bc}(-ip_b + ip_c) + \delta_{db}\delta_{ac}(-ip_a + ip_c)] \\
&- ip_a \cdot [\delta_{ac}\delta_{bd}(-ip_b + ip_d) + \delta_{ad}\delta_{bc}(-ip_b + ip_c) + \delta_{ab}\delta_{cd}(ip_c + ip_d)] \\
&- ip_b \cdot [\delta_{bc}\delta_{ad}(-ip_a + ip_d) + \delta_{bd}\delta_{ac}(-ip_a + ip_c) + \delta_{ba}\delta_{cd}(ip_c + ip_d)]\} \\
&- \frac{M^2}{8F^2}8(\delta_{ab}\delta_{cd} + \delta_{ac}\delta_{bd} + \delta_{ad}\delta_{bc})\Big) \\
= i\Big(&\frac{1}{2F^2}\{\delta_{ab}\delta_{cd}[p_c \cdot (p_a + p_b) + p_d \cdot (p_a + p_b) + p_a \cdot (p_c + p_d) + p_b \cdot (p_c + p_d)] \\
&+ \delta_{ac}\delta_{bd}[p_c \cdot (p_b - p_d) + p_d \cdot (p_a - p_c) + p_a \cdot (p_d - p_b) + p_b \cdot (p_c - p_d)] \\
&+ \delta_{ad}\delta_{bc}[p_c \cdot (p_a - p_d) + p_d \cdot (p_b - p_c) + p_a \cdot (p_c - p_b) + p_b \cdot (p_d - p_a)]\} \\
&- \frac{M^2}{F^2}(\delta_{ab}\delta_{cd} + \delta_{ac}\delta_{bd} + \delta_{ad}\delta_{bc})\Big) \\
= i\Big[&\delta_{ab}\delta_{cd}\frac{s - M^2}{F^2} + \delta_{ac}\delta_{bd}\frac{t - M^2}{F^2} + \delta_{ad}\delta_{bc}\frac{u - M^2}{F^2}\Big].
\end{aligned}$$

In the last step we made use of momentum conservation, $p_a + p_b = p_c + p_d$, and introduced the Mandelstam variables $s = (p_a + p_b)^2$, $t = (p_a - p_c)^2$, and $u = (p_a - p_d)^2$. Comparison with the result of Exercise 3.14 shows that the invariant amplitudes are identical for on-shell pions.

3.17 The first part is given by

$$\frac{1}{48F_0^2}\mathrm{Tr}([\phi,\partial_\mu\phi][\phi,\partial^\mu\phi]) = \frac{1}{48F_0^2}\phi_a\partial_\mu\phi_b\phi_c\partial^\mu\phi_d\mathrm{Tr}(\underbrace{[\lambda_a,\lambda_b]}_{=2if_{abe}\lambda_e}\underbrace{[\lambda_c,\lambda_d]}_{=2if_{cdf}\lambda_f})$$

$$= -\frac{1}{12F_0^2}\phi_a\partial_\mu\phi_b\phi_c\partial^\mu\phi_d\, f_{abe}f_{cdf}\underbrace{\mathrm{Tr}(\lambda_e\lambda_f)}_{=2\delta_{ef}}$$

$$= -\frac{1}{6F_0^2}\phi_a\partial_\mu\phi_b\phi_c\partial^\mu\phi_d\, f_{abe}f_{cde}.$$

For the second part we assume isospin symmetry. The mass matrix is given by

$$\mathcal{M} = \frac{2\hat{m}+m_s}{3}\mathbb{1} + \frac{\hat{m}-m_s}{\sqrt{3}}\lambda_8.$$

We first consider

$$\mathrm{Tr}(\phi^4) = \phi_a\phi_b\phi_c\phi_d\mathrm{Tr}(\lambda_a\lambda_b\lambda_c\lambda_d) = \phi_a\phi_b\phi_c\phi_d\left(\frac{4}{3}\delta_{ab}\delta_{cd} + 2h_{abe}h_{ecd}\right).$$

Since $\phi_a\phi_b\phi_c\phi_d$ is completely symmetric, only the symmetric parts d of h contribute,

$$\cdots = \phi_a\phi_b\phi_c\phi_d\left(\frac{4}{3}\delta_{ab}\delta_{cd} + 2d_{abe}d_{ecd}\right)$$

$$= \phi_a\phi_b\phi_c\phi_d\left[\frac{4}{3}\delta_{ab}\delta_{cd} + \frac{2}{3}(\delta_{ac}\delta_{bd} + \delta_{ad}\delta_{bc} - \delta_{ab}\delta_{cd} + f_{ace}f_{bde} + f_{ade}f_{bce})\right].$$

Again, the f terms do not contribute due to the complete symmetry of $\phi_a\phi_b\phi_c\phi_d$,

$$\cdots = \frac{2}{3}\phi_a\phi_b\phi_c\phi_d(\delta_{ab}\delta_{cd} + \delta_{ac}\delta_{bd} + \delta_{ad}\delta_{bc}) = 2\phi_a\phi_a\phi_b\phi_b.$$

The second term reads

$$\mathrm{Tr}(\phi^4\lambda_8) = \phi_a\phi_b\phi_c\phi_d\mathrm{Tr}(\lambda_a\lambda_b\lambda_c\lambda_d\lambda_8)$$

$$= \phi_a\phi_b\phi_c\phi_d\left(\frac{4}{3}h_{abc}\delta_{d8} + \frac{4}{3}\delta_{ab}h_{cd8} + 2h_{abe}h_{ecf}h_{fd8}\right).$$

The complete symmetry of $\phi_a\phi_b\phi_c\phi_d$ results in

$$\cdots = \frac{4}{3}d_{abc}\phi_a\phi_b\phi_c\phi_8 + \frac{4}{3}d_{cd8}\phi_a\phi_a\phi_c\phi_d$$
$$+ 2\phi_a\phi_b\phi_c\phi_d d_{abe}(d_{ecf}+if_{ecf})(d_{fd8}+if_{fd8}).$$

Upon contraction with $\phi_a\phi_b\phi_c\phi_d$ we can replace:

$$d_{abe}d_{ecf}d_{fd8} = \frac{1}{3}(\delta_{ac}\delta_{bf}+\delta_{af}\delta_{bc}-\delta_{ab}\delta_{cf}+f_{ace}f_{bfe}+f_{afe}f_{bce})d_{fd8}$$
$$\rightarrow \frac{1}{3}\delta_{ac}d_{bd8}+\frac{1}{3}\delta_{bc}d_{ad8}-\frac{1}{3}\delta_{ab}d_{cd8},$$

$$d_{abe}d_{ecf}f_{fd8} = \frac{1}{3}(\delta_{ac}\delta_{bf}+\delta_{af}\delta_{bc}-\delta_{ab}\delta_{cf}+f_{ace}f_{bfe}+f_{afe}f_{bce})f_{fd8}$$
$$\rightarrow \frac{1}{3}\delta_{ac}f_{bd8}+\frac{1}{3}\delta_{bc}f_{ad8}-\frac{1}{3}\delta_{ab}f_{cd8}\rightarrow 0,$$

$$d_{abe}f_{ecf}d_{fd8} = -d_{abe}f_{efc}d_{fd8} = (d_{bfe}f_{eac}+d_{fae}f_{ebc})d_{fd8}\rightarrow 0,$$
$$d_{abe}f_{ecf}f_{fd8} = -d_{abe}f_{efc}f_{fd8} = (d_{bfe}f_{eac}+d_{fae}f_{ebc})f_{fd8}\rightarrow 0,$$

and obtain

$$\cdots = \frac{4}{3}d_{abc}\phi_a\phi_b\phi_c\phi_8 + \frac{4}{3}d_{cd8}\phi_a\phi_a\phi_c\phi_d$$
$$+ 2\phi_a\phi_b\phi_c\phi_d\left(\frac{1}{3}\delta_{ac}d_{bd8}+\frac{1}{3}\delta_{bc}d_{ad8}-\frac{1}{3}\delta_{ab}d_{cd8}\right)$$
$$= \frac{4}{3}\phi_8\phi_a\phi_b\phi_c d_{abc} + 2\phi_a\phi_a\phi_b\phi_c d_{bc8}.$$

We finally obtain

$$\frac{B_0}{24F_0^2}\mathrm{Tr}(\mathcal{M}\phi^4) = \frac{(2\hat{m}+m_s)B_0}{36F_0^2}\phi_a\phi_a\phi_b\phi_b$$
$$+ \frac{(\hat{m}-m_s)B_0}{12\sqrt{3}F_0^2}\left(\frac{2}{3}\phi_8\phi_a\phi_b\phi_c d_{abc}+\phi_a\phi_a\phi_b\phi_c d_{bc8}\right).$$

3.18 (a)

$$D_\mu U = \partial_\mu U - ir_\mu U + iUl_\mu$$
$$\rightarrow \partial_\mu U - i(-e\mathcal{A}_\mu Q)U + iU(-e\mathcal{A}_\mu Q) = \partial_\mu U + ie\mathcal{A}_\mu[Q,U],$$
$$(D^\mu U)^\dagger \rightarrow (\partial^\mu U + ie\mathcal{A}^\mu[Q,U])^\dagger$$
$$= \partial^\mu U^\dagger - ie\mathcal{A}^{\mu\dagger}([Q,U])^\dagger$$
$$= \partial^\mu U^\dagger - ie\mathcal{A}^\mu(-[Q^\dagger,U^\dagger])$$
$$= \partial^\mu U^\dagger + ie\mathcal{A}^\mu[Q,U^\dagger].$$

Under $U \leftrightarrow U^\dagger$

$$D_\mu U = \partial_\mu U + ie\mathscr{A}_\mu[Q,U] \overset{U \leftrightarrow U^\dagger}{\mapsto} \partial_\mu U^\dagger + ie\mathscr{A}_\mu[Q,U^\dagger] = (D_\mu U)^\dagger,$$

$$(D^\mu U)^\dagger \overset{U^\dagger \leftrightarrow U}{\mapsto} D^\mu U.$$

$$\Rightarrow \frac{F_0^2}{4}\mathrm{Tr}[D_\mu U(D^\mu U)^\dagger] \overset{U \leftrightarrow U^\dagger}{\mapsto} \frac{F_0^2}{4}\mathrm{Tr}[(D_\mu U)^\dagger D^\mu U] = \frac{F_0^2}{4}\mathrm{Tr}[D_\mu U(D^\mu U)^\dagger].$$

(b) $\dfrac{F_0^2}{4}\mathrm{Tr}[D_\mu U(D^\mu U)^\dagger]$

$$= \frac{F_0^2}{4}\mathrm{Tr}\{(\partial_\mu U + ie\mathscr{A}_\mu[Q,U])(\partial^\mu U^\dagger + ie\mathscr{A}^\mu[Q,U^\dagger])\}$$

$$= \frac{F_0^2}{4}\mathrm{Tr}(\partial_\mu U\partial^\mu U^\dagger) + \frac{F_0^2}{4}\mathrm{Tr}(\partial_\mu U ie\mathscr{A}^\mu[Q,U^\dagger] + ie\mathscr{A}_\mu[Q,U]\partial^\mu U^\dagger)$$

$$- \frac{F_0^2}{4}e^2\mathscr{A}_\mu\mathscr{A}^\mu\mathrm{Tr}([Q,U][Q,U^\dagger])$$

$$= \cdots + ie\mathscr{A}_\mu\frac{F_0^2}{4}\mathrm{Tr}(\partial^\mu UQU^\dagger - \partial^\mu UU^\dagger Q + QU\partial^\mu U^\dagger - UQ\partial^\mu U^\dagger) + \cdots$$

$$= \cdots - ie\mathscr{A}_\mu\frac{F_0^2}{4}\mathrm{Tr}(-QU^\dagger\partial^\mu U + Q\partial^\mu UU^\dagger - Q\underbrace{U\partial^\mu U^\dagger}_{=-\partial^\mu UU^\dagger} + Q\underbrace{\partial^\mu U^\dagger U}_{=-U^\dagger\partial^\mu U}) + \cdots$$

$$\underbrace{\qquad\qquad\qquad\qquad\qquad\qquad\qquad}_{=2\mathrm{Tr}(Q[\partial^\mu U,U^\dagger])}$$

$$= \frac{F_0^2}{4}\mathrm{Tr}[\partial_\mu U\partial^\mu U^\dagger] - ie\mathscr{A}_\mu\frac{F_0^2}{2}\mathrm{Tr}(Q[\partial^\mu U,U^\dagger]) - e^2\mathscr{A}_\mu\mathscr{A}^\mu\frac{F_0^2}{4}\mathrm{Tr}([Q,U][Q,U^\dagger]).$$

(c)

$$[\partial^\mu U, U^\dagger] = \left[i\frac{\partial^\mu\phi}{F_0} + \cdots, \mathbb{1} - i\frac{\phi}{F_0} + \cdots\right] = \frac{1}{F_0^2}[\partial^\mu\phi, \phi] + \cdots,$$

$$[Q,U][Q,U^\dagger] = \left[Q, \mathbb{1} + i\frac{\phi}{F_0} + \cdots\right]\left[Q, \mathbb{1} - i\frac{\phi}{F_0} + \cdots\right] = \frac{1}{F_0^2}[Q,\phi][Q,\phi] + \cdots.$$

$$\Rightarrow \quad \begin{aligned} \mathscr{L}_2^{A-2\phi} &= -e\mathscr{A}_\mu\frac{i}{2}\mathrm{Tr}(Q[\partial^\mu\phi,\phi]), \\ \mathscr{L}_2^{2A-2\phi} &= -\tfrac{1}{4}e^2\mathscr{A}_\mu\mathscr{A}^\mu\mathrm{Tr}([Q,\phi][Q,\phi]). \end{aligned}$$

(d)

$$[\partial^\mu\phi, \phi]$$

$$= \partial^\mu\begin{pmatrix} \pi^0 + \frac{1}{\sqrt{3}}\eta & \sqrt{2}\pi^+ & \sqrt{2}K^+ \\ \sqrt{2}\pi^- & -\pi^0 + \frac{1}{\sqrt{3}}\eta & \sqrt{2}K^0 \\ \sqrt{2}K^- & \sqrt{2}\bar{K}^0 & -\frac{2}{\sqrt{3}}\eta \end{pmatrix}\begin{pmatrix} \pi^0 + \frac{1}{\sqrt{3}}\eta & \sqrt{2}\pi^+ & \sqrt{2}K^+ \\ \sqrt{2}\pi^- & -\pi^0 + \frac{1}{\sqrt{3}}\eta & \sqrt{2}K^0 \\ \sqrt{2}K^- & \sqrt{2}\bar{K}^0 & -\frac{2}{\sqrt{3}}\eta \end{pmatrix}$$

$$-(\ )\partial^\mu(\ ) = \begin{pmatrix} (\partial^\mu\pi^0 + \frac{1}{\sqrt{3}}\partial^\mu\eta)(\pi^0 + \frac{1}{\sqrt{3}}\eta) + 2\partial^\mu\pi^+\pi^- + 2\partial^\mu K^+ K^- \\ \cdots \\ \cdots \end{pmatrix}$$

$$2\partial^\mu\pi^-\pi^+ + (-\partial^\mu\pi^0 + \frac{1}{\sqrt{3}}\partial^\mu\eta)(-\pi^0 + \frac{1}{\sqrt{3}}\eta) + 2\partial^\mu K^0 \bar{K}^0$$
$$\cdots$$

$$\cdots$$
$$\cdots$$
$$2\partial^\mu K^- K^+ + 2\partial^\mu \bar{K}^0 K^0 + \frac{4}{3}\partial^\mu\eta\eta \Big)$$

$$-\begin{pmatrix} (\pi^0 + \frac{1}{\sqrt{3}}\eta)(\partial^\mu\pi^0 + \frac{1}{\sqrt{3}}\partial^\mu\eta) + 2\pi^+\partial^\mu\pi^- + 2K^+\partial^\mu K^- \\ \cdots \\ \cdots \end{pmatrix}$$

$$\cdots$$
$$2\pi^-\partial^\mu\pi^+ + (-\pi^0 + \frac{1}{\sqrt{3}}\eta)(-\partial^\mu\pi^0 + \frac{1}{\sqrt{3}}\partial^\mu\eta) + 2K^0\partial^\mu\bar{K}^0$$
$$\cdots$$

$$\cdots$$
$$\cdots$$
$$2K^-\partial^\mu K^+ + 2\bar{K}^0\partial^\mu K^0 + \frac{4}{3}\eta\partial^\mu\eta \Big)$$

$$= 2\begin{pmatrix} \partial^\mu\pi^+\pi^- - \pi^+\partial^\mu\pi^- + \partial^\mu K^+ K^- - K^+\partial^\mu K^- & \cdots & \cdots \\ \cdots & \partial^\mu\pi^-\pi^+ - \pi^-\partial^\mu\pi^+ + \partial^\mu K^0 \bar{K}^0 - K^0\partial^\mu\bar{K}^0 & \cdots \\ \cdots & \cdots & \partial^\mu K^- K^+ - K^-\partial^\mu K^+ + \partial^\mu\bar{K}^0 K^0 - \bar{K}^0\partial^\mu K^0 \end{pmatrix}.$$

$$\Rightarrow \mathscr{L}_2^{A-2\phi} = -ie\mathscr{A}_\mu\Big[\frac{2}{3}(\partial^\mu\pi^+\pi^- - \pi^+\partial^\mu\pi^- + \partial^\mu K^+ K^- - K^+\partial^\mu K^-)$$

$$-\frac{1}{3}(\partial^\mu\pi^-\pi^+ - \pi^-\partial^\mu\pi^+ + \partial^\mu K^0 \bar{K}^0 - K^0\partial^\mu\bar{K}^0)$$

$$-\frac{1}{3}(\partial^\mu K^- K^+ - K^-\partial^\mu K^+ + \partial^\mu\bar{K}^0 K^0 - \bar{K}^0\partial^\mu K^0)\Big]$$

$$= -ie\mathscr{A}_\mu(\partial^\mu\pi^+\pi^- - \pi^+\partial^\mu\pi^- + \partial^\mu K^+ K^- - K^+\partial^\mu K^-).$$

$$[Q, \phi] = \begin{pmatrix} \frac{2}{3} & 0 & 0 \\ 0 & -\frac{1}{3} & 0 \\ 0 & 0 & -\frac{1}{3} \end{pmatrix} \begin{pmatrix} \pi^0 + \frac{1}{\sqrt{3}}\eta & \sqrt{2}\pi^+ & \sqrt{2}K^+ \\ \sqrt{2}\pi^- & -\pi^0 + \frac{1}{\sqrt{3}}\eta & \sqrt{2}K^0 \\ \sqrt{2}K^- & \sqrt{2}\bar{K}^0 & -\frac{2}{\sqrt{3}}\eta \end{pmatrix}$$

$$- \begin{pmatrix} \pi^0 + \frac{1}{\sqrt{3}}\eta & \sqrt{2}\pi^+ & \sqrt{2}K^+ \\ \sqrt{2}\pi^- & -\pi^0 + \frac{1}{\sqrt{3}}\eta & \sqrt{2}K^0 \\ \sqrt{2}K^- & \sqrt{2}\bar{K}^0 & -\frac{2}{\sqrt{3}}\eta \end{pmatrix} \begin{pmatrix} \frac{2}{3} & 0 & 0 \\ 0 & -\frac{1}{3} & 0 \\ 0 & 0 & -\frac{1}{3} \end{pmatrix}$$

$$= \begin{pmatrix} 0 & \sqrt{2}\pi^+ & \sqrt{2}K^+ \\ -\sqrt{2}\pi^- & 0 & 0 \\ -\sqrt{2}K^- & 0 & 0 \end{pmatrix},$$

$$[Q, \phi][Q, \phi] = \begin{pmatrix} 0 & \sqrt{2}\pi^+ & \sqrt{2}K^+ \\ -\sqrt{2}\pi^- & 0 & 0 \\ -\sqrt{2}K^- & 0 & 0 \end{pmatrix} \begin{pmatrix} 0 & \sqrt{2}\pi^+ & \sqrt{2}K^+ \\ -\sqrt{2}\pi^- & 0 & 0 \\ -\sqrt{2}K^- & 0 & 0 \end{pmatrix}$$

$$= \begin{pmatrix} -2\pi^+\pi^- - 2K^+K^- & 0 & 0 \\ 0 & -2\pi^-\pi^+ & -2\pi^-K^+ \\ 0 & -2K^-\pi^+ & -2K^-K^+ \end{pmatrix}$$

$$= -2 \begin{pmatrix} \pi^+\pi^- + K^+K^- & 0 & 0 \\ 0 & \pi^-\pi^+ & \pi^-K^+ \\ 0 & K^-\pi^+ & K^-K^+ \end{pmatrix}.$$

$$\Rightarrow \quad \mathcal{L}_2^{2A-2\phi} = -\frac{1}{4}e^2 \mathcal{A}_\mu \mathcal{A}^\mu (-4\pi^+\pi^- - 4K^+K^-)$$

$$= e^2 \mathcal{A}_\mu \mathcal{A}^\mu (\pi^+\pi^- + K^+K^-).$$

(e)

$$\mathcal{M} = [\mp ie\varepsilon'^* \cdot (p+q+p')]\frac{i}{(p+q)^2 - M^2}[\mp ie\varepsilon \cdot (p+ \underbrace{p+q}_{=p'+q'})]$$

$$+ [\mp ie\varepsilon \cdot (p-q'+p')]\frac{i}{(p-q')^2 - M^2}[\mp ie\varepsilon'^* \cdot (p+ \underbrace{p-q'}_{=p'-q})] + 2ie^2\varepsilon'^* \cdot \varepsilon$$

$$= ie^2 \left[2\varepsilon'^* \cdot \varepsilon - \frac{\varepsilon'^* \cdot (p+q+p')\varepsilon \cdot (p+p'+q')}{\underbrace{(p+q)^2}_{=s} - M^2} - \frac{\varepsilon \cdot (p-q'+p')\varepsilon'^* \cdot (p+p'-q)}{\underbrace{(p-q')^2}_{=u} - M^2} \right].$$

The amplitude is the same for π^+ and π^-.

(f) Gauge invariance

$$\mathcal{M} \overset{\varepsilon \to q}{\to} ie^2 \left[2\varepsilon'^* \cdot q - \frac{\varepsilon'^* \cdot (p+q+p')q \cdot \overbrace{(p+p'+q')}^{=p+q}}{2p \cdot q + q^2} \right.$$

$$\left. - \frac{q \cdot (\overbrace{p-q'}^{=p'-q} + p')\varepsilon'^* \cdot (p+p'-q)}{-2p' \cdot q + q^2} \right]$$

$$= ie^2 [2\varepsilon'^* \cdot q - \varepsilon'^* \cdot (p+q+p'-p-p'+q)] = 0.$$

(g) Crossing symmetry

$$\mathcal{M} = ie^2 \left[2\varepsilon'^* \cdot \varepsilon - \frac{\varepsilon'^* \cdot (p+q+p')\varepsilon \cdot (p+p'+q')}{\underbrace{(p+q)^2 - M^2}_{=s}} \right.$$

$$\left. - \frac{\varepsilon \cdot (p-q'+p')\varepsilon'^* \cdot (p+p'-q)}{\underbrace{(p-q')^2 - M^2}_{=u}} \right]$$

$$\mapsto ie^2 \left[2\varepsilon \cdot \varepsilon'^* - \frac{\varepsilon \cdot (p-q'+p')\varepsilon'^* \cdot (p+p'-q)}{\underbrace{(p-q')^2 - M^2}_{=u}} \right.$$

$$\left. - \frac{\varepsilon'^* \cdot (p+q+p')\varepsilon \cdot (p+p'+q')}{\underbrace{(p+q)^2 - M^2}_{=s}} \right]$$

$$= \mathcal{M}.$$

3.19 (a) Using

$$\oint_C dz\, f(z) = \int_{\gamma_1} dz\, f(z) + \lim_{R \to \infty} \int_{\gamma_2} dz\, f(z) = \int_{-\infty}^{\infty} dt\, f(t) + \underbrace{\lim_{R \to \infty} \int_0^{\pi} iRe^{it} dt\, f(Re^{it})}_{=0}$$

$$= 2\pi i \mathrm{Res}[f(z), -(a - i0^+)],$$

we obtain

$$\int_{-\infty}^{\infty} dk_0\, f(k_0) = 2\pi i \mathrm{Res}[f(z), -(a - i0^+)].$$

Determination of the residue:

$$\text{Res}[f(z), -(a - i0^+)] = \lim_{z \to -(a-i0^+)} [z + (a - i0^+)] \frac{1}{[z + (a - i0^+)][z - (a - i0^+)]}$$

$$= -\frac{1}{2} \frac{1}{a - i0^+}.$$

We thus obtain

$$\int_{-\infty}^{\infty} dk_0 f(k_0) = \frac{-i\pi}{\sqrt{\vec{k}^2 + M^2} - i0^+}.$$

(b)

$$\int \frac{d^4 k}{(2\pi)^4} \frac{i}{k^2 - M^2 + i0^+} = \int \frac{d^3 k}{(2\pi)^3} \int_{-\infty}^{\infty} dk_0 \frac{1}{2\pi} \frac{i}{k_0^2 - \vec{k}^2 + M^2 - i0^+}$$

$$= \frac{1}{2} \int \frac{d^3 k}{(2\pi)^3} \frac{1}{\sqrt{\vec{k}^2 + M^2} - i0^+}.$$

(c)

$$\int \frac{d^{n-1} k}{(2\pi)^{n-1}} \frac{1}{\sqrt{\vec{k}^2 + M^2}} = 2 \frac{\pi^{\frac{n-1}{2}}}{\Gamma\left(\frac{n-1}{2}\right)} \frac{1}{(2\pi)^{n-1}} \int_0^{\infty} dr \frac{r^{n-2}}{\sqrt{r^2 + M^2}}$$

$$= \frac{1}{2^{n-2}} \pi^{-\frac{n-1}{2}} \frac{1}{\Gamma\left(\frac{n-1}{2}\right)} \int_0^{\infty} dr \frac{r^{n-2}}{\sqrt{r^2 + M^2}}.$$

(d)

$$\int_0^{\infty} dr \frac{r^{n-2}}{\sqrt{r^2 + M^2}} = \frac{1}{M} \int_0^{\infty} dr \frac{r^{n-2}}{\sqrt{\frac{r^2}{M^2} + 1}}.$$

Substitution $t = r/M$: $r = Mt, dr = M dt$,

$$\cdots = M^{n-2} \int_0^{\infty} dt \frac{t^{n-2}}{\sqrt{t^2 + 1}}.$$

Substitution $t^2 = y$: $dy = 2t dt, dt = dy/(2\sqrt{y})$,

$$\cdots = M^{n-2} \int_0^\infty \frac{dy}{2\sqrt{y}} \frac{y^{\frac{n-2}{2}}}{\sqrt{y+1}}$$

$$= \frac{1}{2} M^{n-2} \int_0^\infty dt \frac{t^{\frac{n-3}{2}}}{(t+1)^{\frac{1}{2}}}.$$

From $x - 1 = \frac{n-3}{2}$ we obtain $x = \frac{n-1}{2}$, from $x + y = \frac{1}{2} = \frac{n-1}{2} + y$ we obtain $y = 1 - \frac{n}{2}$, thus

$$\cdots = \frac{1}{2} M^{n-2} B\left(\frac{n-1}{2}, 1 - \frac{n}{2}\right)$$

$$= \frac{1}{2} M^{n-2} \frac{\Gamma\left(\frac{n-1}{2}\right) \Gamma\left(1 - \frac{n}{2}\right)}{\underbrace{\Gamma\left(\frac{1}{2}\right)}_{= \sqrt{\pi}}}.$$

(e)

$$\int \frac{d^n k}{(2\pi)^n} \frac{i}{k^2 - M^2 + i0^+} = -i\pi \frac{1}{2\pi} i \frac{1}{2^{n-2}} \frac{1}{\pi^{\frac{n-1}{2}}} \frac{1}{\Gamma\left(\frac{n-1}{2}\right)} \frac{1}{2} M^{n-2} \frac{\Gamma\left(\frac{n-1}{2}\right) \Gamma\left(1 - \frac{n}{2}\right)}{\sqrt{\pi}}$$

$$= \frac{1}{(4\pi)^{\frac{n}{2}}} M^{n-2} \Gamma\left(1 - \frac{n}{2}\right).$$

3.20 Let c and s denote $\cos(\theta_1)$ and $\sin(\theta_1)$, respectively, sc denote $\sin(\theta_1)\cos(\theta_2)$ etc., and ssc denote $\sin(\theta_1)\sin(\theta_2)\cos(\theta_3)$ etc.:

$$J = \begin{pmatrix} \frac{\partial l_1}{\partial l} \cdots & \frac{\partial l_1}{\partial \theta_3} \\ \vdots & \vdots \\ \frac{\partial l_4}{\partial l} \cdots & \frac{\partial l_4}{\partial \theta_3} \end{pmatrix} = \begin{pmatrix} c & -ls & 0 & 0 \\ sc & lcc & -lss & 0 \\ ssc & lcsc & lscc & -lsss \\ sss & lcss & lscs & lssc \end{pmatrix}.$$

$\det(J)$

$$= lsss \det \begin{pmatrix} c & -ls & 0 \\ sc & lcc & -lss \\ sss & lcss & lscs \end{pmatrix} + lssc \det \begin{pmatrix} c & -ls & 0 \\ sc & lcc & -lss \\ ssc & lcsc & lscc \end{pmatrix}$$

$$= l\sin(\theta_1)\sin(\theta_2)\sin(\theta_3)\{l\sin(\theta_1)\sin(\theta_2)$$
$$\times [l\cos^2(\theta_1)\sin(\theta_2)\sin(\theta_3) + l\sin^2(\theta_1)\sin(\theta_2)\sin(\theta_3)]$$
$$+ l\sin(\theta_1)\cos(\theta_2)\sin(\theta_3)[l\cos^2(\theta_1)\cos(\theta_2) + l\sin^2(\theta_1)\cos(\theta_2)]\}$$
$$+ l\sin(\theta_1)\sin(\theta_2)\cos(\theta_3)\{l\sin(\theta_1)\sin(\theta_2)$$
$$\times [l\cos^2(\theta_1)\sin(\theta_2)\cos(\theta_3) + l\sin^2(\theta_1)\sin(\theta_2)\cos(\theta_3)]$$
$$+ l\sin(\theta_1)\cos(\theta_2)\cos(\theta_3)[l\cos^2(\theta_1)\cos(\theta_2) + l\sin^2(\theta_1)\cos(\theta_2)]\}$$

$$= l^3 \sin(\theta_1)\sin(\theta_2)\sin(\theta_3)[\sin(\theta_1)\sin^2(\theta_2)\sin(\theta_3) + \sin(\theta_1)\cos^2(\theta_2)\sin(\theta_3)]$$
$$+ l^3 \sin(\theta_1)\sin(\theta_2)\cos(\theta_3)[\sin(\theta_1)\sin^2(\theta_2)\cos(\theta_3) + \sin(\theta_1)\cos^2(\theta_2)\cos(\theta_3)]$$
$$= l^3 \sin^2(\theta_1)\sin(\theta_2)\sin^2(\theta_3) + l^3 \sin^2(\theta_1)\sin(\theta_2)\cos^2(\theta_3)$$
$$= l^3 \sin^2(\theta_1)\sin(\theta_2).$$

Thus

$$dl_1 dl_2 dl_3 dl_4 = l^3 dl \underbrace{\sin^2(\theta_1)\sin(\theta_2)d\theta_1 d\theta_2 d\theta_3}_{d\Omega} .$$

$$\int d\Omega = \int_0^\pi d\theta_1 \sin^2(\theta_1) \int_0^{2\pi} d\theta_2 \sin(\theta_2) \int_0^{2\pi} d\theta_3$$

$$= 2\pi \int_0^\pi d\theta_1 \sin^2(\theta_1) \int_{-1}^1 d\cos(\theta_3)$$

$$= 4\pi \int_0^\pi d\theta_1 \sin^2(\theta_1) = 4\pi \left[\frac{1}{2}\theta_1 - \frac{1}{4}\sin(2\theta_1) \right]_0^\pi = 2\pi^2.$$

3.21 $m = 1$:

$$\int_0^\pi d\theta \sin(\theta) = -\cos(\theta)\Big|_0^\pi = 2, \quad \frac{\sqrt{\pi}\Gamma(1)}{\Gamma(\frac{3}{2})} = 2,$$

because $\Gamma(1) = 1$ and $\Gamma(3/2) = \Gamma(1/2 + 1) = \frac{1}{2}\Gamma(1/2) = \sqrt{\pi}/2$.
Step $m \to m + 1$:

$$\int_0^\pi d\theta \sin^{m+1}(\theta) = \int_0^\pi d\theta \sin(\theta) \sin^m(\theta)$$

$$= \left[-\cos(\theta)\sin^m(\theta) \right]_0^\pi - \int_0^\pi d\theta[-\cos(\theta)]m\sin^{m-1}(\theta)\cos(\theta)$$

$$= m \int_0^\pi d\theta \underbrace{\cos^2(\theta)}_{= 1 - \sin^2(\theta)} \sin^{m-1}(\theta).$$

$$\Rightarrow \quad (m+1) \int_0^\pi d\theta \sin^{m+1}(\theta) = m \int_0^\pi d\theta \sin^{m-1}(\theta).$$

$$\Rightarrow \int_0^\pi d\theta \, \sin^{m+1}(\theta) = \frac{m}{m+1} \int_0^\pi d\theta \, \sin^{m-1}(\theta) d\theta$$

$$= \frac{m}{m+1} \frac{\sqrt{\pi}\,\Gamma\left(\frac{m}{2}\right)}{\Gamma\left(\frac{m+1}{2}\right)}$$

$$= \sqrt{\pi} \frac{\frac{m}{2}\Gamma\left(\frac{m}{2}\right)}{\frac{m+1}{2}\Gamma\left(\frac{m+1}{2}\right)}$$

$$= \sqrt{\pi} \frac{\Gamma\left(\frac{m+2}{2}\right)}{\Gamma\left(\frac{m+3}{2}\right)},$$

where, in the last step, we made use of $x\Gamma(x) = \Gamma(x+1)$.

3.22 (a)

$$\int_{-\infty}^{\infty} dt f(t) = i \int_{-\infty}^{\infty} dt f(it).$$

Thus

$$(k^2)^p = (k_0^2 - \vec{k}^2)^p \rightarrow [(il_1)^2 - (l_2^2 + \cdots + l_n^2)]^p = (-1)^p l^2,$$

$$\frac{1}{(k^2 - M^2 + i0^+)^q} \rightarrow \frac{1}{(-1)^q (l^2 + M^2)^q} = \frac{(-1)^{-q}}{(l^2 + M^2)^q},$$

$$\int \frac{d^n k}{(2\pi)^n} \frac{(k^2)^p}{(k^2 - M^2 + i0^+)^q} = i(-)^{p-q} \int \frac{d^n l}{(2\pi)^n} \frac{(l^2)^p}{(l^2 + M^2)^q}.$$

(b) Perform the angular integration,

$$\cdots = i(-)^{p-q} \frac{2\pi^{\frac{n}{2}}}{\Gamma\left(\frac{n}{2}\right)} \frac{1}{(2\pi)^n} \int_0^\infty dl \frac{l^{2p+n-1}}{(l^2 + M^2)^q}$$

$$= i(-)^{p-q} \frac{2}{(4\pi)^{\frac{n}{2}}\Gamma\left(\frac{n}{2}\right)} \int_0^\infty dl \frac{l^{2p+n-1}}{(l^2 + M^2)^q}.$$

Perform the radial integration,

$$\int_0^\infty dl \frac{l^{n-1}}{(l^2 + M^2)^\alpha} = \frac{1}{2}(M^2)^{\frac{n}{2}-\alpha} \frac{\Gamma\left(\frac{n}{2}\right)\Gamma\left(\alpha - \frac{n}{2}\right)}{\Gamma(\alpha)},$$

with $n \rightarrow 2p + n$ and $\alpha \rightarrow q$,

$$\cdots = i(-)^{p-q} \frac{1}{(4\pi)^{\frac{n}{2}}} (M^2)^{p+\frac{n}{2}-q} \frac{\Gamma\left(p + \frac{n}{2}\right)\Gamma\left(q - p - \frac{n}{2}\right)}{\Gamma\left(\frac{n}{2}\right)\Gamma(q)}.$$

20   3.23 (a)
$$f(z) = a^z$$
$$= \exp[\ln(a)z]$$
$$= \exp\{\ln(a)[\mathrm{Re}(z) + i\mathrm{Im}(z)]\}$$
$$= \exp[\ln(a)\mathrm{Re}(z)] \exp[i\ln(a)\mathrm{Im}(z)]$$
$$= \exp[\ln(a)x]\{\cos[\ln(a)y] + i\sin[\ln(a)y]\}$$
$$\equiv u(x,y) + iv(x,y),$$

i.e.
$$u(x,y) = \exp[\ln(a)x]\cos[\ln(a)y],$$
$$v(x,y) = \exp[\ln(a)x]\sin[\ln(a)y].$$

(b)
$$\partial u/\partial x = \ln(a)\exp[\ln(a)x]\cos[\ln(a)y],$$
$$\partial v/\partial y = \exp[\ln(a)x]\ln(a)\cos[\ln(a)y] = \partial u/\partial x,$$
$$\partial u/\partial y = \exp[\ln(a)x]\ln(a)(-)\sin[\ln(a)y],$$
$$\partial v/\partial x = \ln(a)\exp[\ln(a)x]\sin[\ln(a)y] = -\partial u/\partial y.$$

3.24 (a)
$$L_i^r(\mu) = L_i^r(\mu') + \frac{\Gamma_i}{16\pi^2}\ln\left(\frac{\mu'}{\mu}\right) \quad \Rightarrow \quad \frac{dL_i^r(\mu)}{d\mu} = -\frac{\Gamma_i}{16\pi^2\mu}.$$

(b) Making use of
$$\frac{d}{d\mu}\ln\left(\frac{M^2}{\mu^2}\right) = 2\frac{d}{d\mu}[\ln(M) - \ln(\mu)] = -\frac{2}{\mu},$$

we obtain:
$$\frac{dM_{\pi,4}^2}{d\mu} = \frac{M_{\pi,2}^2}{16\pi^2\mu F_0^2}\left\{\frac{M_{\pi,2}^2}{2}(-2) - \frac{M_{\eta,2}^2}{6}(-2)\right.$$
$$\left. + 16[(2\hat{m} + m_s)B_0(-2\Gamma_6 + \Gamma_4) + \hat{m}B_0(-2\Gamma_8 + \Gamma_5)]\right\}$$
$$= \frac{M_{\pi,2}^2}{16\pi^2\mu F_0^2}\left\{-2B_0\hat{m} + \frac{2}{9}(\hat{m} + 2m_s)B_0\right.$$

$$+ 16\left[(2\hat{m} + m_s)B_0 \underbrace{\left(-2\frac{11}{144} + \frac{1}{8}\right)}_{= -\frac{1}{36}} + \hat{m}B_0 \underbrace{\left(-2\frac{5}{48} + \frac{3}{8}\right)}_{= \frac{1}{6}}\right]\Bigg\}$$

$$= \frac{M_{\pi,2}^2}{16\pi^2 \mu F_0^2}\left\{B_0\hat{m}\left(-2 + \frac{2}{9} - \frac{8}{9} + \frac{8}{3}\right) + B_0 m_s\left(\frac{4}{9} - \frac{4}{9}\right)\right\} = 0.$$

3.25 Expanding U up to and including terms of second order in the fields yields the same functional form for both parameterizations (this is not true for higher orders). Therefore, terms of second order in the fields in the Lagrangian will be the same for both parameterizations (we have seen in Exercise 3.16 that this is in general not true for higher powers of the fields).

Let us begin with $\mathscr{L}_4^{\mathrm{GL}}$. Since $D_\mu U = \partial_\mu U = O(\phi)$ and $D_\mu \chi = \partial_\mu \chi = 0$, the terms potentially generating two powers of fields are l_3 and l_7. Consider

$$\chi U^\dagger \pm U\chi^\dagger = 2\hat{m}B(U^\dagger \pm U).$$

$$U^\dagger + U = \left[2 - \frac{\vec{\phi}^2}{F^2} + O\left(|\vec{\phi}|^4\right)\right]\mathbb{1},$$

$$U^\dagger - U = -2i\frac{\vec{\phi} \cdot \vec{\tau}}{F}\left[1 + O\left(|\vec{\phi}|^2\right)\right].$$

Since $\mathrm{Tr}(\tau_i) = 0$, the l_7 term does not contribute.

$$\left[\mathrm{Tr}\left(\chi U^\dagger + U\chi^\dagger\right)\right]^2 = \left\{2M^2\left[2 - \frac{\vec{\phi}^2}{F^2} + O\left(|\vec{\phi}|^4\right)\right]\right\}^2$$

$$= 16M^4\left[1 - \frac{\vec{\phi}^2}{F^2} + O\left(|\vec{\phi}|^4\right)\right].$$

With respect to the constant term and the term quadratic in the fields, the l_4 term yields the same functional form for both parameterizations of U,

$$\mathscr{L}_4^{\mathrm{GL},2\phi} = -l_3 M^4 \frac{\vec{\phi}^2}{F^2},$$

and an analogous expression in terms of the π_i fields.

Let us now turn to $\mathscr{L}_4^{\text{GSS}}$. In addition we need to investigate

$$\text{Tr}(\partial_\mu U \partial^\mu U^\dagger)\text{Tr}(U^\dagger + U) = 8\frac{\partial_\mu \vec{\phi} \cdot \partial^\mu \vec{\phi}}{F^2} + \cdots.$$

We obtain

$$\mathscr{L}_4^{\text{GSS},2\phi} = -(l_3 + l_4)M^4\frac{\vec{\phi}^2}{F^2} + l_4 M^2\frac{\partial_\mu \vec{\phi} \cdot \partial^\mu \vec{\phi}}{F^2},$$

and an analogous expression in terms of the π_i fields.

3.26 (a) The self-energy diagrams are shown in Fig. 3.14. The tree contribution to $-i\Sigma_{ba}$ is obtained from $\langle p, b | i\mathscr{L}_4^{2\phi} | p, a \rangle$:

$$\text{GL:} - i\Sigma_{4,ba}^{\text{tree}}(p^2) = 2i\left(-l_3 M^4\frac{\delta_{ab}}{F^2}\right) \quad \Rightarrow \quad \Sigma_{4,ba}^{\text{tree}}(p^2) = 2l_3 M^4\frac{\delta_{ab}}{F^2},$$

$$\text{GSS:} - i\Sigma_{4,ba}^{\text{tree}}(p^2) = 2i\left(-(l_3 + l_4)M^4\frac{\delta_{ab}}{F^2}\right) + 2i\left(M^2 l_4\frac{p^2\delta_{ab}}{F^2}\right)$$

$$\Rightarrow \Sigma_{4,ba}^{\text{tree}}(p^2) = 2(l_3 + l_4)M^4\frac{\delta_{ab}}{F^2} - 2l_4 p^2 M^2\frac{\delta_{ab}}{F^2}.$$

The loop contribution for the exponential parameterization is obtained from the Feynman rule of Eq. 3.90 with the replacements $(p_a, a) \to (p, a), (p_b, b) \to (k, c), (p_c, c) \to (p, b)$, and $(p_d, d) \to (k, c)$ (a summation over c is implied):

$$\frac{1}{2}\int \frac{d^4k}{(2\pi)^4} i\left[\underbrace{\delta_{ac}\delta_{bc}}_{=\delta_{ab}}\frac{(p+k)^2 - M^2}{F^2} + \underbrace{\delta_{ab}\delta_{cc}}_{=3\delta_{ab}}\frac{-M^2}{F^2} + \underbrace{\delta_{ac}\delta_{cb}}_{=\delta_{ab}}\frac{(p-k)^2 - M^2}{F^2}\right.$$

$$\left. - \frac{1}{3F^2}\underbrace{(\delta_{ac}\delta_{bc} + \delta_{ab}\delta_{cc} + \delta_{ac}\delta_{cb})}_{=5\delta_{ab}}(2p^2 + 2k^2 - 4M^2)\right]\frac{i}{k^2 - M^2 + i0^+}$$

$$= \frac{1}{2}\delta_{ab}\int \frac{d^4k}{(2\pi)^4}\frac{i}{3F^2}[-4p^2 - 4k^2 + 5M^2]\frac{i}{k^2 - M^2 + i0^+}$$

$$\to \frac{i\delta_{ab}}{6F^2}(-4p^2 + M^2)I(M^2, \mu^2, n),$$

as in Eq. 3.134. In other words

$$\Sigma_{4,ba}^{\text{loop}}(p^2) = \delta_{ab}\left[-\frac{M^2}{6F^2}I(M^2, \mu^2, n) + \frac{2p^2}{3F^2}I(M^2, \mu^2, n)\right].$$

An analogous calculation with the Feynman rule of Exercise 3.16 for the square-root parameterization yields

$$\frac{1}{2}\int \frac{d^4k}{(2\pi)^4}i\left[\underbrace{\delta_{ac}\delta_{bc}}_{=\,\delta_{ab}}\frac{(p+k)^2-M^2}{F^2}+\underbrace{\delta_{ab}\delta_{cc}}_{=\,3\delta_{ab}}\frac{-M^2}{F^2}+\underbrace{\delta_{ac}\delta_{cb}}_{=\,\delta_{ab}}\frac{(p-k)^2-M^2}{F^2}\right]$$

$$\times\frac{i}{k^2-M^2+i0^+}$$

$$=\frac{1}{2}\delta_{ab}\int\frac{d^4k}{(2\pi)^4}\frac{i}{F^2}[2p^2+2k^2-5M^2]\frac{i}{k^2-M^2+i0^+}$$

$$\rightarrow\frac{i\delta_{ab}}{2F^2}(2p^2-3M^2)I(M^2,\mu^2,n).$$

In this representation

$$\Sigma_{4,ba}^{\text{loop}}(p^2)=\delta_{ab}\left[\frac{3M^2}{2F^2}I(M^2,\mu^2,n)-\frac{p^2}{F^2}I(M^2,\mu^2,n)\right].$$

(b) For determination of $M_{\pi,4}^2$ use

$$l_3M^2=\left(l_3^r+\gamma_3\frac{R}{32\pi^2}\right)M^2=\left\{\frac{\gamma_3}{32\pi^2}\left[\bar{l}_3+\ln\left(\frac{M^2}{\mu^2}\right)\right]+\gamma_3\frac{R}{32\pi^2}\right\}M^2.$$

Insert $\gamma_3=-\frac{1}{2}$ and reexpress in terms of integral $I\equiv I(M^2,\mu^2,n)$,

$$\cdots=-\frac{\bar{l}_3M^2}{64\pi^2}-\frac{1}{4}I.$$

Make use of

$$M_{\pi,4}^2=M^2(1+B)+A$$

for
a. GL, exponential:

$$\cdots=M^2\left(1+\frac{2}{3}\frac{I}{F^2}\right)-\frac{1}{6}\frac{M^2}{F^2}I+2l_3\frac{M^4}{F^2}$$

$$=M^2-\frac{\bar{l}_3}{32\pi^2F^2}M^4+\underbrace{\left(\frac{2}{3}-\frac{1}{6}-\frac{1}{2}\right)}_{=\,0}\frac{M^2}{F^2}I,$$

b. GL, square-root:

$$\cdots = M^2\left(1 - \frac{I}{F^2}\right) + \frac{3}{2}\frac{M^2}{F^2}I + 2l_3\frac{M^4}{F^2}$$

$$= M^2 - \frac{\bar{l}_3}{32\pi^2 F^2}M^4 + \underbrace{\left(-1 + \frac{3}{2} - \frac{1}{2}\right)\frac{M^2}{F^2}I}_{= 0},$$

c. GSS, exponential:

$$\cdots = M^2\left(1 + \frac{2}{3}\frac{I}{F^2} - 2l_4\frac{M^2}{F^2}\right) - \frac{1}{6}\frac{M^2}{F^2}I + 2(l_3 + l_4)\frac{M^4}{F^2}$$

$$= M^2 - \frac{\bar{l}_3}{32\pi^2 F^2}M^4,$$

d. GSS, square-root:

$$\cdots = M^2\left(1 - \frac{I}{F^2} - 2l_4\frac{M^2}{F^2}\right) + \frac{3}{2}\frac{M^2}{F^2}I + 2(l_3 + l_4)\frac{M^4}{F^2}$$

$$= M^2 - \frac{\bar{l}_3}{32\pi^2 F^2}M^4.$$

The result for the pion mass is indeed independent of the Lagrangian and parameterization used.

3.27 (a)

$$\mathcal{U}'_{Li}(y) = U'^\dagger(y)\frac{\partial U'(y)}{\partial y^i} = U^\dagger(y)[\mathbb{1} - i\alpha\Delta(x)]\frac{\partial}{\partial y^i}\{[\mathbb{1} + i\alpha\Delta(x)]U(y)\}$$

$$= \mathcal{U}_{Li}(y) + U^\dagger(y)[-i\alpha\Delta(x)]\frac{\partial U(y)}{\partial y^i} + U^\dagger(y)\left\{\frac{\partial}{\partial y^i}[i\alpha\Delta(x)]\right\}U(y)$$

$$\quad + U^\dagger(y)i\alpha\Delta(x)\frac{\partial U(y)}{\partial y^i} + O(\Delta^2)$$

$$= \mathcal{U}_{Li}(y) + U^\dagger(y)\left\{\frac{\partial}{\partial y^i}[i\alpha\Delta(x)]\right\}U(y) + O(\Delta^2).$$

$\Rightarrow$

$$\delta S^0_{\text{ano}} = S^0_{\text{ano}}[U'] - S^0_{\text{ano}}[U]$$

$$= -\frac{in}{240\pi^2}\int_0^1 d\alpha \int d^4 x \varepsilon^{ijklm}\text{Tr}\left[U^\dagger\frac{\partial(i\alpha\Delta)}{\partial y^i}U\mathcal{U}_{Lj}\mathcal{U}_{Lk}\mathcal{U}_{Ll}\mathcal{U}_{Lm}\right.$$

$$+ \mathcal{U}_{Li}U^\dagger\frac{\partial(i\alpha\Delta)}{\partial y^j}U\mathcal{U}_{Lk}\mathcal{U}_{Ll}\mathcal{U}_{Lm} + \mathcal{U}_{Li}\mathcal{U}_{Lj}U^\dagger\frac{\partial(i\alpha\Delta)}{\partial y^k}U\mathcal{U}_{Ll}\mathcal{U}_{Lm}$$

$$\left.+ \mathcal{U}_{Li}\mathcal{U}_{Lj}\mathcal{U}_{Lk}U^\dagger\frac{\partial(i\alpha\Delta)}{\partial y^l}U\mathcal{U}_{Lm} + \mathcal{U}_{Li}\mathcal{U}_{Lj}\mathcal{U}_{Lk}\mathcal{U}_{Ll}U^\dagger\frac{\partial(i\alpha\Delta)}{\partial y^m}U\right].$$

Both ε^{ijklm} and the trace are even under cyclic permutations of the indices $\{i,j,k,l,m\}$. Thus

$$\delta S^0_{\text{ano}} = \frac{n}{48\pi^2} \int_0^1 d\alpha \int d^4x \varepsilon^{ijklm} \text{Tr}\left[U^\dagger \frac{\partial(\alpha\Delta)}{\partial y^i} U \mathcal{U}_{Lj} \mathcal{U}_{Lk} \mathcal{U}_{Ll} \mathcal{U}_{Lm}\right].$$

(b) Make use of integration by parts:

$$\begin{aligned}
\delta S^0_{\text{ano}} = {} & \frac{n}{48\pi^2} \int_0^1 d\alpha \int d^4x \varepsilon^{ijklm} \frac{\partial}{\partial y^i} \text{Tr}\left[U^\dagger(\alpha\Delta) U \mathcal{U}_{Lj} \mathcal{U}_{Lk} \mathcal{U}_{Ll} \mathcal{U}_{Lm}\right] \\
& - \frac{n}{48\pi^2} \int_0^1 d\alpha \int d^4x \varepsilon^{ijklm} \text{Tr}\left[\frac{\partial U^\dagger}{\partial y^i}(\alpha\Delta) U \mathcal{U}_{Lj} \mathcal{U}_{Lk} \mathcal{U}_{Ll} \mathcal{U}_{Lm}\right] \\
& - \frac{n}{48\pi^2} \int_0^1 d\alpha \int d^4x \varepsilon^{ijklm} \text{Tr}\left[U^\dagger(\alpha\Delta) \frac{\partial U}{\partial y^i} \mathcal{U}_{Lj} \mathcal{U}_{Lk} \mathcal{U}_{Ll} \mathcal{U}_{Lm}\right] \\
& - \frac{n}{48\pi^2} \int_0^1 d\alpha \int d^4x \varepsilon^{ijklm} \text{Tr}\left[U^\dagger(\alpha\Delta) U \frac{\partial(\mathcal{U}_{Lj} \mathcal{U}_{Lk} \mathcal{U}_{Ll} \mathcal{U}_{Lm})}{\partial y^i}\right]. \quad \text{(B.3)}
\end{aligned}$$

Consider the individual contributions to the first term of Eq. B.3:
(i) $i = 4$:

$$\begin{aligned}
& \frac{n}{48\pi^2} \int_0^1 d\alpha \int d^4x \varepsilon^{4jklm} \frac{\partial}{\partial\alpha} \text{Tr}\left[U^\dagger(\alpha\Delta) U \mathcal{U}_{Lj} \mathcal{U}_{Lk} \mathcal{U}_{Ll} \mathcal{U}_{Lm}\right] \\
& = \frac{n}{48\pi^2} \int d^4x \underbrace{\varepsilon^{4jklm}}_{=\,\varepsilon^{jklm4}} \text{Tr}\left[U^\dagger \Delta U \mathcal{U}_{Lj} \mathcal{U}_{Lk} \mathcal{U}_{Ll} \mathcal{U}_{Lm}\right] \\
& = \frac{n}{48\pi^2} \int d^4x \varepsilon^{\mu\nu\rho\sigma} \text{Tr}\left[\Delta U \mathcal{U}_{L\mu} \mathcal{U}_{L\nu} \mathcal{U}_{L\rho} \mathcal{U}_{L\sigma} U^\dagger\right] \\
& = \frac{n}{48\pi^2} \int d^4x \varepsilon^{\mu\nu\rho\sigma} \text{Tr}\left[\Delta \underbrace{UU^\dagger}_{=\,1} \underbrace{\partial_\mu UU^\dagger \partial_\nu UU^\dagger \partial_\rho UU^\dagger \partial_\sigma UU^\dagger}_{=\,U\partial_\mu U^\dagger U\partial_\nu U^\dagger U\partial_\rho U^\dagger U\partial_\sigma U^\dagger}\right] \\
& = \frac{n}{48\pi^2} \int d^4x \varepsilon^{\mu\nu\rho\sigma} \text{Tr}(\Delta \mathcal{U}_{R\mu} \mathcal{U}_{R\nu} \mathcal{U}_{R\rho} \mathcal{U}_{R\sigma}). \quad \text{(B.4)}
\end{aligned}$$

(ii) $i = 0$:

$$\frac{n}{48\pi^2} \int_0^1 d\alpha \int d^4x \varepsilon^{0jklm} \frac{\partial}{\partial y^0} \text{Tr}\left[U^\dagger(\alpha\Delta) U \mathcal{U}_{Lj} \mathcal{U}_{Lk} \mathcal{U}_{Ll} \mathcal{U}_{Lm}\right] = 0,$$

because of the boundary conditions $\Delta(\vec{x}, t_1) = \Delta(\vec{x}, t_2) = 0$.

(iii) $i = 1, 2, 3$:

$$\frac{n}{48\pi^2} \int_0^1 d\alpha \int d^4x \sum_{i=1}^{3} \varepsilon^{ijklm} \frac{\partial}{\partial y^i} \text{Tr}\left[U^\dagger(\alpha\Delta) U \mathcal{U}_{Lj} \mathcal{U}_{Lk} \mathcal{U}_{Ll} \mathcal{U}_{Lm} \right] = 0,$$

because of the divergence theorem.

Consider the integrands of the second and third term of Eq. B.3:

$$\alpha\varepsilon^{ijklm}\text{Tr}\left[\Delta\left(U \mathcal{U}_{Lj} \mathcal{U}_{Lk} \mathcal{U}_{Ll} \mathcal{U}_{Lm} \frac{\partial U^\dagger}{\partial y^i} + \frac{\partial U}{\partial y^i} \mathcal{U}_{Lj} \mathcal{U}_{Lk} \mathcal{U}_{Ll} \mathcal{U}_{Lm} U^\dagger \right)\right]$$

$$= \alpha\varepsilon^{ijklm}\text{Tr}\Big[\Delta\big(\underbrace{UU^\dagger}_{= \mathbb{1}} \partial_j UU^\dagger \partial_k UU^\dagger \partial_l UU^\dagger \partial_m U \partial_i U^\dagger$$

$$+ \partial_i UU^\dagger \partial_j UU^\dagger \partial_k UU^\dagger \underbrace{\partial_l UU^\dagger \partial_m UU^\dagger}_{= -\partial_l U \partial_m U^\dagger} \big)\Big]$$

$$= 0,$$

using an even permutation of the indices from (j, k, l, m, i) to (i, j, k, l, m) in the first term.

Finally, the last term of Eq. B.3 vanishes due to permutation symmetry of the trace term. Consider, e.g.,

$$\partial_i \mathcal{U}_{Lj} = \partial_i U^\dagger \partial_j U + U^\dagger \partial_i \partial_j U.$$

Using $UU^\dagger = \mathbb{1}$, the first term can be rewritten as $-\mathcal{U}_{Li} \mathcal{U}_{Lj}$. The second term does not contribute, because the two derivatives are contracted with the epsilon tensor. We obtain

$$\varepsilon^{ijklm}\text{Tr}\left[U^\dagger(\alpha\Delta) U \frac{\partial(\mathcal{U}_{Lj} \mathcal{U}_{Lk} \mathcal{U}_{Ll} \mathcal{U}_{Lm})}{\partial y^i} \right]$$

$$= -\alpha\varepsilon^{ijklm}\text{Tr}\big[U^\dagger \Delta U (\mathcal{U}_{Li} \mathcal{U}_{Lj} \mathcal{U}_{Lk} \mathcal{U}_{Ll} \mathcal{U}_{Lm} + \mathcal{U}_{Lj} \mathcal{U}_{Li} \mathcal{U}_{Lk} \mathcal{U}_{Ll} \mathcal{U}_{Lm}$$

$$+ \mathcal{U}_{Lj} \mathcal{U}_{Lk} \mathcal{U}_{Li} \mathcal{U}_{Ll} \mathcal{U}_{Lm} + \mathcal{U}_{Lj} \mathcal{U}_{Lk} \mathcal{U}_{Ll} \mathcal{U}_{Li} \mathcal{U}_{Lm})\big]$$

$$= 0,$$

because the first two terms as well as the final two terms cancel each other. The final result for δS_{ano}^0 is therefore given by Eq. B.4.

3.28

$$U = \mathbb{1} + i\frac{\phi}{F_0} + \cdots, \qquad \phi = \begin{pmatrix} \pi^0 & 0 & 0 \\ 0 & -\pi^0 & 0 \\ 0 & 0 & 0 \end{pmatrix}.$$

Evaluate the trace:

$$\mathrm{Tr}\big[2Q^2\big(U\partial_\mu U^\dagger - U^\dagger \partial_\mu U\big) - QU^\dagger Q\partial_\mu U + QUQ\partial_\mu U^\dagger\big]$$

$$= \mathrm{Tr}\left[2Q^2\left(-i\frac{\partial_\mu \phi}{F_0} - i\frac{\partial_\mu \phi}{F_0}\right) - Q^2\left(i\frac{\partial_\mu \phi}{F_0}\right) + Q^2\left(-\frac{\partial_\mu \phi}{F_0}\right)\right] + \cdots$$

$$= -\frac{6i}{F_0}\mathrm{Tr}\big(Q^2 \partial_\mu \phi\big) + \cdots$$

$$= -\frac{6i}{F_0}\left[\left(\frac{1}{2N_c} + \frac{1}{2}\right)^2 \partial_\mu \pi^0 + \left(\frac{1}{2N_c} - \frac{1}{2}\right)^2 \big(-\partial_\mu \pi^0\big)\right] + \cdots$$

$$= -\frac{6i\partial_\mu \pi^0}{N_c F_0} + \cdots.$$

Insert into Lagrangian, make use of integration by parts, introduce electromagnetic field-strength tensor, and rename indices:

$$\mathcal{L}_{\pi^0 \gamma\gamma} = i\frac{ne^2}{48\pi^2}\varepsilon^{\mu\nu\rho\sigma}\partial_\nu \mathscr{A}_\rho \mathscr{A}_\sigma \left(-\frac{6i\partial_\mu \pi^0}{N_c F_0}\right)$$

$$= \frac{n}{N_c}\frac{e^2}{8\pi^2}\varepsilon^{\mu\nu\rho\sigma}\partial_\nu \mathscr{A}_\rho \mathscr{A}_\sigma \frac{\partial_\mu \pi^0}{F_0}$$

$$= \text{total derivative} - \frac{n}{N_c}\frac{e^2}{8\pi^2}\varepsilon^{\mu\nu\rho\sigma}\partial_\nu \mathscr{A}_\rho \partial_\mu \mathscr{A}_\sigma \frac{\pi^0}{F_0}$$

$$= -\frac{n}{N_c}\frac{e^2}{32\pi^2}\varepsilon^{\mu\nu\rho\sigma}\mathscr{F}_{\nu\rho}\mathscr{F}_{\mu\sigma}\frac{\pi^0}{F_0}$$

$$= -\frac{n}{N_c}\frac{e^2}{32\pi^2}\varepsilon^{\mu\nu\rho\sigma}\mathscr{F}_{\mu\nu}\mathscr{F}_{\rho\sigma}\frac{\pi^0}{F_0}.$$

We dropped a total derivative and made use of $\varepsilon^{\mu\nu\rho\sigma}\partial_\mu \partial_\nu \mathscr{A}_\rho = 0$.

3.29 Let us define

$$M^{\mu\nu} = -i\frac{\alpha}{\pi F_0}\frac{n}{N_c}\varepsilon^{\mu\nu\rho\sigma}q_{1\rho}q_{2\sigma}, \qquad \alpha = \frac{e^2}{4\pi},$$

where q_1 and q_2 are the four-momenta of the two photons.

$$\sum_{\lambda_1,\lambda_2=1}^{2}|\mathcal{M}|^2 = M_{\mu\nu}M^{\mu\nu*} = \frac{\alpha^2}{\pi^2 F_0^2}\frac{n^2}{N_c^2}\varepsilon_{\mu\nu\alpha\beta}q_1^\alpha q_2^\beta \varepsilon^{\mu\nu\rho\sigma}q_{1\rho}q_{2\sigma}$$

$$= \frac{\alpha^2}{\pi^2 F_0^2}\frac{n^2}{N_c^2}2\big[(q_1 \cdot q_2)^2 - q_1^2 q_2^2\big] = \frac{2\alpha^2}{\pi^2 F_0^2}\frac{n^2}{N_c^2}(q_1 \cdot q_2)^2,$$

because $q_1^2 = q_2^2 = 0$. The decay rate is given by

$$\Gamma = \frac{1}{2M_{\pi^0}}\frac{2\alpha^2}{\pi^2 F_0^2}\frac{n^2}{N_c^2}\int \frac{d^3 q_1}{2\omega_1 (2\pi)^3}\frac{d^3 q_2}{2\omega_2 (2\pi)^3}(2\pi)^4 \delta^4(p - q_1 - q_2)\frac{1}{2}(q_1 \cdot q_2)^2.$$

The factor $1/2$ takes account of two (identical) photons in the final state.

$$
\begin{aligned}
\cdots &= \frac{\alpha^2}{8\pi^4 F_0^2 M_{\pi^0}} \frac{n^2}{N_c^2} \int \frac{d^3 q_1}{2\omega_1} d^4 q_2 \delta(q_2^2) \Theta(\omega_2) \delta^4(p - q_1 - q_2)(q_1 \cdot q_2)^2 \\
&= \frac{\alpha^2}{8\pi^4 F_0^2 M_{\pi^0}} \frac{n^2}{N_c^2} \int \frac{d^3 q_1}{2\omega_1} \delta\left[(p - q_1)^2\right] \Theta(M_{\pi^0} - \omega_1)[q_1 \cdot (p - q_1)]^2 \\
&= \frac{\alpha^2}{8\pi^4 F_0^2 M_{\pi^0}} \frac{n^2}{N_c^2} \frac{1}{2} \underbrace{\int d\Omega_1 \int_0^\infty d\omega_1 \omega_1 \frac{1}{2M_{\pi^0}} \delta\left(\omega_1 - \frac{M_{\pi^0}}{2}\right) \omega_1^2 M_{\pi^0}^2 \Theta(M_{\pi^0} - \omega_1)}_{= \frac{\pi}{8} M_{\pi^0}^4} \\
&= \frac{\alpha^2 M_{\pi^0}^3}{64\pi^3 F_0^2} \frac{n^2}{N_c^2}.
\end{aligned}
$$

Problems of Chapter 4

4.1

$$
\sum_{a=1}^8 \frac{B_a \lambda_a}{\sqrt{2}} = \begin{pmatrix} \frac{B_3}{\sqrt{2}} + \frac{B_8}{\sqrt{6}} & \frac{B_1 - iB_2}{\sqrt{2}} & \frac{B_4 - iB_5}{\sqrt{2}} \\ \frac{B_1 + iB_2}{\sqrt{2}} & -\frac{B_3}{\sqrt{2}} + \frac{B_8}{\sqrt{6}} & \frac{B_6 - iB_7}{\sqrt{2}} \\ \frac{B_4 + iB_5}{\sqrt{2}} & \frac{B_6 + iB_7}{\sqrt{2}} & -\sqrt{\frac{2}{3}} B_8 \end{pmatrix} = \begin{pmatrix} \frac{\Sigma^0}{\sqrt{2}} + \frac{\Lambda}{\sqrt{6}} & \Sigma^+ & p \\ \Sigma^- & -\frac{\Sigma^0}{\sqrt{2}} + \frac{\Lambda}{\sqrt{6}} & n \\ \Xi^- & \Xi^0 & -\frac{2\Lambda}{\sqrt{6}} \end{pmatrix}.
$$

$$
\Sigma^+ = \frac{B_1 - iB_2}{\sqrt{2}}, \qquad \Sigma^0 = B_3, \qquad \Sigma^- = \frac{B_1 + iB_2}{\sqrt{2}},
$$

$$
p = \frac{B_4 - iB_5}{\sqrt{2}}, \qquad n = \frac{B_6 - iB_7}{\sqrt{2}},
$$

$$
\Xi^0 = \frac{B_6 + iB_7}{\sqrt{2}} \qquad \Xi^- = \frac{B_4 + iB_5}{\sqrt{2}},
$$

$$
\Lambda = B_8.
$$

4.2

$$
\begin{aligned}
K(L_1, R_1, R_2 U L_2^\dagger) & K(L_2, R_2, U) \\
&= \sqrt{R_1 R_2 U L_2^\dagger L_1^\dagger}^{-1} R_1 \underbrace{\sqrt{R_2 U L_2^\dagger} \sqrt{R_2 U L_2^\dagger}^{-1}}_{= \mathbb{1}} R_2 \sqrt{U} \\
&= \sqrt{(R_1 R_2) U (L_1 L_2)^\dagger}^{-1} (R_1 R_2) \sqrt{U} \\
&= K[(L_1 L_2), (R_1 R_2), U].
\end{aligned}
$$

4.3

$$u_\mu = i\big[u^\dagger(\partial_\mu - ir_\mu)u - u(\partial_\mu - il_\mu)u^\dagger\big],$$

where

$$u \mapsto u' = V_R u K^\dagger = K u V_L^\dagger, \qquad u^\dagger \mapsto u'^\dagger = K u^\dagger V_R^\dagger = V_L u^\dagger K^\dagger,$$
$$r_\mu \mapsto r'_\mu = V_R r_\mu V_R^\dagger + i V_R \partial_\mu V_R^\dagger, \quad l_\mu \mapsto l'_\mu = V_L l_\mu V_L^\dagger + i V_L \partial_\mu V_L^\dagger.$$

$$u'_\mu = i\Big[u'^\dagger(\partial_\mu - ir'_\mu)u' - u'(\partial_\mu - il'_\mu)u'^\dagger\Big]$$
$$= i\Big[Ku^\dagger V_R^\dagger(\partial_\mu - iV_R r_\mu V_R^\dagger + V_R \partial_\mu V_R^\dagger)V_R u K^\dagger$$
$$- Ku V_L^\dagger(\partial_\mu - iV_L l_\mu V_L^\dagger + V_L \partial_\mu V_L^\dagger)V_L u^\dagger K^\dagger\Big].$$

Consider last term in second line:

$$Ku^\dagger(\partial_\mu V_R^\dagger)V_R u K^\dagger = Ku^\dagger\Big[\partial_\mu(V_R^\dagger V_R u K^\dagger) - V_R^\dagger \partial_\mu(V_R u K^\dagger)\Big]$$
$$= Ku^\dagger\Big[\partial_\mu(u K^\dagger) - V_R^\dagger \partial_\mu(V_R u K^\dagger)\Big]$$
$$= Ku^\dagger \partial_\mu u K^\dagger + K\partial_\mu K^\dagger - Ku^\dagger V_R^\dagger \partial_\mu(V_R u K^\dagger),$$

where we have used $V_R^\dagger V_R = \mathbb{1}$ and $u^\dagger u = \mathbb{1}$. Analogously for the last term in third line $\Rightarrow$

$$u'_\mu = iK\big[u^\dagger(\partial_\mu - ir_\mu)u\big]K^\dagger + iK\partial_\mu K^\dagger - iK\big[u(\partial_\mu - il_\mu)u^\dagger\big]K^\dagger - iK\partial_\mu K^\dagger = Ku_\mu K^\dagger.$$

4.4

$$u^\dagger\partial_\mu u = \left(\mathbb{1} - i\frac{\phi}{2F} - \frac{\phi^2}{8F^2} + \cdots\right)\left(i\frac{\partial_\mu\phi}{2F} - \frac{\partial_\mu\phi\,\phi + \phi\partial_\mu\phi}{8F^2} + \cdots\right)$$
$$= i\frac{\partial_\mu\phi}{2F} + \frac{\phi\partial_\mu\phi}{4F^2} - \frac{\partial_\mu\phi\,\phi + \phi\partial_\mu\phi}{8F^2} + \cdots$$
$$= i\frac{\partial_\mu\phi}{2F} + \frac{\phi\partial_\mu\phi - \partial_\mu\phi\,\phi}{8F^2} + \cdots,$$

$$u\partial_\mu u^\dagger = \left(\mathbb{1} + i\frac{\phi}{2F} - \frac{\phi^2}{8F^2} + \cdots\right)\left(-i\frac{\partial_\mu\phi}{2F} - \frac{\partial_\mu\phi\,\phi + \phi\partial_\mu\phi}{8F^2} + \cdots\right)$$
$$= -i\frac{\partial_\mu\phi}{2F} + \frac{\phi\partial_\mu\phi}{4F^2} - \frac{\partial_\mu\phi\,\phi + \phi\partial_\mu\phi}{8F^2} + \cdots$$
$$= -i\frac{\partial_\mu\phi}{2F} + \frac{\phi\partial_\mu\phi - \partial_\mu\phi\,\phi}{8F^2} + \cdots,$$

$$\Gamma_\mu = \frac{1}{2}\left(u^\dagger \partial_\mu u + u\partial_\mu u^\dagger\right)$$

$$= \frac{1}{2}\left(i\frac{\partial_\mu\phi}{2F} + \frac{\phi\partial_\mu\phi - \partial_\mu\phi\phi}{8F^2} + \cdots - i\frac{\partial_\mu\phi}{2F} + \frac{\phi\partial_\mu\phi - \partial_\mu\phi\phi}{8F^2} + \cdots\right)$$

$$= \frac{1}{8F^2}\left(\phi\partial_\mu\phi - \partial_\mu\phi\phi\right) + \cdots$$

$$= \frac{1}{8F^2}\phi_a\partial_\mu\phi_b(\tau_a\tau_b - \tau_b\tau_a) + \cdots$$

$$= \frac{1}{8F^2}\phi_a\partial_\mu\phi_b[\tau_a,\tau_b] + \cdots$$

$$= \frac{i}{4F^2}\varepsilon_{abc}\phi_a\partial_\mu\phi_b\tau_c + \cdots,$$

$$u_\mu = i\left(u^\dagger\partial_\mu u - u\partial_\mu u^\dagger\right)$$

$$= i\left(i\frac{\partial_\mu\phi}{2F} + \frac{\phi\partial_\mu\phi - \partial_\mu\phi\phi}{8F^2} + \cdots + i\frac{\partial_\mu\phi}{2F} - \frac{\phi\partial_\mu\phi - \partial_\mu\phi\phi}{8F^2} + \cdots\right)$$

$$= -\frac{\partial_\mu\phi}{F} + \cdots.$$

$$\mathscr{L}_{\pi NN}^{(1)} = -\frac{g_A}{2F}\bar{\Psi}\gamma^\mu\gamma_5\partial_\mu\phi_a\tau_a\Psi,$$

$$\Rightarrow$$

$$\mathscr{L}_{\pi\pi NN}^{(1)} = -\frac{1}{4F^2}\varepsilon_{abc}\bar{\Psi}\gamma^\mu\phi_a\partial_\mu\phi_b\tau_c\Psi.$$

4.5 Note that

$$\bar{\Psi}i\!\!\not{D}\Psi = \bar{\Psi}i\!\!\not{\partial}\Psi + \frac{i}{2}\bar{\Psi}\gamma^\mu(u^\dagger\partial_\mu u + u\partial_\mu u^\dagger)\Psi + \frac{1}{2}\bar{\Psi}\gamma^\mu(u^\dagger r_\mu u + u l_\mu u^\dagger)\Psi + \bar{\Psi}\gamma^\mu\Psi v_\mu^{(s)},$$

$$\frac{g_A}{2}\bar{\Psi}\gamma^\mu\gamma_5 u_\mu\Psi = i\frac{g_A}{2}\bar{\Psi}\gamma^\mu\gamma_5(u^\dagger\partial_\mu u - u\partial_\mu u^\dagger)\Psi + \frac{g_A}{2}\bar{\Psi}\gamma^\mu\gamma_5(u^\dagger r_\mu u - u l_\mu u^\dagger)\Psi,$$

$$\mathrm{Tr}[D_\mu U(D^\mu U)^\dagger] = \mathrm{Tr}[(\partial_\mu U - i r_\mu U + i U l_\mu)(\partial^\mu U - i r^\mu U + i U l^\mu)^\dagger]$$

$$= \mathrm{Tr}[(\partial_\mu U - i r_\mu U + i U l_\mu)(\partial^\mu U^\dagger + i U^\dagger r^\mu - i l^\mu U^\dagger)]$$

$$= \mathrm{Tr}(\partial_\mu U\partial^\mu U^\dagger) + 2i\mathrm{Tr}(\partial_\mu U U^\dagger r^\mu) + 2i\mathrm{Tr}(\partial_\mu U^\dagger U l^\mu)$$

$$- 2\mathrm{Tr}(r_\mu U l^\mu U^\dagger) + \mathrm{Tr}(r_\mu r^\mu) + \mathrm{Tr}(l_\mu l^\mu).$$

(a) Electromagnetic interaction:

$$r_\mu = l_\mu = -e\mathscr{A}_\mu\frac{\tau_3}{2}, \quad v_\mu^{(s)} = -\frac{e}{2}\mathscr{A}_\mu.$$

$$u^\dagger r_\mu u + u l_\mu u^\dagger = -e\mathscr{A}_\mu\tau_3 + \cdots,$$

$$u^\dagger\partial_\mu u - u\partial_\mu u^\dagger = i\frac{\partial_\mu\phi_a\tau_a}{F} + \cdots,$$

$$u^\dagger r_\mu u - u l_\mu u^\dagger$$

$$= -\frac{e}{2}\mathscr{A}_\mu(u^\dagger\tau_3 u - u\tau_3 u^\dagger) = -\frac{e}{2}\mathscr{A}_\mu(\tau_3 + [u^\dagger,\tau_3]u - \tau_3 - u[\tau_3,u^\dagger])$$

$$= -\frac{e}{2}\mathscr{A}_\mu([u^\dagger,\tau_3]u + u[u^\dagger,\tau_3]) = -\frac{e}{2}\mathscr{A}_\mu\{[u^\dagger,\tau_3],u\}$$

$$= -\frac{e}{2}\mathscr{A}_\mu\left\{\left[\mathbb{1} - i\frac{\vec{\phi}\cdot\vec{\tau}}{2F} + \cdots,\tau_3\right],\mathbb{1}+\cdots\right\} = -\frac{e}{2}\mathscr{A}_\mu\left[-i\frac{\phi_a\tau_a}{F},\tau_3\right]+\cdots$$

$$= i\frac{e}{2}\mathscr{A}_\mu 2i\varepsilon_{a3b}\frac{\phi_a}{F}\tau_b + \cdots = -\frac{e}{F}\mathscr{A}_\mu\varepsilon_{3ab}\tau_a\phi_b + \cdots,$$

$$\mathrm{Tr}(\partial_\mu U U^\dagger r^\mu) + \mathrm{Tr}(\partial_\mu U^\dagger U l^\mu)$$

$$= -\frac{e}{2}\mathscr{A}^\mu\mathrm{Tr}(\partial_\mu U U^\dagger\tau_3 + \partial_\mu U^\dagger U\tau_3) = -\frac{e}{2}\mathscr{A}^\mu\mathrm{Tr}(\partial_\mu U[U^\dagger,\tau_3])$$

$$= -\frac{e}{2}\mathscr{A}^\mu\mathrm{Tr}\left\{\left(i\frac{\partial_\mu\phi_a\tau_a}{F}+\cdots\right)\left[\mathbb{1}-i\frac{\phi_b\tau_b}{F}+\cdots,\tau_3\right]\right\}$$

$$= -\frac{e}{2F^2}\mathscr{A}^\mu\partial_\mu\phi_a\phi_b\mathrm{Tr}(\tau_a[\tau_b,\tau_3])+\cdots = -\frac{ie}{F^2}\varepsilon_{b3c}\mathscr{A}^\mu\partial_\mu\phi_a\phi_b\mathrm{Tr}(\tau_a\tau_c)+\cdots$$

$$= -2\frac{ie}{F^2}\varepsilon_{3ab}\mathscr{A}^\mu\partial_\mu\phi_a\phi_b+\cdots.$$

$$\mathscr{L}_{\gamma NN} = -e\bar{\Psi}\gamma^\mu\frac{\mathbb{1}+\tau_3}{2}\Psi\mathscr{A}_\mu,$$

$$\Rightarrow \qquad \mathscr{L}_{\pi NN} = -\frac{g_A}{2F}\bar{\Psi}\gamma^\mu\gamma_5\partial_\mu\phi_a\tau_a\Psi,$$

$$\mathscr{L}_{\gamma\pi NN} = -\frac{eg_A}{2F}\varepsilon_{3ab}\bar{\Psi}\gamma^\mu\gamma_5\tau_a\Psi\mathscr{A}_\mu\phi_b,$$

$$\mathscr{L}_{\gamma\pi\pi} = -e\varepsilon_{3ab}\phi_a\partial^\mu\phi_b\mathscr{A}_\mu.$$

(b) In nucleonic Lagrangian use

$$u = \mathbb{1}+\cdots, \quad u^\dagger = \mathbb{1}+\cdots.$$

$$\mathscr{L}_{WNN} = \frac{1}{2}\bar{\Psi}\gamma^\mu l_\mu\Psi - \frac{g_A}{2}\bar{\Psi}\gamma^\mu\gamma_5 l_\mu\Psi$$

$$= -\frac{g}{2\sqrt{2}}\bar{\Psi}\gamma^\mu(1-g_A\gamma_5)(\mathscr{W}_\mu^+ T_+ + \mathscr{W}_\mu^- T_-)\Psi$$

$$= -\frac{g}{2\sqrt{2}}V_{ud}\left[\bar{p}\gamma^\mu(1-g_A\gamma_5)n\mathscr{W}_\mu^+ + \bar{n}\gamma^\mu(1-g_A\gamma_5)p\mathscr{W}_\mu^-\right].$$

In mesonic Lagrangian use

$$\partial_\mu U^\dagger = -i\frac{\partial_\mu \phi_a \tau_a}{F} + \cdots, \quad U = \mathbb{1} + \cdots.$$

$$
\begin{aligned}
\mathcal{L}_{W\pi} &= i\frac{F^2}{2}\mathrm{Tr}\left(-i\frac{\partial_\mu \phi_a \tau_a}{F} l^\mu\right) \\
&= \frac{F}{2}\left(-\frac{g}{\sqrt{2}}\right)V_{ud}\partial^\mu \phi_a \mathrm{Tr}\left[\tau_a\left(\frac{\tau_1 + i\tau_2}{2}\mathcal{W}^+_\mu + \frac{\tau_1 - i\tau_2}{2}\mathcal{W}^-_\mu\right)\right] \\
&= -\frac{g}{\sqrt{2}}V_{ud}\frac{F}{2}\left[(\partial^\mu \phi_1 + i\partial^\mu \phi_2)\mathcal{W}^+_\mu + (\partial^\mu \phi_1 - i\partial^\mu \phi_2)\mathcal{W}^-_\mu\right] \\
&= -gV_{ud}\frac{F}{2}\left(\partial^\mu \pi^- \mathcal{W}^+_\mu + \partial^\mu \pi^+ \mathcal{W}^-_\mu\right).
\end{aligned}
$$

(c) In nucleonic Lagrangian use

$$u = \mathbb{1} + \cdots, \quad u^\dagger = \mathbb{1} + \cdots.$$

Make use of

$$
\begin{aligned}
&\frac{1}{2}(r_\mu + l_\mu) + v^{(s)}_\mu \\
&= \frac{1}{2}\left[e\tan(\theta_W)\mathcal{Z}_\mu\frac{\tau_3}{2} - \frac{g}{\cos(\theta_W)}\mathcal{Z}_\mu\frac{\tau_3}{2} + e\tan(\theta_W)\mathcal{Z}_\mu\frac{\tau_3}{2}\right] + \frac{e\tan(\theta_W)}{2}\mathcal{Z}_\mu \\
&= e\tan(\theta_W)\mathcal{Z}_\mu\frac{\mathbb{1} + \tau_3}{2} - \frac{1}{2}\frac{g}{\cos(\theta_W)}\mathcal{Z}_\mu\frac{\tau_3}{2} \\
&= \frac{g}{\cos(\theta_W)}\left[\sin^2(\theta_W)\mathcal{Z}_\mu\frac{\mathbb{1} + \tau_3}{2} - \frac{1}{2}\mathcal{Z}_\mu\frac{\tau_3}{2}\right] \\
&= \frac{g}{2\cos(\theta_W)}\mathcal{Z}_\mu\left\{\sin^2(\theta_W) + \left[\sin^2(\theta_W) - \frac{1}{2}\right]\tau_3\right\},
\end{aligned}
$$

$$r_\mu - l_\mu = \frac{g}{\cos(\theta_W)}\mathcal{Z}_\mu\frac{\tau_3}{2}.$$

$$
\begin{aligned}
\mathcal{L}_{ZNN} &= \frac{1}{2}\bar{\Psi}\gamma^\mu(r_\mu + l_\mu)\Psi + \bar{\Psi}\gamma^\mu \Psi v^{(s)}_\mu + \frac{g_A}{2}\bar{\Psi}\gamma^\mu \gamma_5(r_\mu - l_\mu)\Psi \\
&= \frac{g}{2\cos(\theta_W)}\bar{\Psi}\gamma^\mu\left(\left\{\sin^2(\theta_W) + \left[\sin^2(\theta_W) - \frac{1}{2}\right]\tau_3\right\} + \frac{g_A}{2}\gamma_5\tau_3\right)\Psi \mathcal{Z}_\mu \\
&= -\frac{g}{2\cos(\theta_W)}\left(\bar{p}\gamma^\mu\left\{\frac{1}{2}[1 - 4\sin^2(\theta_W)] - \frac{g_A}{2}\gamma_5\right\}p\mathcal{Z}_\mu + \bar{n}\gamma^\mu\left(-\frac{1}{2} + \frac{g_A}{2}\gamma_5\right)n\mathcal{Z}_\mu\right).
\end{aligned}
$$

In mesonic Lagrangian use

$$\partial_\mu U = i\frac{\partial_\mu \phi_a \tau_a}{F} + \cdots, \quad U^\dagger = \mathbb{1} + \cdots,$$

$$\partial_\mu U^\dagger = -i\frac{\partial_\mu \phi_a \tau_a}{F} + \cdots, \quad U = \mathbb{1} + \cdots.$$

$$\begin{aligned}
\mathscr{L}_{Z\pi} &= i\frac{F^2}{2}\mathrm{Tr}\left(i\frac{\partial_\mu \phi_a \tau_a}{F}r^\mu - i\frac{\partial_\mu \phi_a \tau_a}{F}l^\mu\right) \\
&= -\frac{F}{2}\partial_\mu \phi_a \mathrm{Tr}[\tau_a(r^\mu - l^\mu)] \\
&= -\frac{g}{2\cos(\theta_W)}\frac{F}{2}\partial_\mu \phi_a \mathscr{Z}^\mu \mathrm{Tr}(\tau_a \tau_3) = -\frac{g}{2\cos(\theta_W)}F\partial_\mu \pi^0 \mathscr{Z}^\mu.
\end{aligned}$$

4.6 Make use of

$$u_\mu = i(u^\dagger \partial_\mu u - u\partial_\mu u^\dagger) = -\frac{\partial_\mu \phi}{F_0} + \cdots,$$

where now $\phi = \phi_a \lambda_a$:

$$-\frac{D}{2}\mathrm{Tr}\left(\bar{B}\gamma^\mu \gamma_5 \{u_\mu, B\}\right) - \frac{F}{2}\mathrm{Tr}\left(\bar{B}\gamma^\mu \gamma_5 [u_\mu, B]\right)$$

$$\rightarrow \frac{1}{4F_0}\bar{B}_b \gamma^\mu \gamma_5 B_a \partial_\mu \phi_c \left[D\underbrace{\mathrm{Tr}(\lambda_b\{\lambda_c, \lambda_a\})}_{=\,4d_{cab}} + F\underbrace{\mathrm{Tr}(\lambda_b[\lambda_c, \lambda_a])}_{=\,4if_{cab}}\right]$$

$$= \frac{1}{F_0}\bar{B}_b \gamma^\mu \gamma_5 B_a \partial_\mu \phi_c (Dd_{cab} + iFf_{cab})$$

$$= \frac{1}{F_0}(d_{abc}D + if_{abc}F)\bar{B}_b \gamma^\mu \gamma_5 B_a \partial_\mu \phi_c$$

$$= \mathscr{L}^{(1)}_{\phi BB}.$$

$$\mathrm{Tr}\left(\bar{B}i\gamma^\mu \frac{1}{2}[u^\dagger \partial_\mu u + u\partial_\mu u^\dagger, B]\right)$$

$$= \frac{i}{4}\bar{B}_b \gamma^\mu B_a \mathrm{Tr}(\lambda_b[[u^\dagger, \partial_\mu u], \lambda_a])$$

$$= \frac{i}{4}\bar{B}_b \gamma^\mu B_a \mathrm{Tr}\left(\lambda_b\left[\left[\mathbb{1} - i\frac{\phi_c \lambda_c}{2F_0} + \cdots, i\frac{\partial_\mu \phi_d \lambda_d}{2F_0} + \cdots\right], \lambda_a\right]\right)$$

$$= \frac{i}{16F_0^2}\bar{B}_b \gamma^\mu B_a \phi_c \partial_\mu \phi_d \mathrm{Tr}(\lambda_b[\underbrace{[\lambda_c, \lambda_d]}_{=\,2if_{cde}\lambda_e}, \lambda_a]) + \cdots$$

$$= -\frac{1}{8F_0^2}\bar{B}_b \gamma^\mu B_a \phi_c \partial_\mu \phi_d f_{cde}\underbrace{\mathrm{Tr}(\lambda_b[\lambda_e, \lambda_a])}_{=\,4if_{eab}} + \cdots$$

$$= -\frac{i}{2F_0^2}\bar{B}_b \gamma^\mu B_a \phi_c \partial_\mu \phi_d f_{cde}f_{eab} + \cdots.$$

$$\Rightarrow \mathscr{L}^{(1)}_{\phi\phi BB} = -\frac{i}{2F_0^2}f_{abe}f_{cde}\bar{B}_b \gamma^\mu B_a \phi_c \partial_\mu \phi_d.$$

4.7 (a)

$$iq_\mu\langle B|A_i^\mu|A\rangle \rightarrow iq_\mu\bar{u}(p')\left[\gamma^\mu G_A(t) + \frac{q^\mu}{2m_N}G_P(t)\right]\gamma_5\frac{\tau_i}{2}u(p)$$

$$= i\bar{u}(p')\left[\slashed{q}\gamma_5 G_A(t) + \frac{q^2}{2m_N}G_P(t)\gamma_5\right]\frac{\tau_i}{2}u(p)$$

$$= i\bar{u}(p')\left[(\slashed{p}'\gamma_5 + \gamma_5\slashed{p})G_A(t) + \frac{t}{2m_N}G_P(t)\gamma_5\right]\frac{\tau_i}{2}u(p)$$

$$= i\bar{u}(p')\left[2m_N G_A(t) + \frac{t}{2m_N}G_P(t)\right]\gamma_5\frac{\tau_i}{2}u(p),$$

$$\hat{m}\langle B|P_i|A\rangle \rightarrow \frac{M_\pi^2 F_\pi}{M_\pi^2 - t}G_{\pi N}(t)i\bar{u}(p')\gamma_5\tau_i u(p),$$

$$2m_N G_A(t) + \frac{t}{2m_N}G_P(t) = 2\frac{M_\pi^2 F_\pi}{M_\pi^2 - t}G_{\pi N}(t).$$

(b)

$$2m_N g_A - \frac{t}{2m_N}\frac{4m^2 g_A}{t - M^2} = 2m g_A\left(1 - \frac{t}{t - M^2}\right) + \mathcal{O}(q^2)$$

$$= 2m g_A\frac{M^2}{M^2 - t} + \mathcal{O}(q^2)$$

$$= 2\frac{M^2 F}{M^2 - t}\frac{m}{F}g_A + \mathcal{O}(q^2),$$

where we have used that $m_N = m + \mathcal{O}(q^2)$, $F_\pi = F[1 + \mathcal{O}(q^2)]$, $M_\pi^2 = M^2[1 + \mathcal{O}(q^2)]$.

4.8 (a) First show that

$$s + t + u = (p + q)^2 + (p' - p)^2 + (p' - q)^2$$
$$= (p + q)^2 + (p' - p)^2 + (p - q')^2$$
$$= 4m_N^2 + 2M_\pi^2 + 2p \cdot (q - p' - q')$$
$$= 4m_N^2 + 2M_\pi^2 - 2p \cdot p$$
$$= 2m_N^2 + 2M_\pi^2,$$

from which $t + u = 2m_N^2 + 2M_\pi^2 - s$ and $s + t = 2m_N^2 + 2M_\pi^2 - u$.

$$2m_N(v - v_B) = \frac{1}{2}(s - u - t + 2M_\pi^2)$$

$$= \frac{1}{2}(2s - 2m_N^2)$$

$$= s - m_N^2,$$

$$-2m_N(v + v_B) = -\frac{1}{2}(s - u + t - 2M_\pi^2)$$
$$= -\frac{1}{2}(-2u + 2m_N^2)$$
$$= u - m_N^2.$$

In the center-of-mass frame, the threshold four-momenta read

$$p^\mu = (m_N, \vec{0}) = p'^\mu, \quad q^\mu = (M_\pi, \vec{0}) = q'^\mu.$$

$$v\big|_{\text{thr}} = \frac{2m_N M_\pi}{2m_N} = M_\pi, \quad v_B\big|_{\text{thr}} = -\frac{M_\pi^2}{2m_N}.$$

(b)
$$T_{ab}(p, q; p', q') = \delta_{ab}T^+(p, q; p', q') - i\varepsilon_{abc}\tau_c T^-(p, q; p', q'),$$
$$T_{ba}(p, -q'; p', -q) = \delta_{ba}T^+(p, -q'; p', -q) - i\varepsilon_{bac}\tau_c T^-(p, -q'; p', -q).$$

Crossing symmetry dictates

$$T^+(p, q; p', q') = T^+(p, -q'; p', -q),$$
$$T^-(p, q; p', q') = -T^-(p, -q'; p', -q). \tag{B.5}$$

Under $q \leftrightarrow -q'$:
$$v \to -v, \quad v_B \to v_B,$$

so that

$$T^\pm(p, -q'; p', -q) = \bar{u}(p')\left[A^\pm(-v, v_B) - \frac{1}{2}(\slashed{q} + \slashed{q}')B^\pm(-v, v_B)\right]u(p).$$

Crossing behavior of Eq. 4.45 then follows from Eq. B.5.

(c)
$$\langle p\pi^0|T|n\pi^+\rangle = \left(\sqrt{\frac{2}{3}}\left\langle\frac{3}{2}, \frac{1}{2}\right| + \frac{1}{\sqrt{3}}\left\langle\frac{1}{2}, \frac{1}{2}\right|\right)T\left(\frac{1}{\sqrt{3}}\left|\frac{3}{2}, \frac{1}{2}\right\rangle - \sqrt{\frac{2}{3}}\left|\frac{1}{2}, \frac{1}{2}\right\rangle\right)$$
$$= \frac{\sqrt{2}}{3}T^{\frac{3}{2}} - \frac{\sqrt{2}}{3}T^{\frac{1}{2}}.$$

Analogously
$$\langle p\pi^0|T|p\pi^0\rangle = \frac{2}{3}T^{\frac{3}{2}} + \frac{1}{3}T^{\frac{1}{2}}, \quad \langle n\pi^+|T|n\pi^+\rangle = \frac{1}{3}T^{\frac{3}{2}} + \frac{2}{3}T^{\frac{1}{2}},$$

and

$$\langle p\pi^0|T|p\pi^0\rangle - \langle n\pi^+|T|n\pi^+\rangle = \frac{1}{3}T^{\frac{3}{2}} - \frac{1}{3}T^{\frac{1}{2}} = \frac{1}{\sqrt{2}}\langle p\pi^0|T|n\pi^+\rangle.$$

4.9

$$\mathcal{M}_s = g_{\pi N}^2 \bar{u}(p')\gamma_5\tau_b\frac{i}{\not{p} + \not{q} - m + i0^+}\gamma_5\tau_a u(p),$$

$$\mathcal{M}_u = g_{\pi N}^2 \bar{u}(p')\gamma_5\tau_a\frac{i}{\not{p} - \not{q}' - m + i0^+}\gamma_5\tau_b u(p).$$

4.10 (a) Only the c_1 term contributes to self energy at tree level:

$$i\langle N|c_1 4M^2\bar{\Psi}\Psi|N\rangle \Rightarrow 4ic_1M^2,$$

$$-i\Sigma_2^{\text{tree}}(\not{p}) = 4ic_1M^2.$$

(b)

$$i\langle N|-\frac{g_A}{2F}\bar{\Psi}\gamma^\mu\gamma_5\partial_\mu\phi_b\tau_b\Psi|N, \phi_a(k)\rangle \Rightarrow -\frac{g_A}{2F}\not{k}\gamma_5\tau_a,$$

$$i\langle N, \phi_b(k')|-\frac{1}{4F^2}\varepsilon_{cde}\bar{\Psi}\gamma^\mu\phi_d\partial_\mu\phi_e\tau_c\Psi|N, \phi_a(k)\rangle \Rightarrow \frac{1}{4F^2}(\not{k} + \not{k}')\varepsilon_{abc}\tau_c.$$

(c) Closing of the pion propagator forces the isospin indices in the Feynman rule to be the same, i.e., $\varepsilon_{aac} = 0$.

(d)

$$-i\Sigma^{\text{loop}}(\not{p}) = \mu^{4-n}\int\frac{d^n k}{(2\pi)^n}\left(-\frac{g_A}{2F}\not{k}\gamma_5\tau_a\right)iS_F(p-k)i\Delta_F(k)\frac{g_A}{2F}\not{k}\gamma_5\tau_a$$

$$= -i\frac{3g_A^2}{4F^2}i\mu^{4-n}\int\frac{d^n k}{(2\pi)^n}\frac{\not{k}\gamma_5(\not{p} - \not{k} + m)\not{k}\gamma_5}{[(p-k)^2 - m^2 + i0^+](k^2 - M^2 + i0^+)}$$

$$= -i\frac{3g_A^2}{4F^2}i\mu^{4-n}\int\frac{d^n k}{(2\pi)^n}\frac{\not{k}(\not{p} - m - \not{k})\not{k}}{[(p-k)^2 - m^2 + i0^+](k^2 - M^2 + i0^+)}.$$

We made use of $\tau_a\tau_a = 3\mathbb{1} = 3$.

(e)

$$\not{k}(\not{p} - m - \not{k})\not{k}$$
$$= (2p\cdot k - \not{p}\not{k} - m\not{k} - k^2)\not{k}$$
$$= 2p\cdot k\not{k} - (\not{p} + m)k^2 - k^2\not{k}$$
$$= -(p-k)^2\not{k} + p^2\not{k} + k^2\not{k} - (\not{p} + m)k^2 - k^2\not{k}$$
$$= -(\not{p} + m)k^2 + (p^2 - m^2)\not{k} - \left[(p-k)^2 - m^2\right]\not{k}$$
$$= -(\not{p} + m)(k^2 - M^2) - (\not{p} + m)M^2 + (p^2 - m^2)\not{k} - \left[(p-k)^2 - m^2\right]\not{k},$$

where we have replaced $k^2 = k^2 - M^2 + M^2$ in the first term of the penultimate line. Inserting this result in Σ^{loop} we obtain the expression of Eq. 4.83.

(f) Contracting

$$p_\mu C = i\mu^{4-n} \int \frac{d^n k}{(2\pi)^n} \frac{k_\mu}{[(p-k)^2 - m^2 + i0^+](k^2 - M^2 + i0^+)}$$

with p^μ we obtain

$$
\begin{aligned}
p^2 C &= i\mu^{4-n} \int \frac{d^n k}{(2\pi)^n} \frac{p \cdot k}{[(p-k)^2 - m^2 + i0^+](k^2 - M^2 + i0^+)} \\
&= i\mu^{4-n} \int \frac{d^n k}{(2\pi)^n} \frac{-\frac{1}{2}[(p-k)^2 - p^2 - k^2]}{[(p-k)^2 - m^2 + i0^+](k^2 - M^2 + i0^+)} \\
&= -\frac{1}{2} i\mu^{4-n} \int \frac{d^n k}{(2\pi)^n} \frac{(p-k)^2 - m^2 - (p^2 - m^2) - (k^2 - M^2) - M^2}{[(p-k)^2 - m^2 + i0^+](k^2 - M^2 + i0^+)},
\end{aligned}
$$

from which we can read off the expression for C.

(g) The expression for $I_{N\pi}^r(-p, 0)$ contains the term $-\frac{1}{16\pi^2}$, which is of $\mathcal{O}(q^0)$. Combined with the factor M^2 multiplying the integral, we see that the self energy contains a term of $\mathcal{O}(q^2)$. Since the power counting predicted the loop contribution to be at least of $\mathcal{O}(q^3)$, we see that renormalization using the $\widetilde{\text{MS}}$ scheme does not result in a consistent power counting.

(h) As $m_N - m \sim \mathcal{O}(q^2)$ and $I_\pi \sim M^2 \sim \mathcal{O}(q^2)$, the last line in Eq. 4.88 is at least of $\mathcal{O}(q^4)$ and can therefore be ignored. We thus only need the expansion of $I_{N\pi}^r$ to first order. As the function $F(\Omega)$ is multiplied by a factor of $\mathcal{O}(q)$, we only need its expansion to lowest order. With $p^2 \approx m^2$ we find for the dimensionless quantity $-1 \le \Omega \le 1$. Therefore,

$$F(\Omega) = \sqrt{1 - \Omega^2} \arccos(-\Omega) = \frac{\pi}{2} + \mathcal{O}(\Omega).$$

Since $p^2 - m^2 - M^2 \sim \mathcal{O}(q^2)$ for on-shell momenta, we can neglect the term containing the logarithm in $I_{N\pi}^r$ and obtain

$$I_{N\pi}^r = \frac{1}{16\pi^2}\left(-1 + \frac{\pi M}{m} + \cdots\right).$$

The loop contribution to the self energy is therefore

$$
\begin{aligned}
\Sigma_r^{\text{loop}}(m_N) &= -\frac{3g_{Ar}^2}{4F_r^2} 2mM^2 \frac{1}{16\pi^2}\left(-1 + \frac{\pi M}{m} + \cdots\right) \\
&= \frac{3g_{Ar}^2}{32\pi^2 F_r^2} M^2 m - \frac{3g_{Ar}^2}{32\pi F_r^2} M^3 + \cdots,
\end{aligned}
$$

where we made use of $m_N - m \sim \mathcal{O}(q^2)$.

(i) Replace $c_{1r} = c_1 + \delta c_1$:

$$m_N = m - 4c_1 M^2 - 4\delta c_1 M^2 + \frac{3g_{Ar}^2}{32\pi^2 F_r^2}M^2 m - \frac{3g_{Ar}^2}{32\pi F_r^2}M^3 + \cdots .$$

The term violating the power counting is absorbed by the choice

$$\delta c_1 = \frac{3g_{Ar}^2}{128\pi^2 F_r^2}m.$$

4.11

$$P_{v\pm}^2 = \frac{1}{2}(\mathbb{1} \pm \not{v})\frac{1}{2}(\mathbb{1} \pm \not{v}) = \frac{1}{4}(\mathbb{1} \pm 2\not{v} + \underbrace{\not{v}\not{v}}_{= v^2 \mathbb{1} = \mathbb{1}}) = \frac{1}{2}(\mathbb{1} \pm \not{v}) = P_{v\pm},$$

$$P_{v\pm}P_{v\mp} = \frac{1}{2}(\mathbb{1} \pm \not{v})\frac{1}{2}(\mathbb{1} \mp \not{v}) = \frac{1}{4}(\mathbb{1} - \not{v}\not{v}) = 0,$$

$$\not{v}e^{+imv\cdot x}P_{v\pm}\Psi = e^{+imv\cdot x}\not{v}\frac{1}{2}(\mathbb{1} \pm \not{v})\Psi = e^{+imv\cdot x}\frac{1}{2}(\not{v} \pm \mathbb{1})\Psi = \pm e^{+imv\cdot x}P_{v\pm}\Psi.$$

4.12 (a) First use integration by parts to rewrite

$$\bar{\Psi}\partial_\mu\Psi = -\partial_\mu\bar{\Psi}\Psi$$

to obtain

$$-\partial_\mu\frac{\partial\mathcal{L}_{\pi N}^{(1)}}{\partial\partial_\mu\bar{\Psi}} = -\partial_\mu(-i\gamma^\mu\Psi) = i\not{\partial}\Psi,$$

$$\frac{\partial\mathcal{L}_{\pi N}^{(1)}}{\partial\bar{\Psi}} = \left[i\left(\not{\Gamma} - i\not{v}^{(s)}\right) - m + \frac{g_A}{2}\gamma^\mu\gamma_5 u_\mu\right]\Psi.$$

EOM follows as

$$\left[i\left(\not{\partial} + \not{\Gamma} - i\not{v}^{(s)}\right) - m + \frac{g_A}{2}\gamma^\mu\gamma_5 u_\mu\right]\Psi = \left(i\not{D} - m + \frac{g_A}{2}\gamma^\mu\gamma_5 u_\mu\right)\Psi = 0.$$

(b) It is sufficient to consider the partial-derivative part acting on the nucleon field:

$$i\not{\partial}\Psi = i\not{\partial}\left[e^{-imv\cdot x}(\mathcal{N}_v + \mathcal{H}_v)\right]$$

$$= e^{-imv\cdot x}\left[m\not{v}(\mathcal{N}_v + \mathcal{H}_v) + i\not{\partial}(\mathcal{N}_v + \mathcal{H}_v)\right]$$

$$= e^{-imv\cdot x}\left[(i\not{\partial} + m)\mathcal{N}_v + (i\not{\partial} - m)\mathcal{H}_v\right].$$

(c) From algebra

$$\not{v}\not{A} = 2v \cdot A - \not{A}\not{v},$$

from which we see that

$$P_{v\pm}\slashed{A} = \frac{1}{2}(\mathbb{1} \pm \slashed{v})\slashed{A} = \pm v \cdot A + \slashed{A} P_{v\mp}.$$

Then

$$
\begin{aligned}
P_{v\pm}\slashed{A}P_{v\pm} &= \pm v \cdot A P_{v\pm} + \slashed{A}P_{v\mp}P_{v\pm} \\
&= \pm v \cdot A P_{v\pm}, \\
P_{v\pm}\slashed{A}P_{v\mp} &= \slashed{A}P_{v\mp}P_{v\mp} \pm v \cdot A P_{v\mp} \\
&= \slashed{A}P_{v\mp} \pm v \cdot A P_{v\mp}.
\end{aligned}
$$

Now use $\mathbb{1} = \slashed{v}\slashed{v}$ and

$$P_{v\pm} = \frac{1}{2}(\mathbb{1} \pm \slashed{v}) = \frac{1}{2}(\slashed{v}\slashed{v} \pm \slashed{v}) = \slashed{v}\frac{1}{2}(\slashed{v} \pm \mathbb{1}) = \pm\slashed{v}P_{v\pm}$$

to obtain

$$
\begin{aligned}
P_{v\pm}\slashed{A}P_{v\mp} &= (\pm v \cdot A + \slashed{A}P_{v\mp})P_{v\mp} \\
&= \pm v \cdot A P_{v\mp} + \slashed{A}P_{v\mp} \\
&= \pm v \cdot A \slashed{v} \underbrace{\slashed{v} P_{v\mp}}_{=\mp P_{v\mp}} + \slashed{A}P_{v\mp} \\
&= -v \cdot A \slashed{v}P_{v\mp} + \slashed{A}P_{v\mp} \\
&= (\slashed{A} - v \cdot A \slashed{v})P_{v\mp} \\
&= \slashed{A}_\perp P_{v\mp}.
\end{aligned}
$$

Use

$$\gamma_5 P_{v\pm} = \gamma_5\frac{1}{2}(\mathbb{1} \pm \slashed{v}) = \frac{1}{2}(\mathbb{1} \mp \slashed{v})\gamma_5 = P_{v\mp}\gamma_5$$

to rewrite the terms containing $\slashed{B}\gamma_5$ in terms of the relations above, e.g.,

$$P_{v\pm}\slashed{B}\gamma_5 P_{v\pm} = P_{v\pm}\slashed{B}P_{v\mp}\gamma_5 = \slashed{B}_\perp P_{v\mp}\gamma_5 = \slashed{B}_\perp\gamma_5 P_{v\pm}$$

and analogously for other expressions.
(d) Project the EOM with $P_{v\pm}$ and use

$$\mathcal{N}_v = P_{v+}\mathcal{N}_v, \quad \mathcal{H}_v = P_{v-}\mathcal{H}_v,$$

e.g.,

$$P_{v+}i\not{D}\mathcal{N}_v = P_{v+}i\not{D}P_{v+}\mathcal{N}_v,$$

then use the relations of Eq. 4.105,

$$\cdots = iv \cdot DP_{v+}\mathcal{N}_v = iv \cdot D\mathcal{N}_v.$$

(e) Formally solve Eq. 4.107 for $\mathcal{H}_v$,

$$\mathcal{H}_v = \left(2m + iv \cdot D - \frac{g_A}{2}\not{u}_\perp \gamma_5\right)^{-1}\left(i\not{D}_\perp - \frac{g_A}{2}v \cdot u\gamma_5\right)\mathcal{N}_v.$$

Insert result into Eq. 4.106.

4.13

$$v \cdot S_v = v_\mu \frac{i}{2}\gamma_5 \sigma^{\mu\nu} v_\nu = -\frac{1}{4}\gamma_5(\not{v}\not{v} - \not{v}\not{v}) = 0.$$

$$\begin{aligned}
\{S_v^\mu, S_v^\nu\} &= \frac{1}{4}\{\gamma_5(\gamma^\mu\not{v} - v^\mu), \gamma_5(\gamma^\nu\not{v} - v^\nu)\} \\
&= \frac{1}{4}\{(\gamma^\mu\not{v} - v^\mu), (\gamma^\nu\not{v} - v^\nu)\} \\
&= \frac{1}{4}(\{\gamma^\mu\not{v}, \gamma^\nu\not{v}\} - \{\gamma^\mu\not{v}, v^\nu\} - \{v^\mu, \gamma^\nu\not{v}\} + \{v^\mu, v^\nu\}) \\
&= \frac{1}{4}(\{\gamma^\mu\not{v}, \gamma^\nu\not{v}\} - 2\gamma^\mu\not{v}v^\nu - 2\gamma^\nu\not{v}v^\mu + 2v^\mu v^\nu).
\end{aligned}$$

Consider

$$\begin{aligned}
\{\gamma^\mu\not{v}, \gamma^\nu\not{v}\} &= \gamma^\mu\not{v}\gamma^\nu\not{v} + \gamma^\nu\not{v}\gamma^\mu\not{v} \\
&= \gamma^\mu\not{v}2v^\nu - \gamma^\mu\not{v}\not{v}\gamma^\nu + \gamma^\nu\not{v}2v^\mu - \gamma^\nu\not{v}\not{v}\gamma^\mu \\
&= 2\gamma^\mu\not{v}v^\nu + 2\gamma^\nu\not{v}v^\mu - \{\gamma^\mu, \gamma^\nu\} \\
&= 2\gamma^\mu\not{v}v^\nu + 2\gamma^\nu\not{v}v^\mu - 2g^{\mu\nu}.
\end{aligned}$$

$$\{S_v^\mu, S_v^\nu\} = \cdots = \frac{1}{2}(v^\mu v^\nu - g^{\mu\nu}).$$

Evaluate left-hand side of commutator,

$$\begin{aligned}
[S_v^\mu, S_v^\nu] &= \frac{1}{4}[\gamma_5(\gamma^\mu\not{v} - v^\mu), \gamma_5(\gamma^\nu\not{v} - v^\nu)] \\
&= \frac{1}{4}[\gamma^\mu\not{v} - v^\mu, \gamma^\nu\not{v} - v^\nu] \\
&= \frac{1}{4}[\gamma^\mu\not{v}, \gamma^\nu\not{v}] \\
&= \frac{1}{4}(\gamma^\mu\not{v}\gamma^\nu\not{v} - \gamma^\nu\not{v}\gamma^\mu\not{v})
\end{aligned}$$

$$= \frac{1}{4}[\gamma^\mu \not{p}(-\not{p}\gamma^\nu + 2v^\nu) - \gamma^\nu \not{p}(-\not{p}\gamma^\mu + 2v^\mu)]$$

$$= \frac{1}{4}(-\gamma^\mu\gamma^\nu + \gamma^\nu\gamma^\mu + 2v^\nu\gamma^\mu\not{p} - 2v^\mu\gamma^\nu\not{p}).$$

Rewrite the right-hand side using

$$\gamma_5\sigma^{\sigma\tau} = -\frac{i}{2}\varepsilon^{\sigma\tau\alpha\beta}\sigma_{\alpha\beta},$$

$$\varepsilon^{\mu\nu\rho}{}_\sigma\varepsilon^{\sigma\tau\alpha\beta} = \det\begin{pmatrix} g^{\mu\tau} & g^{\mu\alpha} & g^{\mu\beta} \\ g^{\nu\tau} & g^{\nu\alpha} & g^{\nu\beta} \\ g^{\rho\tau} & g^{\rho\alpha} & g^{\rho\beta} \end{pmatrix}.$$

$$i\varepsilon^{\mu\nu\rho}{}_\sigma v_\rho S^\sigma_\nu = i\varepsilon^{\mu\nu\rho}{}_\sigma v_\rho \frac{i}{2}\gamma_5\sigma^{\sigma\tau}v_\tau$$

$$= \frac{i}{4}\varepsilon^{\mu\nu\rho}{}_\sigma\varepsilon^{\sigma\tau\alpha\beta}v_\rho\sigma_{\alpha\beta}v_\tau$$

$$= -\frac{1}{8}v_\rho[\gamma_\alpha,\gamma_\beta]v_\tau\left[g^{\mu\tau}\left(g^{\nu\alpha}g^{\rho\beta} - g^{\rho\alpha}g^{\nu\beta}\right) - \cdots\right]$$

$$= -\frac{1}{8}(v^\mu[\gamma^\nu,\not{p}] - [\not{p},\gamma^\nu]v^\mu - [\gamma^\mu,\not{p}]v^\nu + [\gamma^\mu,\gamma^\nu] + [\not{p},\gamma^\mu]v^\nu - [\gamma^\nu,\gamma^\mu])$$

$$= -\frac{1}{4}(v^\mu[\gamma^\nu,\not{p}] + [\not{p},\gamma^\mu]v^\nu + [\gamma^\mu,\gamma^\nu])$$

$$= -\frac{1}{4}(v^\mu(2\gamma^\nu\not{p} - 2v^\nu) + (2v^\mu - 2\gamma^\mu\not{p})v^\nu + [\gamma^\mu,\gamma^\nu])$$

$$= -\frac{1}{4}(2v^\mu\gamma^\nu\not{p} - 2v^\nu\gamma^\mu\not{p} + [\gamma^\mu,\gamma^\nu]).$$

4.14 First note that

$$\bar{\mathcal{N}}_v = \mathcal{N}_v^\dagger\gamma_0$$

$$= \Psi^\dagger e^{-imv\cdot x}\frac{1}{2}(\mathbb{1} + \not{p})^\dagger\gamma_0$$

$$= e^{-imv\cdot x}\Psi^\dagger\frac{1}{2}(\mathbb{1} + \gamma_0\not{p}\gamma_0)\gamma_0$$

$$= e^{-imv\cdot x}\Psi^\dagger\gamma_0\frac{1}{2}(\mathbb{1} + \not{p}) = e^{-imv\cdot x}\bar{\Psi}P_{v+}.$$

From $P_{v+}\not{p} = P_{v+}$ it then follows that

$$\bar{\mathcal{N}}_v = \bar{\mathcal{N}}_v\not{p}.$$

$$\bar{\mathcal{N}}_v\gamma_5\mathcal{N}_v = \bar{\mathcal{N}}_v\gamma_5\not{p}\mathcal{N}_v = -\bar{\mathcal{N}}_v\not{p}\gamma_5\mathcal{N}_v = -\bar{\mathcal{N}}_v\gamma_5\mathcal{N}_v \quad \Rightarrow \quad \bar{\mathcal{N}}_v\gamma_5\mathcal{N}_v = 0.$$

$$\bar{\mathcal{N}}_v \gamma^\mu \mathcal{N}_v = \bar{\mathcal{N}}_v \gamma^\mu \not v \mathcal{N}_v = 2v^\mu \bar{\mathcal{N}}_v \mathcal{N}_v - \bar{\mathcal{N}}_v \not v \gamma^\mu \mathcal{N}_v = 2v^\mu \bar{\mathcal{N}}_v \mathcal{N}_v - \bar{\mathcal{N}}_v \gamma^\mu \mathcal{N}_v$$
$$\Rightarrow \quad \bar{\mathcal{N}}_v \gamma^\mu \mathcal{N}_v = v^\mu \bar{\mathcal{N}}_v \mathcal{N}_v.$$

Use $\gamma^\mu \gamma_5 \not v = -\gamma_5 \gamma^\mu \not v = 2S_v^\mu - v^\mu \gamma_5$,

$$\bar{\mathcal{N}}_v \gamma^\mu \gamma_5 \mathcal{N}_v = \bar{\mathcal{N}}_v \gamma^\mu \gamma_5 \not v \mathcal{N}_v = 2\bar{\mathcal{N}}_v S_v^\mu \mathcal{N}_v - v^\mu \bar{\mathcal{N}}_v \gamma_5 \mathcal{N}_v = 2\bar{\mathcal{N}}_v S_v^\mu \mathcal{N}_v.$$

Use

$$\sigma^{\mu\nu} = -\frac{i}{2}\varepsilon^{\mu\nu\rho\sigma}\gamma_5 \sigma_{\rho\sigma} = \frac{1}{4}\varepsilon^{\mu\nu\rho\sigma}\gamma_5[\gamma_\rho, \gamma_\sigma] = \frac{1}{2}\varepsilon^{\mu\nu\rho\sigma}\gamma_5 \gamma_\rho \gamma_\sigma$$

to reduce the problem to the previous one:

$$\bar{\mathcal{N}}_v \sigma^{\mu\nu} \mathcal{N}_v = \frac{1}{2}\varepsilon^{\mu\nu\rho\sigma} \bar{\mathcal{N}}_v \gamma_5 \gamma_\rho \gamma_\sigma \mathcal{N}_v$$

$$= \frac{1}{2}\varepsilon^{\mu\nu\rho\sigma} \bar{\mathcal{N}}_v \gamma_5 \gamma_\rho \gamma_\sigma \not v \mathcal{N}_v$$

$$= \frac{1}{2}\varepsilon^{\mu\nu\rho\sigma} \left(-\bar{\mathcal{N}}_v \gamma_5 \gamma_\rho \not v \gamma_\sigma \mathcal{N}_v + 2v_\sigma \bar{\mathcal{N}}_v \gamma_5 \gamma_\rho \mathcal{N}_v \right)$$

$$= \frac{1}{2}\varepsilon^{\mu\nu\rho\sigma} \left(\bar{\mathcal{N}}_v \gamma_5 \not v \gamma_\rho \gamma_\sigma \mathcal{N}_v - 2v_\rho \bar{\mathcal{N}}_v \gamma_5 \gamma_\sigma \mathcal{N}_v + 2v_\sigma \bar{\mathcal{N}}_v \gamma_5 \gamma_\rho \mathcal{N}_v \right)$$

$$= \frac{1}{2}\varepsilon^{\mu\nu\rho\sigma} \left(-\bar{\mathcal{N}}_v \not v \gamma_5 \gamma_\rho \gamma_\sigma \mathcal{N}_v - 4v_\rho \bar{\mathcal{N}}_v \gamma_5 \gamma_\sigma \mathcal{N}_v \right)$$

$$= \frac{1}{2}\varepsilon^{\mu\nu\rho\sigma} \left(-\bar{\mathcal{N}}_v \gamma_5 \gamma_\rho \gamma_\sigma \mathcal{N}_v + 4v_\rho \bar{\mathcal{N}}_v \gamma_\sigma \gamma_5 \mathcal{N}_v \right)$$

$$= -\bar{\mathcal{N}}_v \sigma^{\mu\nu} \mathcal{N}_v + 2\varepsilon^{\mu\nu\rho}{}_\sigma v_\rho \bar{\mathcal{N}}_v \gamma^\sigma \gamma_5 \mathcal{N}_v$$

$$\Rightarrow \bar{\mathcal{N}}_v \sigma^{\mu\nu} \mathcal{N}_v = \varepsilon^{\mu\nu\rho}{}_\sigma v_\rho \bar{\mathcal{N}}_v \gamma^\sigma \gamma_5 \mathcal{N}_v = 2\varepsilon^{\mu\nu\rho}{}_\sigma v_\rho \bar{\mathcal{N}}_v S_v^\sigma \mathcal{N}_v.$$

$$\bar{\mathcal{N}}_v \sigma^{\mu\nu} \gamma_5 \mathcal{N}_v = \frac{i}{2}\bar{\mathcal{N}}_v(\gamma^\mu \gamma^\nu - \gamma^\nu \gamma^\mu)\gamma_5 \mathcal{N}_v$$

$$= \frac{i}{2}\bar{\mathcal{N}}_v \not v(\gamma^\mu \gamma^\nu - \gamma^\nu \gamma^\mu)\gamma_5 \mathcal{N}_v$$

$$= \frac{i}{2}\bar{\mathcal{N}}_v(-\gamma^\mu \not v \gamma^\nu + 2v^\mu \gamma^\nu + \gamma^\nu \not v \gamma^\mu - 2v^\nu \gamma^\mu)\gamma_5 \mathcal{N}_v$$

$$= \frac{i}{2}\bar{\mathcal{N}}_v(\gamma^\mu \gamma^\nu \not v - 2\gamma^\mu v^\nu + 2v^\mu \gamma^\nu - \gamma^\nu \gamma^\mu \not v + 2\gamma^\nu v^\mu - 2v^\nu \gamma^\mu)\gamma_5 \mathcal{N}_v$$

$$= -\frac{i}{2}\bar{\mathcal{N}}_v(\gamma^\mu \gamma^\nu - \gamma^\nu \gamma^\mu)\gamma_5 \not v \mathcal{N}_v$$

$$\quad + 2i(v^\mu \bar{\mathcal{N}}_v \gamma^\nu \gamma_5 \mathcal{N}_v - v^\nu \bar{\mathcal{N}}_v \gamma^\mu \gamma_5 \mathcal{N}_v)$$

$$= -\bar{\mathcal{N}}_v \sigma^{\mu\nu} \gamma_5 \mathcal{N}_v + 4i(v^\mu \bar{\mathcal{N}}_v S_v^\nu \mathcal{N}_v - v^\nu \bar{\mathcal{N}}_v S_v^\mu \mathcal{N}_v)$$

$$\Rightarrow \quad \bar{\mathcal{N}}_v \sigma^{\mu\nu} \gamma_5 \mathcal{N}_v = 2i(v^\mu \bar{\mathcal{N}}_v S_v^\nu \mathcal{N}_v - v^\nu \bar{\mathcal{N}}_v S_v^\mu \mathcal{N}_v).$$

4.15 (a) Using the Feynman parameterization, the denominator is given by

$$
\begin{aligned}
&\left[(k^2 - 2p \cdot k + (p^2 - m^2) + i0^+)z + (k^2 - M^2 + i0^+)(1 - z)\right]^2 \\
&= \left[k^2 - 2p \cdot kz + (p^2 - m^2)z - M^2(1 - z) + i0^+\right]^2 \\
&\rightarrow \left[(k + zp)^2 - 2p \cdot (k + zp)z + (p^2 - m^2 + M^2)z - M^2 + i0^+\right]^2 \\
&= \left[k^2 + 2p \cdot kz + z^2p^2 - 2p \cdot kz - 2z^2p^2 + (p^2 - m^2 + M^2)z - M^2 + i0^+\right]^2 \\
&= \left[k^2 - z^2p^2 + z(p^2 - m^2 + M^2) - M^2 + i0^+\right]^2 \\
&= \left[k^2 - A(z) + i0^+\right]^2.
\end{aligned}
$$

(b) Use the given equation with the exponents $p = 0$ and $q = 2$. The integral is then given by

$$
\begin{aligned}
H(p^2, m^2, M^2; n) &= -i \int_0^1 dz \frac{i}{(4\pi)^{\frac{n}{2}}} \frac{\Gamma\left(\frac{n}{2}\right)\Gamma\left(2 - \frac{n}{2}\right)}{\Gamma\left(\frac{n}{2}\right)\Gamma(2)}[A(z) - i0^+]^{\frac{n}{2}-2} \\
&= \frac{1}{(4\pi)^{\frac{n}{2}}}\Gamma\left(2 - \frac{n}{2}\right) \int_0^1 dz[A(z) - i0^+]^{\frac{n}{2}-2}.
\end{aligned}
$$

(c) At threshold $p_{\text{thr}}^2 = (m + M)^2 \Rightarrow$

$$
\begin{aligned}
A(z) &= z^2(m + M)^2 - z\left[(m + M)^2 - m^2 + M^2\right] + M^2 \\
&= z^2(m + M)^2 - 2z(m + M)M + M^2 = [z(m + M) - M]^2.
\end{aligned}
$$

For $0 \leq z \leq 1, A(z)$ is never negative, therefore can drop small imaginary part. Integrand is zero for $z_0 = \frac{M}{m+M}$, and splitting of integral allows to rewrite $[A(z)]^{\frac{n}{2}-2}$ as

$$
\begin{aligned}
\int_0^1 dz[A(z)]^{\frac{n}{2}-2} &= \int_0^{z_0} dz[M - z(m + M)]^{n-4} + \int_{z_0}^1 dz[z(m + M) - M]^{n-4} \\
&= -\frac{1}{(n - 3)(m + M)}[M - z(m + M)]^{n-3}\Big|_0^{z_0} \\
&\quad + \frac{1}{(n - 3)(m + M)}[z(m + M) - M]^{n-3}\Big|_{z_0}^1 \\
&= \frac{1}{(n - 3)(m + M)}(M^{n-3} + m^{n-3})
\end{aligned}
$$

for $n > 3$.

(d) Expansion in small quantities corresponds to expansion of $1/(m+M)$ in M:

$$\frac{1}{m+M} = \frac{1}{m}\frac{1}{1+\frac{M}{m}} = \frac{1}{m}\left(1 - \frac{M}{m} + \frac{M^2}{m^2} + \cdots\right),$$

i.e., only nonnegative integer powers of M are generated. Therefore the infrared-singular part has an expansion

$$\sim M^{n-3}\left(1 - \frac{M}{m} + \frac{M^2}{m^2} + \cdots\right) = M^{n-3} - \frac{M^{n-2}}{m} + \frac{M^{n-1}}{m^2} + \cdots,$$

while for the infrared-regular part

$$\sim m^{n-3}\left(1 - \frac{M}{m} + \frac{M^2}{m^2} + \cdots\right) = m^{n-3} - Mm^{n-4} + M^2 m^{n-5} + \cdots.$$

4.16 (a)

$$\Omega_{\text{thr}} = \frac{(m+M)^2 - m^2 - M^2}{2mM} = \frac{m^2 + 2mM + M^2 - m^2 - M^2}{2mM} = 1,$$

$$\begin{aligned}
D_{\text{thr}}(x) &= 1 - 2\Omega_{\text{thr}}x + x^2 + 2\alpha x(\Omega_{\text{thr}}x - 1) + \alpha^2 x^2 \\
&= 1 - 2x + x^2 + 2\alpha x(x-1) + \alpha^2 x^2 = (x-1)^2 + 2\alpha x(x-1) + \alpha^2 x^2 \\
&= [(x-1) + \alpha x]^2 = [(1+\alpha)x - 1]^2,
\end{aligned}$$

$$\begin{aligned}
I_{\text{thr}} &= \kappa(m;n)\alpha^{n-3}\int_0^\infty dx [D_{\text{thr}}(x) - i0^+]^{\frac{n}{2}-2} \\
&= \kappa(m;n)\alpha^{n-3}\int_0^\infty dx\left\{[(1+\alpha)x - 1]^2 - i0^+\right\}^{\frac{n}{2}-2}.
\end{aligned}$$

Integral potentially divergent for large x, where integrand behaves as $x^{n-4} \Rightarrow$ convergent for $n < 3$.

(b)

$$\begin{aligned}
&\frac{(1+\alpha)x - 1}{(1+\alpha)(n-4)}\frac{d}{dx}\left\{[(1+\alpha)x - 1]^2 - i0^+\right\}^{\frac{n}{2}-2} \\
&= \frac{(1+\alpha)x - 1}{(1+\alpha)(n-4)}\frac{n-4}{2}\left\{[(1+\alpha)x - 1]^2 - i0^+\right\}^{\frac{n}{2}-3} 2[(1+\alpha)x - 1](1+\alpha) \\
&= [(1+\alpha)x - 1]^2\left\{[(1+\alpha)x - 1]^2 - i0^+\right\}^{\frac{n}{2}-3} \\
&= \left\{[(1+\alpha)x - 1]^2 - i0^+\right\}^{\frac{n}{2}-2}.
\end{aligned}$$

$$\int_0^\infty dx \left\{ [(1+\alpha)x - 1]^2 - i0^+ \right\}^{\frac{n}{2}-2}$$

$$= \int_0^\infty dx \frac{(1+\alpha)x - 1}{(1+\alpha)(n-4)} \frac{d}{dx} \left\{ [(1+\alpha)x - 1]^2 - i0^+ \right\}^{\frac{n}{2}-2}$$

$$= \left[\frac{(1+\alpha)x - 1}{(1+\alpha)(n-4)} \left\{ [(1+\alpha)x - 1]^2 - i0^+ \right\}^{\frac{n}{2}-2} \right]_0^\infty$$

$$- \int_0^\infty dx \left\{ [(1+\alpha)x - 1]^2 - i0^+ \right\}^{\frac{n}{2}-2} \underbrace{\frac{d}{dx} \frac{(1+\alpha)x - 1}{(1+\alpha)(n-4)}}_{= \frac{1}{n-4}}.$$

(c) The second expression on the right is proportional to the original integral. Bringing it to the left we obtain

$$\frac{n-3}{n-4} \int_0^\infty dx \left\{ [(1+\alpha)x - 1]^2 - i0^+ \right\}^{\frac{n}{2}-2}$$

$$= \left[\frac{(1+\alpha)x - 1}{(1+\alpha)(n-4)} \left\{ [(1+\alpha)x - 1]^2 - i0^+ \right\}^{\frac{n}{2}-2} \right]_0^\infty.$$

The expression on the right vanishes at the upper limit for $n < 3$, and at the lower limit yields $1/[(1+\alpha)(n-4)]$, so that

$$\int_0^\infty dx \left\{ [(1+\alpha)x - 1]^2 - i0^+ \right\}^{\frac{n}{2}-2} = \frac{1}{(n-3)(1+\alpha)}.$$

With $\alpha = M/m$ and $\kappa(m; n)$ given in Eq. 4.125

$$I_{\text{thr}} = \kappa(m; n)\alpha^{n-3} \frac{1}{(n-3)(1+\alpha)} = \frac{\Gamma\left(2 - \frac{n}{2}\right)}{(4\pi)^{\frac{n}{2}}(n-3)} \frac{M^{n-3}}{m+M}.$$

(d)

$$C_{\text{thr}}(z) = z^2 - 2\alpha z(1-z) + \alpha^2(1-z)^2 = [z - \alpha(1-z)]^2 = [z(1+\alpha) - \alpha]^2,$$

$$R_{\text{thr}} = -\kappa(m; n) \int_1^\infty dz [C_{\text{thr}}(z) - i0^+]^{\frac{n}{2}-2}$$

$$= -\kappa(m; n) \int_1^\infty dz \left\{ [z(1+\alpha) - \alpha]^2 - i0^+ \right\}^{\frac{n}{2}-2}$$

$$= -\kappa(m; n) \frac{1}{m^{n-4}} \int_1^\infty dz [z(m+M) - M]^{n-4}$$

$$= -\kappa(m; n) \frac{1}{m^{n-4}} \left[\frac{1}{(n-3)(m+M)} [z(m+M) - M]^{n-3} \right]_1^\infty$$

$$\overset{n<3}{=} \frac{\Gamma\left(2 - \frac{n}{2}\right)}{(4\pi)^{\frac{n}{2}}(n-3)} \frac{m^{n-3}}{m+M}.$$

4.17 (a)

$$F\left(1, 2 - \frac{n}{2}; 4 - n; -\Delta\right) = 1 + \frac{2 - \frac{n}{2}}{4 - n}(-\Delta) + \frac{2(2 - \frac{n}{2})(3 - \frac{n}{2})}{(4 - n)(5 - n)} \frac{\Delta^2}{2} + \cdots$$

$$= 1 - \frac{\Delta}{2} + \frac{6 - n}{5 - n} \frac{\Delta^2}{4} + \cdots.$$

$$\varepsilon = 4 - n \quad \Rightarrow \quad \begin{cases} 6 - n = 2 + \varepsilon \\ 5 - n = 1 + \varepsilon \end{cases}$$

$$\cdots = 1 - \frac{\Delta}{2} + \frac{2 + \varepsilon}{1 + \varepsilon} \frac{\Delta^2}{4} + \cdots$$

(b)

$$\cdots = 1 - \frac{\Delta}{2} + (2 + \varepsilon)(1 - \varepsilon)\frac{\Delta^2}{4} + \cdots$$

$$= 1 - \frac{\Delta}{2} + (2 - \varepsilon)\frac{\Delta^2}{4} + \cdots.$$

(c)

$$\frac{\Gamma(2 - \frac{n}{2})}{n - 3} = \frac{\Gamma(3 - \frac{n}{2})}{2 - \frac{n}{2}} \frac{1}{n - 3}$$

$$= \frac{\Gamma(1 + \frac{\varepsilon}{2})}{\frac{\varepsilon}{2}} \frac{1}{1 - \varepsilon}$$

$$= \frac{2}{\varepsilon} + 2 + \Gamma'(1) + O(\varepsilon).$$

(d)

$$\left[-\frac{\Delta}{2} + (2 - \varepsilon)\frac{\Delta^2}{4}\right]\left[\frac{2}{\varepsilon} + 2 + \Gamma'(1) + \cdots\right]$$

$$= -\Delta\left[\frac{1}{\varepsilon} + 1 + \frac{\Gamma'(1)}{2}\right] + \Delta^2\left[\frac{1}{\varepsilon} + \frac{1}{2} + \frac{\Gamma'(1)}{2}\right] + \cdots.$$

(e)

$$(-\Delta)^{n-3} = (-\Delta)(-\Delta)^{-\varepsilon} = (-\Delta)e^{-\varepsilon \ln(-\Delta)}$$

$$= (-\Delta)[1 - \varepsilon \ln(-\Delta) + \cdots]$$

$$= -\Delta + \varepsilon \Delta \ln(-\Delta) + \cdots.$$

(f)

$$F\left(\frac{n}{2}-1, n-2; n-2; -\Delta\right) = 1 + \left(\frac{n}{2}-1\right)(-\Delta) + \cdots$$
$$= 1 - \left(1 - \frac{\varepsilon}{2}\right)\Delta + \cdots.$$

$$\Gamma\left(\frac{n}{2}-1\right)\Gamma(3-n) = \Gamma\left(1-\frac{\varepsilon}{2}\right)\Gamma(-1+\varepsilon)$$
$$= \Gamma\left(1-\frac{\varepsilon}{2}\right)\frac{\Gamma(\varepsilon)}{-1+\varepsilon}$$
$$= -\Gamma\left(1-\frac{\varepsilon}{2}\right)\frac{\Gamma(1+\varepsilon)}{\varepsilon(1-\varepsilon)}$$
$$= -\frac{1}{\varepsilon}(1+\varepsilon+\cdots)\left[1-\frac{\varepsilon}{2}\Gamma'(1)+\cdots\right]\left[1+\varepsilon\Gamma'(1)+\cdots\right]$$
$$= -\frac{1}{\varepsilon} - \left[1+\frac{\Gamma'(1)}{2}\right] + \cdots.$$

$$(-\Delta)^{n-3}\Gamma\left(\frac{n}{2}-1\right)\Gamma(3-n)F\left(\frac{n}{2}-1, n-2; n-2; -\Delta\right)$$
$$= (-\Delta)[1 - \varepsilon \ln(-\Delta) + \cdots]\left[-\frac{1}{\varepsilon} - \left(1+\frac{\Gamma'(1)}{2}\right)+\cdots\right]\left[1-\left(1-\frac{\varepsilon}{2}\right)\Delta+\cdots\right]$$
$$= \Delta\left[\frac{1}{\varepsilon}+1+\frac{\Gamma'(1)}{2}\right] - \Delta\ln(-\Delta) - \Delta^2\left[\frac{1}{\varepsilon}+\frac{1}{2}+\frac{\Gamma'(1)}{2}\right] + \Delta^2\ln(-\Delta) + \cdots.$$

Comparison with (d) shows that terms proportional to Δ and Δ^2 precisely cancel.

4.18 (a) Equation of motion for $A = -1$:

$$\frac{\partial \mathscr{L}_{\frac{3}{2}}}{\partial \bar{\Psi}_\mu} - \partial_\rho \frac{\partial \mathscr{L}_{\frac{3}{2}}}{\partial \partial_\rho \bar{\Psi}_\mu} = \Lambda^{\mu\nu}(A)\Psi_\nu$$

$$\overset{A=-1}{=} \quad -(i\partial\!\!\!/ - m_\Delta)\Psi^\mu + i(\gamma^\mu\partial^\nu + \partial^\mu\gamma^\nu)\Psi_\nu - i\gamma^\mu\partial\!\!\!/\gamma^\nu\Psi_\nu - m_\Delta\gamma^\mu\gamma^\nu\Psi_\nu = 0.$$

(b) Use $\gamma_\mu\gamma^\mu = 4$,

$$\gamma_\mu\cdots = -i\gamma_\mu\partial\!\!\!/\Psi^\mu + m_\Delta\gamma_\mu\Psi^\mu + i\gamma_\mu\gamma^\mu\partial^\nu\Psi_\nu$$
$$\quad + i\partial\!\!\!/\gamma^\nu\Psi_\nu - i\gamma_\mu\gamma^\mu\partial\!\!\!/\gamma^\nu\Psi_\nu - m_\Delta\gamma_\mu\gamma^\mu\gamma^\nu\Psi_\nu$$
$$= -i\gamma_\mu\partial\!\!\!/\Psi^\mu - 3m_\Delta\gamma_\mu\Psi^\mu + 4i\partial_\mu\Psi^\mu - 3i\partial\!\!\!/\gamma_\mu\Psi^\mu$$
$$= -2i\partial_\mu\Psi^\mu + i\partial\!\!\!/\gamma_\mu\Psi^\mu + 4i\partial_\mu\Psi^\mu - 3i\partial\!\!\!/\gamma_\mu\Psi^\mu - 3m_\Delta\gamma_\mu\Psi^\mu$$
$$= 2i\partial_\mu\Psi^\mu - 2i\partial\!\!\!/\gamma_\mu\Psi^\mu - 3m_\Delta\gamma_\mu\Psi^\mu = 0 \qquad (B.6)$$

for solutions of the EOM.
(c)

$$\partial_\mu \cdots = -(i\slashed{\partial} - m_\Delta)\partial_\mu \Psi^\mu + i(\slashed{\partial}\partial_\mu \Psi^\mu + \Box \gamma_\mu \Psi^\mu) - i\Box \gamma_\mu \Psi^\mu - m_\Delta \slashed{\partial} \gamma_\mu \Psi^\mu$$
$$= m_\Delta \partial_\mu \Psi^\mu - m_\Delta \slashed{\partial} \gamma_\mu \Psi^\mu = 0 \tag{B.7}$$

for solutions of the EOM. For $m_\Delta \neq 0$,

$$\partial_\mu \Psi^\mu = \slashed{\partial} \gamma_\mu \Psi^\mu$$

for solutions of the EOM.
(d) Insert into Eq. B.6 $\Rightarrow$

$$-3m_\Delta \gamma_\mu \Psi^\mu = 0$$

and thus $(m_\Delta \neq 0)$

$$\gamma_\mu \Psi^\mu = 0 \tag{B.8}$$

for solutions of the EOM.
(e) Insert Eq. B.8 into Eq. B.7 $\Rightarrow$

$$\partial_\mu \Psi^\mu = 0 \tag{B.9}$$

for solutions of the EOM.
(f) Insert Eqs. B.8 and B.9 into EOM $\Rightarrow$

$$(i\slashed{\partial} - m_\Delta)\Psi^\mu = 0.$$

4.19 In the following, we neglect terms of order ε^2.

$$\begin{aligned}
D_{ij}(V) &= \frac{1}{2}\mathrm{Tr}(\tau_i V \tau_j V^\dagger) \\
&= \frac{1}{2}\mathrm{Tr}\left[\tau_i\left(\mathbb{1} - i\varepsilon_a\frac{\tau_a}{2}\right)\tau_j\left(\mathbb{1} + i\varepsilon_b\frac{\tau_b}{2}\right)\right] \\
&= \frac{1}{2}\mathrm{Tr}(\tau_i \tau_j) + i\frac{\varepsilon_a}{4}\mathrm{Tr}(\tau_i \tau_j \tau_a) - i\frac{\varepsilon_a}{4}\mathrm{Tr}(\tau_i \tau_a \tau_j) \\
&= \delta_{ij} + i\frac{\varepsilon_a}{4}\mathrm{Tr}(\tau_a \underbrace{[\tau_i, \tau_j]}_{= 2i\varepsilon_{ijk}\tau_k}) \\
&= \delta_{ij} + i\frac{\varepsilon_a}{4}4i\varepsilon_{ijk}\delta_{ak} \\
&= \delta_{ij} - i\varepsilon_a(-i\varepsilon_{aij}) \\
&= \delta_{ij} - i\varepsilon_a t_{a,ij}^{\mathrm{ad}}.
\end{aligned}$$

4.20 (a)

$$P_{\frac{1}{2}} = \left| \left(1\tfrac{1}{2}\right)\tfrac{1}{2}, -\tfrac{1}{2} \right\rangle \left\langle \left(1\tfrac{1}{2}\right)\tfrac{1}{2}, -\tfrac{1}{2} \right| + \left| \left(1\tfrac{1}{2}\right)\tfrac{1}{2}, \tfrac{1}{2} \right\rangle \left\langle \left(1\tfrac{1}{2}\right)\tfrac{1}{2}, \tfrac{1}{2} \right|$$

$$= \left(-\sqrt{\tfrac{2}{3}}|1,-1\rangle \left|\tfrac{1}{2}, \tfrac{1}{2}\right\rangle + \tfrac{1}{\sqrt{3}}|1,0\rangle \left|\tfrac{1}{2}, -\tfrac{1}{2}\right\rangle \right)$$

$$\times \left(-\sqrt{\tfrac{2}{3}}\langle 1,-1| \left\langle\tfrac{1}{2}, \tfrac{1}{2}\right| + \tfrac{1}{\sqrt{3}}\langle 1,0| \left\langle\tfrac{1}{2}, -\tfrac{1}{2}\right| \right)$$

$$+ \left(-\tfrac{1}{\sqrt{3}}|1,0\rangle \left|\tfrac{1}{2}, \tfrac{1}{2}\right\rangle + \sqrt{\tfrac{2}{3}}|1,1\rangle \left|\tfrac{1}{2}, -\tfrac{1}{2}\right\rangle \right)$$

$$\times \left(-\tfrac{1}{\sqrt{3}}\langle 1,0| \left\langle\tfrac{1}{2}, \tfrac{1}{2}\right| + \sqrt{\tfrac{2}{3}}\langle 1,1| \left\langle\tfrac{1}{2}, -\tfrac{1}{2}\right| \right)$$

$$= \tfrac{1}{3}|1,0\rangle\langle 1,0| \left(\left|\tfrac{1}{2}, \tfrac{1}{2}\right\rangle\left\langle\tfrac{1}{2}, \tfrac{1}{2}\right| + \left|\tfrac{1}{2}, -\tfrac{1}{2}\right\rangle\left\langle\tfrac{1}{2}, -\tfrac{1}{2}\right| \right)$$

$$- \tfrac{\sqrt{2}}{3}(|1,0\rangle\langle 1,1| + |1,-1\rangle\langle 1,0|) \left|\tfrac{1}{2}, \tfrac{1}{2}\right\rangle\left\langle\tfrac{1}{2}, -\tfrac{1}{2}\right|$$

$$- \tfrac{\sqrt{2}}{3}(|1,1\rangle\langle 1,0| + |1,0\rangle\langle 1,-1|) \left|\tfrac{1}{2}, -\tfrac{1}{2}\right\rangle\left\langle\tfrac{1}{2}, \tfrac{1}{2}\right|$$

$$+ \tfrac{2}{3}\left(|1,1\rangle\langle 1,1| \left|\tfrac{1}{2}, -\tfrac{1}{2}\right\rangle\left\langle\tfrac{1}{2}, -\tfrac{1}{2}\right| + |1,-1\rangle\langle 1,-1| \left|\tfrac{1}{2}, \tfrac{1}{2}\right\rangle\left\langle\tfrac{1}{2}, \tfrac{1}{2}\right| \right).$$

(b)

$$|1,0\rangle\langle 1,0| = \begin{pmatrix} 0 \\ 1 \\ 0 \end{pmatrix} (0 \quad 1 \quad 0) = \begin{pmatrix} 0 & 0 & 0 \\ 0 & 1 & 0 \\ 0 & 0 & 0 \end{pmatrix},$$

$$|1,0\rangle\langle 1,1| + |1,-1\rangle\langle 1,0| = \begin{pmatrix} 0 \\ 1 \\ 0 \end{pmatrix} (1 \quad 0 \quad 0) + \begin{pmatrix} 0 \\ 0 \\ 1 \end{pmatrix} (0 \quad 1 \quad 0) = \begin{pmatrix} 0 & 0 & 0 \\ 1 & 0 & 0 \\ 0 & 1 & 0 \end{pmatrix},$$

$$|1,1\rangle\langle 1,0| + |1,0\rangle\langle 1,-1| = \begin{pmatrix} 1 \\ 0 \\ 0 \end{pmatrix} (0 \quad 1 \quad 0) + \begin{pmatrix} 0 \\ 1 \\ 0 \end{pmatrix} (0 \quad 0 \quad 1) = \begin{pmatrix} 0 & 1 & 0 \\ 0 & 0 & 1 \\ 0 & 0 & 0 \end{pmatrix},$$

$$|1,1\rangle\langle 1,1| = \begin{pmatrix} 1 \\ 0 \\ 0 \end{pmatrix} (1 \quad 0 \quad 0) = \begin{pmatrix} 1 & 0 & 0 \\ 0 & 0 & 0 \\ 0 & 0 & 0 \end{pmatrix},$$

$$|1,-1\rangle\langle 1,-1| = \begin{pmatrix} 0 \\ 0 \\ 1 \end{pmatrix} (0 \quad 0 \quad 1) = \begin{pmatrix} 0 & 0 & 0 \\ 0 & 0 & 0 \\ 0 & 0 & 1 \end{pmatrix}.$$

(c)

$$\left|\frac{1}{2},\frac{1}{2}\right\rangle\left\langle\frac{1}{2},\frac{1}{2}\right| + \left|\frac{1}{2},-\frac{1}{2}\right\rangle\left\langle\frac{1}{2},-\frac{1}{2}\right| = \begin{pmatrix} 1 & 0 \\ 0 & 1 \end{pmatrix} = \mathbb{1},$$

$$\left|\frac{1}{2},\frac{1}{2}\right\rangle\left\langle\frac{1}{2},-\frac{1}{2}\right| = \begin{pmatrix} 0 & 1 \\ 0 & 0 \end{pmatrix} = \frac{1}{2}(\tau_1 + i\tau_2),$$

$$\left|\frac{1}{2},-\frac{1}{2}\right\rangle\left\langle\frac{1}{2},\frac{1}{2}\right| = \begin{pmatrix} 0 & 0 \\ 1 & 0 \end{pmatrix} = \frac{1}{2}(\tau_1 - i\tau_2),$$

$$\left|\frac{1}{2},-\frac{1}{2}\right\rangle\left\langle\frac{1}{2},-\frac{1}{2}\right| = \begin{pmatrix} 0 & 0 \\ 0 & 1 \end{pmatrix} = \frac{1}{2}(\mathbb{1} - \tau_3),$$

$$\left|\frac{1}{2},\frac{1}{2}\right\rangle\left\langle\frac{1}{2},\frac{1}{2}\right| = \begin{pmatrix} 1 & 0 \\ 0 & 0 \end{pmatrix} = \frac{1}{2}(\mathbb{1} + \tau_3).$$

$$\Rightarrow \quad \xi_{\text{sph}}^{\frac{1}{2}} = \frac{1}{3}\begin{pmatrix} 0 & 0 & 0 \\ 0 & 1 & 0 \\ 0 & 0 & 0 \end{pmatrix} \otimes \mathbb{1}$$

$$-\frac{\sqrt{2}}{3}\begin{pmatrix} 0 & 0 & 0 \\ 1 & 0 & 0 \\ 0 & 1 & 0 \end{pmatrix} \otimes \frac{\tau_1 + i\tau_2}{2} - \frac{\sqrt{2}}{3}\begin{pmatrix} 0 & 1 & 0 \\ 0 & 0 & 1 \\ 0 & 0 & 0 \end{pmatrix} \otimes \frac{\tau_1 - i\tau_2}{2}$$

$$+\frac{2}{3}\begin{pmatrix} 1 & 0 & 0 \\ 0 & 0 & 0 \\ 0 & 0 & 0 \end{pmatrix} \otimes \frac{\mathbb{1} - \tau_3}{2} + \frac{2}{3}\begin{pmatrix} 0 & 0 & 0 \\ 0 & 0 & 0 \\ 0 & 0 & 1 \end{pmatrix} \otimes \frac{\mathbb{1} + \tau_3}{2}$$

$$= \frac{1}{3}\begin{pmatrix} \mathbb{1} - \tau_3 & -\frac{1}{\sqrt{2}}(\tau_1 - i\tau_2) & 0 \\ -\frac{1}{\sqrt{2}}(\tau_1 + i\tau_2) & \mathbb{1} & -\frac{1}{\sqrt{2}}(\tau_1 - i\tau_2) \\ 0 & -\frac{1}{\sqrt{2}}(\tau_1 + i\tau_2) & \mathbb{1} + \tau_3 \end{pmatrix}.$$

Problems of Chapter 5

5.1 (a) For contact interaction without pion fields, we can replace the covariant derivative $D_\mu \to \partial_\mu$, and

$$\partial_\mu \mathcal{N}_\nu = \partial_\mu e^{imv\cdot x} P_{v+}\Psi = imv^\mu \mathcal{N}_\nu + e^{imv\cdot x} P_{v+}\partial_\mu\Psi$$
$$\to imv^\mu \mathcal{N}_\nu - ip^\mu \mathcal{N}_\nu = -ik_p^\mu \mathcal{N}_\nu.$$

$$-i\Sigma^{\text{tree}}(p) = i\frac{k_p^2}{2m} + i4c_1 M^2.$$

(b) Expand u_μ and Γ_μ as in Exercise 4.4.

$$\left\langle N \left| -i\frac{g_A}{F} \bar{\mathcal{N}}_v S_v^\mu \partial_\mu \phi_b \tau_b \mathcal{N}_v \right| N, \phi_a(q) \right\rangle \Rightarrow -i\frac{g_A}{F} S_v^\mu (-iq_\mu) \delta_{ab} \tau_b = -\frac{g_A}{F} S_v \cdot q \tau_a.$$

$$\left\langle N, \phi_b(q') \left| -i\frac{v^\mu}{4F^2} \bar{\mathcal{N}}_v \varepsilon_{dec} \phi_d \partial_\mu \phi_e \tau_c \mathcal{N}_v \right| N, \phi_a(q) \right\rangle$$

$$\Rightarrow -i\frac{v^\mu}{4F^2}[\varepsilon_{abc}(iq'_\mu) + \varepsilon_{bac}(-iq_\mu)]\tau_c = \frac{v \cdot (q + q')}{4F^2} \varepsilon_{abc} \tau_c.$$

(c)

$$-i\Sigma^{\text{loop}}(p)$$

$$= \mu^{4-n} \int \frac{d^n q}{(2\pi)^n} \frac{g_A}{F} S_v \cdot q \tau_a \frac{i}{q^2 - M^2 + i0^+} \frac{i}{v \cdot (k_p + q) + i0^+} \left(-\frac{g_A}{F}\right) S_v \cdot q \tau_a$$

$$= -i\frac{3g_A^2}{F^2} S_v^\mu S_v^\nu i \mu^{4-n} \int \frac{d^n q}{(2\pi)^n} \frac{q_\mu q_\nu}{(q^2 - M^2 + i0^+)[v \cdot (k_p + q) + i0^+]}.$$

(d) Contracting with $v^\mu v^\nu$, using $v^2 = 1$:

$$C_{20}(\omega, M^2) + C_{21}(\omega, M^2) = i\mu^{4-n} \int \frac{d^n q}{(2\pi)^n} \frac{(v \cdot q)(v \cdot q)}{(q^2 - M^2 + i0^+)(v \cdot q + \omega + i0^+)}$$

$$= i\mu^{4-n} \int \frac{d^n q}{(2\pi)^n} \frac{(v \cdot q + \omega - \omega)v \cdot q}{(q^2 - M^2 + i0^+)(v \cdot q + \omega + i0^+)}$$

$$= \underbrace{i\mu^{4-n} \int \frac{d^n q}{(2\pi)^n} \frac{v \cdot q}{q^2 - M^2 + i0^+}}_{= 0}$$

$$- \omega i\mu^{4-n} \int \frac{d^n q}{(2\pi)^n} \frac{v \cdot q + \omega - \omega}{(q^2 - M^2 + i0^+)(v \cdot q + \omega + i0^+)}$$

$$= -\omega i\mu^{4-n} \int \frac{d^n q}{(2\pi)^n} \frac{1}{q^2 - M^2 + i0^+}$$

$$+ \omega^2 i\mu^{4-n} \int \frac{d^n q}{(2\pi)^n} \frac{1}{(q^2 - M^2 + i0^+)(v \cdot q + \omega + i0^+)}$$

$$= -\omega I_\pi(0) + \omega^2 J_{\pi N}(0; \omega).$$

Contracting with $g^{\mu\nu}$, using $g^{\mu\nu}g_{\mu\nu} = n$:

$$C_{20}(\omega, M^2) + nC_{21}(\omega, M^2) = i\mu^{4-n} \int \frac{d^n q}{(2\pi)^n} \frac{q^2}{(q^2 - M^2 + i0^+)(v \cdot q + \omega + i0^+)}$$

$$= i\mu^{4-n} \int \frac{d^n q}{(2\pi)^n} \frac{q^2 - M^2 + M^2}{(q^2 - M^2 + i0^+)(v \cdot q + \omega + i0^+)}$$

$$= i\mu^{4-n} \underbrace{\int \frac{d^n q}{(2\pi)^n} \frac{1}{v \cdot q + \omega + i0^+}}_{= 0}$$

$$+ M^2 i\mu^{4-n} \int \frac{d^n q}{(2\pi)^n} \frac{1}{(q^2 - M^2 + i0^+)(v \cdot q + \omega + i0^+)}$$

$$= M^2 J_{\pi N}(0; \omega).$$

(e) Since $S_v \cdot v = 0$, only $C_{21}(\omega, M^2)$ contributes. Solve system of equations to obtain

$$C_{21}(\omega, M^2) = \frac{1}{n-1} \left[(M^2 - \omega^2) J_{\pi N}(0; \omega) + \omega I_\pi(0) \right],$$

i.e., factor $1 - n$ from $S_v^\mu S_v^\nu g_{\mu\nu} = S_v^2$ cancels.

(f) Insert the expressions given for the loop integrals, extract a factor of $4\pi^2$ to give the denominator $(4\pi F)^2$, and collect the remaining terms.

(g)

$$p^\mu = m_N v^\mu = m v^\mu + k_p^\mu \quad \Rightarrow \quad k_p^\mu = (m_N - m) v^\mu$$

and with $v^2 = 1$ we obtain $\omega = v \cdot k_p = m_N - m$. Neglecting the tree-level term $\sim (m_N - m)^2$ and setting $\omega = 0$ in the loop contribution, we find

$$m_N = m - 4c_1 M^2 - \frac{3g_A^2}{(4\pi F)^2} M^3 \underbrace{\arccos(0)}_{= \frac{\pi}{2}} = m - 4c_1 M^2 - \frac{3\pi g_A^2 M^3}{2(4\pi F)^2}.$$

5.2

$$Q_{Ab}(x_0) = \int d^3 x\, q^\dagger(x_0, \vec{x}) \gamma_5 \frac{\lambda_b}{2} q(x_0, \vec{x}),$$

$$\mathcal{H}_{sb}(x) = \bar{q}(x) \mathcal{M} q(x).$$

Make use of Eq. 1.103 with $\Gamma_1 = \gamma_5$, $\Gamma_2 = \gamma_0$, $F_1 = \frac{\lambda_b}{2}$, and $F_2 = \mathcal{M}$:

$$\Gamma_1 \Gamma_2 = \gamma_5 \gamma_0 = -\gamma_0 \gamma_5 = -\Gamma_2 \Gamma_1.$$

$$[Q_{Ab}(x_0), \mathcal{H}_{sb}(x)] = -q^\dagger(x) \gamma_0 \gamma_5 \left\{ \frac{\lambda_b}{2}, \mathcal{M} \right\} q(x).$$

Make use of Eq. 1.103 with $\Gamma_1 = \gamma_5$, $\Gamma_2 = \gamma_0 \gamma_5$, $F_1 = \frac{\lambda_a}{2}$, and $F_2 = \left\{ \frac{\lambda_b}{2}, \mathcal{M} \right\}$:

$$\Gamma_1 \Gamma_2 = \gamma_5 \gamma_0 \gamma_5 = -\gamma_0 = -\Gamma_2 \Gamma_1.$$

$$\sigma_{ab}(x) = [Q_{Aa}(x_0), [Q_{Ab}(x_0), \mathscr{H}_{sb}(x)]]$$

$$= -\left[Q_{Aa}(x_0), q^\dagger(x)\gamma_0\gamma_5\left\{\frac{\lambda_b}{2}, \mathscr{M}\right\}q(x)\right]$$

$$= q^\dagger(x)\gamma_0\left\{\frac{\lambda_a}{2}, \left\{\frac{\lambda_b}{2}, \mathscr{M}\right\}\right\}q(x)$$

$$= \bar{q}(x)\left\{\frac{\lambda_a}{2}, \left\{\frac{\lambda_b}{2}, \mathscr{M}\right\}\right\}q(x).$$

$$\mathscr{M} = \begin{pmatrix} \hat{m} & 0 & 0 \\ 0 & \hat{m} & 0 \\ 0 & 0 & m_s \end{pmatrix},$$

$$\{\lambda_1, \mathscr{M}\} = \begin{pmatrix} 0 & 1 & 0 \\ 1 & 0 & 0 \\ 0 & 0 & 0 \end{pmatrix}\begin{pmatrix} \hat{m} & 0 & 0 \\ 0 & \hat{m} & 0 \\ 0 & 0 & m_s \end{pmatrix} + \begin{pmatrix} \hat{m} & 0 & 0 \\ 0 & \hat{m} & 0 \\ 0 & 0 & m_s \end{pmatrix}\begin{pmatrix} 0 & 1 & 0 \\ 1 & 0 & 0 \\ 0 & 0 & 0 \end{pmatrix}$$

$$= \begin{pmatrix} 0 & \hat{m} & 0 \\ \hat{m} & 0 & 0 \\ 0 & 0 & 0 \end{pmatrix} + \begin{pmatrix} 0 & \hat{m} & 0 \\ \hat{m} & 0 & 0 \\ 0 & 0 & 0 \end{pmatrix}$$

$$= 2\hat{m}\lambda_1,$$

$$\{\lambda_1, \lambda_1\} = 2\begin{pmatrix} 1 & 0 & 0 \\ 0 & 1 & 0 \\ 0 & 0 & 0 \end{pmatrix},$$

$$\frac{1}{4}\{\lambda_1, \{\lambda_1, \mathscr{M}\}\} = \hat{m}\begin{pmatrix} 1 & 0 & 0 \\ 0 & 1 & 0 \\ 0 & 0 & 0 \end{pmatrix},$$

$$\{\lambda_2, \mathscr{M}\} = 2\hat{m}\lambda_2, \quad \{\lambda_2, \lambda_2\} = 2\begin{pmatrix} 1 & 0 & 0 \\ 0 & 1 & 0 \\ 0 & 0 & 0 \end{pmatrix}, \quad \frac{1}{4}\{\lambda_2, \{\lambda_2, \mathscr{M}\}\} = \hat{m}\begin{pmatrix} 1 & 0 & 0 \\ 0 & 1 & 0 \\ 0 & 0 & 0 \end{pmatrix},$$

$$\{\lambda_3, \mathscr{M}\} = 2\hat{m}\lambda_3, \quad \{\lambda_3, \lambda_3\} = 2\begin{pmatrix} 1 & 0 & 0 \\ 0 & 1 & 0 \\ 0 & 0 & 0 \end{pmatrix}, \quad \frac{1}{4}\{\lambda_3, \{\lambda_3, \mathscr{M}\}\} = \hat{m}\begin{pmatrix} 1 & 0 & 0 \\ 0 & 1 & 0 \\ 0 & 0 & 0 \end{pmatrix},$$

$$\{\lambda_4, \mathscr{M}\} = (\hat{m} + m_s)\lambda_4, \quad \{\lambda_4, \lambda_4\} = 2\begin{pmatrix} 1 & 0 & 0 \\ 0 & 0 & 0 \\ 0 & 0 & 1 \end{pmatrix},$$

$$\frac{1}{4}\{\lambda_4,\{\lambda_4,\mathcal{M}\}\} = \frac{\hat{m}+m_s}{2}\begin{pmatrix} 1 & 0 & 0 \\ 0 & 0 & 0 \\ 0 & 0 & 1 \end{pmatrix},$$

$$\{\lambda_5,\mathcal{M}\} = (\hat{m}+m_s)\lambda_5, \quad \{\lambda_5,\lambda_5\} = 2\begin{pmatrix} 1 & 0 & 0 \\ 0 & 0 & 0 \\ 0 & 0 & 1 \end{pmatrix},$$

$$\frac{1}{4}\{\lambda_5,\{\lambda_5,\mathcal{M}\}\} = \frac{\hat{m}+m_s}{2}\begin{pmatrix} 1 & 0 & 0 \\ 0 & 0 & 0 \\ 0 & 0 & 1 \end{pmatrix},$$

$$\{\lambda_6,\mathcal{M}\} = (\hat{m}+m_s)\lambda_6, \quad \{\lambda_6,\lambda_6\} = 2\begin{pmatrix} 0 & 0 & 0 \\ 0 & 1 & 0 \\ 0 & 0 & 1 \end{pmatrix},$$

$$\frac{1}{4}\{\lambda_6,\{\lambda_6,\mathcal{M}\}\} = \frac{\hat{m}+m_s}{2}\begin{pmatrix} 0 & 0 & 0 \\ 0 & 1 & 0 \\ 0 & 0 & 1 \end{pmatrix},$$

$$\{\lambda_7,\mathcal{M}\} = (\hat{m}+m_s)\lambda_7, \quad \{\lambda_7,\lambda_7\} = 2\begin{pmatrix} 0 & 0 & 0 \\ 0 & 1 & 0 \\ 0 & 0 & 1 \end{pmatrix},$$

$$\frac{1}{4}\{\lambda_7,\{\lambda_7,\mathcal{M}\}\} = \frac{\hat{m}+m_s}{2}\begin{pmatrix} 0 & 0 & 0 \\ 0 & 1 & 0 \\ 0 & 0 & 1 \end{pmatrix},$$

$$\frac{1}{4}\{\lambda_8,\{\lambda_8,\mathcal{M}\}\} = \lambda_8^2\mathcal{M} = \frac{1}{3}\begin{pmatrix} \hat{m} & 0 & 0 \\ 0 & \hat{m} & 0 \\ 0 & 0 & 4m_s \end{pmatrix},$$

$$\frac{1}{4}\{\lambda_3,\{\lambda_8,\mathcal{M}\}\} = \lambda_3\lambda_8\mathcal{M} = \frac{\hat{m}}{\sqrt{3}}\lambda_3.$$

5.3

$$\begin{aligned}
\frac{\partial E(\lambda)}{\partial \lambda} &= \frac{\partial}{\partial \lambda}(E(\lambda)\langle\alpha(\lambda)|\alpha(\lambda)\rangle) \\
&= \frac{\partial}{\partial \lambda}\langle\alpha(\lambda)|H(\lambda)|\alpha(\lambda)\rangle \\
&= \frac{\partial\langle\alpha(\lambda)|}{\partial\lambda}H(\lambda)|\alpha(\lambda)\rangle + \left\langle\alpha(\lambda)\left|\frac{\partial H(\lambda)}{\partial\lambda}\right|\alpha(\lambda)\right\rangle + \langle\alpha(\lambda)|H(\lambda)\frac{\partial|\alpha(\lambda)\rangle}{\partial\lambda} \\
&= E(\lambda)\frac{\partial}{\partial\lambda}\underbrace{\langle\alpha(\lambda)|\alpha(\lambda)\rangle}_{=1} + \left\langle\alpha(\lambda)\left|\frac{\partial H(\lambda)}{\partial\lambda}\right|\alpha(\lambda)\right\rangle \\
&= \left\langle\alpha(\lambda)\left|\frac{\partial H(\lambda)}{\partial\lambda}\right|\alpha(\lambda)\right\rangle.
\end{aligned}$$

5.4

$$m_N = m + k_1 M^2 + k_2 M^3 + k_3 M^4 \ln\left(\frac{M}{m}\right) + k_4 M^4 + \mathcal{O}(M^5).$$

Make use of

$$\frac{\partial M^3}{\partial M^2} = \frac{\partial (M^2)^{\frac{3}{2}}}{\partial M^2} = \frac{3}{2}(M^2)^{\frac{1}{2}} = \frac{3}{2}M,$$

$$\frac{\partial}{\partial M^2}\ln\left(\frac{M}{m}\right) = \frac{1}{2}\frac{\partial}{\partial M^2}\ln\left(\frac{M^2}{m^2}\right) = \frac{1}{2M^2}.$$

$$\sigma = \sigma_1 M^2 + \sigma_2 M^3 + \sigma_3 M^4 \ln\left(\frac{M}{m}\right) + \sigma_4 M^4 + \mathcal{O}(M^5)$$

$$= M^2 \frac{\partial m_N}{\partial M^2}$$

$$= M^2 \left[k_1 + \frac{3}{2}k_2 M + 2k_3 M^2 \ln\left(\frac{M}{m}\right) + \frac{k_3}{2}M^2 + 2k_4 M^2 + \mathcal{O}(M^3) \right].$$

$\Rightarrow$

$$\sigma_1 = k_1, \quad \sigma_2 = \frac{3}{2}k_2, \quad \sigma_3 = 2k_3, \quad \sigma_4 = \frac{k_3}{2} + 2k_4.$$

5.5

$$(\delta^2 - M^2)^{\frac{3}{2}} = \delta^3 (1 - x^2)^{\frac{3}{2}} = \delta^3 \left[1 - \frac{3}{2}x^2 + \mathcal{O}(x^4) \right],$$

$$\ln\left(\frac{\delta - \sqrt{\delta^2 - M^2}}{M}\right) = \ln\left[\frac{\delta - \delta(1 - x^2)^{\frac{1}{2}}}{M}\right]$$

$$= \ln\left\{\frac{\delta - \delta\left[1 - \frac{1}{2}x^2 - \frac{1}{8}x^4 + \mathcal{O}(x^6)\right]}{M}\right\}$$

$$= \ln\left\{\frac{x}{2}\left[1 + \frac{x^2}{4} + \mathcal{O}(x^4)\right]\right\}$$

$$= \ln\left(\frac{x}{2}\right) + \ln\left[1 + \frac{x^2}{4} + \mathcal{O}(x^4)\right]$$

$$= \ln\left(\frac{x}{2}\right) + \frac{x^2}{4} + \mathcal{O}(x^4).$$

$\Rightarrow$

$$(\delta^2 - M^2)^{\frac{3}{2}}\ln\left(\frac{\delta - \sqrt{\delta^2 - M^2}}{M}\right) = \delta^3 \left[1 - \frac{3}{2}x^2 + \mathcal{O}(x^4)\right]\left[\ln\left(\frac{x}{2}\right) + \frac{x^2}{4} + \mathcal{O}(x^4)\right]$$

$$= \delta^3 \left[\ln\left(\frac{x}{2}\right) - \frac{3}{2}x^2 \ln\left(\frac{x}{2}\right) + \frac{x^2}{4} + \mathcal{O}(x^4)\right].$$

5.6 (a) From $\mathscr{L}_{\pi N}^{(1)}$:

$$\left\langle N(p') \left| -ie\bar{\Psi}\mathscr{A}_\mu\gamma^\mu\frac{1}{2}(\mathbb{1}+\tau_3)\Psi \right| N(p), \gamma(\varepsilon,q) \right\rangle \Rightarrow -ie\varepsilon_\mu\gamma^\mu\frac{1}{2}(\mathbb{1}+\tau_3).$$

From $\mathscr{L}_{\pi N}^{(2)}$, we only need c_6 and c_7 terms. With choice of external fields we find

$$f_{L/R\mu\nu} = -e(\partial_\mu\mathscr{A}_\nu - \partial_\nu\mathscr{A}_\mu)\frac{\tau_3}{2} + \cdots,$$
$$f_{\mu\nu}^+ = -e(\partial_\mu\mathscr{A}_\nu - \partial_\nu\mathscr{A}_\mu)\tau_3 + \cdots,$$

where the ellipsis stands for terms containing a larger number of fields.

$$\left\langle N(p') \left| -i\frac{e}{2}\bar{\Psi}\sigma^{\mu\nu}\left(c_6\tau_3 + \frac{c_7}{2}\mathbb{1}\right)(\partial_\mu\mathscr{A}_\nu - \partial_\nu\mathscr{A}_\mu)\Psi \right| N(p), \gamma(\varepsilon,q) \right\rangle$$
$$\Rightarrow -ie\frac{1}{2}\sigma^{\mu\nu}\left(c_6\tau_3 + \frac{c_7}{2}\mathbb{1}\right)[-iq_\mu\varepsilon_\nu - (-iq_\nu)\varepsilon_\mu]$$
$$= e\varepsilon_\mu\sigma^{\mu\nu}q_\nu\frac{1}{2}(2c_6\tau_3 + c_7\mathbb{1}).$$

In $\mathscr{L}_{\pi N}^{(3)}$, we can set $D_\mu\Psi = \partial_\mu\Psi$ as all other terms contain more fields. Then

$$i\left\langle N(p') \left| \left(\frac{-ie}{2m}\bar{\Psi}(d_6\tau_3 + 2d_7\mathbb{1})\partial^\mu(\partial_\mu\mathscr{A}_\nu - \partial_\nu\mathscr{A}_\mu)\partial^\nu\Psi + \text{H.c.}\right) \right| N(p), \gamma(\varepsilon,q) \right\rangle$$
$$\Rightarrow \frac{e}{2m}(d_6\tau_3 + 2d_7\mathbb{1})(-ip'^\nu - ip^\nu)(q\cdot\varepsilon q_\nu - q^2\varepsilon_\nu)$$
$$= ie\varepsilon_\mu(q^2 P^\mu - q^\mu q\cdot P)\left(\frac{1}{2m}d_6\tau_3 + \frac{1}{m}d_7\mathbb{1}\right),$$

where $P^\mu = p'^\mu + p^\mu$.

(b) Contributions from first- and second-order Lagrangian can be read off as the Feynman rule is simply $\mathscr{M}$. For third order use

$$\begin{aligned}
\bar{u}(p')P^\mu u(p) &= \bar{u}(p')(p'^\mu + p^\mu)u(p) \\
&= \bar{u}(p')(p'_\nu g^{\nu\mu} + g^{\mu\nu}p_\nu)u(p) \\
&= \bar{u}(p')[p'_\nu(\gamma^\nu\gamma^\mu + i\sigma^{\nu\mu}) + (\gamma^\mu\gamma^\nu + i\sigma^{\mu\nu})p_\nu]u(p) \\
&= \bar{u}(p')(p\!\!\!/'\gamma^\mu + \gamma^\mu p\!\!\!/ - i\sigma^{\mu\nu}q_\nu)u(p) \\
&= \bar{u}(p')(2m_N\gamma^\mu - i\sigma^{\mu\nu}q_\nu)u(p)
\end{aligned}$$

and

$$q\cdot P = (p'-p)\cdot(p'+p) = p'^2 - p^2 = 0$$

for on-shell momenta.

(c) Take Lagrangian $\mathscr{L}_{\gamma\pi NN}$ from solution to Exercise 4.5,

$$\left\langle N(p'), \phi_a(k) \left| -ie\frac{g_A}{2F}\varepsilon_{3bc}\bar{\Psi}\gamma^\mu\gamma_5\tau_b\Psi\mathscr{A}_\mu\phi_c \right| N(p), \gamma(\varepsilon,q) \right\rangle$$
$$\Rightarrow ie\varepsilon_\mu\frac{g_A}{2F}\gamma^\mu\gamma_5\varepsilon_{3ab}\tau_b.$$

(d) Apply Feynman rules

$$\mathscr{M} = \bar{u}(p')\mu^{4-n}\int\frac{d^nk}{(2\pi)^n}\left(-\frac{g_A}{2F}\slashed{k}\gamma_5\tau_a\right)i\Delta_F(k)iS_F(p'-k)\left(ie\varepsilon_\mu\frac{g_A}{2F}\gamma^\mu\gamma_5\varepsilon_{3ab}\tau_b\right)u(p),$$

and make use of $\slashed{k}\gamma_5 = -\gamma_5\slashed{k}$ and $\varepsilon_{3ab}\tau_a\tau_b = 2i\tau_3$.
(e)

$$\slashed{k} = -(\slashed{p}'-\slashed{k}-m) + (\slashed{p}'-m)$$
$$= -S_F^{-1}(p'-k) + (\slashed{p}'-m)$$

leads to the simplification

$$\bar{u}(p')\gamma_5\slashed{k}S_F(p'-k)\Delta_F(k)\gamma^\mu\gamma_5 u(p)$$
$$= \bar{u}(p')\gamma_5\left[-\Delta_F(k) + (\slashed{p}'-m)S_F(p'-k)\Delta_F(k)\right]\gamma^\mu\gamma_5 u(p)$$
$$= \bar{u}(p')\gamma_5\left\{-\Delta_F(k) - (m_N+m)\frac{\slashed{p}'-\slashed{k}+m}{[(p'-k)^2-m^2+i0^+]}\Delta_F(k)\right\}\gamma^\mu\gamma_5 u(p)$$
$$= \bar{u}(p')\gamma_5\left\{-\Delta_F(k) + \frac{(m_N^2-m^2) + (m_N+m)\slashed{k}}{[(p'-k)^2-m^2+i0^+]}\Delta_F(k)\right\}\gamma^\mu\gamma_5 u(p)$$
$$= \bar{u}(p')\left\{1 - \frac{m_N^2-m^2}{[(p'-k)^2-m^2+i0^+]} + \frac{(m_N+m)\slashed{k}}{[(p'-k)^2-m^2+i0^+]}\right\}\Delta_F(k)\gamma^\mu u(p).$$

(f)

$$\mathscr{M} = -ie\varepsilon_\mu\bar{u}(p')\frac{g_A^2}{2F^2}\tau_3\left\{I_\pi - (m_N^2-m^2)I_{N\pi}(-p',0)\right.$$
$$\left. + \frac{m_N+m}{2m_N^2}[I_N - I_\pi + (m_N^2-m^2+M^2)I_{N\pi}(-p',0)]\slashed{p}'\right\}\gamma^\mu u(p)$$
$$= -ie\varepsilon_\mu\bar{u}(p')\frac{g_A^2}{2F^2}\tau_3\left\{I_\pi - (m_N^2-m^2)I_{N\pi}\right.$$
$$\left. + \frac{m_N+m}{2m_N}[I_N - I_\pi + (m_N^2-m^2+M^2)I_{N\pi}]\right\}\gamma^\mu u(p),$$

where $I_{N\pi} = I_{N\pi}(-p',0)|_{p'^2=m_N^2}$.

Problems of Appendix B

B.1

$$\frac{\delta f(y)}{\delta f(x)} = \lim_{\varepsilon \to 0} \frac{F_y[f + \varepsilon \delta_x] - F_y[f]}{\varepsilon}$$

$$= \lim_{\varepsilon \to 0} \frac{\int d^n z \delta^n(z - y)[f(z) + \varepsilon \delta^n(z - x)] - \int d^n z \delta^n(z - y) f(z)}{\varepsilon}$$

$$= \int d^n z \delta^n(z - y) \delta^n(z - x)$$

$$= \delta^n(y - x).$$

Index

A

Adler-Gilman relation, 240

Anomalous
  action, 134, 136, 138
  Ward identity, 121, 133

Anomalous magnetic moment, 170, 231, 236, 241

Anomaly, 23, 27, 36, 41, 45, 46, 121

Anticommutation relations
  fermion fields, 20
  Gell-Mann matrices, 3
  Pauli matrices, 251

Axial-vector coupling constant, 152, 158, 159, 227, 229

Axial-vector current, 23, 85
  divergence of, 26, 46

B

B_0, 87–89, 92, 94, 100

Baryon mass, 154, 243

Baryon number conservation, 69

C

Cabibbo-Kobayashi-Maskawa matrix, 45, 97, 153

Canonical quantization, 16, 29, 30

Charge conjugation, 42, 43, 91, 94, 95, 158, 241

Charge operator, 17, 18, 20, 24, 25, 28, 30, 61, 70, 72, 74, 75, 222

χ, 92–94

$\chi_\pm$, 169

Chiral algebra, 24

Chiral connection, 150, 157, 163, 208, 237

Chiral extrapolation, 230, 239

Chiral limit, 9, 24, 41, 69, 71, 72, 74, 76, 80, 90, 129, 152, 168
  QCD Lagrangian in, *see* QCD Lagrangian in chiral limit

Chiral logarithm, 64, 129, 139, 229

Chiral symmetry, 1, 12

Chiral transformation properties (local)
  baryons, 154
  Δ, 207, 208
  external fields, 44
  Goldstone bosons U, 91, 150
  nucleon, 150
  quarks, 43

Chiral unitary approach, 243

Chiral vielbein, 151, 156, 184, 208

Chiral-symmetry-breaking scale, 9, 118, 180, 213, 243

Clebsch-Gordan coefficient, 28, 102, 205–207

Coleman theorem, 71

Commutation relations
  angular momentum, 16
  boson fields, 20
  charge operators, 18, 24, 25
  equal-time, 16, 25, 29, 30, 38, 73
  Gell-Mann matrices, 3
  Pauli matrices, 251

Complex-mass scheme, 243

Compton scattering
  nucleon, 36, 240
  pion, 104–105, 140, 141
  virtual, 241

Constraints, 201, 209, 210

S. Scherer and M. R. Schindler, *A Primer for Chiral Perturbation Theory*,
Lecture Notes in Physics 830, DOI: 10.1007/978-3-642-19254-8,
© Springer-Verlag Berlin Heidelberg 2012